Anaerobic Sewage Treatment

Optimization of Process and Physical Design of Anaerobic and Complementary Processes

Anaerobic Sewage Treatment

Optimization of Process and Physical Design of Anaerobic and Complementary Processes

Adrianus van Haandel and
Jeroen van der Lubbe

IWA PUBLISHING

Published by **IWA Publishing**
Alliance House
12 Caxton Street
London SW1H 0QS, UK
Telephone: +44 (0)20 7654 5500
Fax: +44 (0)20 7654 5555
Email: publications@iwap.co.uk
Web: www.iwapublishing.com

First published 2019

British Library Cataloguing in Publication Data
A CIP catalogue record for this book is available from the British Library

ISBN: 9781780409610 (Paperback)
ISBN: 9781780409627 (eBook)
ISBN: 9781780409634 (ePub)

Contents

Preface

In this book the authors seek to present relevant aspects of the theory of anaerobic digestion and use them for optimized design and operation of anaerobic sewage treatment systems. The Upflow Anaerobic Sludge Bed (UASB) reactor is identified as the most appropriate solution for anaerobic treatment of municipal sewage. The book has been written for students at MSc or PhD level, as well as for engineers in consulting firms and environmental protection agencies.

Since its conception almost half a century ago, the UASB reactor has evolved as the most widely applied configuration for anaerobic wastewater treatment. This is due to its high efficiency in removing suspended solids and organic material at short retention time in the reactor. Anaerobic digestion had been applied for centuries as a treatment process for wastes and wastewater, but its efficiency was low, even with long retention times. The breakthrough of the UASB and other modern anaerobic reactors only became possible when finally it was understood than any biological system can only be efficient if two pre-conditions are satisfied: (1) there must be intensive contact between the organic material to be removed and the biomass in the reactor; and (2) a high concentration of biomass must be kept in the treatment system. In the UASB reactor both pre-conditions are automatically satisfied: the influent is introduced at the bottom of the reactor and passes upward through a dense sludge bed and then through a gas–liquid–solids separator, thereby collecting the biogas produced, retaining the biomass by settling, and discharging an effluent essentially free from solids.

A significant part of the book is dedicated to the presentation of guidelines for rational design of the UASB reactor itself, in which valuable experience from many full-scale UASB systems has been incorporated. Both authors have a long history in anaerobic treatment and have been personally involved with the design, engineering and construction of large UASB systems with multiple reactors.

In anaerobic treatment the influent organic material is divided into three fractions: (1) the largest part is collected as biogas, in which methane is the main component; (2) a second fraction is the sludge mass generated in the biological process; and finally (3) a fraction is not affected by the treatment and is discharged with the effluent. In the book it is shown that the division between these three fractions

Editor(s): Adrianus van Haandel and Jeroen van der Lubbe
doi: 10.2166/9781780409627_xi

depends basically on two variables: the temperature and the sludge age. If the process value of these variables increases, so will the transformation of influent organic material into methane, and consequently less will end up in the other two fractions. Secondly, a more stabilized excess sludge is obtained, which after drying can be readily applied for soil improvement and fertilization. On the other hand, a longer sludge age requires a larger reactor volume, so an optimal value must be determined where high digestion efficiency is combined with low investment cost. Insofar as temperature is concerned, a low value is detrimental, but the energy generated from the biogas can help to increase the digestion efficiency, even at low temperatures.

The book contains chapters on the treatment, disposal and/or utilization of each of the three basic organic material fractions. Biogas is collected and can be flared, or used for generation of electrical energy. The excess sludge produced must be dewatered and dried before its use as organic manure or for energy generation by combustion. Alternatively it may be discharged into landfills. Sludge drying beds, when properly designed and operated, are very suitable for application in regions with a warm climate.

Finally, the effluent of anaerobic treatment can be used for irrigation or for industrial or urban reuse. Irrigation is feasible if remaining pathogens, such as thermotolerant bacteria and helminth eggs, are removed, which can be easily achieved in polishing ponds. For reuse in industry or in urban areas, nutrients must be removed as well. Nitrogen removal is possible in shallow polishing ponds, where the pH increases due to photosynthetic carbon dioxide consumption. The high pH is also instrumental for precipitation of phosphate. Alternatively phosphate may be removed by chemical precipitation followed by clarification, which can produce a high-quality effluent.

In this book the two most suitable alternatives for post-treatment of anaerobic effluent will be discussed. The first is the activated sludge system, which is very applicable as a high-rate supplementary treatment step to remove organic material. The second is the polishing pond, which requires much more area, but allows for both pathogen and nutrient removal, and produces a high-quality effluent. The superiority of sequential batch operations in terms of treatment efficiency and ease of operation will be demonstrated.

Authors

Adrianus van Haandel (1948) holds an MSc degree from the Technical University of Eindhoven in the Netherlands and a PhD from the University of Cape Town, South Africa. He has worked at the University of Campina Grande in Brazil since 1971, where he coordinates research on biological wastewater treatment. He has extensive experience as an independent consultant and is involved with several international expert committees. Together with other authors he has written several books about different aspects of wastewater treatment including the *Handbook of Biological Wastewater Treatment – Design and Optimization of Activated Sludge Systems* (co-authored with Jeroen van der Lubbe) and *Advanced Biological Treatment Processes for Industrial Wastewaters – Principles and Applications.* Adrianus can be contacted at adrianusvh@gmail.com.

Jeroen van der Lubbe (1971) graduated in 1995 at the Environmental Department of the Wageningen University, Netherlands, and since then has been involved extensively in the design, engineering and operation of both industrial and municipal wastewater treatment plants. Currently he is Technology Manager at Biothane, which is a part of Veolia Water Solutions and Technologies. Biothane focuses on anaerobic wastewater and biogas treatment, mainly for the industrial market. Product development has always been an important activity and as such he was for example project manager for the development of the Upthane® – Veolia's municipal UASB reactor.

Acknowledgements

This book reflects the experience of the authors with different aspect of biological wastewater treatment. Research was mainly carried out in Brazil as part of a cooperative program of several Universities, PROSAB, financed by the World Bank and several Brazilian government agencies. The authors want to thank especially Professors Pedro Alem, Silvânia Lucas dos Santos and Francisco Viera Paiva.

The experimental investigations were carried out over a period of more than thirty years with four PhD and twenty MSc students, as well as many undergraduates and internship students.

A lot of practical feedback came from visits to and return of experience from full-scale UASB-based sewage treatment plants located all over the world. Thanks are due to Jeroen's colleagues from Veolia Brazil (Luis Abrahão), VWT UAE (Ashok Tailor) and VWT Egypt (Johnathan Sharrock), who made these visits possible. And a special word of thanks goes to Barry Heffernan (at the time Biothane, now Oxymem), who faithfully travelled all over the world with Jeroen and was instrumental in developing the Upthane® municipal UASB reactor. Finally Frank van der Zee (Biothane) was responsible for assuring us that the status with respect to anaerobic microbiology was such that we could indeed still use our old microbial reaction schemes....

Also many thanks to the people who have read the drafts of the book and given many suggestions to improve it, especially Willy Verstraete (Ghent University) and Juan Garrido (University of Santiago de Compostela).

Finally, special words of thanks are due to the authors' wives, Paula Frassinetti Feitosa Cavalcanti and Lotje Jacqueline van de Poll, for their unfailing support during the long incubation period in which this book was conceived and written.

Symbols and Abbreviations

The variables used in the book can have a prefix and a suffix. As an example, the load of organic material in a system is expressed as MS_{ti} where M indicates the prefix "daily load", S is the variable substrate (COD), t refers to total and i to influent. Variables and parameters are listed in Table A and the suffixes and prefixes are in Table B, as well as the variables with a Greek letter.

Table A Symbols for variables and abbreviations.

A = area	TP = transfer pond
Ac = acidity	TSS = total suspended sludge
Alk = alkalinity	U = humidity
B = mass balance, buffer index	VSS = volatile suspended solids
BOD = biological oxygen demand	V = volume
C = concentration, load	VFA = volatile fatty acids concentration
COD = chemical oxygen demand	W = width, weight
D = diameter, dispersion number	WSP = waste stabilization pond
E = daily excess sludge production, spacing	X = generic symbol for sludge
GLS = gas-liquid-solid separator	Y = yield coefficient
H = height	b = decay rate constant
K = constant	d = distribution
L = length	f = fraction
N = number of transferred electrons, moles, number or reactors, nitrogen	g = gravitation constant
OUR = oxygen uptake rate	h = water level, head
P = weight, drying bed productivity, phosphorus	p = pressure
PP = polishing pond	v = velocity
Q = flow rate	x = number of carbon atoms
R = retention time, rate, ratio, gas constant	y = number of hydrogen atoms
S = substrate concentration	z = number of oxygen atoms
T = temperature	
TOC = total organic carbon	

Table B Suffixes, prefixes and variables with a Greek symbol.

Suffix	Prefix
a = active sludge, storage of material	d = differential
ae = aerobic	m = generic prefix for "per unit mass of"
an = anaerobic	p = −log
b = biodegradable, decay	M = mass in the reactor or influent mass per day
c = critical, cycle, carbonic	
cv = conversion COD to VSS	**Greek symbols**
d = digested, desorption	α = angle (°)
dig = digestion	β = angle (°)
e = effluent, endogenous, excess sludge	η = efficiency
end = endogenous	θ = temperature coefficient
exo = exogenous	μ = growth rate constant
g = growth, gas	ω = contraction coefficient
gb = gas box	Δ = difference (finite)
h = hydraulic (retention time)	
hi = hydrolysis	
i = influent, inert	
l = liquid	
m = maximum, methanized	
max = maximum	
min = minimum	
n = non-biodegradable, nitrogen	
na = non-active	
o = oxidized, organic material, overflow	
p = particulate, phosphorus	
r = reactor, rate, productivity	
rel = relative	
s = soluble, sludge, saturation, solids load	
sb = sludge batch	
t = total	
u = utilization, UASB, upwards	
v = volatile	
x = sludge (solid)	
H = Henry constant	
T = temperature	

Contents in Brief

This book deals with anaerobic sewage treatment systems including complementary treatment (effluent post-treatment, biogas handling and sludge treatment) and is composed of ten chapters that can be summarized as follows.

***Chapter 1: Sewage Characteristics and Treatment Systems*:** In this chapter the undesirable components of sewage are presented as well as the problems they cause when treatment is not adequate. After a brief discussion about effluent standards and treatment objectives, the main treatment systems are presented and compared. Basic thermodynamics are used to evaluate the energy of organic material in wastewaters and to select the most appropriate parameter to measure the mass of organic material in a wastewater.

***Chapter 2: Organic Material and Bacterial Metabolism*:** The metabolism of the micro-organisms that develop in anaerobic treatment system is presented as a dual process in which organic material is used for both energy generation and cell growth: catabolism and anabolism. Anaerobic and aerobic metabolisms are compared, and the positive and negative aspects of both are briefly discussed. The different processes and associated populations of micro-organisms for anaerobic digestion and competing processes are presented. The influence of sewage characteristics and operational conditions are discussed, and expressions for the kinetics of the different anaerobic processes are derived.

***Chapter 3: Process Model for the UASB Reactor*:** The conceptual flaws in classic anaerobic systems are shown in order to explain how these systems cannot perform well, because contact between sludge and influent substrate is not ensured and/or sludge retention is inefficient. Modern anaerobic treatment systems do not have these failures and perform much better. In a comparison of five modern anaerobic reactor configurations the UASB reactor showed a superior performance, and for that reason is selected as the preferred anaerobic system.

The biological and mechanical properties of UASB sludge are presented as functions of operational conditions, which includes both start-up and steady-state performance. Apart from the sewage characteristics in terms of concentration and composition, the performance of the UASB reactor is shown to depend fundamentally on two variables: the sludge age and the temperature. The sludge age that can be applied depends on the quality of the phase separator and the settling properties of the sludge, while the sewage temperature in the coldest month must be taken for design of the UASB reactor.

Chapter 4: Design of UASB Reactors and Components: A well performing UASB reactor requires proper design and operation. Insofar as process design aspects are concerned, important issues are the sulphate concentration in the wastewater and the return for stabilization of aerobic sludge from post-treatment. For adequate physical design of UASB reactor components, the following aspects are of vital importance: choice of adequate construction materials, design of the influent flow distribution and effluent collection system, as well as collection and utilization of the produced biogas. Where operational aspects are concerned, important issues are the appropriate pre-treatment of the wastewater and adequate provision for scum removal and excess sludge. A reactor design procedure is presented with special attention to the design of the GLS separators and the deflectors. At the end of the chapter, several alternatives to the design of conventional UASB reactors are discussed, which are very suitable for decentralized treatment at the level of the individual household.

Chapter 5: pH Value and Stability: It will be shown that in the case of sewage treatment the pH value of the UASB reactor depends almost entirely on the carbonic system and is determined by changes in alkalinity and acidity due to the anaerobic treatment process. A calculation procedure for alkalinity and volatile fatty acid concentrations using the data from a simple titration is presented. An expression to calculate the reactor pH is presented. The implications of the two-phase gas–liquid equilibrium and the three-phase gas–liquid–solids equilibrium are discussed in relation to the biogas composition and the corrosion potential of the reactor concrete.

Chapter 6: Biogas Generation in Anaerobic Digesters: An iterative calculation procedure to determine the production and composition of biogas from anaerobic sewage treatment systems is developed, resulting in calculations of the biogas flow and the partial pressures of methane, carbon dioxide, water, nitrogen, and hydrogen sulphide. Biogas collection and potential for energy production are discussed.

Chapter 7: Dewatering, Dehydration and Drying of Sludge: The production and composition of sludge from anaerobic sewage treatment systems and the influence of the solids fraction (humidity) on its physical characteristics are discussed. Three mechanical dehydration processes are presented. The attractive alternative of natural drying on sludge drying beds is discussed in detail and a procedure for rational design based on the water evaporation rate is presented for different types of drying beds. A procedure to quantify the influence of rainwater events on the productivity of drying beds is developed. Expressions are developed to estimate the energy production potential from combustion of dried sludge.

Chapter 8: Aerobic Post-Treatment of Anaerobic Effluents: It is shown that the best high-rate alternative for aerobic post-treatment is the activated sludge process and more precisely the sequential batch reactor (SBR). It is demonstrated that the steady-state IWA Activated Sludge Model is perfectly able to describe the sludge production, oxygen consumption and final effluent quality. Based on this model an optimized design procedure for the SBR is presented. The large advantages of integrated anaerobic–aerobic sewage treatment for reduction of required investment (reactor volume) and operational costs (energy) are quantified, but it is shown that no important reduction of sludge production is accomplished. Stabilization of excess aerobic sludge production in the UASB reactor is perfectly feasible, but only if the solids from the reactor are separated before the effluent is fed to post-treatment.

***Chapter 9: Post-Treatment in Polishing Ponds*:** The combination of a UASB reactor with polishing ponds (PP) is a huge improvement over obsolete waste stabilization ponds (WSP) systems. It results in a reduction of investment costs (volume and area), and an improved effluent quality and quantity as well as the environmental and aesthetic advantages of PP. Equations are presented to calculate the required retention time for coliform removal in different pond configurations, and the advantage of the sequential batch regime for thermotolerant coliform removal is shown. A model is developed that is able to predict the behaviour of the pH in a polishing pond, as a result of photosynthesis and desorption of carbon dioxide and ammonia. It is shown that nutrient removal (ammonia and phosphate) in PP is perfectly feasible, if the depth is shallow, in contrast to WSP.

***Chapter 10: Spreadsheet Design of Anaerobic and Complementary Treatment Plants*:** In this final chapter, spreadsheets are presented that will help the design engineer make optimal selections for the design of a UASB reactor. The consequences of setting the values of variables such as applied load, temperature, depth and number of GLS separators are readily shown. The spreadsheet also calculates the composition and production of biogas, as well as the sludge production and the required drying bed area. The dimensions of both aerobic and polishing pond options are calculated as well.

Chapter 1

Sewage characteristics and treatment systems

1.1 INTRODUCTION

Wastewater is water that after being used in some human activity has acquired characteristics that mean it is no longer fit for its intended purpose and for this reason is being discarded. Most countries have set in place effluent limits to which wastewater must conform: its disposal takes place in accordance with the norms and criteria established by such legislation. For this reason, it will be mostly be necessary to apply some form of treatment to remedy that wastewater's undesirable characteristics. In many cases the treatment can generate an effluent with characteristics such that it can be fed into processes that reuse this treated wastewater. In other cases, it will be preferred to discharge the treated effluent into water bodies, eventually reaching the sea.

There are three main origins of wastewater: (1) domestic wastewater or sewage, (2) industrial wastewater and (3) agricultural wastewater. The proportion of water used for domestic, industrial and agricultural purposes (and hence of the produced wastewaters) depends greatly on local conditions. Factors such as climate and economic and social development have a strong influence, but in general domestic usage is lowest, while agricultural usage is highest.

Domestic wastewater is usually collected from residences by public sewer networks, into which other wastewaters from urban activities such as commerce, public offices, schools, bars and hotels are also discharged. Depending on the nature of the collection network (combined or separate) it may also serve to convey rainwater. The municipal wastewater thus produced is called sewage. Owing to the large variations in urban drainage water flow, nowadays separate sewage networks are used instead of combined networks, but owing to illegal connections there is usually a significant amount of sewage discharged into the drainage network while there are drainage waters in the sewage network as well. Depending on the nature of the industries the network can also be used for industrial wastewaters, provided that a pre-treatment is applied that enables the wastewater to acquire features that allow this option. However, in many cases the amount of water and the concentration of undesirable constituents in industrial effluents is such that a specific

Editor(s): Adrianus van Haandel and Jeroen van der Lubbe
doi: 10.2166/9781780409627_0001

treatment system is justified for an industry or for a group of industries, and only after this can the treated effluent be reused or discharged into the public network. In agricultural uses of water, a major aim is to only apply enough water to meet the demands of the plants being cultivated. Owing to contact with the soil and fertilizers applied thereto, a wastewater rich in salts is created. Usually there is no specific structure to collect the wastewater and this therefore ends up in the groundwater reserves or drains to surface water.

In the subsequent sections of this chapter the problems caused by inadequate sewage treatment are discussed first. Then the main treatment steps of sewage treatment systems are briefly discussed. Finally, characteristics of municipal sewage are discussed as well, including the definition and measurement of its content of organic material.

1.2 IMPACT OF INADEQUATE SEWAGE TREATMENT

The primary objective of wastewater treatment is to correct its undesirable characteristics in such a way that its reuse or final disposal may occur in accordance with the standards and criteria defined by legislation. For this reason, the treatment will include reducing the concentration of various undesirable constituents found in the water. In Table 1.1, the main undesirable constituents of sewage are shown, as well as the main problems they cause and alternative solutions to remove these components or reduce their concentrations.

Suspended solids are the most noticeable component of sewage and can be determined experimentally by gravimetric tests after filtration and drying. Suspended solids may be present as visible macroscopic particles or as colloidal particles. Colloids cannot be seen by the naked eye but give sewage its typical turbidity. If sewage is not treated, these solids can accumulate in the receiving water body, forming banks of decomposing solids, which eventually may occupy much of its volume in a process called silting. This is a problem in numerous water bodies, from small dams to large rivers.

The solution to the problem is simple: by simple sedimentation the settleable material is removed before the discharge of the sewage, thus largely avoiding deposits formation. The retention of the solids by settling is not enough: without continuously or periodically removing solids from the treatment system, accumulation will unavoidably result in silting of the sedimentation system itself. Unfortunately, there are numerous examples of silted treatment systems, particularly stabilization ponds.

Organic material is invariably present in sewage, causing organic pollution: it serves as food for micro-organisms, in the process consuming oxygen dissolved in the water of the receiving body. Depending

Table 1.1 Main components of sewage, the problems associated with them and the treatment methods often applied for their removal.

Component	Problems Observed	Solution
Suspended solids	Silting of the receiving water body	Primary treatment: separation/ digestion
Organic material	Pollution: oxygen consumption, death of macro organisms generation of bad odours	Biological treatment (anaerobic and/or aerobic)
Pathogens	Contamination Disease transmission micro-organisms: virus, bacteria, protozoa helminth eggs	Elimination by: natural death (time) or imposed death (chemical or physical methods)
Nutrients	Eutrophication: excessive aquatic life growth (algal blooms)	Nitrogen & phosphorus removal or use in agriculture

on the intensity of the organic pollution, the oxygen can be consumed completely, making it impossible for macro organisms to survive in the water. Without oxygen an anaerobic environment can develop which, in addition to the simple absence of oxygen, is characterized by bad odours (like rotten eggs) and a black colour, giving the water a bad aesthetic appearance. Removal of the organic material is possible, using the same methods found in nature: biological treatment, applying anaerobic and/or aerobic micro-organisms to eliminate organic material. Quantitatively there are standard tests available which measure the concentration of organic material.

In aerobic systems the oxygen required for the oxidation of organic material can be transferred from the air or generated by the treatment system itself. In the case of the activated sludge process there is transfer of oxygen from the atmosphere to the liquid phase of the treatment system by mechanical aerators. In stabilization ponds photosynthesis by algae produces oxygen which is subsequently used by other micro-organisms, mainly bacteria, to oxidize organic material in the sewage. In aerobic systems the organic material is oxidized and decomposed into stable inorganic compounds, mainly carbon dioxide and water. In anaerobic systems organic material decomposes, in a biological process called anaerobic digestion, producing biogas, which is composed mainly of methane and carbon dioxide.

Receiving water bodies can be contaminated due to the introduction of pathogenic organisms present in the sewage. Many different micro-organisms (viruses, bacteria, protozoa) as well as helminth eggs exist in sewage that can cause several diseases in humans and animals which come into contact with water or drink it. Waterborne diseases are an important public health problem in developing countries and often are one of the most frequent causes for illness or even death. Lack of sanitation is also among the leading causes of child mortality. Methods for the removal of pathogens can be divided into natural methods which retain the organisms in the treatment system for a period long enough to cause their natural death or applications of specific chemical or physical disinfection methods which accelerate the decay rate of pathogenic organisms. To evaluate the hygienic effluent quality of treatment systems there are two important indicators: the number of thermotolerant bacteria per unit volume (formerly called faecal coliforms) and the concentration of helminth eggs.

Surface water eutrophication is a direct result of effluent discharge of nutrients (notably the macronutrients nitrogen and phosphorus) to an excessive concentration in the receiving water body. Eutrophication is the exacerbated growth of aquatic organisms, especially algae, which give the water an intense green colour and additionally give it a characteristic and unpleasant odour and taste. Eutrophication is more serious in standing (stagnant) or lentic waters like dams. It is possible to remove nutrients from the wastewater before discharge, but the treatment processes are complicated and costly.

The alternative is not to discharge the treated sewage into surface water, but to use it as irrigation water in agriculture, where the presence of nutrients is advantageous, replacing fertilizer which otherwise would have to be applied to increase the harvest. However this is not always feasible, especially if agriculture is a seasonal activity in which part of the year there is no demand for nutrients. If agricultural reuse is not feasible, there is the alternative of biological nitrogen removal through the sequential processes of nitrification and denitrification. Nitrogen in sewage is predominantly present in the form of ammonia, and nitrification is the oxidation of ammonia to nitrate by specific bacteria that use the energy released in oxidation for their vital functions. Denitrification is the reduction of nitrate to molecular nitrogen that is released from the liquid phase. This reduction is performed by bacteria that use nitrate instead of oxygen when it is not available, consuming organic material in the process. Phosphorus can be removed by the incorporation of phosphate into the microbiologic sludge mass, or it can be removed by addition of metal salts that form insoluble precipitates (chemical precipitation).

Worldwide there is a tendency to reduce the quantity of sewage to be treated in centralized systems by the introduction of decentralized systems nearer to the origin of the sewage. The objective of these local systems is to partially treat the generated sewage, aiming at reuse in more basic applications such as external floor washing and toilet flushing, as well as garden irrigation. One possibility is to separate wastewater flows at source. In practice, black waters (from the sanitary basin), grey water (from the kitchen and shower) and yellow water (urine) are distinguished. Specifically the treatment of grey water, which comprises one-third to half of the total sewage flow, has a very promising future. To take advantage of using treated water in toilet flushing, for example, there must be two water networks installed: one for all uses including drinking water and the other for specific uses of recovered water.

1.3 EFFLUENT STANDARDS AND TREATMENT OBJECTIVES

In many countries, there are federal, regional and municipal institutions that are responsible for protection of the environment through specific legislation and administrative actions. Legislatively, these institutions issue resolutions defining the quality of surface waters and any effluents discharged into them. Regarding the implementation of that legislation, there are bodies for federal and regional control in many countries that are tasked with: (1) verifying if the quality of the water resources is within the limits set by the legal standards, and (2) verifying if the legislation is being respected by all sources of effluents. To do this, the institutions must have the necessary human and material resources to allow them to exercise this control efficiently. Basically, what legislation prescribes are standards for maintaining in all surface or groundwater a quality compatible with its most significant use. For this purpose, a classification of the waters in a country may be made which defines the quality of the water that is to be maintained. A second aspect of this is the definition of the quality of any effluents that can be discharged into the water bodies. In addition to the undesirable constituents of Table 1.1, there are others whose limitations may also impose effluent quality restrictions. Normally this quality is defined by regional environmental control agencies.

1.4 SEWAGE TREATMENT SYSTEMS

Sewage treatment systems are intended to reduce or remove one or more of the undesirable constituents of Table 1.1 and thus produce an effluent compatible with the imposed standards. Figure 1.1 schematically shows the sequence of processes that normally occur in sewage treatment systems. The treatment systems basically consist of two obligatory components – preliminary treatment and biological treatment – and two optional systems – primary treatment and post-treatment. There is almost always a third component to treat and dispose of the waste sludge that is generated in the treatment system. This waste sludge is composed of inorganic and organic solids, where in the organic solids it is possible to further distinguish between three fractions: (1) a mass of living and dead micro-organisms (2) biodegradable organic material and (3) non-biodegradable organic material.

The sludge treatment system can have two components: (1) a biological component to stabilize the sludge, reducing the organic fraction and the concentration of live micro-organisms, and (2) a physical component to increase the concentration of the sludge solids, thus facilitating its final disposal.

1.4.1 Preliminary treatment

The preliminary treatment step seeks to remove fractions of sewage that strictly should not even be present or that can lead to malfunction of parts of the treatment system. There are three main fractions: (1) gross solids and floating materials, (2) sand and other granular inorganic materials, and (3) oils and greases.

Gross solids and floating materials are made up of various materials that should be disposed of as solid waste, but have been dumped into the sewer: leftovers, paper, plastics, branches and leaves, diapers, condoms, etc. The need to remove gross solids is clear: their presence in the treatment system itself would

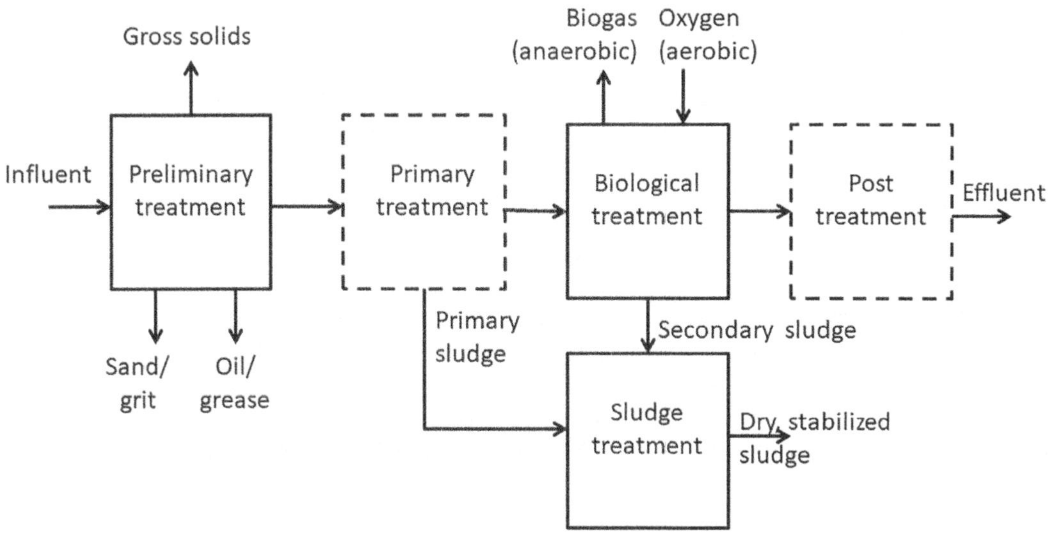

Figure 1.1 Basic flowsheet containing the main treatment steps in a municipal sewage treatment plant.

cause major operational stability problems because together with fat they would inevitably cause clogging or breakdown of, for example, pumps in the subsequent biological treatment system. Also, many of these materials are non-biodegradable and would accumulate in the treatment system, taking up useful space. Usually the removal of coarse solids occurs by mechanical methods: screen bars, possibly aided by static or rotating sieves.

Efficient removal by preliminary treatment can greatly contribute to reducing the organic load that the subsequent biological treatment system receives, both in quantitative and qualitative terms. Quantitatively, more than 30% of the organic sewage material can be retained, depending on the type of sieve holes, etc. It is important to note that the retention of organic material greatly improves the biodegradability of the sieved sewage by retaining two fractions: non-biodegradable particulate material and biodegradable particulate material. If the non-biodegradable or inert particulate material is not retained, it accumulates in the biological treatment system and is then only discharged with excess sludge.

The particulate and biodegradable material must be hydrolysed prior to being metabolized, but this hydrolysis may not be complete, especially in the case of anaerobic treatment. Thus, in the treatment of raw sewage a large part of the volume of a biological reactor will be occupied by particulate material, biodegradable or not. Removal of this material in the preliminary treatment releases the volume of the biological reactor to be occupied mainly by the bacterial mass which is active in the metabolism of the material not retained in the preliminary treatment. On the other hand, it creates a new problem: a large volume of non-stabilized, malodorous primary sludge that needs processing before final disposal. Hence in practice bar screens with 2–3 cm bar separation followed by 4–6 mm fine screens are often used.

For removal of sand, etc., the inorganic solids (with a much higher density than organic material) settle with a much more rapidly, allowing their separation in special units. The removal of oil and grease is by the same principle: oils and greases, being immiscible and lighter than water, tend to float on water in an undisturbed environment and can thus be separated and removed. Preliminary treatment, although vitally important to the operational stability of treatment systems, will not be considered in this text, but is discussed in detail in several books (e.g. Metcalf & Eddy, 2013).

1.4.2 Primary treatment

When the size of cities increased in the nineteenth century, the enormous negative impact of the discharge of raw sewage into surface waters became increasingly apparent. When later in that century it was also shown that this practice was also the cause of the spread of potentially lethal diseases such as cholera and typhus, a program of collection and treatment of sewage began. Around the same time, water consumption (and therefore sewage production) increased enormously by the introduction of large-scale piped water supply, which made sewage treatment even more important.

The first community sewage treatment systems where built in the late nineteenth and early twentieth centuries in several European countries (Mouras in France, Cameron in England, Imhoff in Germany) and in the United States (Philbrick) (McCarty, 1981). Although different in some specific respects, what these systems had in common was that they were designed to remove the most visible part of the pollution: the settleable solids. Removal was performed in masonry tanks through simple sedimentation, now known as primary treatment. The separated material was partially transformed into biogas by anaerobic digestion, thereby reducing the accumulation of solids in the tank and opening the possibility of operation without interruption for long periods. The "liquefied" effluent from the tanks was infiltrated into the ground or discharged into surface waters.

When it was found that the discharge of effluents from primary treatment did not significantly improve the water quality in the receiving bodies, it became clear that the mere removal of the settling solids was insufficient for adequate protection of the surface waters. Initially this was attributed mainly to the presence of organic material in the sewage after sedimentation and led to the application of large-scale biological treatment systems from the first decades of the last century onwards. Today primary treatment systems like septic tanks and Imhoff tanks still have wide application in single-family treatment systems. However, the quality of the effluent is considered insufficient for application in community systems. In Chapter 3, Section 3.2.1 discusses the septic tank, while more recent alternatives for anaerobic treatment in single-family systems are presented in Section 4.6.

In the primary treatment process the settleable solids are separated from the liquid phase by simple settling; the primary sludge which is formed is then discharged into a digestion unit for stabilization of the biodegradable material and subsequently into a dehydration unit, producing a final dry and stable sludge. Initially primary treatment was operated as the whole treatment system, but it is now known that primary treatment is insufficient to produce a satisfactory effluent quality. If it is applied, it is intended to reduce the load on the subsequent biological treatment unit and only when the biological treatment process is aerobic. In this case the reduction of the load translates into a decrease in oxygen demand and therefore in the cost of aeration.

1.4.3 Secondary treatment

Secondary treatment systems are so called because they were developed after it was found that the treatment provided by primary systems was only partial and insufficient for adequate environmental protection. Secondary treatment systems almost invariably use biological methods and were initially developed to reduce the content of organic material. In the secondary or biological treatment step, the same principles that play a role in the removal of organic material in nature apply: degradation by micro-organisms, notably bacteria. Two basic mechanisms for the removal of organic material are distinguished: (1) the aerobic pathway, which results in the oxidation of the material into mineral end products (CO_2 and water), and (2) the anaerobic pathway, which develops when there is no suitable oxidant (e.g. nitrate or oxygen) available. An important third system, which has found ample application in developing countries and other places with a hot climate, are stabilization ponds, where both aerobic and anaerobic mechanisms are used.

In aerobic systems, oxygen is introduced into the wastewater by mechanical aeration. This oxygen sustains a bacterial mass that uses the organic material as a source material for its growth or as a source of energy, using oxygen for oxidation. The bacterial mass may be fixed within the treatment system on a support material, or it may be present in the form of a suspended slurry, which is retained in the system by a final settler separating the solids (containing the bacterial mass) from the liquid effluent before its final discharge. The solids are pumped back to the biological reactor as return sludge. Figure 1.2 shows a schematic representation of an activated sludge system, the aerobic system of greatest application in practice.

Waste stabilization ponds are shallow tanks, where wastewater is maintained for a considerable time (20–30 days), allowing the growth of algae that produce oxygen by the process of photosynthesis that develops in the presence of sunlight. For this reason, the oxygen forms mainly in the upper part of the liquid layer, where the penetration of sunlight is greater. Oxidation of the organic material by the bacteria present in the lagoon then occurs, using in this process oxygen of photosynthetic production. In regions where oxygen is lacking, anaerobic digestion tends to develop, which decomposes organic material through the anaerobic pathway. Figure 1.3 shows a schematic representation of a waste stabilization pond system, consisting of anaerobic and facultative lagoons followed by maturation ponds (this concept will be discussed in detail in Chapter 9).

In anaerobic digestion organic material decomposes into gaseous end products, predominantly methane and carbon dioxide, which are released from the liquid phase into the atmosphere as biogas. At the beginning of the construction of community treatment systems (in the first decades of the twentieth century) anaerobic secondary treatment had much application. Later aerobic treatment became the predominant option, because of the better final effluent quality that could be obtained. But in the last decades of the 20th century, owing to a better understanding of the functioning of anaerobic systems, modern systems of anaerobic treatment with a large organic material removal capacity were developed and the number of anaerobic systems has increased, especially in hot-climate regions, where anaerobic digestion develops at a higher rate. Also, in industrial wastewater treatment anaerobic technologies are widely applied, as the

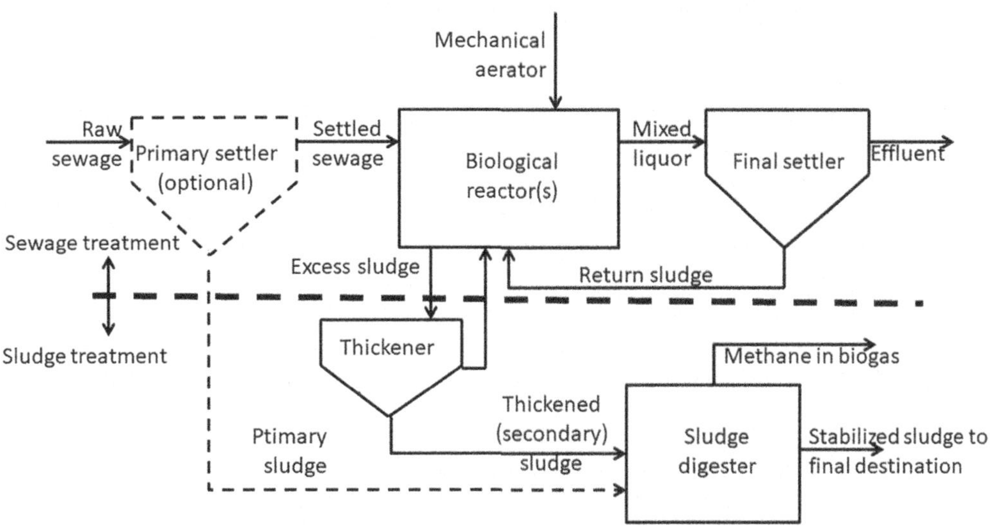

Figure 1.2 Typical aerobic treatment system in activated sludge configuration with a mechanically aerated biological reactor and units to treat the generated waste sludge streams.

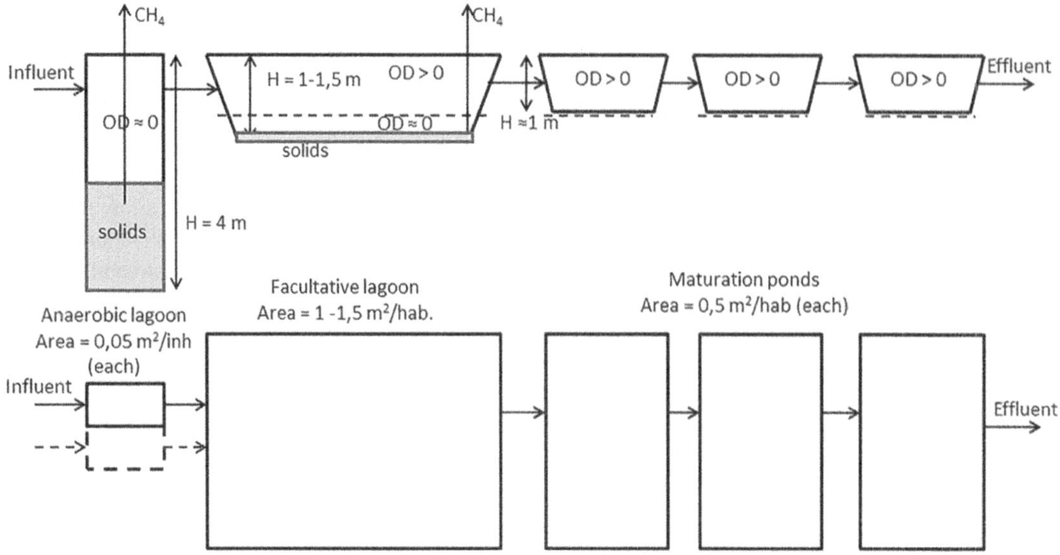

Figure 1.3 Schematic representation of a waste stabilization pond system for sewage treatment.

wastewater is often produced at much higher temperatures and, if not, may contain a high concentration of organic material that upon digestion allows for the heating of the wastewater with the produced biogas.

A large sewage treatment plant based on anaerobic technology (the Upflow Anaerobic Sludge Blanket (UASB) reactor) is shown in Figure 1.4. Having now identified three types of wastewater treatment systems, it is necessary to evaluate which one is optimal for sewage treatment. In this context not only the performance in terms of the removal efficiency of the undesirable components of sewage must be evaluated, but also several other aspects, which are summarized in Table 1.2 and are discussed in more detail below:

(1) ***Performance***
All the variants of the basic systems have good removal of organic material, with the activated sludge system tending to be more efficient than the others. The activated sludge system also has superior removal efficiency of suspended solids and can be designed for a high degree of nutrient removal efficiency, which is normally not possible in stabilization ponds and anaerobic treatment units. On the other hand, in terms of hygienic quality, the activated sludge system is less effective: it removes helminth eggs (caught in the sludge), but the removal of micro-organisms is insufficient. However, with a specific post-treatment (disinfection) the micro-organisms can be removed. Hence it is concluded that in terms of the removal of the undesirable constituents listed in Table 1.2, the activated sludge system has the best performance.

(2) ***Size***
The very large size required for stabilization pond systems weighs against this alternative, especially since there are usually serious odour problems for at least part of the year. For this reason, the lagoon must be built a good distance from the urban centre that it serves. The need for a long sewage outfall often has a higher cost than the waste stabilization pond itself. On the other hand the activated sludge system and anaerobic treatment units are much more compact and when well constructed can be inserted without problem in an urban region and without causing

Figure 1.4 Modern anaerobic system for sewage treatment, the Upflow Anaerobic Sludge Blanket (UASB) reactors at the largest system in Brazil: Onça Treatment Plant in Belo Horizonte.

Table 1.2 Important aspects of the three main biological treatment systems.

Parameter	Stabilization Pond	Activated Sludge	UASB
Removal efficiency			
BOD	90%	95%	80%
COD	80%	90%	70%
TSS	80%	90%	75%
Nitrogen	20%	90%	20%
Phosphorus	20%	90%	20%
Thermo-tolerant bacteria	99.9%	90%	90%
Helminth eggs	99.9%	50%	50%
Size			
Retention time (d)	25	1	0.25
Area ($m^2.hab^{-1}$)	3	0.03	0.01
Depth (m)	1	4	5
Costs			
Construction (US$.$hab^{-1}$)	100	200	50
Operation (US$.$hab^{-1}.year^{-1}$)	1	5	1
Simplicity			
Construction	Positive	Negative	Positive
Operation	Positive	Negative	Positive
Stability	Negative	Negative	Positive
Negative aspects	Methane emission odour, insects, visibility	Sludge, aerosols	Odour Sewage temp. >15–18°C
Positive aspects			Segmentation of the collection network Energy production

discomfort for the population. Due to the large area requirement and the high cost of the sewage pipeline, the stabilization pond is a solution only for small urban areas and is not suitable for large cities. In contrast, large activated sludge systems and UASB reactors for more than 1 million inhabitants have been implemented in many cities around the globe.

(3) ***Costs***
The cost of treatment systems tends to depend heavily on the specific conditions of the site where the system is to be built. However, experience so far clearly shows that anaerobic units tend to have a lower cost than the alternative solutions. In terms of operation the activated sludge systems have a much higher cost, because they consume energy for aeration and produce excess sludge, whose destination is another important cost factor. Also activated sludge systems require specialized personnel for operation, which in practice is usually the biggest factor of operational cost.

(4) ***Simplicity***
All stabilization ponds and anaerobic treatment systems are simple to construct and operate without motors or moving parts. Generally, construction materials and personnel are available near the area where they are operating, and the operation is limited to actions like the removal of blockages to ensure that the sewer flow enters the treatment system, following the expected path. On the other hand, in the activated sludge system there are daily actions such as the discharge of excess sludge and evaluation of settleability, as well as preventive and corrective maintenance activities of the electromechanical parts of the system, which suffer continuous wear.

(5) ***Operational stability***
Operational stability is directly related to the degree of interference necessary for proper performance of the treatment system. Since activated sludge has a more intensive operation it is predictable that there may be more frequent operational problems. In UASB reactors the operation is limited to sludge discharges when the accumulated mass is too large, and sludge is discharged along with the effluent. The lagoon system, because it is a unit that depends heavily on the weather, can show signs of instability when the climate changes or when there are prolonged periods of atypical or adverse climatic conditions. The fact that required pond operations are minimal should not be confused with the absence of operation and maintenance, as is generally the case in practice: there are dozens of ponds that are not functioning because the responsible – or rather irresponsible – company has allowed the accumulation of solids that caused silting

(6) ***Negative aspects***
Although the population may be aware of the benefits that sewage collection and treatment bring to public health and the environment, it is generally not willing to readily accept the presence of treatment systems near its residence (NIMBY). In particular, there is often resistance against the construction of waste stabilization ponds that are not only very visible, but can also be problematic because of odours and mosquito breeding.

A particularly negative aspect is the production of methane gas in the anaerobic lagoons, which escapes into the atmosphere and thus contributes to global warming. Methane gas has an impact 21 times higher than carbon dioxide on the greenhouse effect (Yvon-Durocher *et al.* 2014). Activated sludge systems can have major problems with the generation of waste sludge: its final disposal is becoming increasingly problematic and costly. Anaerobic systems may have odour problems, particularly when they are not well constructed to prevent gas leakage.

(7) ***Positive externalities***
In the past it was generally considered advantageous for a city to have a single system for treatment of all the generated wastewater. With the growth of city size, the cost of sewage collection in built sewage networks has grown more than proportionately and it has become economically feasible to

have more than one system in large cities. In fact, today almost all major cities have more than one treatment system. Another cost factor of centralized systems is the need for the pumping stations that become necessary to transport sewage over large distances to the single treatment system. These stations are also a high cost factor for investment and operation for sanitation companies. For these reasons the division of sewage treatment into geographical segments each with its own treatment system becomes increasingly attractive. In many cases the segmentation of the sewage network will reduce the cost so much that the treatment systems themselves can be constructed from the cost savings of the network construction.

Anaerobic reactors are especially suitable for use in this segmented treatment approach, and this possibility has been adopted in several large cities in Brazil, such as the metropolitan areas of Curitiba and Campinas where the local companies operate several anaerobic systems. An important consideration of anaerobic treatment is that the effluent obtained will have so few solids in suspension that it can be considered equivalent to raw water and can be pumped in pipelines under pressure to the post-treatment sites, which greatly reduces the cost. Of course, it is not always possible to have several treatment systems: in the case of stabilization ponds for example, odour problems will often lead to the need to use a single system. In the case of activated sludge systems, the operational requirement for advanced human and material resources can also make it impossible to divide the sewage collection into several segments.

Another positive aspect of anaerobic treatment is the production of energy in the form of biogas, rich in methane, which can be used to generate electricity in gas motors. However, in the case of treatment of wastewaters as dilute as municipal sewage, the potential for energy production is limited and is only worth implementing in large systems, from 500,000 to 1,000,000 equivalent inhabitants. This means that for smaller systems the biogas is often flared.

Considering the different aspects discussed above, it is concluded that all three basic types of treatment have problematic aspects. The "best" system depends on several factors such as:

(1) Legislation that defines the minimum quality of the final effluent: normally the legislation is such that the allowable concentration of nitrogen and phosphorus nutrients cannot be achieved using stabilization ponds or anaerobic systems, so that a post-treatment system is necessary. In countries like Brazil there are also problems of governmental interference in the legislation to support plans or policies. For example, in the recent housing expansion program the government, in order to avoid the high construction cost of nutrient removal systems, in particular for nitrogen, simply increased the permissible limit in the effluent by decree. The effluent discharge limits in Brazil, which are now very relaxed compared to those imposed in other countries, will have to be revised in the future. Stricter limits will be needed to impose to protect water resources adequately, particularly with respect to eutrophication. On the other hand, new processes of treatment are continuously being developed that will allow a good quality of effluents to be obtained at reduced costs, as is the case of Anammox (Mulder *et al.* 1995) for nitrogen removal, although stable operation at lower temperatures for municipal applications still not has been demonstrated.

(2) The local situation can impose limits on system size and thus render unfeasible treatment systems like stabilization ponds. A clear example is the situation of the city of Campina Grande in Brazil. The city has a stabilization pond system to treat its sewage and the effluent is released into the Bodocongó stream, eventually reaching the Acauã dam that supplies several cities in the central region of Paraíba State. The great nutrient load causes calamitous eutrophication of the water in the reservoir. This situation will only improve if a nutrient treatment system is in place or if all sewage is treated for agricultural reuse.

(3) The intended use of the effluent is another important factor. In the case of agricultural reuse, it is advantageous to maintain nutrients in the treated wastewater, but it is important that the evaporation of water is limited, not only to avoid losses of the precious liquid, but also prevent an excessive increase in the salinity that reduces the value of water for irrigation. If the final effluent is reused in the industry or discharged into surface water, it will be necessary to remove nutrients.

It is only after a detailed analysis of these aspects that a conclusion can be drawn as to which system should be deployed. In many cases, the optimal solution will not be one of the three basic systems, but rather a combined solution in which full advantage is taken of the positive aspects of each unit and the negative aspects are minimized. Examples are combinations with a UASB reactor followed by an activated sludge system, or a UASB reactor with a complementary polishing pond; such combinations are discussed in Chapters 8 and 9, respectively. The combined anaerobic–aerobic system is smaller than an aerobic system without anaerobic pre-treatment and also will have lower energy demand and sludge production, but it may not be possible to remove nutrients in quantity. When the UASB is combined with a polishing pond, this combination requires a much smaller area than a traditional stabilization pond and has no odour problems, nor does it emit methane into the atmosphere.

1.4.4 Tertiary treatment

In the second half of the last century it became evident that the discharge of effluents from secondary treatment systems could result in deterioration of the water quality of the receiving body, even when removal of suspended solids and organic material was very efficient. The deterioration could be attributed to the presence of the macro-nutrients nitrogen and phosphorus. Ammonium nitrogen (usually predominant in sewage) exerts a demand for oxygen, since in the aquatic environment it can be oxidized to nitrate in a process called nitrification. However, even when nitrification is performed in the treatment system itself, the discharge of effluent can still impair the quality of the water of the receiving body due to the introduction of nitrogen (nitrate) and phosphorus (as phosphate) in excess.

The availability of these elements tends to cause an excessive growth of aquatic life, which needs these elements for cell synthesis, especially autotrophic organisms like algae, which use inorganic carbon (CO_2) instead of organic carbon for cell synthesis. In this way an explosive development of (autotrophic) biomass can occur in water when nitrogen and phosphorus are available in abundance. This biomass can produce dissolved oxygen (DO) during the day by the process of photosynthesis, but at night there will be oxygen consumption and the concentration of DO may reach levels too low to sustain the life of other (macro) organisms. The phenomenon of water quality deterioration due to the excessive discharge of nutrients, called eutrophication, led to the development of tertiary treatment systems: in these, nitrogen and phosphorus are removed, as well as settleable solids (primary treatment) and organic material (secondary treatment). Biological treatment systems were developed for the removal of nitrogen through the sequential processes of nitrification and denitrification (the biological reduction of nitrate to molecular nitrogen using organic material), and of phosphate by creating a favourable environment for the development of a bacterial mass with high content of phosphorus (Rabinowitz & Marais, 1981)). The removal of phosphorus occurs by incorporating phosphate into the bacterial mass, which is then removed from the system as excess sludge.

Currently the removal of nitrogen through denitrification by heterotrophic bacteria that use nitrate as an oxidant for organic material is the most applied process in practice. Recently a process that opens great possibilities for the treatment of wastewater with deficiency of organic material was developed at the University of Delft (Mulder *et al.* 1995): the Anammox process, which has been applied in full-scale systems. It is the discovery of a type of autotrophic bacteria, which uses nitrite to oxidize ammonia, transforming the two forms of nitrogen into molecular nitrogen that desorbs from the liquid phase. In this

Table 1.3 Some examples of post-treatment methods for secondary or tertiary effluents.

Post-treatment	Objective	Goal
Filtration – granular	Improve hygienic quality	Environmental protection
Clarification + filtration of pond effluents	Algae removal	Use of algae (digestion, biodiesel, animal fodder) and non-agricultural reuse
Clarification of anaerobic or aerobic effluent	Improve aesthetics	Reuse: washings, agricultural reuse
Softening	Hardness removal	Industrial reuse
Desalinization by membranes	Removal of the ionic matrix	Industrial reuse

way nitrogen removal is possible without the mediation of heterotrophs and without the consumption of organic material. In its simplest form the process is written as:

$$NH_4^+ + NO_2^- \rightarrow N_2 + 2H_2O$$

The Anammox process has two important advantages compared with the traditional process of nitrification to nitrate and denitrification with organic material: (1) it considerably reduces the demand for oxygen, and (2) it does not require organic material. The discovery of the Anammox process allowed the development of several systems to treat effluents rich in nitrogen and lacking in organic material such as effluent from sludge digesters or effluent from the anaerobic treatment of (industrial) wastewater. These can be two-step systems, having an aerobic zone (nitritation) and an anoxic zone (Anammox), separated in space or time. Carrying out autotrophic nitrogen removal in two steps implies that both can be optimized and controlled individually. However, it soon became apparent that the cost of the investment and the difficulty of balancing the two steps are reduced when operating a single-stage system where the two processes are developed simultaneously (Vlaeminck *et al.*, 2012). Because of these advantages, large-scale applications are tending toward single-stage configuration.

Today it is possible to operate tertiary treatment systems and discharge an effluent essentially free of suspended solids, organic material and nutrients. The post-treatment can also be applied to correct aspects of the effluent to make it more suitable for reuse for certain purposes. Some applications of post-treatment for certain purposes can be observed in Table 1.3. The problem with these systems is that the cost of construction and operation is much higher than for the secondary treatment systems.

1.5 SEWAGE CHARACTERISTICS

The most important constituents in sewage treatment are those that give wastewater its undesirable physical, chemical and biological properties: suspended solids, organic material, nutrients and pathogenic organisms. The composition and concentration of these constituents in sewage depends greatly on the socio-economic customs of the contributing population and the nature of the collecting network that transports the sewage to the treatment system. This network can be the same one that also drains the rainwater (combined network), but more modern collection networks try to separate the rainwater from sewage (separated network). Because of variations in population contributions and per capita flow, the concentration of the components in sewage can vary considerably in different places.

Table 1.4 shows the composition of sewage as function of the economic situation of the contributing population in Brazil. The data reflect the experience obtained by the authors and several colleagues in different regions of Brazil (Pacheco & Pessoa, 2017). Three social classes are considered: upper, middle and lower class. The last column of Table 1.4 shows the average contribution per capita in the US as reported by Metcalf and Eddy (2013). From the data in Table 1.4 the following conclusions can be drawn:

(1) In low- to middle-income countries like Brazil there is a strong quantitative and qualitative correlation between income and sewage production:
   (a) Populations with a high income produce much more sewage;
   (b) The sewage has a higher per capita load of organic material, suspended solids and nutrients if the contributor has a higher income;
   (c) The TKN/COD and P/COD ratios are higher for high-income groups, presumably because of the diet being richer in proteins.
(2) Comparing the Brazilian and US data one can see that:
   (a) The flow of sewage in the US is much greater than in Brazil (for any income class);
   (b) The per capita contribution of all undesirable constituents is much larger in the US than in Brazil;
   (c) The TKN/COD and P/COD ratios are high, indicating lots of proteins in the US diet.

It is interesting to note that the composition and concentration of sewage tends to change over time, especially in emerging countries such as Brazil, where great changes in the economic conditions and consequently the consumption of food by the population have occurred. Notably the concentrations of nutrients nitrogen and phosphorus have increased greatly in recent decades, presumably due to a large increase in protein consumption in the Brazilian diet.

In addition to the constituents that are directly related to the discharge of waste, the composition of sewage is also influenced by the composition of the drinking water supply, which eventually will end up in the produced sewage. In this context the alkalinity and concentration of mineral dissolved solids, i.e. the inorganic salts, are particularly important. The alkalinity determines to a large extent the pH value and stability in the sewage and the receiving treatment system. The difference of alkalinity between sewage produced and drinking water supplied is mainly due to the discharge of organic nitrogen that tends to decompose in the sewage network and in the biological wastewater treatment system, forming ammonia, which is a base. Because of this ammonification process there is a difference between the alkalinity in the sewage and the public water supply (in $meq \cdot L^{-1}$) that is, to good approximation, equal to the concentration of ammonia in the sewage (also in $meq \cdot L^{-1}$).

Table 1.4 Main constituents of predominantly municipal sewage in Brazil and in the US (adapted from Metcalf & Eddy, 2013; Pacheco & Pessoa, 2017; Vlaeminck *et al.* 2012).

Constituent	Unit	Income Class in Brazil			US
		High	Medium	Low	Average
Per capita contribution	$l \cdot d^{-1}$	180	120	75	483
COD load	$g \cdot hab^{-1} \cdot d^{-1}$	120	80	60	192
BOD load	$g \cdot hab^{-1} \cdot d^{-1}$	60	40	30	85
Suspended solids load	$g \cdot hab^{-1} \cdot d^{-1}$.	60	40	30	91
Nitrogen load	$g\,N \cdot hab^{-1} \cdot d^{-1}$	14	8	5	17
Phosphorus load	$g\,P \cdot hab^{-1} \cdot d^{-1}$	3	2	1	3

The total concentration of salts is important because it determines the electrical conductivity: a high conductivity contributes to corrosion of metals. The presence of a high concentration of sulphate tends to cause corrosion problems for concrete structures in the treatment system and can be converted into toxic and noxious H_2S gas under anaerobic conditions. Another important aspect is that it reduces the applicability of treated sewage as irrigation water as it increases salinity. For these reasons it is important to limit the addition of ions during the treatment and to avoid evaporation of water in the treatment system, since the loss of water vapour increases the salinity of the remaining water.

1.5.1 Flow rate

The flow of sewage is determined primarily by the size of the contributing population and the per capita contribution, which in turn strongly depends on the economic situation (Table 1.4). The nature and quality of the sewage network also has a strong influence. In the past, a single network was built for the two purposes, sewage and rainwater collection. Since there was usually no treatment system, the raw sewage was discharged together with the rainwater into the receiving water body. At present usually there are separate networks for the different water qualities.

In practice there is always an influence of rainwater on the sewage flow, even in separated systems, because the separation is never perfect: part of the rainwater is discharged by clandestine connections to the sewerage network and irregular sewage discharges into the rainwater network also occur. Even considering only sewage production, there is typically a strong variability of the flow rate received at the treatment plant, due to the daily cyclic nature of activities of the population and the associated water use and sewage production. In Figure 1.5 a typical incoming sewage flow profile can be observed. In this flow the extent of the variation is already attenuated because the sewage is introduced at different points in the network and it takes different periods of time to reach the end of the network, where the treatment system is located. Figure 1.6 shows the accumulated (total) sewage volume that is received at the treatment system. In Example 1.1 it will be demonstrated that with the installation of an equalization basin it is possible to eliminate to a large extent the variations in sewage flow and obtain an almost constant flow in the treatment system, which improves the operational stability of the treatment system. However, equalization tanks are generally not used in treatment systems because the cost and difficulty of operating such a basin is greater than the advantages it potentially brings.

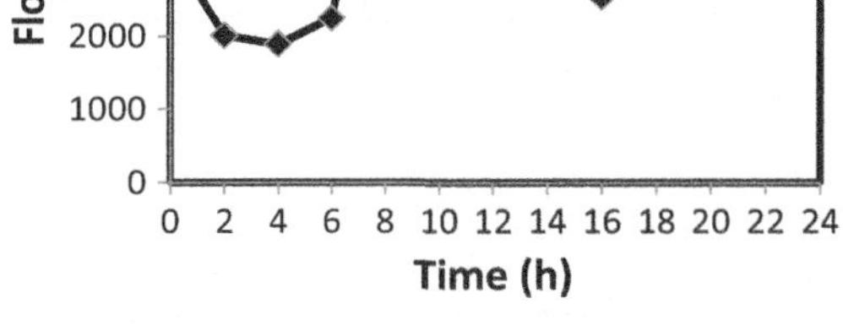

Figure 1.5 Typical profile of the variation of the incoming sewage flow.

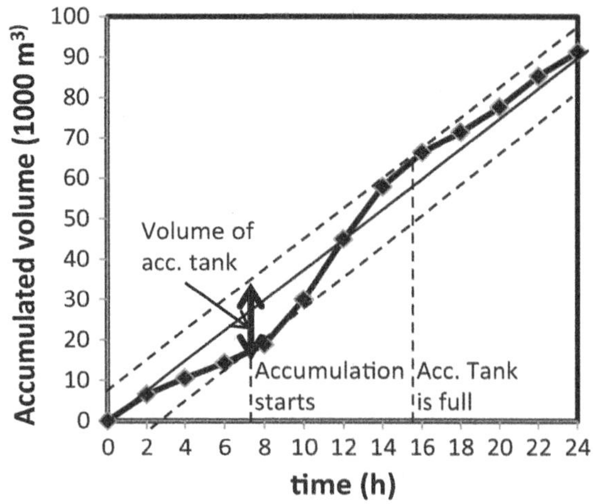

Figure 1.6 Accumulated daily sewage volume received in treatment plant.

EXAMPLE 1.1

Estimate for the data in Table 1.5 (shown in Figures 1.5 and 1.6) the minimum volume of an equalization basin required to be able to process sewage at essentially constant flow rate.

Solution

Using the data in Table 1.5, the Figures show the instantaneous flow rate and the accumulated daily volume of sewage as a function of time. The procedure for calculating the volume of the equalization basin is summarized as follows:

(1) Draw a straight line from the accumulated volume at the starting point (0 h) to the volume at the end of the day (24 h)
(2) Draw two lines tangent to the volume–time curve and parallel to the line drawn in step 1
(3) Estimate the volume of the equalization basin considering that the basin must be empty at the lower tangent point (when accumulation begins) and full at the upper tangent point (when emptying begins)
(4) In this way, for the case of Figure 1.6, a volume of approximately 16,000 m^3 or equal to 17% of the daily flow is calculated (resulting in a hydraulic residence time of 5.7 h). This volume is rather large, considering that the retention time of a UASB reactor may be of the same order of magnitude.

It is interesting to observe in Table 1.5 that the organic load varies much more than the flow or COD concentration of the sewage, because in general the maximum and minimum values of both the flow and the COD concentration occur more or less at the same time.

For this reason, the average load (2842 $kg \cdot d^{-1}$ in Table 1.5), calculated as the average of the products of instantaneous flow and COD concentration, is much higher than the product of the average flow and the average COD ($3801 \times 675 = 2566$ $kg \cdot d^{-1}$). Of course, the value to be used to estimate the average load received by the treatment system should be the average of the products of flow and COD concentration (i.e. the higher value).

Table 1.5 Values of sewage flow rate (column 2) and COD concentration (column 3) received at a treatment system as function of the time of day (column 1). The values of the accumulated volume in one day and the instantaneous organic load are also indicated (columns 4 and 5).

(1) Time (h)	(2) Flow ($m^3 \cdot h^{-1}$)	(3) COD ($mg \cdot L^{-1}$)	(4) Totalized flow (m^3)	(5) Organic load kg $COD \cdot h^{-1}$
0:00	3240	325	0	1053
2:00	2010	517	6480	1039
4:00	1890	439	10,500	830
6:00	2250	672	14,280	1512
8:00	5590	821	18,780	4589
10:00	7470	930	29,960	6947
12:00	6510	1102	44,900	7174
14:00	4200	892	57,920	3746
16:00	2550	935	66,320	2384
18:00	3110	661	71,420	2056
20:00	3810	428	77,640	1631
22:00	2980	384	85,260	1144
24:00	3240	325	91,220	1053
Average	3801	675		2842

1.5.2 Measurement of organic material

Organic compounds in wastewater are of particular interest to the sanitary engineer: a wide variety of micro-organisms (which may be present in the wastewater or the receiving body) interact with the organic material, using it as a source of energy or as source of material to synthesize new cells. The use of organic material by micro-organisms is called metabolism. It consists of two processes, the first being catabolism, the consumption of organic material to obtain energy, while the second, anabolism, is the synthesis of cellular material from the organic material. These processes are discussed in more detail in Chapter 2.

To describe the metabolic processes in treatment systems, it is necessary to be able to quantitatively characterize the concentration of the organic material in the sewage. Given the enormous variety of organic compounds in sewage, it was quickly realized that it is totally impractical (if not impossible) to determine them all individually. Therefore, a parameter was needed to evaluate the concentration of organic compounds, which uses one of two properties that almost all organic substances have in common: (1) organic material contains organic carbon and (2) organic material can be oxidized to inorganic substances.

In sanitary engineering there are two standardized tests that are based on the oxidation of organic material: the Biochemical Oxygen Demand (BOD) and the Chemical Oxygen Demand (COD) tests. In both there is destruction of the organic material by an oxidant and the concentration of organic material is determined from the consumption of the oxidant. The essential differences between the BOD and COD tests are the nature of the oxidant and the operating conditions prevailing during the test.

It should be noted that in an anaerobic environment and in the presence of a mass of suitable micro-organisms, organic material can be converted into biogas, a mixture of gaseous products, mainly methane and carbon dioxide. In this process, known as anaerobic digestion, there is in fact no destruction of the organic material: it is converted into an equivalent mass of methane, which is released, thereby effecting the physical removal of the organic material from the liquid phase.

The quantification of organic carbon led to the development of the Total Organic Carbon (TOC) test as an alternative to quantify the concentration of the organic material. Considering the great importance of organic material in sewage treatment, the tests are considered in more detail below.

1.5.2.1 The COD test

In the COD test, a sample of wastewater is mixed with a strong oxidant: a mixture of dichromate and sulfuric acid. Silver sulphate is added as catalyst (accelerating the oxidation of fats and acetate) and the mixture is heated to its boiling point (>150°C). The result is that after a period of two hours (the duration of the test) the oxidation of the organic substances is practically complete. This can be verified by using a solution of a pure organic substance of known concentration.

The theoretical COD of such a solution is calculated from the stoichiometry of its complete oxidation. The theoretical value can be compared with the experimental value obtained. For the general structural formula $C_xH_yO_z$, the oxidation reaction with oxygen is expressed as:

$$C_xH_yO_z + \tfrac{1}{4}(4x + y - 2z)O_2 \rightarrow xCO_2 + (y/2)H_2O \tag{1.1}$$

With the aid of the known molar weights of H (1 g), C (12 g) and O (16 g) it is calculated that the oxidation of 1 "mol" of organic material, i.e., a mass of $x + 12y + 16z$ g, requires an oxygen mass of ¼ $(4x + y - 2z)$ moles or $8(4x + y - 2z)$ g O_2. Therefore, the theoretical COD of a component $C_xH_yO_z$ can be expressed as:

$$COD_t = 8 \cdot (4x + y - 2z)/(12x + y + 16z) \text{ mg COD} \cdot \text{mg}^{-1}\ C_xH_yO_z. \tag{1.2}$$

EXAMPLE 1.2

Calculate the theoretical COD concentration of a solution with 1200 $mg \cdot L^{-1}$ of sodium acetate.

Solution

Sodium acetate ($CH_3COONa \cdot 3H_2O$) is a salt of acetic acid. The oxidation of the acetate converts the acetic acid ($x = 2$, $y = 4$, $z = 2$), while the sodium ion remains in solution:

$$CH_3COOH + \tfrac{1}{4}(4 \times 2 + 4 - 2 \times 2)\ O_2 \rightarrow 2\ CO_2 + (4/2)H_2O$$

or

$$CH_3COOH + 2\ O_2 \rightarrow 2\ CO_2 + 2\ H_2O$$

Therefore, the theoretical COD value of acetic acid $COD_t = 8(4 \times 2 + 4 - 2 \times 2)/(2 \times 12 + 12 + 16 \times 2) =$ 1.067 mg COD $\cdot$ mg^{-1} CH_3COOH. Since the COD of 1 mol of sodium acetate (136 g) is equal to the COD of 1 mol of acetic acid (60 g), the COD value of the solution of 1200 $mg \cdot L^{-1}$ of sodium acetate is equivalent to $1200 \times 60/136 = 529$ $mg \cdot L^{-1}$ acetic acid. Thus, the theoretical COD of a solution of 1200 $mg \cdot L^{-1}$ of sodium acetate (equivalent to 529 mg HAc $\cdot L^{-1}$ of acetic acid) is $529 \times 1.067 = 565$ $mg \cdot L^{-1}$.

EXAMPLE 1.3

Calculate the theoretical COD of a mixture of 1200 mg·L^{-1} of sodium acetate and 0.5 g·L^{-1} of glucose ($C_6H_{12}O_6$).

Solution

The COD concentration of 1200 mg·l^{-1} of sodium acetate was calculated above as 565 mg COD·L^{-1}. Glucose ($x=6$; $y=12$ and $z=6$) has a theoretical COD value equal to $8(4x+y-2z)/(12x+y+16z)=1.067$ mg COD·mg^{-1} $C_6H_{12}O_6$. Hence a solution of 0.5 g·L^{-1} of glucose has a COD concentration of $0.5\times1000\times1.067=533$ mg·L^{-1}. Therefore, the mixture has a COD concentration of $565+533=1098$ mg·L^{-1}.

In Table 1.6 the theoretical COD values per unit mass were calculated for different compounds of the type $C_xH_yO_z$. It is interesting to note that COD per unit mass may have very different values for different compounds. In the case of reduced compounds such as methane (# 18 in the list of Table 1.6) the theoretical COD value will be high. Substituting for the methane values (i.e. $x=1$, $y=4$ and $z=0$) in Eq. (1.2), a value of $COD_t=8(4\times1+4-2\times0)/(12\times1+4+16\times0)=4$ mg COD·mg^{-1} CH_4 is calculated. In the case of a much more oxidized compound such as oxalic acid, $(COOH)_2$ (# 1 of the list) the COD value per unit mass is much lower. In this case $x=2$, $y=2$ and $z=4$, so that for oxalic acid $COD_t=8(4\times2+2-2\times4)/(12\times2+2+16\times4)=0.18$ mg COD·mg^{-1} $(COOH)_2$.

Table 1.6 Stoichiometric values of COD and TOC per mass unit for pure single substances.

No.	1	2			3	4	5
	Substance	Formula $C_xH_yO_z$			g COD/g $C_xH_yO_z$	g TOC/g $C_xH_yO_z$	g COD/g TOC
		x	y	z			
1	Oxalic acid	2	2	4	0.18	0.267	0.67
2	Formic acid	1	2	2	0.35	0.261	1.33
3	Citric acid	6	8	7	0.75	0.375	2.00
4	Glucose	6	12	6	1.07	0.400	2.67
5	Lactic acid	3	6	3	1.07	0.400	2.67
6	Acetic acid	2	4	2	1.07	0.400	2.67
7	Glycerine	3	8	3	1.22	0.391	3.11
8	Phenol	6	6	1	2.38	0.766	3.11
9	Ethylene glycol	2	6	2	1.29	0.387	3.33
10	Benzene	6	6	0	3.08	0.923	3.33
11	Acetone	3	6	1	2.21	0.621	3.56
12	Palmate acid	16	32	2	2.88	0.750	3.83
13	Cyclohexane	6	12	0	3.43	0.857	4.00
14	Ethylene	2	4	0	3.43	0.857	4.00
15	Ethanol	2	6	1	2.09	0.522	4.00
16	Methanol	1	4	1	1.50	0.375	4.00
17	Ethane	2	6	0	3.73	0.800	4.67
18	Methane	1	4	0	4.00	0.750	5.33

From the above it is clear that the expression "mass of organic material" as used throughout this book – and more broadly in the field of wastewater treatment – actually means "mass of oxygen required to oxidize the organic material present." Thus, 1 g of organic material (expressed as COD) may be equal to 1/4 = 0.25 g of methane (CH_4) or 1/0.18 = 5.6 g of oxalic acid $(COOH)_2$. Both these masses require a mass of 1 g of oxygen for their complete oxidation.

Another very important conclusion is that when oxygen is used as an oxidant for organic material, the mass of consumed oxygen will by definition be exactly equal to the oxidized mass of COD_t. The following general observations about the COD test can be made:

(1) Even though the dichromate–sulphuric acid mix is a very potent oxidant, it does not oxidize organic nitrogen or ammonium.
(2) In contrast, it will oxidize sulphur compounds to the most oxidized compound: sulphate.
(3) Not all organic compounds can react with the reagents of COD analysis. The most notable exceptions are acetate (which can be overcome with a silver acetate catalyst) and organic nitrogen compounds such as trimethyl glycine (betaine, often present in sugar beets) and methylamines.
(4) The COD test yields reliable and reproducible results for soluble low- to medium-strength wastewaters, but less so for high-strength wastewaters or high-solids slurries.

1.5.2.2 The BOD test

The BOD test is like the COD test in that oxidant consumption is used to quantify the organic material. However, in the BOD test the oxidant is the dissolved oxygen itself and for the redox reaction with organic material to materialize, mediation of bacteria is needed. If bacteria are not present in enough concentration in the sample to be tested, they must be added (seeded) together with nutrients, so that their concentration does not represent a limiting factor for the oxidation of organic material (APHA, AWWA and WEF, 2017). Whereas in the COD test the oxidation of organic material is essentially complete in less than two hours, the biological oxidation of organic material is a much slower process and takes several weeks to complete. As it is impractical to wait so long for the test result, a duration period of 5 days was agreed for the test, determining only a partial consumption in this period. To exclude the influence of temperature on the oxidation rate, a constant temperature of 20°C is maintained during the test. Therefore, unless otherwise specified, the BOD of a wastewater represents the biological consumption of oxygen over a 5 day incubation period at a temperature of 20°C. To avoid photosynthetic oxygen production, the sample is kept out of reach of the light during the test.

Due to the nature of the BOD test the accuracy is lower than that of the COD test: while the COD test can be performed with a maximum error of 5%, the BOD may easily have a 10–20% error. In the case of wastewater with a wide range of organic compounds, an additional difficulty for BOD as a quantitative parameter is the fact that the oxidation rate of the organic material depends on the nature and size of its molecules. Small molecules can be used easily by bacteria, but the macromolecules of colloidal and particulate material can only be metabolized after the necessary preparatory steps. Thus, it becomes impossible to establish a theoretical relation with general validity relating the BOD after 5 days of incubation and the ultimate BOD, i.e. after several weeks of incubation. However, an empirical relationship between these two parameters was presented by Phelps (1944) for raw sewage:

$$BOD_t = BOD_u(1 - \exp(-k_p t)) \tag{1.3}$$

where

BOD_t = BOD value after a period "t" of incubation
BOD_u = ultimate (maximum) BOD concentration
k_p = empirical proportionality constant = 0.23 d^{-1} at 20°C.

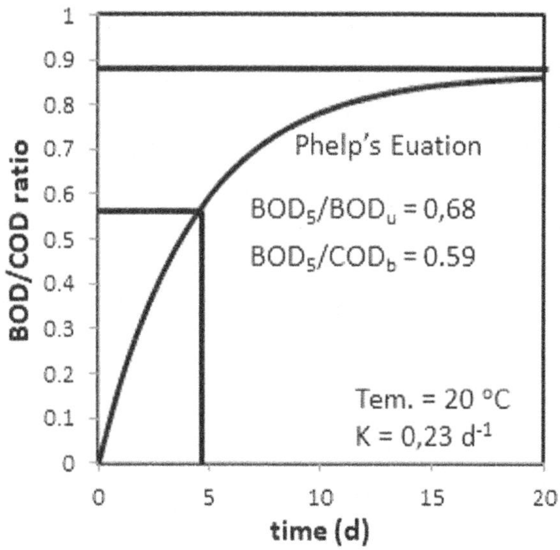

Figure 1.7 BOD_5/COD and COD_u/COD_b as a function of incubation time for biodegradable material.

Figure 1.7 is a graphical representation of Eq. (1.3). At 20°C the BOD after 5 days of incubation is equal to a factor $1 - \exp(-0.23 \times 5) = 0.68$ of the ultimate BOD of a sewage sample. Therefore, the ultimate BOD is around 1.5 times greater than the value it has at 5 days. It should be noted that Eq. (1.3) is an empirical expression that is based on experimental observations of BOD of wastewater as municipal sewage. The validity of the expression is therefore limited to this type of wastewater.

In Figure 1.7 it can be observed that the ultimate BOD value deviates from the COD value. This is because a fraction of the organic material is not oxidized but converted into cellular material, even after a long incubation period. This material is called the endogenous residue (Van Haandel & Van der Lubbe, 2012). The fraction of metabolized organic material that ultimately is converted into endogenous residue is about 13% (Marais & Ekama, 1976), so that the ultimate BOD is only 87% of the COD concentration if all the material of the sample is biodegradable. It is important to know that not all organic compounds can be oxidized under the conditions prevailing in the BOD test. Some organic compounds do not exert any measurable oxygen demand during the BOD test. In this way a distinction is made between the biodegradable material (which can be oxidized during the BOD test) and non-biodegradable material (which is not affected by the test). In most wastewater the organic material is a mixture of biodegradable and non-biodegradable material. For the biodegradable material there is a proportional relation between BOD and COD. Since the COD after 5 days is equal to 68% of the ultimate BOD value, which is equal to 87% of COD concentration, it follows that:

$$BOD_5 = 0.68 \times BOD_u = 0.68 \times 0.87 \times COD_b = 0.59 \times COD_b \text{ or } COD_b/BOD_5 = 1/0.59 = 1.70 \tag{1.4}$$

where

COD_b = biodegradable COD concentration.

In general, the COD/BOD ratio in wastewater exceeds the minimum value of 1.70, which represents a wastewater of which the organic material would be completely degradable. This is because there is almost always non-biodegradable material present. Thus, for example, in the case of domestic sewage the

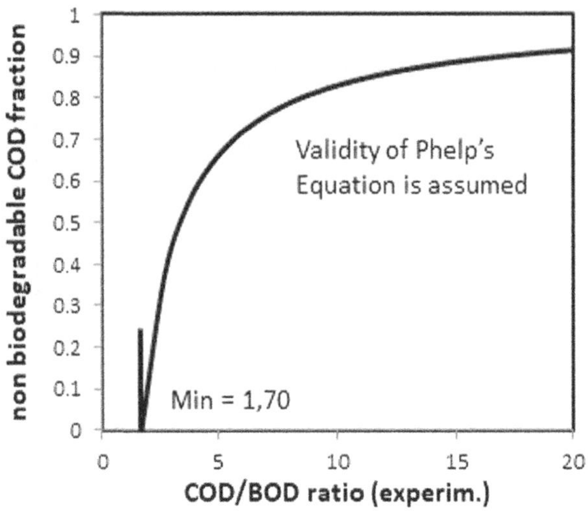

Figure 1.8 Fraction of non-biodegradable material as a function of the COD/BOD ratio.

COD/BOD ratio usually lies in the range of 1.8–2.2. For any value "*R*" of the COD/BCOD ratio, the non-biodegradable fraction of the organic material can be estimated as:

$$f_n = COD_n/COD = (R - 1.70)/R \tag{1.5}$$

where

f_n = non-biodegradable fraction of the organic material
COD_n = non-biodegradable COD of the organic material
R = experimentally determined ratio between COD and BOD.

Figure 1.8 shows the non-biodegradable fraction as a function of the COD/BOD ratio. To qualify and quantify as accurately as possible the organic material that is present in wastewater it is interesting to know both its BOD and COD value. In practice it is generally too laborious to perform both tests routinely. Normally COD is performed more frequently, because this test leads to a more accurate result in a much shorter period. The BOD test is performed less frequently, aiming to establish an empirical relationship between BOD and COD concentrations. Once the ratio between the two values is known, the BOD value is estimated from the COD test result. The COD test has another very significant advantage: in Section 1.4.3 it is shown that, when using this parameter, it is possible to make a mass balance of organic material. This is important because it allows verification of the experimental procedures used to measure the organic material in its various forms. Also a mass balance is very useful to rationally design a wastewater treatment plant, as will be discussed in later chapters.

EXAMPLE 1.4

A BOD test with a solution of 1200 $mg \cdot L^{-1}$ of sodium acetate (Example 1.2) yields 300 $mg \cdot L^{-1}$. Is this result compatible with theory? Consider that acetate is fully biodegradable.

Solution

The theoretical COD concentration of the solution is 565 mg · L^{-1} (as shown in Example 1.2). The theoretical BOD would therefore be $0.59 \times 565 = 333$ mg · L^{-1}. Hence the result of the test of 300 mg · L^{-1} is a fraction of $300/333 = 0.90$ of the theoretical value. The difference of 10% can be attributed to experimental error and therefore the result is compatible with its theoretical value.

1.5.2.3 *The anaerobic biodegradability test*

The biodegradability of organic material is evaluated in the BOD test for an aerobic environment, but the result obtained is not necessarily indicative for the anaerobic environment, since the anaerobic digestion is mediated by other micro-organisms. Biodegradability in the anaerobic environment can be experimentally determined by measuring the production of methane obtained from the organic material in a sample of wastewater when it is brought into contact with a batch of anaerobic sludge under previously selected and stable conditions. In Appendix A1 the test setup and protocol, the operational conditions and the calculation procedures to determine anaerobic biodegradability are presented in detail.

In analogy with Eq. (1.1) to express the oxidation of a compound $C_xH_yO_z$, one can develop a generic equation for anaerobic digestion:

$$C_xH_yO_z + (4x - y - 2z)/4\ H_2O \rightarrow (4x - y + 2z)/8\ CO_2 + (4x + y - 2z)/8\ CH_4 \quad (1.1a)$$

1.5.2.4 *Total Organic Carbon – TOC*

The TOC test uses the property that all organic substances contain organic carbon. Organic carbon is measured as carbon dioxide after incineration of the organic material present in a sample of wastewater. Correction should be made for the inorganic carbon originally present in the sample. The theoretical value of TOC is calculated from stoichiometry:

$$1\ \text{mol}\ C_xH_yO_z = (12x + y + 16z)\ \text{g} \rightarrow x\ \text{moles}\ CO_2 = 12x\ \text{g C} \quad (1.8)$$

Therefore, the theoretical TOC is calculated as:

$$TOC_t = 12x/(12x + y + 16z)\ \text{g TOC} \cdot \text{g}^{-1}\ C_xH_yO_z \quad (1.9)$$

Using Eqs. (1.2 and 1.9), the following relationship is established between the theoretical values of COD and TOC:

$$COD/TOC = 8(4x + y - 2z)/(12x). \quad (1.10)$$

EXAMPLE 1.5

A TOC test reports a value of 205 mg $C_{org} \cdot L^{-1}$ for a solution of 1200 mg · L^{-1} of sodium acetate. Is this value compatible with stoichiometry?

Solution

For acetate it can be determined with Eq. (1.10) that the theoretical COD/TOC ratio $= 8(4x + y - 2z)/(12x) = 2.67$. In Example 1.2, a COD value of 565 mg · L^{-1} was measured for the solution, so that the value of TOC = 565/2.67 = 211 mg $C \cdot L^{-1}$. Alternatively, one may consider that the solution is equivalent to a solution of 529 $mg \cdot L^{-1}$ of acetic acid (Example 1.2) or 529/60 = 8.8 mmol $HAc \cdot L^{-1}$. Therefore, a result of 2 × 8.8 mmol $C_{org} \cdot L^{-1}$ (acetate has two C atoms) or 12 × 2 × 8.8 = 211 mg $C_{org} \cdot L^{-1}$ is to be expected. Therefore, the obtained practical result of 205 mg $C_{org} \cdot L^{-1}$ differs by only 3% of the stoichiometric value.

The theoretical values of COD (column 3) and TOC (column 4) as well as the ratio between COD and TOC (column 5) for different compounds are all shown in Table 1.6. The data in Table 1.6 clearly show that the COD/TOC ratio is very different for different substances, so it must be concluded that if one of the two parameters is adequate to quantify the organic material concentration, the other one is not. In Section 1.6 it will be demonstrated that it is COD which is the more appropriate parameter: TOC can only be used in special cases when there is a known empirical relationship between the two parameters, so that the TOC value can be converted to the equivalent COD value.

The use of TOC is convenient when it is desired to continuously monitor the concentration of organic material, for example in the effluent of a treatment system. The TOC test, unlike the COD test, can be automated, does not require the presence of an operator or analyst, and the result is obtained in a few minutes, and can be transmitted automatically to a computer or telephone. However, it must be remembered that the TOC test like any online analyser will be prone to clogging/fouling and will thus require regular intervention for maintenance. In addition, the TOC analyser is an expensive piece of equipment.

Table 1.7 shows important aspects of the four tests used to quantify the concentration of organic material in wastewater. These aspects are: (1) the test principle, (2) their duration, (3) the precision that can be expected, (4) the minimum organic material concentration that can be determined with the test, (5) whether the test has an environmental impact, (6) whether there is a relation of the test to the chemical energy content contained in the organic material, and (7) whether it is feasible to use the results to derive a mass balance of the organic material for the treatment system. The table shows that with respect to the last two important aspects, the COD and ANB tests have a significant advantage.

Table 1.7 Comparison between the tests to determine the concentration of organic material: chemical oxygen demand (COD), biochemical oxygen demand (BOD), anaerobic biodegradability (ANB) and total organic carbon (TOC)

Parameter	**COD**	**BOD**	**ANB**	**TOC**
Test Principle	**Oxidation with $Cr_2O_7^{2-}/H_2SO_4$**	**Biological Oxidation w/O_2**	**Fermentation**	**Thermal Oxidation w/O_2**
Duration	**Hours**	**Days**	**Days**	**Minutes**
Precision	good (±5%)	medium (±15%)	good (±5%)	excellent (±3%)
Minimum concentration	±30 $mg \cdot L^{-1}$	>±3 $mg \cdot L^{-1}$	±30 $mg \cdot L^{-1}$	±3 $mg \cdot L^{-1}$
Cost	Medium	Low	Low	High
Environmental impact	yes (Hg, Cr, Ag)	No	No	No
Relationship with energy content	Direct	No	Direct	No
Mass balance possible?	Yes	No	Yes	No

1.6 COMPARISON OF MEASUREMENT PARAMETERS

Organic material in wastewater is of interest because it serves as a source of energy for micro-organisms present in wastewater or in the treatment system. For this reason, it is important to evaluate the energy content of the organic material in the wastewater and, if possible, to correlate this parameter with one of the parameters used to quantify the organic material concentration (COD, BOD or TOC).

From the point of view of thermodynamics, it is relevant to introduce the concept of free energy, which is the useful energy that is released when a chemical reaction occurs. Thus, if a bacterium oxidizes organic material, then the free energy from such oxidation becomes available to the bacteria and can in principle be used, for example, for anabolic processes. In practice, the bacteria will not be able to use all the energy provided by catabolism, so that a part will be lost as heat to the aquatic phase in which the biochemical reaction occurs. This can result in measurable increase in liquid phase temperature if the rate of the catabolic process and the released free energy per unit mass of catabolized material are high enough.

The numerical values of the released free energy have tabulated for many chemical reactions, or they can be calculated from the formation energy. In Table 1.8 the free energy released upon oxidation of 1 mole of different organic compounds (the ones listed in Table 1.7) was calculated. The values of the free energy per mole of the different compounds vary widely. However, when the free energy released per

Table 1.8 Values of the free energy released per gram of oxidized COD for different organic compounds and the number of electrons transferred per C-atom (N_{el}*).

N°	Compound	*x*	*y*	*z*	Free Energy (kJ/mol)	Free Energy (kJ/g COD)	N_{el} (e^- per C)*
1	Oxalic acid	2	2	4	342	21.39	1
2	Formic acid	1	2	2	283	17.89	2
3	Citric acid	6	8	7	3817	14.93	3
4	Glucose	6	12	6	2858	14.89	4
5	Lactic acid	3	6	3	1358	14.14	4
6	Acetic acid	2	4	2	863	13.47	4
7	Glycerine	3	8	3	1613	14.39	4.67
8	Phenol	6	6	1	3013	13.43	4.67
9	Ethylene glycol	2	6	2	1171	14.64	5
10	Benzene	6	6	0	3171	13.22	5
11	Acetone	3	6	1	1708	13.34	5.33
12	Palmate acid	16	32	2	9742	13.26	5.75
13	Cyclohexane	6	12	0	3754	13.01	6
14	Ethylene	2	4	0	1308	13.64	6
15	Ethanol	2	6	1	1300	13.55	6
16	Methanol	1	4	1	688	14.34	6
17	Ethane	2	6	0	1433	12.80	7
18	Methane	1	4	0	796	12.43	8

gram of theoretical COD is calculated, it is noted that its value remains practically constant for almost all substances: in Table 1.8, the free energy per g COD does not deviate more than 10% from an average value of 13.7 $kJ \cdot g^{-1}$ COD, except for the two compounds on top of the list: oxalic acid (21.4 $kJ \cdot g^{-1}$ COD) and formic acid (17.9 $kJ \cdot g^{-1}$ COD).

The large deviations of these two compounds can be related to their oxidation state, i.e. the number of electrons transferred during the oxidation of these compounds. The number of electrons transferred during oxidation can be deduced with ease from the equation reaction. For a compound $C_xH_yO_z$ the half-reaction of the oxidation is written as:

$$C_xH_yO_z + (4x - 2z)\ H_2O \rightarrow x\ CO_2 + (4x + y - 2z) + H^+ + (4x + y - 2z)\ e^- \quad (1.13)$$

From Eq. (1.13) it can be observed that the oxidation of 1 mol $C_xH_yO_z$ or x moles of carbon results in a transfer of $(4x + y - 2z)$ electron equivalents, so that the number of transferred electrons per C-atom, N_{el}, is equal to:

$$N_{el} = (4x + y - 2z)/x = 4 + (y - 2z)/x \quad (1.14)$$

The N_{el} value is indicative for the oxidation state of the compound. The N_{el} values are listed in Table 1.8 for the different compounds (last column).

It can be observed that the number of electrons transferred per C atom varies from 1 for oxalic acid to 8 electrons/C atom for methane. In the list in Table 1.8 there are some N_{el} values that are not integers. This does not mean that there are parts of electrons being transferred, but that the number of electrons transferred is not the same for all carbon atoms.

In the case of oxalic acid ($N_{el} = 1$) and formic acid ($N_{el} = 2$) the number of transferred electrons per C atom is lower than for other compounds. Thus, the high free-energy release of these two compounds can

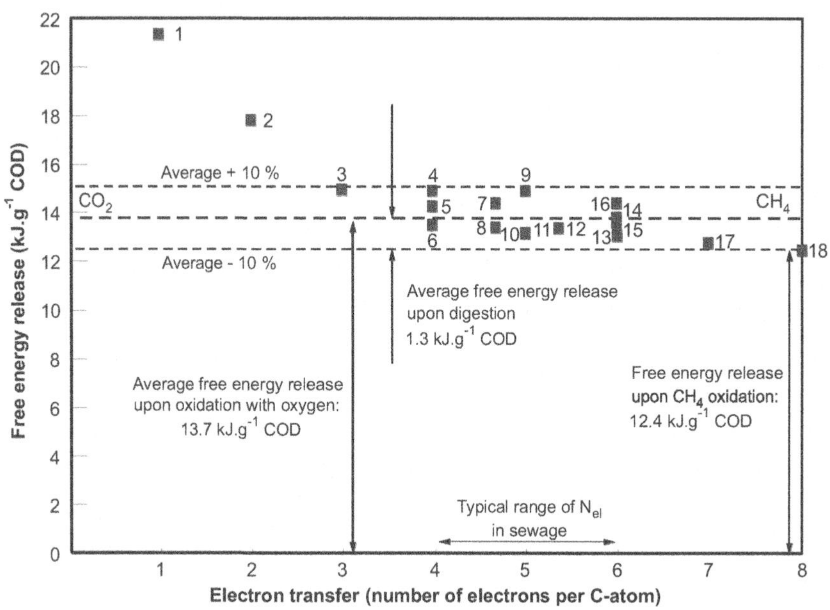

Figure 1.9 Free energy released upon oxidation of organic compounds per g of COD as a function of the number of transferred electrons per C-atom.

be attributed to their high oxidation state. It is concluded that as the oxidation state of a compound is high (and N_{el} is low), the transferred energy is higher.

This conclusion is also evident in Figure 1.9. In this diagram the free energy of oxidation per unit mass of COD is plotted as a function of the number of electrons transferred per atom of carbon for different compounds $C_xH_yO_z$. Note that there is a tendency for this energy to decrease as the number of electrons increases from a minimum of 1 to a maximum of 8 electrons/C atom. However, when the N_{el} value is 3 or more, the free energy released in the oxidation becomes practically constant and deviates only marginally from the average of 13.7 $kJ \cdot g^{-1}$ COD.

In practice, the number of electrons transferred in most of the organic compounds present in wastewater varies from 4 (for carbohydrates and most proteins) to 6 (for lipids). Therefore, it is concluded that the COD is a suitable parameter to evaluate the potential of available chemical energy in wastewater.

The free energy released in anaerobic digestion can now be estimated by considering that the free energy release upon oxidation of methane is 12.4 $kJ \cdot g^{-1}$ COD (refer to the last line of Table 1.8). As the average free energy of oxidation of organic material is 13.7 $kJ \cdot g^{-1}$ COD, the average free energy for anaerobic digestion can now be estimated as $13.7 - 12.4 = 1.3$ $kJ \cdot g^{-1}$ COD.

As it is known that the ratio between COD and TOC varies considerably for different compounds (Table 1.6), it is concluded that the free energy released upon oxidation of 1 g of TOC must vary a lot depending on the nature of the oxidized compound, because the energy per unit mass of COD is approximately constant. Therefore, the TOC parameter is inadequate to evaluate the available chemical energy in a wastewater. In practice, it is much more important to estimate the chemical energy of a wastewater than the organic carbon content. Thus, it is concluded that TOC is an intrinsically inferior parameter to quantify the concentration of organic material in wastewater.

It is important to note that free energy is a measure of the useful energy released in a given reaction, but says nothing about the quantity of energy absorbed by the micro-organism that mediates the reaction. Thus, it is quite possible that two reactions mediated by different micro-organisms release the same amount of free energy, but that the energy used by organisms is very different.

Chapter 2

Organic material & bacterial metabolism

2.1 BACTERIAL METABOLISM

One of the most important objectives in wastewater treatment is the removal of organic material, a task almost invariably performed by biological processes because these are more reliable and economical. Organic material in wastewater is unstable, i.e. it tends to decompose into simpler organic or inorganic compounds. In biological wastewater treatment systems, micro-organisms are responsible for the degradation or stabilization of organic material. Micro-organisms can be divided into two groups according to the structuring of their genetic material within the cells: prokaryotes and eukaryotes (Bitton, 2005). Prokaryotes in turn can be divided into two groups, bacteria and archaea, which are the most important micro-organisms in treatment systems. It was only relatively recently that archaea were recognized as a specific group of micro-organisms and in many publications archaea are mistaken for bacteria. However, for purposes of design and operation of treatment systems the distinction is of no importance.

Bacteria are micro-organisms that can be classified according to the method by which they obtain the energy needed for their vital functions: from sunlight (phototrophic bacteria), from organic material (organotrophic bacteria) or from certain inorganic substances (lithotrophic bacteria). A further distinction can be made between bacteria that use organic material as a source material for their cell synthesis (heterotrophs) and those that use carbon dioxide (autotrophs).

- In aerobic treatment systems heterotrophic bacteria are responsible for the metabolism of organic material, while autotrophic bacteria are responsible for the oxidation of ammonium nitrogen.
- In anaerobic systems the micro-organisms have a more complex metabolism, with a syntrophic cooperation between bacteria and archaea, both of which are necessary for the overall transformation of organic material into biogas. Several groups of bacteria transform complex organic material into materials that can be used as substrates of the methanogenic archaea, which in turn complete the

Editor(s): Adrianus van Haandel and Jeroen van der Lubbe
doi: 10.2166/9781780409627_0029

transformation of organic material into biogas. Without the bacteria archaea cannot use most of the organic compounds in wastewater, and without the archaea there is no removal of organic material from the liquid phase in an anaerobic environment, i.e. anaerobic digestion does not develop.

As mentioned earlier, heterotrophic bacteria use organic material; this is for two purposes. One is as a material source, in a process called anabolism, when a part of the organic material in the wastewater is converted into cell mass; the other is as a source of energy, in a process called catabolism, when organic material is transformed into stabilized end products and energy.

The processes of anabolism and catabolism occur simultaneously, and part of the catabolic energy released is used by the bacteria to perform the anabolic reaction. Therefore, anabolism and catabolism are interdependent: catabolism provides the necessary energy for anabolism, and anabolism is necessary for growth, so that the bacterial mass remains, even if there are losses due to predators and decay. Figure 2.1 shows the overall scheme of bacterial metabolism.

Figure 2.1 also demonstrates another aspect of bacterial metabolism. Bacterial decay happens: over time micro-organisms cease to exist as living material. Some of the decayed cell material may be metabolized again, but there is a residue of particulate material that is either not biodegradable or only biodegrades very slowly. This material accumulates in the system as endogenous residue until it is discharged as excess sludge or together with the effluent. This chapter discusses aspects related to the stoichiometry and the kinetics of the metabolic processes that develop in wastewater treatment systems. These aspects are:

(1) The ratio between the fractions of organic material used in anabolism and catabolism;
(2) Thermodynamic considerations to assess whether catabolic processes can develop;
(3) Kinetics, which deals with the rate of processes that develop; and
(4) The decay rate of living micro-organisms and the generation of the endogenous residue.

Biological treatment systems can be divided into two categories according to the nature of catabolism. In aerobic systems the organic material is oxidized to products like carbon dioxide and water, that is, the organic material is destroyed, having its electrons transferred to the oxidant (which is usually oxygen). In anaerobic systems there is fermentation of organic compounds with some end products that are more oxidized and others more reduced than the original material, in the sense that there are products with greater or lesser potential of electron transfer than the original material. Often anaerobic degradation of organic material does not involve only one fermentation, but a sequence of fermentations.

At the end of all possible fermentations, the formed products are the most oxidized or the most reduced compounds. The most oxidized is CO_2, which has no more electron transfer capacity, and the most

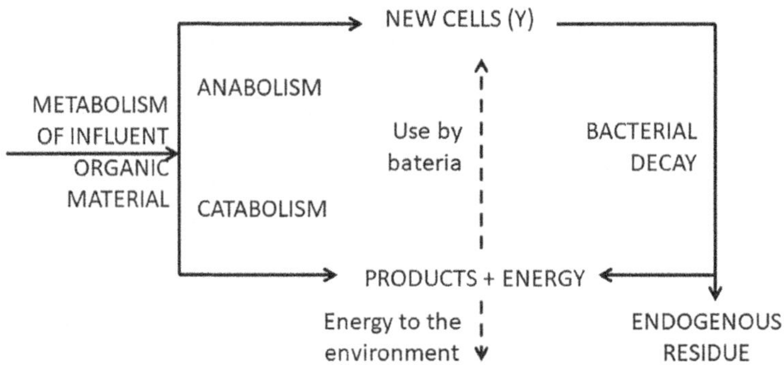

Figure 2.1 Schematic representation of metabolism by heterotrophic bacteria.

reduced is CH_4, having the greatest potential for electron transfer: eight electrons per carbon atom. The set of fermentations that leads to the production of CO_2 and CH_4 is the process of anaerobic digestion. It is concluded that in anaerobic digestion the final products are gaseous and tend to desorb from the liquid phase, forming biogas. This feature is an important advantage in wastewater treatment: biological processes which transform any biodegradable organic material into biogas, which separates from the water to be treated and can readily be used as a fuel or otherwise.

EXAMPLE 2.1

The sewage of a city of 400,000 inhabitants with a per capita contribution of 100 $g \cdot d^{-1}$ of COD is treated in an anaerobic system. The system has a COD removal efficiency of 80% while 10% of the removed COD is used for sludge production. Estimate the methane production, knowing that 1 g of CH_4 is generated by the digestion of 4 g of COD.

Solution

The city generates $400{,}000 \times 0.1 = 40{,}000\ kg \cdot d^{-1}$ of sewage COD. If 20% of the COD ends up in the effluent and if the production of sludge takes up 10% of the COD, it is calculated that a fraction of 70% of the COD is digested, producing $0.7 \times 40{,}000/4 = 7000\ kg \cdot d^{-1}$ of CH_4.

To evaluate the proportion between anabolism and catabolism, it is necessary to determine quantitatively the effects of the two processes: anabolism that results in an increase in sludge (or bacterial) mass and catabolism that causes oxygen consumption in the case of the oxidative mechanism and methane production in the case of anaerobic digestion.

2.1.1 Anabolism

Anabolism can be measured directly as the increase of the bacterial mass, commonly expressed in terms of volatile solids in suspension (VSS). It is well established that there is proportionality between the generated sludge mass and the metabolized COD mass:

$$Y = \Delta X_v / \Delta S_{met} \tag{2.1}$$

where

Y = yield coefficient
ΔX_v = generated bacterial mass (mass of volatile sludge) and
ΔS_{met} = metabolized COD mass (organic matter or substrate).

In the case of metabolism in an aerobic environment, experimental data from many researchers indicate that the value of the yield coefficient is to good approximation a constant and does not depend on the nature of the organic material.

Based on their own research and results from other researchers, Marais & Ekama (1976) suggested a value for the aerobic yield coefficient Y_{ae} of 0.45 gram of VSS per gram of COD metabolized. This value has since been found by many other researchers and is widely used in aerobic process design. In contrast, the value of the yield coefficient in an anaerobic environment (Y_{an}) does depend on the nature of the organic material. This is because anaerobic digestion is a complex process, developed in stages and with various populations of bacteria and archaea involved. If the organic material to be digested is particulate (macromolecules), it will initially have to be solubilized, and subsequently be transformed into intermediate substances, that end up forming substrates for the archaea and finally these transform them into biogas. In most wastewaters, particularly in sanitary sewage, the organic material is a mixture: one (usually small) part is present as a substrate that can be used directly by the methanogenic archaea; another part is present as soluble substances and the remainder as particulate material. For this reason, in the case of anaerobic digestion, two extremes can be distinguished:

(1) If the organic material in the wastewater is composed only of acetate (a substrate for methanogens), there will be an increase in the population of methanogenic archaea, with a low value of the yield coefficient $Y_{an,min} = 0.02$ g VSS · g^{-1} COD metabolized
(2) If the organic material is composed of macromolecules, all the populations that transform the macromolecule into methane will develop, with a much higher value of the overall yield coefficient: $Y_{an,max} = 0.12$ g VSS · g^{-1} COD (Henze & Harremoës, 1983).

In the specific case of domestic sewage, several researchers found values for the yield coefficient in anaerobic environment Y_{an} ranging from 0.04 to 0.06 g VSS · g^{-1} COD converted, with $Y_{an} = 0.05$ g VSS · g^{-1} COD as an average. Figure 2.2 schematically represents the aerobic and anaerobic metabolic processes with numerical values for yield factor, energy effect of catabolism, decay constant and endogenous residue fraction.

2.1.2 Oxidative catabolism

Obligate aerobic organisms can only use oxygen for the oxidation of organic material. Facultative organisms (abundant in aerobic wastewater treatment systems) can use both dissolved oxygen (DO) and nitrate as the oxidant, but they use nitrate only when there is no DO available. Sulphate reducing bacteria that develop only in the absence of DO also have an oxidative catabolism. Oxidative catabolism is also called respiration, which can be aerobic respiration when oxygen is the oxidant and anoxic- or anaerobic respiration when nitrate or sulphate are the electron acceptors of the organic material. The half reactions for the reduction of the three oxidants can be written as:

$$e + H^+ + \tfrac{1}{4}O_2 \rightarrow \tfrac{1}{2}H_2O \tag{2.2}$$

$$e + \tfrac{6}{5}H^+ + \tfrac{1}{5}NO_3^- \rightarrow \tfrac{1}{10}N_2 + \tfrac{3}{5}H_2O \tag{2.3}$$

$$e + H^+ + \tfrac{1}{8}SO_4^{2-} \rightarrow \tfrac{1}{8}S^{2-} + \tfrac{1}{2}H_2O \tag{2.4}$$

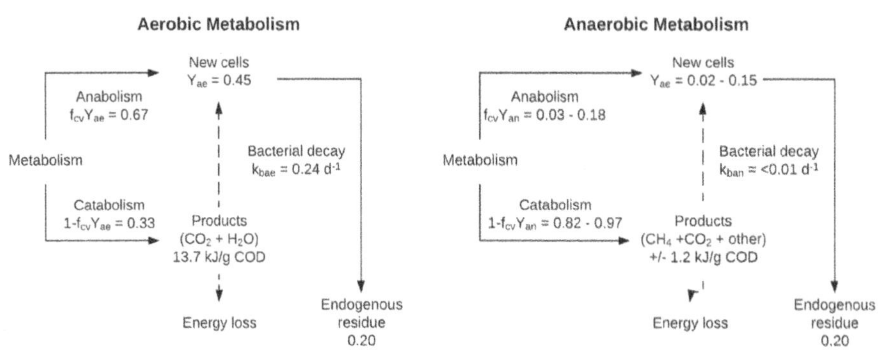

Figure 2.2 Fractions of the organic material utilized for bacterial catabolism and anabolism in aerobic (left) and anaerobic environments (right).

The half-equations show that in oxidimetric terms 1/4th of a mole of oxygen (8 g O_2) is equivalent to 1/5th of a mole of nitrate (14/5 = 2.8 g N) or to 1/8th of a mole of sulphate (32/8 = 4 g S): all can accept one electron equivalent, which means that they can oxidize organic material with a mass of 8 g COD. Therefore, stoichiometrically 1 g COD can be oxidized by 1 g O_2 (by definition) or by 2.8/8 = 0.35 g nitrate-N or by 4/8 = 0.5 g sulphate-S. Conversely the oxidimetric equivalents for 1 mg NO_3-N and 1 mg SO_4-S (reduction to S^{2-}!) are 2.86 and 2.0 mg O_2 respectively.

2.1.3 Fermentative catabolism or anaerobic digestion

The fermentative catabolism is a process that results in intramolecular transfer (as opposed to intermolecular transfer in oxidative catabolism) in such a way that the catabolized compound decomposes in at least two smaller parts. If the catabolic products are different, one will be more oxidized and the other will be more reduced than the original compound. The most known fermentation is probably the formation of ethanol:

$$C_6H_{12}O_6 \rightarrow 2CO_2 + 2C_2H_5OH \tag{2.5}$$

In this process carbohydrates such as glucose and fructose ($N_{el} = 4$, Table 1.8) are transformed into ethyl alcohol ($N_{el} = 6$) and carbon dioxide ($N_{el} = 0$). Note that the number of transferable electrons does not change because of the fermentation: before fermentation glucose (with six carbon atoms per mol) had several transferable electron equivalents equal to $6 \times N_{el} = 24$ equivalents per mol glucose. After fermentation the two moles of ethanol (with two carbon atoms per mol of ethanol) can transfer $2 \times 2 \times 6 = 24$ equivalents per mol glucose.

Anaerobic digestion is a specific fermentation as it is characterized by the production of methane, which is the smallest organic compound ($N_{el} = 8$) and of carbon dioxide ($N_{el} = 0$), which is a product that cannot be further oxidized.

In this sense anaerobic digestion can be considered as the ultimate fermentation: the products of anaerobic digestion are stable in the sense that they are not susceptible to further fermentations. For the general formula the $C_xH_yO_z$ anaerobic digestion can be expressed as:

$$C_xH_yO_z + \tfrac{1}{4}(4x - y - 2z)\,H_2O \rightarrow \tfrac{1}{8}(4x - y + 2z)\,CO_2 + \tfrac{1}{8}(4x + y - 2z)\,CH_4 \tag{2.6}$$

EXAMPLE 2.2

If ethanol is subjected to anaerobic digestion, what will be the composition of the biogas that is generated in the catabolic process?

Solution

For ethanol, C_2H_5OH, one has $x = 2$, $y = 6$ and $z = 1$, so that the number of moles of CO_2 in Equation (2.6) will be: $N_{CO2} = (4x - y + 2z)/8 = (4 \cdot 2 - 6 + 2 \cdot 1)/8 = 0.5$ mol·mol^{-1} ethanol. Similarly, the number of moles of CH_4 formed will be $N_{CH4} = (4 \cdot 2 + 6 - 2 \cdot 1)/8 = 1.5$ mol·mol^{-1} ethanol. Therefore, the composition of the biogas will be $N_{CH4}/N_{CO2} = 1.5{:}0.5 = 3{:}1$. The produced biogas will contain 75% methane and 25% carbon dioxide (disregarding other components and dissolution in the liquid phase).

It is very important to note that anaerobic digestion is a process of transformation and not of destruction of organic material: since there is no oxidant used in the process, the electron transfer capacity of the organic material remains intact in the produced methane. Stated differently, as there is no oxidation, the theoretical COD value of methane will be equal to the COD value of the organic material digested to form it. Knowing that the theoretical COD of methane is equal to 4 g $COD \cdot g^{-1}CH_4$, it is concluded that for the formation of 1 g of methane, a mass of 4 g COD of organic material is needed. This ratio is stoichiometric and independent of the nature of the organic material.

The energy production of oxidative catabolism (notably when oxygen or nitrate is used to oxidize the component) is much greater than that for fermentative catabolism. This difference is shown more clearly by an example: Table 1.8 shows that in the case of oxidation of acetic acid (compound number 6) with oxygen, the free energy release is equal to 796 $kJ \cdot Mol^{-1}$. This reaction is written as:

$$C_2H_4O_2 + 2O_2 \rightarrow 2CO_2 + 2H_2O + 796 \text{ kJ} \cdot \text{mol}^{-1} \quad (2.7a)$$

When acetic acid is digested to methane and carbon dioxide the reaction is as follows:

$$C_2H_4O_2 \rightarrow CH_4 + CO_2 + E_d \quad (2.7b)$$

The free energy released during the digestion E_d can be calculated knowing that if methane is oxidized by oxygen, the released free energy will be 734 $kJ \cdot mol^{-1}$ as shows the last line of Table 1.8

$$CH_4 + 2O_2 \rightarrow CO_2 + 2H_2O + 734 \text{ kJ} \cdot \text{mol}^{-1} \quad (2.7c)$$

As the overall result of Equations (2.7b and c) is a reaction identical to Equation (2.7a), the sum of the released free energies from the last two reactions is equal to that of the first reaction so that $E_d + 734 \text{ kJ} \cdot \text{mol}^{-1} = 796 \text{ kJ} \cdot \text{mol}^{-1}$, or $E_d = 62 \text{ kJ} \cdot \text{mol}^{-1}$. It is concluded that the energy released in the anaerobic digestion of acetic acid is only a small fraction of the available free energy: i.e. 62/769 = 8%. The remaining 92% remains in the methane released from wastewater.

The above example shows that in anaerobic digestion only a small part of the chemical energy of the organic material is released: most of it remains in the generated methane and can be used, for example, for combustion. The small energetic benefit of fermentations has an important consequence: it is necessary that relatively much organic material is converted in the catabolic processes to generate the energy required for the anabolic processes.

2.2 CONVERSION PROCESSES IN BIOLOGICAL TREATMENT

The essence of biological wastewater treatment is the conversion of organic material. Metabolism of organic material through anabolism and catabolism converts organic material into sludge and stable products as shown in the previous chapter. However, in Chapter 1 it was also established that not all organic material is biodegradable: a fraction will not be converted by physical or biological processes and will hence be discharged unaffected in the effluent. Appendix A1 shows the experimental procedure to determine the biodegradable fraction of the organic material in an anaerobic environment.

The fractions of organic material that end up in the effluent or are converted into sludge can be determined directly using standard tests: determination of COD to measure the organic material in the effluent and determination of either VSS (volatile suspended solids) or particulate COD to measure the organic material in the excess sludge. The fraction of organic material converted into stable end products can also be determined experimentally: in aerobic systems the fraction of oxidized organic material is by definition equal to the oxygen consumption for that oxidation (i.e. oxygen demand for nitrification should be ignored), and in anaerobic systems the methane production is ¼ of the digested organic material, expressed as COD.

To estimate the production and composition of biological sludge and the catabolic products as well as non-biodegradable organic material in wastewater treatment systems, it is necessary to evaluate the interaction between the bacterial mass and the influent organic material. Marais & Ekama (1976) considered that the organic material in wastewater is composed of four fractions, which can be distinguished because of different interactions with the bacterial mass in treatment systems:

(1) *Non-biodegradable particulate fraction* – this fraction is adsorbed by the bacterial mass and forms the inert organic fraction of the sludge;
(2) *Non-biodegradable soluble fraction* – this is characterized by the fact that it is not removed by biological or physical–chemical action of the bacterial mass and is discharged together with the effluent;
(3) *Biodegradable soluble fraction* – this fraction can be readily metabolized by bacteria or archaea in both aerobic and anaerobic environments, forming intermediate (in the case of anaerobic treatment) or final metabolic products;
(4) *Biodegradable particulate fraction* – this fraction is typically slowly degradable and after hydrolysis is biologically converted by the bacterial mass into substrate suitable to be metabolized, and hence this material is converted into soluble biodegradable material.

In subsequent sections the removal of organic material and the consequent production of sludge and catabolic products are discussed in both aerobic and anaerobic treatment systems.

2.2.1 Conversions in aerobic systems

Figure 2.3 shows a basic representation of the interactions between the organic material in the influent and the sludge in aerobic treatment systems. There are several processes that occur simultaneously which lead to the removal of organic material, the production of excess sludge and oxygen consumption.

The following five basic processes can be identified in aerobic treatment systems:

(1) *Flocculation*: non-biodegradable and particulate material is removed from the liquid phase and becomes part of the suspended sludge flocs, forming the inert sludge fraction X_i. This process can be considered as instantaneous.
(2) *Adsorption*: the biodegradable particulate material is adsorbed onto the biological sludge as stored material. This process also develops at high rate.
(3) *Hydrolysis*: the stored particulate material is converted into a soluble material, due to the action of extracellular enzymes, excreted by the bacteria. This process is slower, but the rate is sufficiently high to prevent accumulation of stored degradable material, unless the system is overloaded.
(4) *Metabolism*: biodegradable soluble organic material either present in the influent or originating from hydrolysis of stored slowly degradable material is used by micro-organisms for anabolism and catabolism.
(5) *Decay*: upon decay of the active sludge (living organisms), a fraction of inert endogenous residue is generated, while the rest of the decayed sludge will be used as substrate for the remaining active sludge.

When Figure 2.3 is analysed the following important conclusions can be drawn:

- In the aerobic system the organic material of the influent is divided over three fractions:
  - A part (mS_{xv}) is transformed into sludge, composed of generated active and inert sludge (anabolic process), endogenous residue and possibly some stored non-hydrolysed degradable material, which eventually is discharged from the system as excess sludge;
  - A part (mS_o) is oxidized in two main processes: (1) exogenous respiration of the degradable material from the influent (catabolic process) and (2) endogenous respiration to oxidize the biodegradable part of the decayed active sludge;

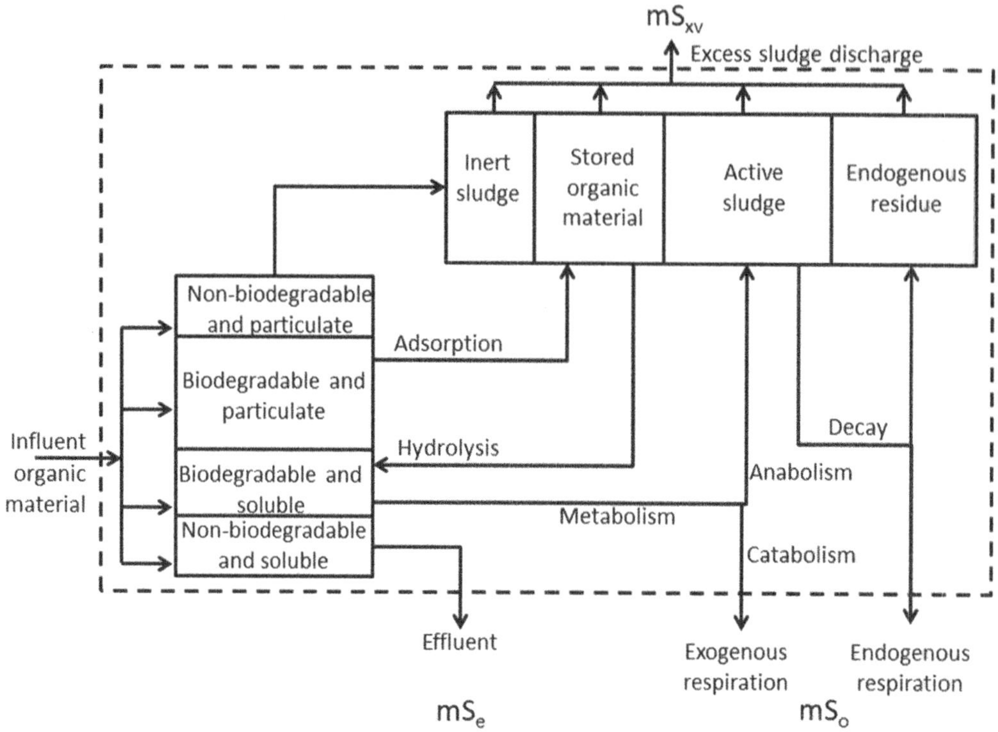

Figure 2.3 Simplified model of the interaction between influent organic material and sludge in aerobic treatment systems.

- ◦ A part of the organic material (mS_e) is not affected by the biochemical or physical–chemical processes in the sludge and is discharged with the effluent;
- ◦ The values of these three fractions can be used to generate a mass balance over the system, as will be demonstrated in Section 2.6.
- The key variable governing the behaviour of aerobic systems is the sludge age, which is defined as the average retention time of the solids in the system and can be expressed as the ratio between the sludge mass in the system and the daily discharged mass of solids. At longer sludge age the fraction ending up in the waste sludge (mS_{xv}) will become smaller while the oxidized fraction (mS_o) will decrease. The value of mS_e is typically not affected above a certain (low) minimum value of the sludge age.
- The values of the fractions mS_{xv} and mS_o defined above depend on two variables: (i) the composition of the organic material, notably the non-biodegradable soluble and particulate fractions, and (ii) the sludge age that is applied. It is important to know the value of these two fractions because they directly affect the main parameters of aerobic treatment.
- The fraction mS_{xv} defines the mass of excess sludge produced and thus also defines the required volume of the biological reactor.
- The fraction mS_o defines the oxygen demand and therefore the capacity of the aerators to be installed, the main contributor to energy demand. The fraction mS_e defines the residual concentration of organic material in the effluent and thus the effluent quality.

The conversion efficiency of the biodegradable material, including the preceding processes of adsorption and hydrolysis, depends on the nature of the wastewater and the operational conditions, notably the temperature and the sludge age. In regions with a hot climate the processes that eliminate the biodegradable organic material are essentially complete for a sludge age above a minimum value of 2 to 3 days. However, in general a minimum value of 5 days is recommended to improve the stability of the aerobic system and generate a better-quality effluent – mainly because of the establishment of higher micro-organisms that predate on single bacteria present outside the sheltered environment of the sludge flocs. In contrast, the sludge decay process requires much more time to be essentially complete. Marais & Ekama (1976) suggested a first-order equation to describe the active sludge decay rate:

$$(dX_a/dt)_d = -b_{ae} \cdot X_a \tag{2.8}$$

where

X_a = concentration of active sludge (live bacterial mass)
b_{ae} = decay constant = $0.01 \times (1.04)^{(T-20)}$ (h^{-1}), with T = temperature (°C)

Therefore, at a temperature of 20°C the active sludge decays at a rate of 1% per hour. To achieve a significant degree of decay, the sludge age must be in the range of 20 to 30 days or 10 times greater than the sludge age to have essentially complete removal of the biodegradable material. Under normal conditions the sludge age in the system will be shorter, so decay will be partial, and another unit will be required to complete the decay: a sludge digester, which may be aerobic or anaerobic.

The three fractions of the organic material can be *experimentally* determined by operating an activated sludge system under steady conditions (i.e. with constant organic and hydraulic load):

$$mS'_e = S_e/S_{ti} \tag{2.9a}$$

$$mS'_{xv} = f_{cv} \cdot X_v (R_h/R_s)/S_{ti} \tag{2.9b}$$

$$mS'_o = OUR \cdot R_h/S_{ti} \tag{2.9c}$$

where

mS'_e = experimentally determined fraction of COD discharged with the effluent
mS'_{xv} = experimentally determined fraction of COD in excess volatile sludge
mS'_o = experimentally determined fraction of COD that is oxidized
X_v = concentration of volatile sludge
OUR = rate of oxygen consumption
R_h = retention time of the liquid = V_r/Q_i
R_s = retention time of solids = sludge age
Q_i = influent flow and V_r = reactor volume.

Equation (2.9a, b and c) show that the three fractions can be determined experimentally, because they depend on measurable variables and constants. On the other hand, Van Haandel and Van der Lubbe (2012) also derived *theoretical* expressions to calculate the values of the above fractions, under the condition that the removal of biodegradable organic material in the aerobic system is essentially complete:

$$mS_{te} = f_{ns} \tag{2.10a}$$

$$mS_{xv} = f_{cv} (1 - f_{ns} - f_{np}) (1 + f \cdot b_{ae} \cdot R_s) Y_{ae}/(1 + b_{ae}R_s) + f_{np} \tag{2.10b}$$

$$mS_o = (1 - f_{ns} - f_{np}) [1 - f_{cv}Y_{ae} + f_{cv} (1-f) \cdot b_{ae}Y_{ae} \cdot R_s/(1 + b_{ae}R_s)] \tag{2.10c}$$

where

f_{ns} = soluble non-biodegradable fraction of influent organic material
f_{np} = particulate non-biodegradable fraction of influent organic material
Y_{ae} = yield coefficient = 0.45 mg$X_a \cdot$mg^{-1} COD (Marais & Ekama, 1976)
f_{cv} = ratio of COD to VSS in organic biomass = 1.5 mg COD$\cdot$mg^{-1} COD
f = fraction remaining as endogenous residue upon decay of active sludge = 0.2 mg VSS$\cdot$mg^{-1} VSS (Marais & Ekama, 1976).

In many papers by different researchers there is always an excellent correlation between the experimentally determined values mS'_{te}, mS'_{xv} and mS'_o and their theoretical equivalents of mS_{te}, mS_{xv}, and mS_o for any value of the fundamental variable describing activated sludge behaviour: the sludge age. It is concluded that the model presented in Figure 2.3 adequately describes the division of the organic material of the influent into the three fractions in aerobic systems: (1) organic material in the effluent, (2) organic material converted into sludge, and (3) oxidized organic material. Since the mass parameters f, f_{cv} and Y are constant, and b_{ae} only depends on temperature, it is possible to calculate these fractions from Eqs. (2.10a to c), if the non-biodegradable fractions f_{ns} and f_{np} are known or can be estimated. Figure 2.4 shows the theoretical curves of the three fractions as a function of the most important variable, the sludge age. The results justify the assumption that in aerobic systems the metabolism of the biodegradable material is complete. Figure 2.4 also shows the division of the feed COD into fractions that are transformed into inert sludge, active sludge and endogenous residue, also in this case expressed as mg COD$\cdot$mg^{-1} COD. Theoretical equations of these fractions were presented by Van Haandel and Van der Lubbe (2012)

$$mS_{xi} = f_{np} \tag{2.11}$$

$$mS_{xa} = (1 - f_{ns} - f_{np}) \cdot f_{cv} \cdot Y_{ae}/(1 + b_{ae} \cdot R_s) \tag{2.12}$$

$$mS_{xe} = f \cdot b_{ae} \cdot R_s \cdot mS_{xa} \tag{2.13}$$

$$mS_{xv} = mS_{xi} + mS_{xa} + mS_{xe} = (1 - f_{ns} - f_{np}) \cdot f_{cv} \cdot Y_{ae} \cdot (1 + f \cdot b_{ae} \cdot R_s)/(1 + b_{ae} \cdot R_s) + f_{np} \tag{2.14}$$

2.2.2 Conversions in anaerobic systems

Having established a coherent model to describe the fate of organic material in aerobic systems, we now attempt to repeat this strategy for anaerobic treatment systems. In anaerobic systems the organic material of the influent is also divided into three fractions:

mS_e = organic material discharged in the effluent
mS_d = digested organic material (CH_4) resulting from fermentative catabolism
mS_x = organic material transformed into sludge by various anabolic and physical–chemical processes.

Figure 2.5 shows the results of an experimental investigation that is described in more detail in Chapter 5 and presents the three fractions as a function of sludge age. When comparing the division of the feed COD into various fractions in aerobic and anaerobic systems (Figures 2.4 and 2.5) the following differences are evident:

(1) The value of the sludge age that is required for decent COD conversion in anaerobic systems is much higher than in aerobic systems. In the latter the metabolism of the biodegradable material is already complete when the sludge age is no more than 2 to 3 days (at 25°C), whereas in anaerobic systems digestion with associated methane production does not develop below a minimum sludge

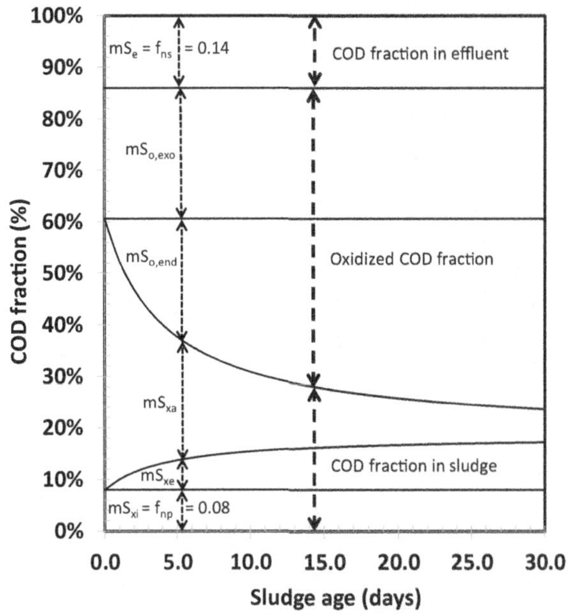

Figure 2.4 COD fractions in the effluent, oxidized and in the sludge as function of the sludge age in aerobic treatment systems.

age of approximately 6 days (at 25°C). Above the minimum sludge age, the efficiency of anaerobic digestion gradually increases, but to obtain a significant removal of biodegradable material the sludge age must be at least 50 days (at 25°C), a value rarely applied in aerobic systems.

(2) The fraction of organic material that is converted into biological sludge tends to be smaller for anaerobic systems than for aerobic systems, but only when the applied anaerobic sludge age is sufficiently high (>40 days at 25°C). Below this value the adsorbed biodegradable material tends to contribute significantly to sludge production, whereas in aerobic systems this component is usually insignificant.

(3) The concentration of biodegradable and soluble material in anaerobic systems is much higher than in aerobic systems, unless a very long sludge age (>100 days at 25°C) is applied.

It is concluded that the main difference between aerobic systems and anaerobic systems with respect to the conversion of organic material in wastewater is that in aerobic systems the metabolism of biodegradable material is essentially complete even at short sludge age, whereas in anaerobic systems it is only partial over a wide range of sludge ages and substantially complete removal only occurs at high values. Therefore, the behaviour of the anaerobic systems cannot be described without considering the reaction kinetics, or the rate of the processes that occur. This issue will be discussed in detail in the following sections.

Gujer and Zehnder (1983) proposed a model for the anaerobic digestion of complex organic material. Anaerobic digestion of organic material is a very complex process in which there are several bacterial and archaeal populations that cooperatively (syntrophic) transform organic compounds into their final products: methane and carbon dioxide. Depending on the composition of the organic material, other compounds such as water, ammonium, and hydrogen sulphide may also be produced. Gaseous intermediates such as carbon

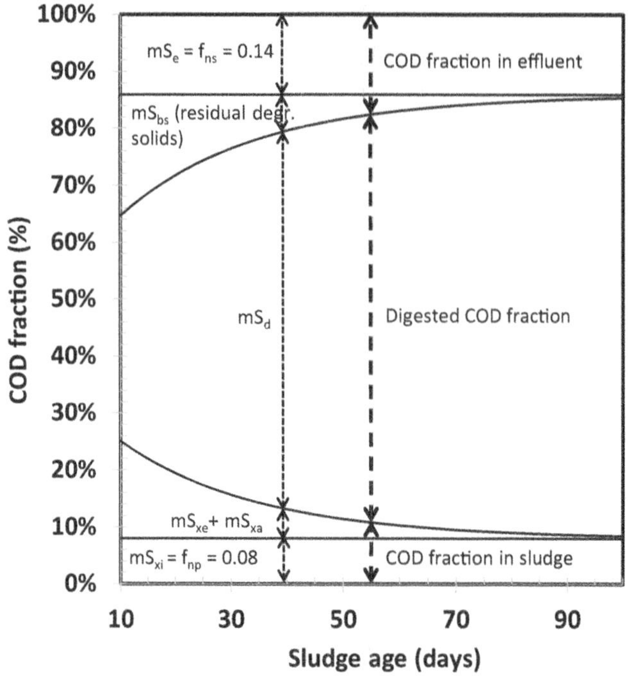

Figure 2.5 COD fractions in the effluent, digested into methane and in the sludge as function of the sludge age in anaerobic treatment systems.

monoxide and hydrogen may be present at low concentrations in biogas, while low concentrations of incompletely degraded materials such as alcohols and fatty acids may be present in the effluent. Figure 2.6 shows the traditional anaerobic digestion scheme with the various microbial populations involved in the treatment of complex organic material, for example in domestic sewage.

The process of degradation of complex organic material via fermentative catabolism involves four populations of different micro-organisms, of which three are classified as bacteria while the last is composed of archaea. The bacterial processes of hydrolysis, acidogenesis and acetogenesis result in the transformation of organic material into the precursors required for the final process of methanogenesis, which is mediated by archaea.

The hydrolysis of particulate organic material is carried out by specific hydrolytic bacteria excreting extracellular enzymes solubilizing and breaking down the particulate matter, consisting of carbohydrates, proteins and lipids. In this process primary metabolites are produced, which are mainly long-chain sugars, amino acids and long-chain fatty acids. These primary metabolites become the secondary metabolites in the process of acidogenesis, again with a population of specific bacteria. These secondary metabolites are short-chain fatty acids (propionic, butyric, valeric) and other soluble compounds like alcohols. In the third stage of the digestion process, acetogenesis, these secondary metabolites are transformed into substrates (acetate, hydrogen, methanol) for the last stage of anaerobic digestion: methanogenesis.

This is the most important process of the anaerobic digestion process because now the organic material that was produced after the preparatory steps is transformed into biogas that desorbs from the liquid phase.

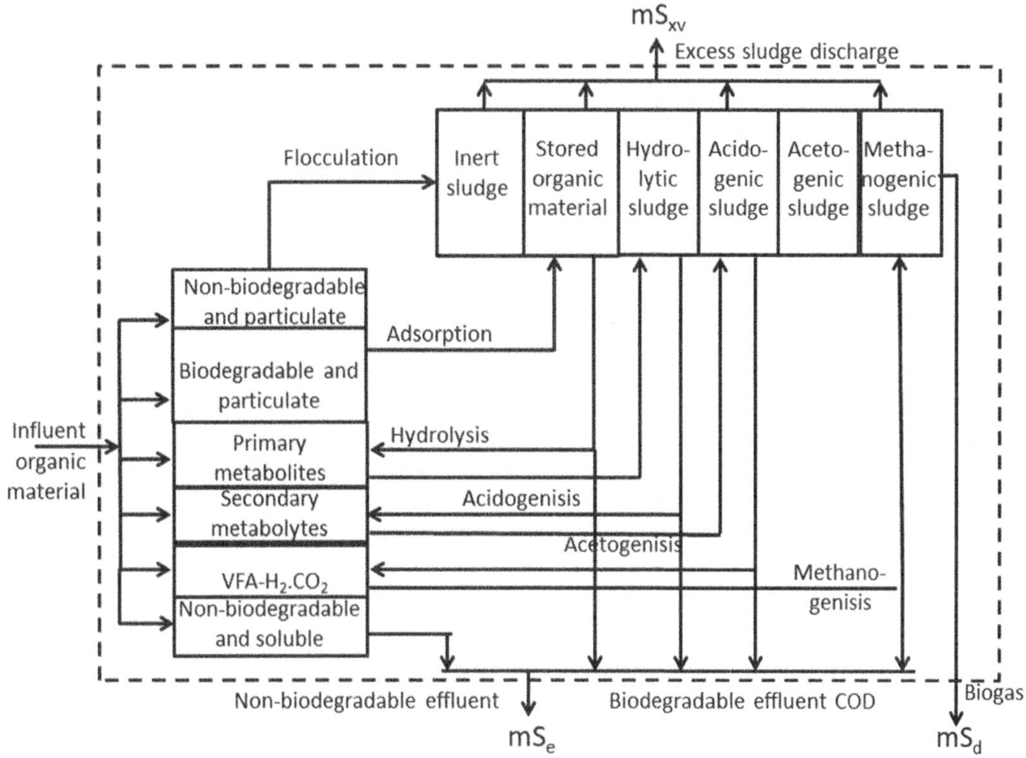

Figure 2.6 Schematic representation of the sequence of processes of anaerobic digestion of complex organic material (Gujer & Zehnder, 1983).

Methanogenesis is carried out by several groups of archaea, of which two are particularly important: the acetoclastic archaea that produce methane from the breakdown of acetate and the hydrogenotrophic archaea that reduce carbon dioxide with hydrogen.

It is important to note that in the processes mediated by the bacterial populations there is only transformation and not removal of organic material, so the COD does not change. In the methanogenesis process there is also no removal of organic material, but the product formed (methane) is gaseous and desorbs from the liquid phase in the anaerobic reactor, so that organic material is physically removed.

An interesting aspect is that the Gujer and Zehnder model does not explicitly consider the decay of bacterial populations and archaea. In fact, the decay occurs, but the rate is much lower in anaerobic systems than in aerobic systems. Therefore, if this process is ignored it will not result in a large deviation from true behaviour.

When the schematic representation of aerobic treatment systems as in Figure 2.3 is compared with that of anaerobic systems as in Figure 2.6, it can be noted that the model for the anaerobic digestion of organic material is much more complex. First there are several processes in series that need to develop, each having a specific population for its realization. A second aspect is that in anaerobic digestion most of the processes of the specific conversions are partial, whereas in aerobic processes the removal of biodegradable organic material is substantially complete for sludge ages of more than a few days. For this reason, it becomes essential to describe and quantify the kinetics of these anaerobic processes, that is, to consider the rate at which the different processes develop and the residual substrate concentration that remains in each process.

The sequence of processes that results in anaerobic digestion with biogas production has different models to describe each step. Hydrolysis is generally described as a first-order process, being proportional to the concentration of particulate organic material and to the concentration of hydrolytic bacteria. For the conversion processes of dissolved organic material, i.e. in the processes of acidogenesis, acetogenesis and methanogenesis, a widely used model is the Monod equation (Monod, 1950). These two basic models will be are briefly described later (Sections 2.3.1 and 2.3.4.1) in this chapter.

2.3 ANAEROBIC DIGESTION OF ORGANIC MATERIAL

The analysis of the stoichiometry of biological reactions in the previous chapter showed clear advantages of anaerobic treatment over aerobic. In anaerobic treatment there is no need for the introduction of an oxidant, the process results in the production of less sludge and, as an additional benefit, produces methane that can be used as fuel. However, the intrinsically favourable stoichiometry of anaerobic digestion alone does not make it an adequate alternative for the removal of organic material in sewage. Basically there are two other important factors to consider: (1) the efficiency of organic material removal should be high, so that a low concentration of residual organic material in the effluent of the treatment system can be obtained; and (2) the organic material removal rate must be high, so that the removal is possible in a reactor with a short retention time, i.e. a reactor with a small volume. Both factors are linked to the removal kinetics of organic material, which describes the rate of bacterial metabolism, and to the operational and environmental conditions in the sewage treatment system.

An anaerobic treatment system will tend to develop a bacterial population compatible with the nature of the organic material and the applied organic and hydraulic loads. In a "mature" or steady-state treatment system (i.e. having a population compatible with the load and composition of the organic material in the feed), the following factors are important for the removal efficiency of biodegradable organic material:

(1) The nature of the organic material to be digested;
(2) The existence of adequate environmental factors for anaerobic digestion;
(3) Bacterial population size (sludge retention efficiency in the system);
(4) Intensity of contact between the organic influent material and bacterial populations;
(5) The sewage retention time in the treatment system.

Factors 1 and 2 are basically determined by sewage characteristics, while factors 3 and 4 depend on the design and operation of the treatment system. Factor 5 is a dependent variable: the first four factors determine the environmental and operational conditions in the digester and therefore the retention time required for the desired removal efficiency of organic biodegradable material.

The transformation into biogas of the complex organic macromolecules present in the sewage requires the mediation of several different groups of micro-organisms. Figure 2.7 shows a schematic representation, suggested by Gujer and Zehnder (1983), of the different processes required for the anaerobic digestion of complex substrates such as sewage. For anaerobic digestion of proteins, carbohydrates and lipids (most particulate organic material belongs to these categories) four different stages can be distinguished in the overall conversion process: hydrolysis, acidogenesis, acetogenesis and methanogenesis. These processes will now be briefly reviewed.

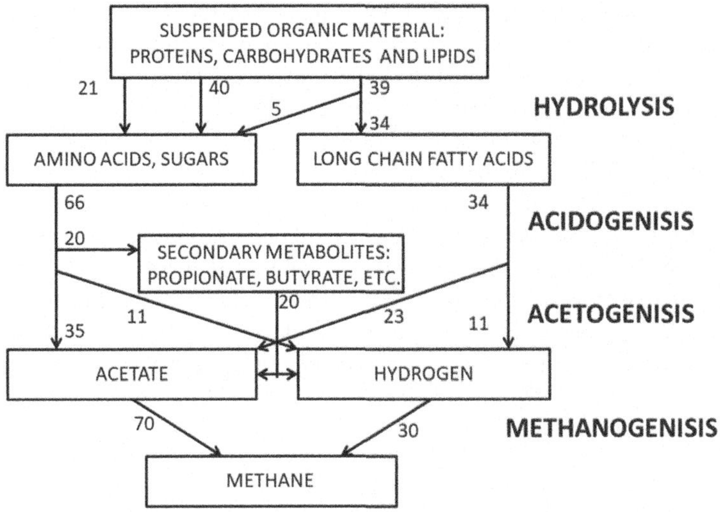

Figure 2.7 Schematic representation of the sequential steps in anaerobic digestion of complex macromolecules (solids) as present in municipal sewage (Gujer & Zehnder, 1983). The relative importance of the different pathways is indicated with a typical division in percentages.

2.3.1 Hydrolysis

Since bacteria are unable to metabolize particulate organic material directly, owing to the size of the particles, the first step in anaerobic degradation is hydrolysis of macromolecules and solid particles present in wastewater. The hydrolysis occurs through the action of enzymes that are excreted from the bacteria and results in the solubilisation of the polymeric particles, producing smaller molecules that can cross the membrane of the bacterial cell. In the process of enzymatic hydrolysis, the proteins are hydrolysed to amino acids, polysaccharides to monomeric sugars, and lipids to long-chain fatty acids like stearate, palmitate or oleate.

Several researchers (Dold *et al.*, 1980) have modelled the hydrolysis in an aerobic environment as a two-process sequence, the first being the storage of slowly biodegradable material and the second the solubilisation of this stored material. Dold *et al.* (1980) described the processes for aerobic systems as:

$$\text{Storage: } r_a = K_a \cdot X \cdot S_{bp} \cdot (K_{ap} - S_{ba}/X_h) \quad (2.15)$$

$$\text{Hydrolysis: } r_{hi} = K_{mp} \cdot X \cdot S_{pa}/(S_{pa} + K_{sp} \cdot X_h) \quad (2.16)$$

where

r_a = rate of storage of particulate and biodegradable organic material
K_a, K_{ap} = kinetic constants
X_h = concentration of hydrolytic micro-organisms
S_{bp} = concentration of particulate organic material
S_{ba} = concentration of stored/adsorbed organic material
r_{hi} = hydrolysis rate of stored/adsorbed organic material
K_{mp}, K_{sp} = kinetic constants.

Hydrolysis in the anaerobic environment follows the same model as the aerobic one, with one difference: the anaerobic hydrolytic bacteria form a specific population and hydrolysis products are used by other populations. A major problem is that the values of the kinetic constants of the processes are not known *a priori*: they tend to vary according to the operating conditions, such as temperature and the degree of disintegration by biological, mechanical or thermal preparatory processes. Thus, the efficiency of organic material hydrolysis in anaerobic digesters cannot really be predicted in advance of operating the reactor.

Generally, hydrolysis is the process that limits the overall degradation rate of complex organic material. Apparently, the rate of hydrolysis is not limited by the availability of enzymes, but by the available area on which the enzymes can act (Angelidaki & Sanders, 2004; Vavilin *et al.*, 2008). Furthermore, the hydrolysis process is heavily temperature dependent. Based on the results of O'Rourke (1968) and Gujer and Zehnder (1983), an Arrhenius-type dependency with a coefficient of 1.067 is adopted, i.e.,

$$r_{hT} = r_{h25} \cdot 1.067^{(T-25)} \tag{2.17}$$

where

r_{hT} = hydrolysis rate at a temperature T
r_{h25} = hydrolysis rate at 25°C
T = temperature in °C.

In the case of sanitary sewage, the difficulty in applying Equation (2.17) is that the hydrolysis rate is generally unknown at the standard temperature, since this depends on factors such as the time that the wastewater was generated, contact with micro-organisms in the sewer network, and pumping impact on particle size, which are generally not known.

2.3.2 Acidogenesis

During the process of acidogenesis, the hydrolysis products are transformed into even smaller molecules by fermentation processes. Acidogenesis is a very common reaction and is carried out by a large group of hydrolytic and non-hydrolytic micro-organisms. The main acidogenesis products are volatile fatty acids (VFA), which are short-chain carboxylic acids such as acetate, propionate, butyrate and valerate.

A variety of other organic and inorganic compounds are also formed, such as lactic acid, alcohols, as well as H_2, CO_2, and ammonia. As an example, different fermentation reactions with glucose as the source material can be observed in Table 2.1. These reactions (and their subsequent follow-up reactions are shown in Figure 2.8 (Section 2.3.2).

Glucose in turn is the main product of the hydrolysis of carbohydrates such as cellulose and starch. The table shows that there are several possible fermentations, and in anaerobic systems several of these reactions will occur simultaneously. A fermentation that often occurs in practice (for example, in silage) is the transformation of glucose into lactic acid (#7 in Table 2.1). Ethanol (#6) is another intermediate compound which forms when the concentration of glucose is high ($>1\ g \cdot L^{-1}$) and the pH is low (<4.5). Butyric acid (#5) can be formed and results in the simultaneous production of hydrogen and carbon dioxide. In contrast, propionic acid formation (#1 and 2) is a process that develops when hydrogen pressure is high, and this fermentation leads to hydrogen consumption. The consequences from hydrogen production and consumption processes will be discussed in more detail in Sections 2.3.5 and 2.4.

The acidogenesis process received its name for a reason: it results in the formation of acids that tend to lower the pH value, which thus may easily decrease to below the ideal range for methanogenesis. As a result, the entire process of anaerobic digestion may fail.

Table 2.1 Alternative pathways for the fermentation of glucose. The electron transfer and the free energy release (under standard conditions, $\Delta G°$) are also indicated.

N°	Reaction Equation			$\Delta G°$ (kJ/mol)
1	$3C_6H_{12}O_6$ 3×24 e$^-$	→	$2\ CH_3COOH + 4\ CH_3CH_2COOH + 2CO_2 + 2H_2O$ 16 e$^-$ (acetic acid) + 56 e$^-$ (propionic acid)	−310
2	$C_6H_{12}O_6 + 2\ H_2$ 24 e$^-$ + 4 e$^-$	→	$2\ CH_3CH_2COOH + 2H_2O$ 28 e$^-$ (propionic acid)	−358
3	$C_6H_{12}O_6$ 24 e$^-$	→	$3\ CH_3COOH$ 24 e$^-$ (acetic acid)	−310
4	$C_6H_{12}O_6 + 4\ H_2O$ 24 e$^-$	→	$2\ CH_3COOH + 2\ CO_2^- + 4\ H_2$ 16 e$^-$ (acetic acid) + 8 e$^-$ (hydrogen)	−206
5	$C_6H_{12}O_6 + 2H_2O$ 24 e$^-$	→	$CH_3CH_2CH_2COOH + 2\ CO_2 + 2\ H_2$ 20 e$^-$ (butyric acid) + 4 e$^-$ (hydrogen)	−254
6	$C_6H_{12}O_6$ 24 e$^-$	→	$2\ C_2H_5OH + 2\ CO_2$ 24 e$^-$ (ethanol)	−225
7	$C_6H_{12}O_6$ 24 e$^-$	→	$2\ CH_3CHOHCOOH$ 24 e$^-$ (lactic acid)	−198

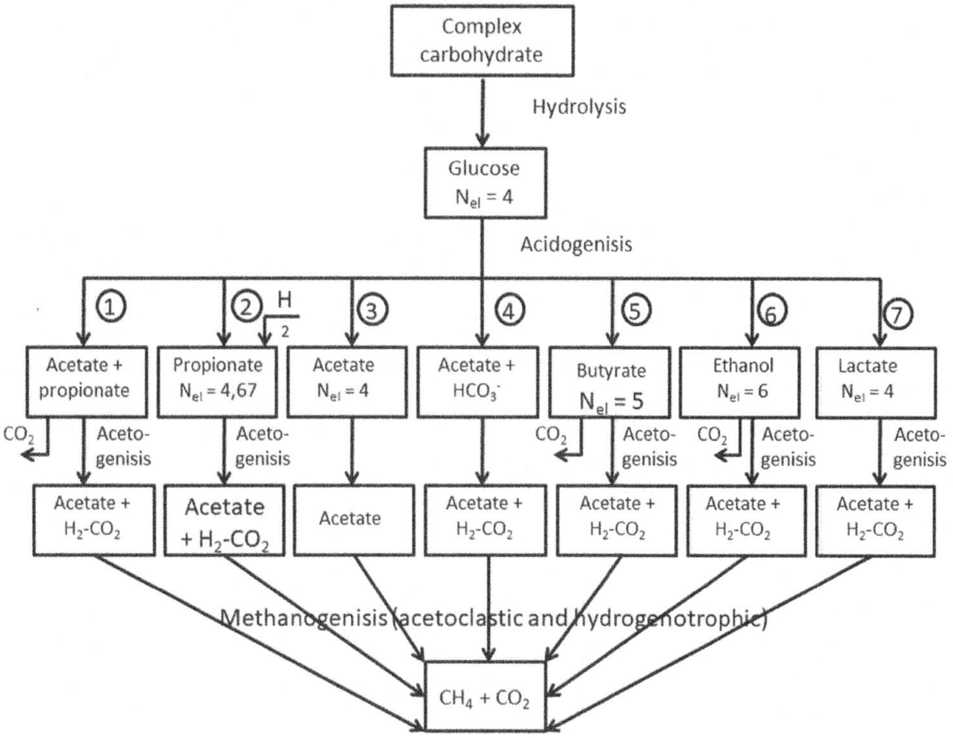

Figure 2.8 Different pathways of the acidogenic and acetogenic processes in the transformation of glucose to biogas (methane + carbon dioxide).

Table 2.2 Inhibitory concentration for methanogenesis for several acids formed in acidogenesis.

Acid	Formula (dissociated)	Acidogenic pH	Inhibitory Concentration ($g \cdot L^{-1}$)
Formic	$HCOO^-$	5.0–6.0	0.1–1
Acetic	CH_3COO^-	4.0–5.0	5–10
Propionic	$CH_3CH_2COO^-$	5.0–6.0	1.0
Butyric	$CH_3CH_2CH_2COO^-$	5.0–6.0	5–10
Caproic	$CH_3CH_2CH_2CH_2CH_2COO^-$	5.0–6.0	1.0

Acidogenesis generally is the fastest step in the sequence of processes that develop in the anaerobic conversion of complex organic material. For this reason, anaerobic reactors may be subject to sudden pH drops, for example when reactors are overloaded or disturbed by toxic compounds. In the first case, the rate of acidogenic processes (acid producers) is higher than the maximum rate of methanogenic processes (acid consumers) and the pH will fall, reducing the rate of methanogenesis further and thus aggravating the problem. In the second case the toxic material may affect methanogenesis more than the other processes, with the same results.

In wastewater a pH buffer is present in the form of alkalinity, which protects against abrupt changes in pH when acid accumulates in the reactor. In the specific case of anaerobic sewage treatment, because it is a wastewater with low concentration of organic material, the alkalinity is normally enough to avoid instability of the acidogenic process. In the case of more concentrated wastewater, pH control is of fundamental importance to maintain stable operation of the anaerobic system.

It is important to note that in addition to the effect on pH, the formation of acids due to acidogenesis can reach levels that are toxic to the methanogens, even though these will use the products of acidogenesis and acetogenesis. In Table 2.2 the range of concentrations resulting in the inhibition of methanogenic archaea can be observed, as well as the pH value that will develop when acidogenesis proceeds unchecked.

2.3.3 Acetogenesis

Acetogenesis is the conversion of the products of acidogenesis into compounds that form the substrates to produce methane: acetate and hydrogen + carbon dioxide. Gujer and Zehnder (1983) estimated that in the case of sewage treatment a fraction of approximately 70% of the COD originally present is converted into acetic acid, while the remainder is converted into hydrogen. However, this depends heavily on the type of substrate used and in our experiments there is an enormous variation of the ratio between acetoclastic and hydrogenotrophic methanogenic capacity. It is possible to have 70% of methanogenesis occurring via the hydrogenotrophic pathway.

Depending on the oxidation state of the organic material to be digested, the formation of acetic acid may be accompanied by the production of carbon dioxide or hydrogen, as follows from the following equations of reaction:

When $y < 2 \cdot z$ ($N_{el} < 4$):

$$C_xH_yO_z + {}^1/_4 \cdot (4x - y - 2z)\, H_2O \rightarrow {}^1/_8 \cdot (4x + y - 2z)\, CH_3COOH + {}^1/_4 \cdot (2z - y)\, CO_2. \quad (2.23)$$

When $y > 2 \cdot z$ ($N_{el} \geq 4$), which is usually the case:

$$C_xH_yO_z + (x - z)\, H_2O \rightarrow x/2\, CH_3COOH + {}^1/_2 \cdot (y - 2z)\, H_2. \quad (2.24)$$

In a mixture of organic pollutants, such as in domestic sewage, it is possible that both the above processes occur simultaneously, but there will always be a tendency to form more hydrogen than carbon dioxide. This is because the average number of transferable electrons (N_{el}) always tends to be greater than four per atom of C: for carbohydrates $N_{el} = 4$, but for lipids $N_{el} > 4$. In this way, the conversion of the compounds to acetic acid (which has a N_{el} value of 4, see Table 1.8) tends to be an oxidative process with the consequent hydrogen generation to maintain the balance of transferable electrons.

EXAMPLE 2.3

Calculate the composition of acidogenesis and acetogenesis of long-chain fatty acids like palmitic acid ($C_{16}H_{32}O_2$).

Solution

The acetogenic process produces the substrates for methanogenesis. The reaction equation can be written as:

$$C_xH_yO_z + (x - z)\,H_2O \rightarrow x/2\;CH_3COOH + ½\,(y - 2z)\,H_2$$

Thus, with $x = 16$; $y = 32$ and $z = 2$:

$$C_{16}H_{32}O_2 + 14H_2O \rightarrow 8CH_3COOH + 18H_2$$

Thus 1 mol of palmitic acid has the potential of producing 8 mols of acetic acid and 18 mols of hydrogen. The production of methane is equal $8 + 4.5 = 12.5$ mol CH_4 (8 mols of acetoclastic and 4.5 mols of hydrogenotrophic methane, because 4 mols of H_2 can produce 1 mol of CH_4 by reducing CO_2). The acetoclastic process produces 8 mols of CO_2 of which 4.5 are used reduction by hydrogen. Thus:

$$8CH_3COOH \rightarrow 8CH_4 + 8CO_2 \text{ and}$$
$$18H_2 + 4.5CO_2 \rightarrow 4.5CH_4 + 9H_2O$$

The overall acetogenic and methanogenic reactions are:

$$C_{16}H_{32}O_2 + 14H_2O \rightarrow 8CH_3COOH + 18H_2 \rightarrow 12.5CH_4 + 3.5CO_2 + 9H_2O$$

A limited number of bacteria can ferment secondary metabolites such as ethanol, propionate, butyrate, caproate (all with $N_{el} > 4$) into acetate. In this case the association with hydrogenotrophic methanogenic archaea is necessary because the formation of acetate is accompanied by the formation of hydrogen that must be removed to maintain the thermodynamic viability of acetogenesis (see Section 2.3.3).

2.3.4 Methanogenesis

Methanogenic archaea carry out the final stage of the anaerobic conversion of organic material into methane (the product that cannot be further reduced) and carbon dioxide (the product that cannot be further oxidized). During this fourth and last phase of anaerobic degradation of organic matter, a group of methanogenic archaea both reduces carbon dioxide, using hydrogen as electron donor, while another breaks the acetate molecule to form CH_4. It is only at this stage that the influent COD is converted into gaseous form and released from the liquid phase. Methanogens are obligate anaerobes, with few substrates they can use: acetate, methanol, formate, and H_2/CO_2 or CO (Bitton, 2005; Holmes & Smith, 2016; Kotsyurbenko *et al.*, 2004). Methanogens are classified into two major groups: those that convert acetate or acetoclastic methanogens and those that use hydrogen or hydrogenotrophic archaea.

Acetoclastic methanogens: $CH_3COOH \rightarrow CH_4 + CO_2$

Hydrogenotrophs: $4H_2 + CO_2 \rightarrow CH_4 + 2H_2O$

Figure 2.8 shows the pathways of the seven acidogenic processes available for the fermentation of glucose as discussed in Table 2.1, followed by acetogenic processes that generate the substrates for methanogenesis: acetate and H_2+CO_2. It can be clearly seen that there are many pathways that all lead to the same result: the transformation of glucose (obtained from complex carbohydrates) into biogas, the equimolar mixture of CH_4 and CO_2.

Gujer & Zehnder (1983) considered that about 70% of the methane produced originates from acetate as the main precursor (Figure 2.7), but in sewage treatment the actual proportion of hydrogenotrophic methane production is higher and may actually exceed the acetoclastic methane production. The growth rate of acetoclastic methanogens is very slow, with a doubling time of several days. The low growth rates explain why anaerobic reactors require a very long start-up time if the inoculum is poorly adapted to the substrate. Archaea have a much higher maximum growth rate than acetoclastic methanogens, with doubling times of 4 to 12 hours. Due to this characteristic, oscillations in the hydrogen partial pressure can quickly be absorbed and thus guarantee the continued viability of the acetoclastic methanogens. For this reason, high-rate anaerobic reactors can exhibit remarkable stability, even when operated under varying conditions.

2.3.4.1 Monod model for methanogenesis

Acetate and hydrogen are the substrates used by archaea to produce biogas. Unlike in the case of particulate matter, the modelling of the kinetics of processes that transform the dissolved organic material into biogas using the Monod model is possible. Monod (1950), when investigating the kinetics of alcoholic fermentation of grape juice to wine, found an empirical equation relating the concentration of the substrate (sugars) to that of the micro-organisms (yeast):

$$(dX/dt)_g = \mu \cdot X = \mu_m \cdot S/(S + K_s) \cdot X \tag{2.18}$$

where

$(dX/dt)_g$ = biomass growth rate of yeast due to sugar fermentation (g VSS $\cdot L^{-1} \cdot d^{-1}$)
μ = specific growth rate of the micro-organism (d^{-1})
μ_m = maximum growth rate constant of the micro-organism (when $S >> K_s$) (d^{-1})
X = micro-organism concentration (g VSS $\cdot d^{-1}$)
S = substrate concentration (mg S $\cdot L^{-1}$)
K_s = Monod or half-saturation constant (mg S $\cdot L^{-1}$)

Subsequently the Monod model has been successfully used to describe a wide range of microbiological processes, including the acidogenesis, acetogenesis and methanogenesis processes of anaerobic wastewater digestion. As an example, in Figure 2.9 the Monod model is used to represent the utilization rate of acetate due to growth of micro-organisms as a function of the acetate concentration. It can be observed that:

- The specific growth rate μ increases as a function of the substrate concentration: from $\mu=0$ for $S=0$ mg$\cdot L^{-1}$ to a maximum value (referred to as μ_m) when the substrate concentration is not growth-limiting anymore. Without substrate limitation, Equation (2.18) simplifies to $(dX/dt)_g=\mu_m \cdot X$, so that $X=X_o \cdot \exp(\mu_m \cdot t)$. Therefore, under these conditions exponential growth of the mass of micro-organisms will occur. Thus, for example, when $\mu_m=1.0\ d^{-1}$ the mass of micro-organisms after 1 day of uninterrupted growth without substrate limitation increases with $\exp(1)=2.71$ times.
- When the growth rate is equal to half the maximum rate, then by definition the substrate concentration is equal to the so-called half-saturation constant value. This is the reason for the denomination of "half-saturation constant". Thus, in the Monod model the growth rate of micro-organism X, using a

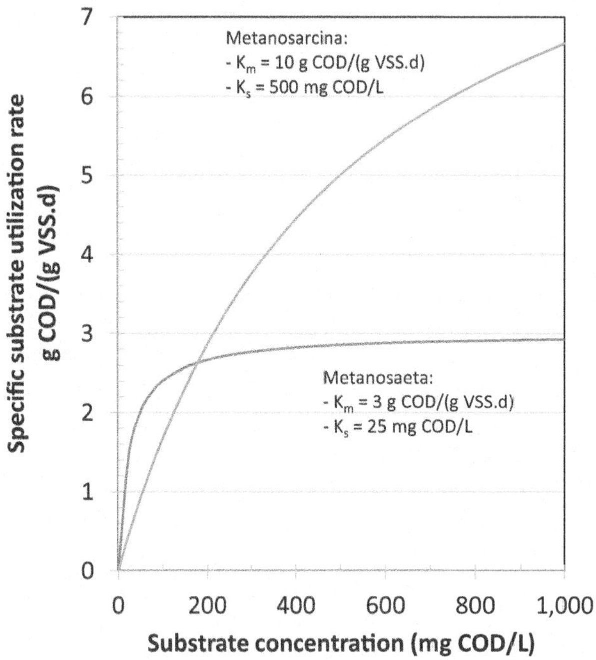

Figure 2.9 Specific substrate utilization rate as a function of the acetate concentration according to the Monod equation.

particular substrate S, is determined by two constants: the maximum growth rate constant and the half-saturation constant.

The Monod model is complemented with three other considerations:

(1) The growth rate of the micro-organisms is proportional to the substrate utilization rate:

$$(\mathrm{d}X/\mathrm{d}t)_{\mathrm{c}} = Y \cdot (\mathrm{d}S/\mathrm{d}t)_{\mathrm{u}} \tag{2.19}$$

where

$(\mathrm{d}X/\mathrm{d}t)_{\mathrm{g}}$ = growth rate of the micro-organisms (g VSS · L^{-1} · d^{-1})
$(\mathrm{d}S/\mathrm{d}t)_{\mathrm{u}}$ = substrate utilization rate (g COD · L^{-1} · d^{-1})
Y = yield factor (g VSS · g^{-1} COD).

Therefore, the growth rate of the bacteria is always a factor Y of the utilization rate of the substrate. The utilization constant, K_{m}, which denotes the maximum rate of substrate metabolism can therefore be defined as $K_{\mathrm{m}} = \mu_{\mathrm{m}}/Y$ (g COD · g^{-1} VSS · d^{-1}).

(2) The micro-organisms decay in a process with first-order kinetics:

$$(\mathrm{d}X/\mathrm{d}t)_{\mathrm{d}} = -b_{\mathrm{an}} \cdot X \tag{2.20}$$

where

$(\mathrm{d}X/\mathrm{d}t)_{\mathrm{d}}$ = decay rate of the micro-organisms (g VSS · L^{-1} · d^{-1})
b_{an} = decay rate constant (d^{-1}).

(3) In a treatment system there is loss of sludge by unintentional discharge with the effluent or by intentional discharge as excess sludge. This loss can be described as:

$$(dX/dt)_e = X/R_s \tag{2.21}$$

where

$(dX/dt)_e$ = sludge discharge rate (g VSS · L^{-1} · d^{-1})
R_s = sludge age (d^{-1}).

Figure 2.9 shows the specific substrate utilization rate (the COD utilization rate per unit sludge mass) as a function of concentration for two types of methanogens: *Methanosarcina* with $K_m = \mu_m/Y = 10$ g COD · g^{-1} VSS · d^{-1} and $K_s = 500$ mg · L^{-1} and *Methanosaeta* with $K_m = 3$ g COD · g^{-1} VSS · d^{-1} and $K_s = 25$ mg · L^{-1}. These values are typical for these archaea (De Vrieze *et al.*, 2012). Figure 2.9 illustrates well that, although it is the same process (the conversion of acetate into methane), the digestion rates for the two types of methanogens show very different profiles: at high acetate concentrations *Methanosarcina* utilize acetate at a higher rate than the *Methanosaeta*, but at low concentrations the opposite is true. Therefore, it is expected that in anaerobic treatment systems, which normally have a low concentration of acetate, *Methanosaeta* will predominate over *Methanosarcina*, which is proven in practice. *Methanosarcina* may predominate only if the concentration is high, i.e. when the digester is systemically overloaded or when a regime of batch feeding is applied.

By equating the sum of the sludge loss rates from decay and sludge discharge to the growth rate in a completely mixed reactor under steady conditions (i.e. constant feed flow with constant substrate concentration and composition), the residual substrate concentration S_e can be calculated:

$$S_e = K_s \cdot (b + 1/R_s)/[(\mu_m - (b + 1/R_s)] \tag{2.22}$$

Equation (2.22) shows that the residual concentration of a substrate being metabolized by micro-organism under steady state conditions depends on the values of three kinetic constants (μ_m, K_s and b) and an operational variable, the sludge age R_s. Figure 2.10 is a graphical representation of Equation (2.22) for the same bacteria of Figure 2.9 where $\mu_m = K_m \cdot Y$ and considering that the decay rate is very low: $b_{an} \approx 0.00$. It can be noted that for all sludge ages *Methanosaeta* produces a lower residual acetate concentration than *Methanosarcina*. Therefore, if the main objective of the anaerobic treatment is the efficient removal of the organic material, *Methanosaeta* is preferable. However, if the main objective of the treatment is to have high substrate utilization rate with consequently a high methane production rate, *Methanosarcina* should be preferred, especially when the residual concentration of organic material in the effluent is not very important.

An important aspect of Monod kinetics is that if sequential processes develop (e.g. acidogenesis, acetogenesis and methanogenesis), the residual concentration of the substrate (which in this case will be expressed in mg COD · L^{-1}) is the sum of the residual concentrations of the three processes. This is also borne out in Figure 2.6 where the effluent COD is composed of a non-biodegradable fraction and several biodegradable fractions related to the biological processes. Therefore, if one of the processes is not efficient, the aggregate COD concentration in the effluent will be high.

Whether a particular step in a sequence of several reactions (as in the series of acidogenesis, acetogenesis and methanogenesis) develops efficiently can be evaluated by the analytical measurement of the generated products. For example, if the sludge age is reduced and it is noted that the COD of the effluent increases, but not the concentration of acetate, it is concluded that in this case methanogenesis is not the process that limits the efficiency of anaerobic digestion.

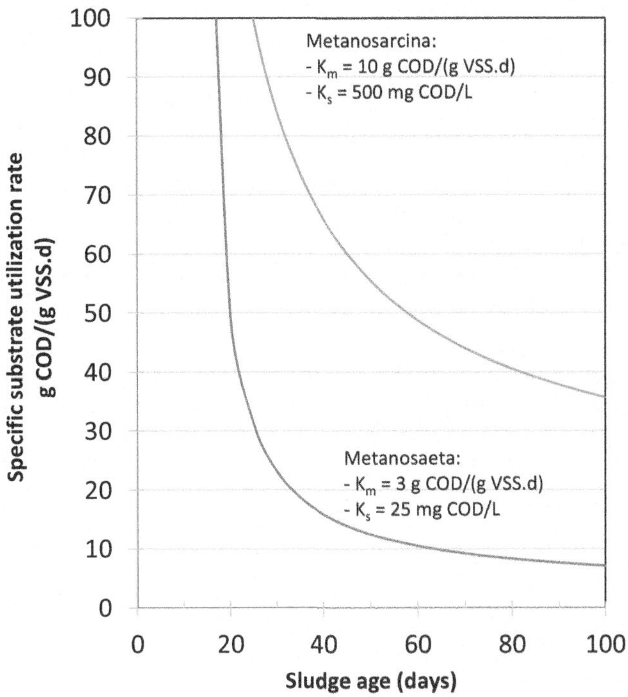

Figure 2.10 Residual substrate concentration as a function of the sludge age for the conditions of Figure 2.9.

2.3.5 Thermodynamic considerations

The viability of a biological reaction depends on the question whether it releases energy for the micro-organisms, that is, if the reaction is exergonic. Assume the following generic reaction equation:

$$a[\mathrm{A}] + b[\mathrm{B}]\ldots \rightarrow c[\mathrm{C}] + d[\mathrm{D}]\ldots \tag{2.25}$$

The free energy of this reaction can be calculated by the Nernst equation which is expressed as:

$$\Delta G = \Delta G^{o} + R \cdot T \cdot \ln\{(c[\mathrm{C}]\ d[\mathrm{D}])/(a[\mathrm{A}]\ b[\mathrm{B}])\} \tag{2.26}$$

where

[A], [B] = concentrations of reactants
[C], [D] = concentrations of products formed in the reaction
a, b, c, d = coefficients in the reaction equation
ΔG = free energy released in reaction
ΔG^{o} = free energy released under standard conditions i.e. when [A] = [B] = [C] = [D] = 1 (concentration in $\mathrm{mol \cdot kg^{-1}}$ except for gases which are expressed in atm.)
T = absolute temperature in Kelvin
R = universal gas constant = $8.31 \times 10^{-3}\ \mathrm{kJ \cdot mol^{-1} \cdot K^{-1}}$

If the released free energy is positive, the reaction is thermodynamically feasible; if it is negative the reaction cannot develop. In the case of anaerobic digestion there is a particularity with respect to hydrogen. Depending on the reaction, hydrogen can either be a product or a reactant in the transformation of glucose to biogas. For example, in the acetogenic reaction of acetate production from propionic acid (reaction 1), hydrogen is a product, but hydrogen is a reactant in hydrogenotrophic methane production (reaction 2).

Reaction 1: $H_3CH_2COO^- + 3H_2O \rightarrow CH_3COO^- + HCO_3^- + H^+ + 3H_2$

Reaction 2: $4H_2 + HCO_3^- + H^+ \rightarrow CH_4 + 3H_2O$

From the Nernst equation the energy released in the production of methane increases with increasing hydrogen pressure, but in the case of propionic acid acetogenesis the free energy decreases with increasing hydrogen pressure. This situation is shown in Figure 2.11. It may be noted that for an H_2 pressure less than 10^{-6} atm (0.101 Pa) the hydrogenotrophic production of methane becomes impossible, while at pressures greater than 10^{-4} atm (about 10.1 Pa) the generation of acetate from propionic acid becomes infeasible.

For the proper functioning of an anaerobic treatment system it is necessary that the acetogenesis of propionic acid can develop, because propionate will always be formed as an intermediate product.

Hence to ensure the feasibility of acetogenesis it is necessary to limit the hydrogen pressure in the formed biogas, but this is only possible if the hydrogenotrophic archaea efficiently eliminate hydrogen, transforming it into methane. Therefore, to have stability in the anaerobic digestion of organic material in treatment systems, cooperation between hydrogen-producing bacteria and hydrogen-consuming archaea is necessary.

Another case where the importance of the hydrogen partial pressure is shown is the anaerobic digestion of lipids. Long-chain fatty acids may be saturated, when of animal origin such as palmitate ($C_{16}H_{32}O_2$,

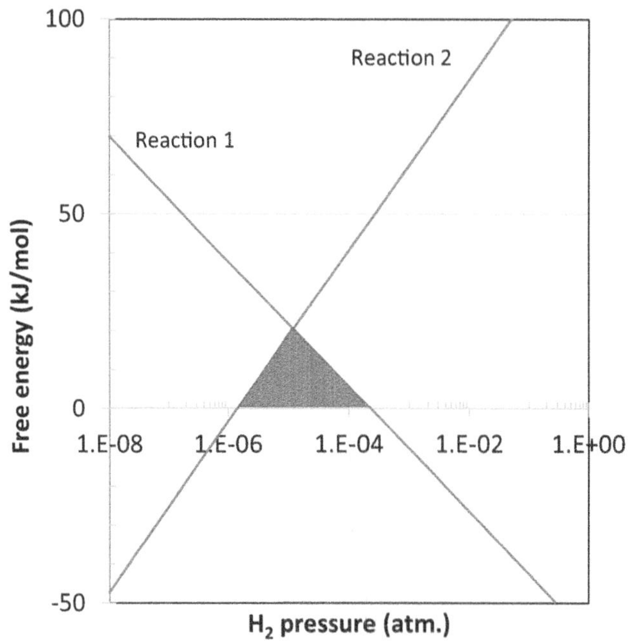

Figure 2.11 Relation between the released free energy and the partial hydrogen pressure in the gas phase for acetogenesis from propionic acid (reaction 1) and the production of hydrogenotrophic methane (reaction 2).

$N_{el} = 5.75$) or stearate ($C_{18}H_{36}O_2$, $N_{el} = 5.78$), or they may be unsaturated when of vegetable origin such as oleate ($C_{18}H_{34}O_2$, $N_{el} = 5.67$). They can be transformed into volatile (short-chain) fatty acids, but this requires a decrease in the number of transferable electrons per atom of carbon, i.e. a partial oxidation, which can be affected by the release of hydrogen. For example, as shown in Example 2.3 for palmitate:

$$C_{16}H_{32}O_2 + nH_2O \rightarrow (n/2)\, C_2H_4O_2 + C_{16\text{-}2n}H_{28\text{-}4n}O_2 + nH_2$$

The further the palmitate is decomposed, the more hydrogen will be produced, and the reaction will become thermodynamically unviable unless there is simultaneous removal of hydrogen by another mechanism, such as hydrogenotrophic methanogenesis. It is concluded that it is not possible to carry out anaerobic digestion of long-chain fatty acids in a two-stage system, separating acid fermentation (acidogenesis and acetogenesis) from methanogenic fermentation: without hydrogenotrophic methanogenesis, acid fermentation is not thermodynamically feasible.

2.4 COMPETING PROCESSES

There are a variety of micro-organisms that can develop in anaerobic reactors and compete with the microbiological methanogenic community. When electron acceptors such as dissolved oxygen or nitrate are introduced, the aerobic and anoxic organisms with their oxidative catabolism have a much greater energy liberation than those of fermentative catabolism as shown in Table 1.8. For this reason, they will tend to prevail over the methanogenic micro-organisms until all the oxidant is consumed. In the same way there are bacteria that use sulphate as oxidant of organic material in a process that has a ΔG value only marginally higher than that of methanogens when using acetate as a substrate. However, these sulphate-reducing bacteria along with methanogenic substrates also use other substrates, notably from the acidogenic and acetogenic processes (Lens & Hulshof Pol, 2000) and for that reason tend to prevail. Many wastewaters contain sulphate. In the case of domestic sewage, the source of the sulphate may be the water itself, but in the case of industrial wastewater sulphate and less oxidized forms of sulphur (sulphite, thiosulphate) may appear in high concentrations. Although these compounds themselves are not toxic to methanogens, the problem is that in anaerobic digesters they tend to be reduced to sulphide by sulphate-reducing bacteria (SRB), which also use sulphite and thiosulphate. Figure 2.12 shows the processes that take place when complex substrates are used for both methane production and sulphate reduction.

Knowing that only acetate and hydrogen are used for methanogenesis, but that SRB may also use other intermediates and that hydrogen is preferentially used by SRB, it is to be expected that in anaerobic reactors fed with complex substrates the SRB will outcompete methanogens and methanogenesis only develops when sulphate reduction is complete. The literature shows that this is usually what occurs. The influence of sulphate reduction on anaerobic digestion will be discussed in more detail in Chapters 3 and 6.

The problem with sulphate reduction is not only the reduction of the methane production: the product of sulphate reduction and notably the non-dissociated form of H_2S in the liquid phase has a toxic effect on the methanogens. Several researchers (Rinzema, 1988) have found that for a non-dissociated H_2S concentration of 50 $mg \cdot L^{-1}$ there is already a 50% reduction of methanogenic activity, and for a concentration of 200 $mg \cdot L^{-1}$ methanogenesis will be severely inhibited.

In Chapter 6 the division between sulphur in the gas phase and in the liquid phase is discussed as well as the division between the dissociated and non-dissociated sulphur in the liquid phase and hydrogen sulphide in the gas phase.

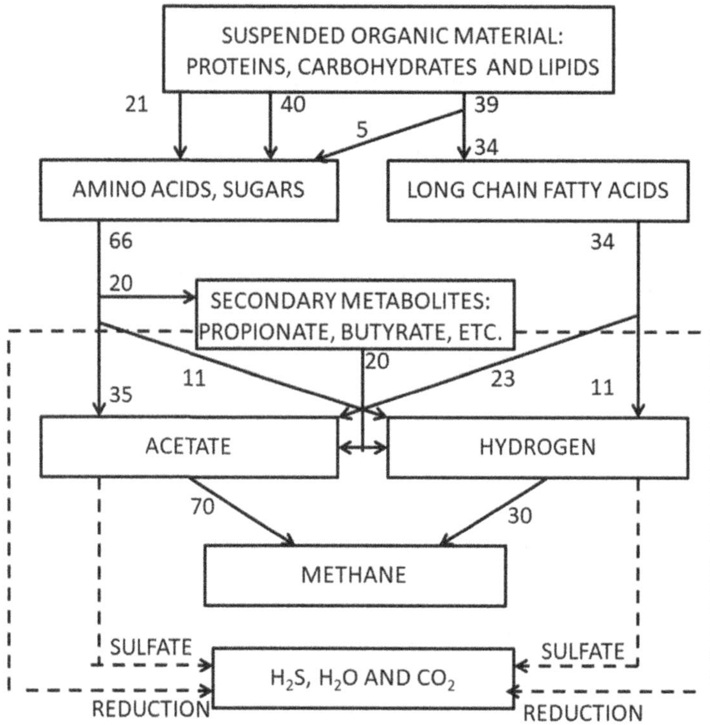

Figure 2.12 Schematic representation of the processes that develop in an anaerobic environment, when sulphate reduction and anaerobic digestion both occur (Lens & Hulshof Pol, 2000).

2.5 INFLUENCE OF ENVIRONMENTAL FACTORS

Important environmental factors affecting the anaerobic digestion process are temperature, pH, presence of essential nutrients and absence of excessive concentrations of toxic compounds, either in the influent or generated in the reactor. In the case of municipal sewage, the last three factors usually do not need consideration:

- An adequate and stable pH is defined by the presence of the carbonic system and normally chemicals to correct pH are not necessary (see also Chapter 4);
- Nutrients: both macro-nutrients like nitrogen, phosphorus, sulphur etc. and micro-nutrients (trace metals) are abundantly available in the sewage;
- Compounds that may exert a toxic influence on micro-organisms are generally absent in domestic wastewater.

2.5.1 Temperature

Anaerobic digestion, like other biological processes, depends heavily on temperature. Regarding the conversion rate of digestion processes, there is a maximum between 35 to 40°C for the mesophilic range and between 52 to 55°C for the thermophilic range.

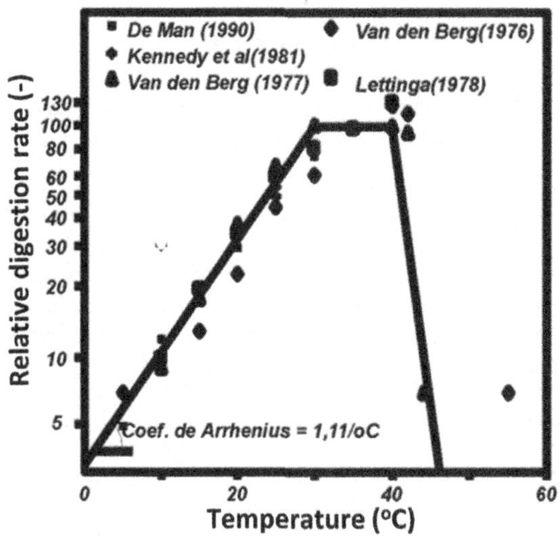

Figure 2.13 Influence of the temperature on the mesophilic digestion rate (Henze & Harremoes, 1983).

For treatment of municipal sewage only the mesophilic region is relevant. The influence of temperature on the rate and extent of anaerobic digestion has been the subject of much research. Figure 2.13 shows a graphical representation of data from several authors relating the relative rate of anaerobic digestion as a function of temperature. The optimal range for anaerobic sewage treatment is between 30 and 40°C: and for temperatures below the ideal range the digestion rate decreases about 11% per °C, or according to Arrhenius's expression:

$$r_t = r_{30} \cdot (1.11)^{(T-30)} \tag{2.26}$$

where

T = temperature in °C
r_t, r_{30} = digestion rate at temperature T and 30°C respectively

With the aid of Equation (2.26) the digestion rates at 20 and 10°C are calculated at about 35% and 12% respectively, relative to the rate at 30°C. The influence of temperature is not limited to the rate of anaerobic digestion. The extent of anaerobic digestion is also affected, as found by O'Rourke (1968) and Van der Last (1991). Figure 2.14 shows the extent of digestion achieved for settled sludge from sewage (primary sludge) as function of the digestion time at different temperatures, according to the results of O'Rourke (1968). This diagram clearly shows a strong temperature dependency for the digestion efficiency of solids: the reduction at lower temperature can be attributed to a lower hydrolysis rate. In practical terms, this means that organic suspended matter can be removed from the liquid phase at low temperatures, even when it is not metabolized, because it can be trapped in the sludge bed. Of course, the quality of the sludge will decrease as it will contain more non-hydrolysed material. Chapter 3 discusses anaerobic digestion at low temperatures.

For concentrated wastewaters the operating temperature to some extent can be considered as a process variable for anaerobic treatment systems, because the methane produced can be used to heat the wastewater.

This is however not the case for wastewater of low concentration such as sewage, because the heat obtained from the combustion of the produced methane is insufficient to cause a significant increase of the temperature. However, it is possible to remove settleable solids, for example by gravity, and concentrate these at a much lower flow rate (1% of the sewage flow). Under these conditions it is possible to apply

EXAMPLE 2.4

Estimate the maximum temperature increase of wastewater that can be obtained by transferring heat from methane combustion, if the concentration of the digestible COD is 600 mg·L^{-1}.

Solution

The production of methane can be estimated at 600/4 = 150 mg CH_4·L^{-1}. The combustion heat of methane is 12.4 kJ·g^{-1} COD or 3 kCal·g^{-1} COD (Table 1.7 and Figure 1.8). Therefore, the maximum heat that can be produced from the combustion of 150 mg CH_4·L^{-1} is 0.6·3 = 1.8 kCal·L^{-1}. Per definition, 1 kCal increases the temperature of 1 L of water by 1°C, so that a temperature increase of 1.8°C is theoretically possible, but this maximum value is only reached when organic biodegradable material is completely converted to methane, no methane remains dissolved in the liquid phase and all the released heat is used to raise the sewage temperature. Even then, this increase in temperature is insignificant.

It is concluded that digestion temperature in practice will be equal to the sewage temperature, which is invariably less than the ideal temperature for anaerobic digestion. The possibility of separating a large part of the incoming COD by gravity and concentrating this material in a small fraction of the volume is discussed in Chapter 3.

EXAMPLE 2.5

What should be the minimum COD digested concentration to raise the temperature of a wastewater from 10°C to 35°C?

Solution

In the previous example it was calculated that a temperature increase of 1.8°C was theoretically possible for a COD concentration of 600 mg·L^{-1}, or 1.8/0.6 = 3°C·g^{-1} COD·L^{-1}. If the required temperature increase is 35 − 10 = 25°C, then the digested COD concentration should be at least equal to 25/3 = 8.3 g COD·L^{-1}.

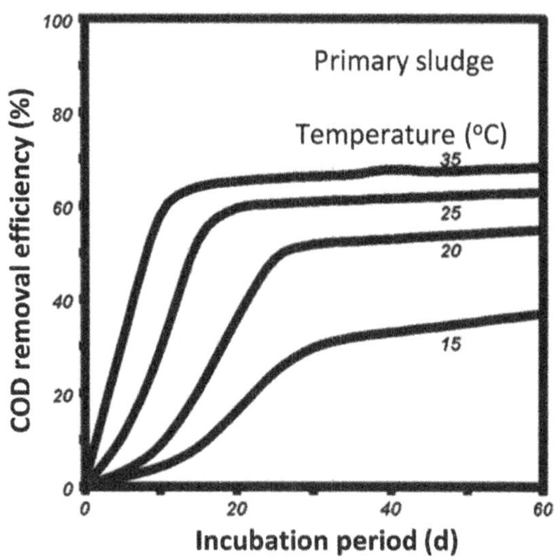

Figure 2.14 Influence of the temperature on the extent and the rate of anaerobic digestion of primary sludge (O'Rourke, 1968).

thermal energy to heat the separated solids in a hydrolytic reactor and thus accelerate hydrolysis for further digestion along with non-settled material. This is discussed in Chapter 3.

2.5.2 pH in the reactor

The pH value in an anaerobic reactor is established after ionic equilibrium is reached of the different acid/base systems present in the system. Weak acid/base systems have a great influence; in particular the carbonic system is often dominant because its concentration generally substantially exceeds that of other systems combined, such as phosphate, ammonia or sulphide. This subject will be discussed in detail in Chapter 4. It is important to note that the pH of a wastewater can be very different from the pH value the anaerobic digester, even without applying any corrections. In the case of low pH wastewater such as vinasse, the pH in the digester may rise because methanogenesis converts acetic acid, which is present at high concentration in the influent.

On the other hand, in the case of alkaline wastewater, as in paper production, CO_2 produced in methanogenesis can be absorbed and converted into bicarbonate, thereby reducing the pH value. For this reason hardly any wastewater requires acid addition to lower its pH for anaerobic digestion.

Among possible chemicals used to supply alkalinity are lime or hydrated lime (CaO or $Ca(OH)_2$), but these may cause precipitation of calcium carbonate in the anaerobic reactor and on the anaerobic biomass. Caustic soda ($NaOH$), soda (Na_2CO_3) or urea ($CO(NH_2)_2 \rightarrow$ ammonification!) may be used. In Chapter 4 a method is presented to estimate the demand for alkalinity, which depends on the characteristics of the influent.

The hydrolysis and acid fermentation processes occur already at high rate when the pH is above 4.5. In contrast, methanogenesis requires a much narrower pH range near the neutral value of 7.0 to develop at the maximum rate: values outside the range $6.3 < pH < 7.8$ lead to a severe reduction of methanogenic activity. In the case of fermentation of carbohydrates there is a tendency to acidification. The acido- and methanogenic fermentations of carbohydrates can be written as:

$$(CH_2O)_n \rightarrow n/2\ CH_3COOH \text{ and} \tag{2.27}$$

$$n/2\ CH_3COOH \rightarrow n/2\ CH_4 + n/2\ CO_2 \tag{2.28}$$

In the case of fermentation of proteins, which occurs for example in the anaerobic digestion of waste sludge or manure, there is a partial compensation of acid souring due to ammonia formation. Using the generic formula $C_5H_7O_2N$ for the bacterial mass (Van Haandel & Van der Lubbe, 2012), for acid fermentation the following reaction formula applies:

$$C_5H_7O_2N + 3H_2O \rightarrow 2.5CH_3COOH + NH_3 \rightarrow 2.5CH_3COOH^- + 1.5H^+ + NH_4^+ \tag{2.29}$$

For complete anaerobic digestion:

$$C_5H_7O_2N + 3H_2O + H^+ \rightarrow 2.5CH_4 + 2.5CO_2 + NH_4^+ \tag{2.29}$$

In this case there is production of alkalinity or consumption of acidity and anaerobic digestion can be sustained. However, there is a potential danger: the mineralization of organic nitrogen can increase the concentration of ammonium (+ ammonia) to the point where it becomes toxic to methanogens, as will be discussed in the subsequent section. This can be a real problem in the anaerobic digestion of high-solids waste streams such as primary sludge or manure.

2.5.3 Toxic compounds and inhibitors

In contrast to nutrients that are indispensable to maintain high sludge activity, there are toxic compounds and inhibitors which, if present, reduce or even stop the activity of the micro-organism. There is often a differentiation made between reversible (inhibition) and irreversible toxicity, the difference being that anaerobic activity recovers quickly in the case of inhibition, whereas the effects of irreversible toxicity (resulting in cell decay) take much longer to remedy – often the biomass must be replaced or regrown.

In addition to the concentration of hydrogen ions already discussed in Section 2.3.5, several other compounds can affect the rate of anaerobic digestion. Potentially toxic compounds that may be present in sewage (or be generated in the anaerobic reactor itself) include oxygen, nitrate/nitrate, ammonium/ammonia and sulphide. However, the presence of these compounds in inhibitory concentrations in sewage is unlikely.

Several other components are very toxic to anaerobic biomass but are under normal conditions not present in municipal sewage, unless there are sizeable industrial contributions to the sewage. Examples are mineral oil (e.g. used as lubricants!), all compounds containing quaternary ammonia and many other kinds of cleaning agents and disinfectants. A comprehensive list of toxic compounds can be found in Speece (2008).

Some oxygen or nitrate/nitrate can be introduced into the influent distribution system, but it will quickly be used for the oxidative metabolism in the acidogenesis process. Thus, dissolved oxygen and nitrate/nitrite will not be present in the anaerobic reactor, so that its introduction will be irrelevant to the performance of the reactor, apart from a slight reduction in methane yield.

Sulphide can be formed in the anaerobic reactor due to the occurrence of sulphate reduction. However, the sulphide concentration that normally can be expected in anaerobic sewage treatment systems (i.e. up to 20 $mg \cdot L^{-1}$) is much lower than the minimum concentration for appreciable toxicity. Therefore, toxicity will not normally be a problem in anaerobic sewage treatment systems. There are cases however where the sulphate concentration in the sewage is high, for example because of high sulphate levels in the water supply or infiltration of seawater in the sewer system. Values of up to 500–600 $mg\ SO_4 \cdot L^{-1}$ are regularly reported in the Middle East and this may severely limit the biogas production, even if the produced sulphide is not toxic to the methanogens.

As to ammonium toxicity, this compound is in equilibrium with dissolved ammonia (NH_3) gas. Like with H_2S, it is this gaseous component, called free ammonia, which is by far the more toxic one, so the toxicity effect is much higher at increased values of the pH value. Being a weak acid, the following reaction equation applies for the dissociation of ammonium in water:

$$NH_4^+ \leftrightarrow NH_3 + H^+ \tag{2.30}$$

There is already significant toxicity at a free ammonia concentration of 100 $mg\ N \cdot L^{-1}$ and almost complete inhibition at 200 $mg\ N \cdot L^{-1}$, although some degree of adaptation may occur. The equilibrium moves significantly to the right-hand side at higher temperatures, which is why ammonia toxicity is much more severe in thermophilic reactors. This is shown in Figure 2.15, where the free ammonia fraction as function of the pH is given for both mesophilic and thermophilic conditions. The actual free ammonia concentration that develops for a total $NH_4^+ + NH_3$ of 1000 $mg\ N \cdot L^{-1}$ is also indicated. It can be observed that significant inhibition starts only at a pH of around 8.0 under mesophilic conditions, but already at 7.5 under thermophilic conditions. For high-solids digesters, a pH value in the reactor in the range of 7.8–8.2 is not at all unlikely.

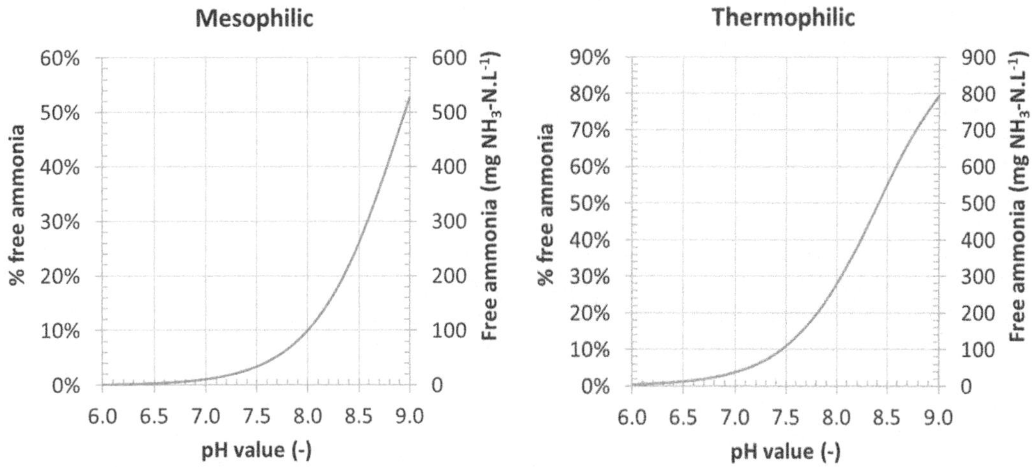

Figure 2.15 Free NH_3 concentration as function of the pH under mesophilic (left) and thermophilic conditions (right), for $pK_a = 0.09018 + 2730/(T + 273)$ and $NH_4\text{-}N + NH_3\text{-}N = 1000\ \text{mg N} \cdot L^{-1}$.

2.5.4 Nutrients

Nutrients are needed for the synthesis of new cells of micro-organisms. A unit of volatile solids of bacterial mass contains approximately 10% nitrogen and 2.5% of phosphorus. Assuming a production of approximately 0.1 g $VSS \cdot g^{-1}$ COD metabolized, about 0.01 g N and 0.0025 g P per g of metabolized COD is required. These concentrations are generally available in sewage and most wastewaters, but if they are not, they should be added in sufficient quantity to avoid inhibition of bacterial growth. The presence of nitrogen and phosphorus in the form of ammonia and phosphate respectively together with calcium and magnesium ions allows the formation of struvite and apatite precipitates in addition to calcium carbonate which can all cause scaling problems:

$$NH_4^+ + Mg^{2+} + PO_4^{3-} \rightarrow MH_4MgPO_4\downarrow \text{ (struvite)} \quad (2.30a)$$

$$4Ca^{2+} + 3PO_4^{3-} + H^+ \rightarrow Ca_4H(PO_4)_3\downarrow \text{ (octacalcium phosphate)} \quad (2.30b)$$

$$Ca^{2+} + CO_3^{2-} \rightarrow CaCO_3\downarrow \text{ (calcium carbonate)} \quad (2.30c)$$

In addition to the macro-nutrients N, P and K there are several minerals that are required in smaller amounts. If these micro-nutrients are not available, the rate of digestion in anaerobic systems can be severely inhibited (Speece, 2008). Again, this is unlikely in municipal sewage treatment.

2.6 MASS BALANCE OF ORGANIC MATERIAL

As discussed in the previous sections, there are four pathways for organic material fed to wastewater treatment systems:

(1) Conversion into sludge (anabolism or absorption);
(2) Conversion into methane (fermentative catabolism);
(3) Mineralization by oxidation (oxidative catabolism);
(4) Discharge with the liquid phase.

If there is no accumulation of organic material in the treatment system (steady state), the daily mass of organic material in the effluent must be equal to the sum of the daily masses of the organic material leaving the system as methane, sludge or present in the effluent and the daily mass of destroyed (oxidized) organic material. Thus, the following mass balance of organic material can be established:

$$MS_{ti} = MS_{te} + MS_{xv} + MS_d + MS_o \tag{2.31}$$

where

MS_{ti} = daily flux of organic material in the influent (kg COD·d^{-1})
MS_{te} = daily flux of organic material in the effluent (kg COD·d^{-1})
MS_{xv} = daily flux of organic material in the excess sludge (kg COD·d^{-1})
MS_d = daily flux of digested organic material (kg COD·d^{-1})
MS_o = daily flux of oxidized organic material (kg COD·d^{-1})

If the mass fluxes in Equation (2.31) are expressed in measurable parameters, then the validity of the equation can be verified experimentally. However, due to experimental errors in the determinations of the different terms of Equation (2.31) and to the fact that sewage treatment systems never operate under strictly stationary conditions, the equation generally is not rigorously valid. For this reason, the recovery factor of organic material was introduced, which is the ratio between the fluxes of the different forms of organic material and flux in the influent:

$$B_o = (MS_{te} + MS_{xv} + MS_d + MS_o)/MS_{ti} \tag{2.32a}$$

or

$$B_o = (MS_{te} + f_{cv} \cdot ME_v + MS_d + MS_o)/MS_{ti} \tag{2.32b}$$

where

B_o = value of the recovery factor of organic material (expressed as COD)
ME_v = organic excess sludge production = daily mass of produced sludge (kg VSS·d^{-1})
f_{cv} = ratio between organic sludge mass and its COD value
= 1.5 g COD·g^{-1} VSS (Van Haandel & Van der Lubbe, 2012).

Theoretically, the value of the recovery should be exactly equal to unity but due to experimental errors the value of B_o will tend to deviate from this theoretical value. The magnitude of the deviation is indicative of the quality of the experimental data. When calculating the recovery factor over a long period of operation of the treatment system, then much of the oscillations are eliminated. This allows an evaluation of the reliability of the experimental data: if the B_o deviation is much greater than could be expected based on the experimental errors in the analyses, then it must be assumed that some systematic error is being made in the analytical procedures or that some measuring device (e.g. flow or reactor volume) is not working properly. The fact that the COD test allows the calculation of mass balance and BOD test does not, is the main reason that today the first parameter is much more used than the second one.

Chapter 3

Anaerobic sewage treatment systems: process design and performance

3.1 INTRODUCTION

The designs of the first anaerobic treatment systems, implemented in the late nineteenth century, were not really adequate for efficient removal of organic material. The systems were designed to retain the solids present in sewage by applying simple settling, possibly with the idea that this would be enough to obtain a good-quality effluent (McCarty, 1981). Although settleable solids make up the most visible fraction of the impurities present in sewage, they represent no more than approximately one-third of the organic material in raw sewage, while another one-third is due to colloids (which confers turbidity to the sewage) and the remaining part is due to dissolved material. Much of the colloids and dissolved material was not removed, because there was little contact between the organic material fed to the system and the anaerobic sludge already present. Hence, the organic material removal efficiency of these systems was only of the order of 30 to 50%. When it was found that the aerobic systems that were developed soon thereafter (Arden & Lockett, 1914) performed much better in this respect, the erroneous belief was established that anaerobic systems had much lower treatment capacity and efficiency than aerobic systems, and that for that reason their retention time had to be long while their efficiency remained low.

There are two important prerequisites to be met for any biological treatment system to combine a good organic material removal efficiency with a short retention time: (1) enough contact between the influent organic material and the bacterial mass in the treatment system, and (2) presence of a large bacterial mass in the system. However, among the first treatment systems, neither of these two prerequisites for success were met. In these initial projects, the wastewater was flowing horizontally over the settled sludge mass without intentional mixing, hence contact between sludge and non-settleable organic material was impossible or at least unlikely. In contrast, in aerobic systems the sludge is kept in suspension through vigorous agitation by the aerators, so that contact with the substrate is always ensured. Also, in aerobic systems the sludge mass that could be maintained was large, either by adhering micro-organisms to

Editor(s): Adrianus van Haandel and Jeroen van der Lubbe
doi: 10.2166/9781780409627_0061

support material present in the system (attached biofilms) or by settling and recirculating sludge from a final settler. None of these arrangements were foreseen in those early anaerobic systems. Although now the above conditions for good performance of biological systems may seem obvious, for many years in the design of anaerobic systems they were ignored. It was not until the 1960s that the first anaerobic treatment systems with adequate designs were constructed, and soon it was shown that anaerobic systems actually had an even higher treatment capacity than aerobic systems.

Young and McCarty (1969) successfully operated an anaerobic filter treating wastewater from rum distillate with a high concentration of soluble organic material. The anaerobic sludge was retained in the system by filling the reactor with stones, which created a fixed bed with a large surface area for biofilm attachment.

Furthermore, the mixing of influent and sludge was induced by the imposition of an upward flow direction, so that the influent that was distributed over the bottom of the reactor would pass through the fixed bed. The improvements resulted in a treatment capacity of more than 10 kg $COD \cdot m^{-3} \cdot d^{-1}$, much more than the 3–4 kg $COD \cdot m^{-3} \cdot d^{-1}$ that had hitherto been the most possible in aerobic systems. Thus, it was conclusively demonstrated that the poor performance of classical anaerobic treatment systems was not due to the "inherent weakness" of anaerobic treatment, but rather to inadequate design. As a result, the anaerobic filter has found widespread application in the treatment of industrial and domestic wastewaters.

A very important subsequent event was the development of the Upflow Anaerobic Sludge Bed reactor by Professor Lettinga and his collaborators at the University of Wageningen in Holland (Lettinga *et al.*, 1980). Originally developed for concentrated agro-industrial effluents, its potential for the treatment of sewage, especially in regions with a hot climate was soon demonstrated (Draaijer *et al.*, 1992; Haskoning, 1989; Schellinkhout & Collazos, 1992; Vieira & Garcia, 1992).

The fact that anaerobic systems have a higher capacity than aerobic systems is mainly due to the much larger sludge mass that can be retained in the former. This greater mass in turn is due to two factors: (1) the anaerobic sludge has a higher settleability, and (2) in anaerobic systems there is not much need for agitation, whereas in aerobic systems the oxygen demand is supplied by mechanical aeration, which causes higher turbulence in the reactor. It is still not known why the anaerobic sludge has a much higher settleability than aerobic sludge, but it may be due to the shear stress on the sludge flocs resulting from aeration. Flocculant UASB reactors can operate at sludge concentrations of 20–40 g $TSS \cdot L^{-1}$ or even up to 50–100 g $TSS \cdot L^{-1}$ in the case of the granular UASB systems applied on mostly soluble industrial wastewater. In contrast, aerobic systems operate at a typical concentration of 3–6 g $TSS \cdot L^{-1}$. However, when comparing the metabolic capacity per unit sludge mass or specific capacity (expressed in g $COD \cdot g^{-1}$ VSS), the aerobic sludge is superior to the sludge from anaerobic systems. Overall, this higher specific capacity does not compensate for the much lower sludge concentration and for this reason the treatment capacity in terms of kg $COD \cdot m^{-3} \cdot d^{-1}$ of the anaerobic reactors tends to be larger than that of aerobic systems.

In order to compare the different types of anaerobic systems available on the market today with each other, it is generally considered that the removal efficiency of a treatment system will be higher as the hydraulic retention time (R_h) of the wastewater increases, so often the efficiency is evaluated as a function of R_h.

Later in this chapter (Section 3.9) it will be shown that in fact the important operational and design parameter is not the retention time but rather the sludge age. Systems with equal retention time can have very different efficiencies because of different sludge masses. The problem is that in the design stage the value of the sludge age is not yet known, because its value depends not only on reactor volume and feed flow, but also on the efficiency of sludge retention and the sludge characteristics, especially its settleability. For this reason, even knowing well the limitations of retention time as an independent variable, it is used in the first part of this chapter to compare different anaerobic reactor types in terms of performance.

Another important aspect is the definition of the term "treatment efficiency". In general terms, as was shown in Chapter 2, in aerobic systems the organic material of the influent is divided into three parts: (1) discharged into the effluent, mS_e (2) transformed into sludge, mS_{xv} and (3) transformed by oxidation, mS_o. As the objective of the anaerobic treatment system is the removal of organic material from the influent, it can be considered that in anaerobic systems the efficiency can be given as either (1) the fraction of the influent COD that is removed from the liquid phase, or (2) the influent COD fraction that is transformed into methane. Therefore:

$$\eta_1 = 1 - mS_e \tag{3.1a}$$

and

$$\eta_2 = mS_d = 1 - mS_e - mS_{xv} \tag{3.1b}$$

In practice the fraction mS_{xv} is generally not well defined, because it is not clear how much of the organic material in the effluent is composed of settleable material (which should be considered as sludge) and how much is really non-settleable (dissolved and colloidal). Researchers generally do not even consider sludge discharge, and when they do, they often do not quantify the discharge. Thus, it will generally be considered that efficiency is the fraction removed from the organic material present in the influent:

$$\eta = 1 - S_{ti}/S_{te} \tag{3.1c}$$

where

S_{ti} = influent COD concentration
S_{te} = effluent COD concentration

3.2 CLASSIC ANAEROBIC TREATMENT SYSTEMS

Classic anaerobic sewage treatment systems were developed at the end of nineteenth century by Mouras in France in 1872, and Cameron in 1896 and Travis 1903 in England. All are a settler–digester type of system, in which the settled solids are retained and digested at the bottom of the unit (Figure 3.1). A typical aspect of these classic systems is that wastewater enters at a horizontal trajectory in the upper part of the system, while sludge is deposited at the bottom of the tank. Therefore, no specific effort is made to ensure contact between the influent organic material and the biological sludge at the bottom. These classic systems found widespread application until the 1930s, when they began to lose popularity due to the application of aerobic sewage treatment systems, but they persist until today for the treatment of small units (on-site or local treatment). The maximum efficiency in the classic anaerobic systems does not exceed more than 50% of the biodegradable material and depends on the nature of the sewage, the operating conditions (temperature!) and the settling efficiency.

3.2.1 The septic tank

Septic tanks are treatment units, where the primary treatment (solids settling) is coupled to the anaerobic digestion of settled sludge. A septic tank can be built as a single unit or can be divided into two parts: an upper chamber for settling and a lower part for digestion. The latter configuration is called the Imhoff tank, after the name of its inventor in 1906 in Germany. The advantage of separating the two functions is that the biogas bubbles generated will avoid re-suspension of the settled solids. For this reason, the retention time in units with a single chamber (12–24 hours) needs to be longer than in the Imhoff tank (2 hours in the settling section). This shorter retention time is a considerable advantage, since the volume reduction is proportional. The removal efficiency of organic material and suspended solids in the sewage is in the order of 50%.

Septic tanks are probably the most widely applied system in the world for the treatment of sewage from individual households. This is mainly due to their simplicity and reliability, even though the treatment efficiency is relatively low. The septic tank can be prefabricated in fiberglass, plastic, or built onsite in masonry or concrete. Typically, they are constructed partially underground. The effluent is infiltrated into the soil or discharged to the surface (gutters) or receiving water, sometimes after some form of post-treatment, usually an anaerobic filter. Septic tanks for individual families usually have only one chamber.

In Brazil, the design of these units is regulated by a national standard: NBR 7229, which determines that the chamber should have a volume such that the retention time is 24 h for flows of up to 6000 $L \cdot d^{-1}$, decreasing gradually to a minimum of 12 h for flows of more than 14,000 $L \cdot d^{-1}$. The minimum volume of these tanks is 1500 L.

In the case of sewage treatment, the biological sludge will grow in the septic tank without the necessity of inoculation, due to the presence of the necessary bacteria in the influent.

Over time, non-biodegradable solids accumulate in the tank and after some time they will occupy almost the entire useful volume of the tank, reducing the liquid retention time and the removal efficiency of settleable solids. For this reason, the accumulated sludge must be periodically removed from septic tanks, usually with the aid of a properly equipped (vacuum) truck. The interval between of sludge removal can vary considerably, from one to several years.

Despite their widespread application, the septic tank has a fatal flaw in its design: it does not allow good contact between the organic material in the sewage and the sludge at the bottom of the tank and so it has a low efficiency, even though the retention time is long. Along with the need for periodic sludge discharge, these are the main disadvantages of septic tanks. Chapter 4 presents modern alternatives to the use of septic tanks for household (localized) sewage treatment, with a much higher performance despite having a much smaller volume. It is concluded that the septic tank is obsolete and ready to take its place in a technology museum.

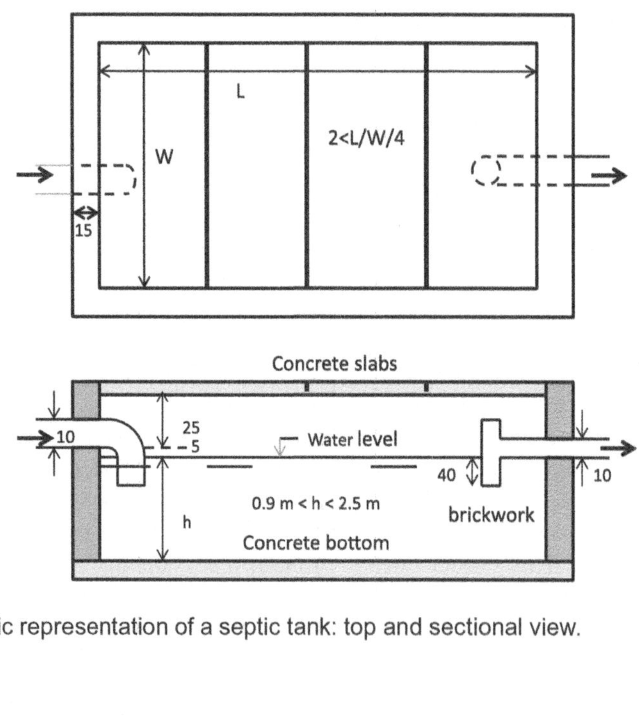

Figure 3.1 Schematic representation of a septic tank: top and sectional view.

3.2.2 The anaerobic lagoon

Another system with a horizontal flow pattern passing over accumulated sludge at the bottom of the reactor is the anaerobic lagoon or pond, which is widely used for the treatment of communal wastewater, normally combined with other lagoon systems forming what is called a stabilization pond system (Mara, 1975; Mara & Pearson, 1987): the facultative pond and maturation lagoons.

The anaerobic lagoon is not normally used as a stand-alone single-unit treatment system. In conventional stabilization pond systems, the anaerobic pond receives the sewage after a preliminary treatment step (Chapter 1) and removes part of the organic material by anaerobic digestion and by settling of particulate material. The treatment is complemented with facultative and maturation ponds, where additional organic material is removed together with pathogens (helminth eggs and thermotolerant bacteria). This configuration is widely applied in small and medium-sized communities, especially in developing countries. Ponds are usually constructed as shallow earth dams with reinforced material on the slopes to prevent erosion. Plastic lining or a layer of clay is placed on the bottom to prevent infiltration.

The mechanism for the removal of organic material in anaerobic ponds is the same as in the septic tank: the liquid flows through the pond and the influent settleable material accumulates at the bottom where the biodegradable fraction is digested by the mass of anaerobic sludge present there. Due to the lack of contact between sludge and organic influent material, the retention time to obtain a good removal efficiency is long (usually 2–5 days) and consequently the required area is considerable, so it is usually not covered. The biogas produced therefore escapes to the atmosphere and causes aesthetic and even public health problems owing to the bad – and at times horrible – odour that can be perceived at great distances from the lagoon. The odour problem is mainly caused by hydrogen sulphide (H_2S), a gas that forms due to anaerobic protein decomposition or by reduction of sulphate, which is present in the wastewater.

Another problem related to the release of biogas is the emission of methane into the atmosphere. It is well known that methane is a major contributor to global warming (greenhouse effect): on a molar basis, it has an effect 21 times stronger than carbon dioxide (Yvon-Durocher *et al.*, 2014). In a few cases the anaerobic lagoon is covered to retain and use the biogas that is generated. In these cases, the cost of covering the lagoon should be weighed against benefits in the form of biogas use, for example for electricity generation and eventual carbon credit payments. Because utilization of biogas from pond systems in practice is very rare, it is deduced that only in special cases are the benefits of covering ponds greater than the costs. On the other hand, it will certainly reduce odour and methane emissions to the environment.

The sewage retention time in anaerobic ponds is often higher than that in a primary treatment system such as the septic tank. Therefore, the removal efficiency of the organic material tends to be higher. Mara (1975) reports 50–70% BOD removal efficiency from sewage in anaerobic lagoons operating at hydraulic retention times of 1 to 5 days. The results of several researchers on the BOD removal efficiency in regions with a hot climate, all collected before 1995, are plotted in Figure 3.2 (on a log–log scale) and show that the efficiency clearly improves with increasing liquid retention time. From the experimental data set (all with temperatures >19°C), it was possible to derive the following empirical relation:

$$\eta = 1 - 2.4 \cdot R_h^{-0.50} \tag{3.2}$$

where

η = organic material removal efficiency (expressed in BOD)
R_h = hydraulic retention time (h)

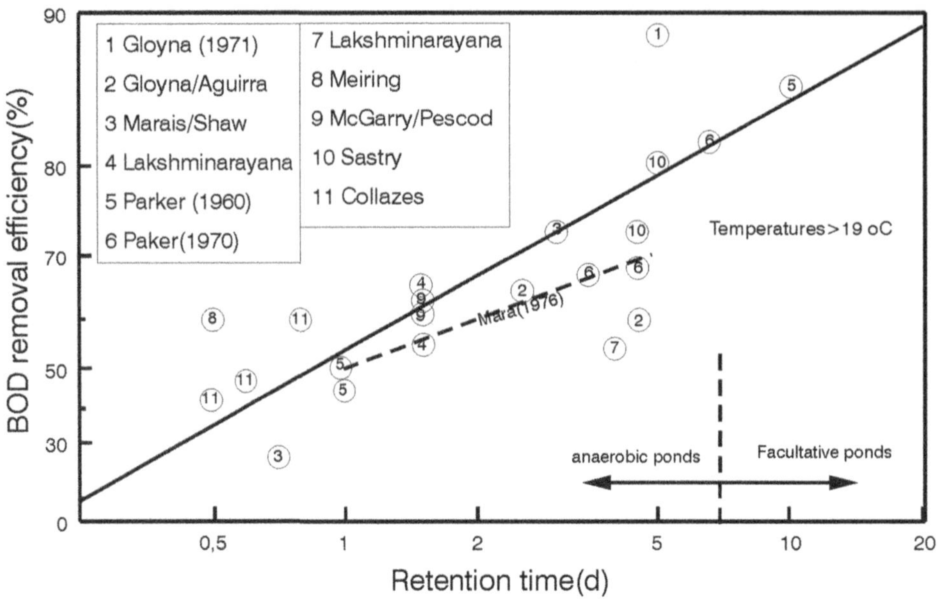

Figure 3.2 Experimental data of several authors for the BOD removal efficiency from municipal sewage in anaerobic ponds.

Efficient BOD removal (more than 80%) can only be achieved with a long hydraulic retention time of about 6 d. This is only possible if some of the non-settleable organic material present in the sewage is also removed, probably as a result of some degree of contact between this material and the sludge at the bottom of the pond. Mixing of the liquid phase (with a typical depth of 2–5 m) may occur due to the agitation introduced by rising bubbles of biogas, and by the effects of wind and sun (mechanical and thermal mixing). If the organic loading rate is low enough – below a loading rate of about 1000 kg $BOD \cdot ha^{-1} \cdot d^{-1}$ or 0.1 kg $BOD \cdot m^{-2} \cdot d^{-1}$ – a pond tends to be facultative: i.e. with an aerobic rather than an anaerobic top liquid layer. For a common pond depth of 2.5 m and for a typical sewage influent BOD (250 $mg \cdot L^{-1}$ or 0.25 $kg \cdot m^{-3}$), a load of 0.1 kg $BOD \cdot m^{-2} \cdot d^{-1}$ is achieved for a retention time of $0.25 \times 2.5/0.1 = 6.25$ d. Therefore, a retention time of less than about 6 d is required to ensure complete anaerobic conditions in the pond.

EXAMPLE 3.1

Use Eq. (3.2) to estimate the retention time and per capita area requirement of an anaerobic pond to obtain a BOD removal efficiency of 80%. Assume a per capita sewage production of 120 $L \cdot d^{-1}$ and a pond depth of 4 m.

Solution

For a BOD removal efficiency of 80%, a required (minimum) retention time of 144 h or 6 d is calculated from Eq. (3.2). Therefore, the required per capita volume for a contribution of 0.12 $m^3 \cdot d^{-1}$ is $6 \times 0.12 = 0.72$ m^3 and for a depth of 4 m the per capita area is thus equal to $0.72/4 = 0.18$ m^2. It will be shown (Chapter 9) that this is relatively small compared to the area of the stabilization pond system, but much larger than a UASB reactor (see Section 3.3.4). Again using Eq. (3.2), it can be estimated that when settled solids occupy half of the pond volume, efficiency drops to 71%.

The disadvantages of anaerobic ponds are so great that their application should be avoided. When it is desired to use ponds for sewage treatment, the alternative of using a high-rate anaerobic reactor as main treatment step for the removal of organic load (as described in Section 3.3) with a complementary polishing pond is much preferred. This alternative is presented in detail in Chapter 9. In this case the complementary polishing lagoon may be much smaller than in the conventional lagoon system. The advantages of anaerobic treatment are fully materialized, but there are none of the disadvantages of the anaerobic lagoon: (1) strong odour, (2) release of methane into the atmosphere, (3) no need to empty the pond for periodic sludge removal. In addition, owing to the smaller pond volume, (4) water loss is reduced and (5) increased salinity due to evaporation is less. A final important advantage is that the sewage network can be divided into two or more parts each with its treatment system, which can be close to urban regions, since there is no longer an odour problem. This can lead to an enormous construction cost reduction.

It is concluded that in general the application of conventional systems of stabilization ponds is not advisable: it is an obsolete system that has much more efficient modern alternatives, as will be seen in Section 3.3 and in Chapter 9.

3.3 MODERN ANAEROBIC TREATMENT SYSTEMS

The introduction of modern anaerobic treatment systems in the second half of last century was characterized by the development of new configurations with adequate contact between substrate and sludge and with mechanisms to retain the sludge in the reactor or to return it after phase separation of the effluent.

Initially these systems were primarily used for the treatment of concentrated industrial effluents, where the use of aerobic systems was not attractive due to the high energy consumption for aeration and large sludge production. Soon the applicability of anaerobic sewage treatment became evident, especially in regions with a warm climate. However, worldwide the fraction of anaerobic reactors used for sewage treatment is only a small part of the total installed base, the proportion used for industrial systems is several times higher. Also, the total sewage flow treated in anaerobic systems is much smaller than that in aerobic systems, with a few exceptions (mainly Brazil and India).

Despite the sludge retention mechanism, the storage capacity in the reactors is limited. Thus, if no action is taken, the reactor will inevitably become "full" of sludge in the sense that no sludge produced beyond this point can be accommodated. Thereafter the anaerobic reactor will expel sludge at the same rate it is being produced. The presence of sludge particles in the effluent will increase the concentration of suspended solids and organic material. Therefore, the sludge must be discharged continuously or periodically from the reactor, or a separation device (for example, a settler) should be introduced to separate discharged sludge particles in the effluent from the liquid phase.

If the discharged sludge is not removed, the quality of the effluent will deteriorate, which still may be acceptable if it is considered in the design phase that post-treatment units must have enough capacity to handle this additional load. As the produced anaerobic sludge is usually stable, there is no need for further digestion (e.g. anaerobic sludge digestion) and can be processed directly by sludge dewatering and drying units to remove the water.

3.3.1 The anaerobic contact process (ACP)

The first high-rate anaerobic treatment systems operated similarly to conventional activated sludge systems, but then without aeration (Figure 3.3), the so-called anaerobic contact process, having a biological reactor with mechanical mixing and a settler for separating sludge from the effluent (Schroepfer *et al.*, 1955). These systems met with only moderate success, not only because the local environmental conditions

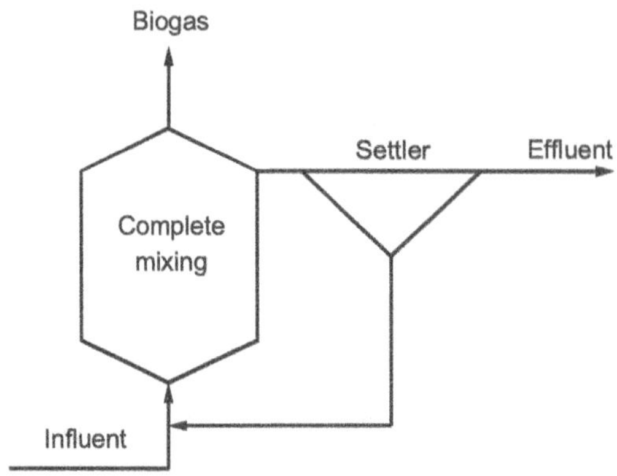

Figure 3.3 Schematic representation of the anaerobic contact process (ACP).

(Germany) were not adequate, but mainly because there was a difficulty in phase separation, due to the continued production of biogas in the settler, which tended to cause sludge flotation. Another problem was that even a relatively small sludge loss with the effluent could already lead to process instability due to the slow growth of anaerobic sludge and especially that of methanogenic archaea.

The use of the contact process for sewage treatment was never considered, possibly because at that time it was believed that anaerobic treatment was only feasible for very concentrated influents. When it became clear that treatment of less concentrated influents was also possible in anaerobic systems, new alternatives that were superior to the anaerobic contact process had already been developed.

3.3.2 The anaerobic filter (AF)

The second high-rate anaerobic treatment system was the anaerobic filter, developed after the pioneering work of Young and McCarty (1969). As indicated in Figure 3.4, in the anaerobic filter the biological sludge is attached to the surface of support material (the medium) which fills the biological reactor, while the influent flows upward through the voids of the medium. At the same time there is also sludge present in suspended form in the interstices between the filling material. In the anaerobic filter the thickness of the fixed film layer of sludge on the support material remains essentially constant, because on the one hand there is some

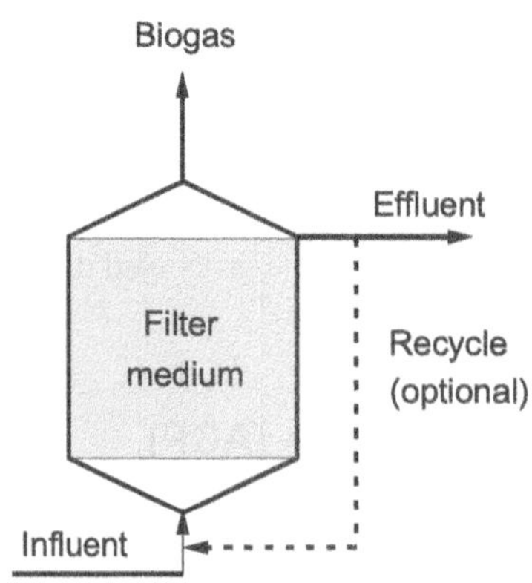

Figure 3.4 Schematic representation of an anaerobic filter (AF).

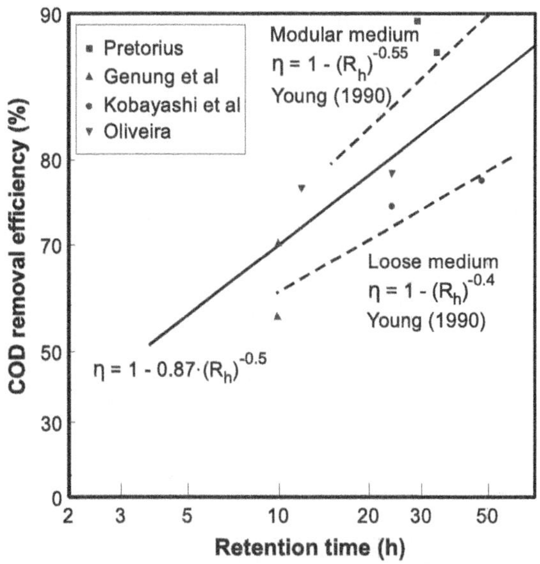

Figure 3.5 COD removal efficiency as a function of the retention time for sewage treatment in AF systems.

increase due to the growth of the biological sludge and adsorption of influent particles, and on the other hand there is a decrease due to erosion of the film by the bubbles of biogas and scour of the liquid phase.

Originally the medium of the anaerobic filter was typically composed of 5 cm diameter stones. Recent studies have shown the feasibility of using alternative materials such as bamboo rings, hollow bricks, or pieces of plastic conduits. They are relatively easily available on the market, lighter and with good specific surface area for bacterial film adhesion (Andrade Neto, 2004). Later it was shown that the use of plastic support materials such as Pall or Raschig rings with a high specific area led to a much higher organic material removal efficiency and applicable load.

On the other hand, the cost of the filling media increased, and eventually could even be higher than the cost of the reactor itself (Speece, 1983), so that it has found no application in the practice of municipal sewage treatment.

The anaerobic filter was the first system that – for mainly soluble wastewaters – showed the feasibility of applying organic loads of up to 10–15 kg $COD \cdot m^{-3} \cdot d^{-1}$, much larger than those of high-rate activated sludge systems which were in the range of 3–4 kg $COD \cdot m^{-3} \cdot d^{-1}$. This greatly increased the applicability of anaerobic digestion for wastewater treatment. However, when the anaerobic filter was applied to treat influents with a significant fraction of the organic material in particulate form, it was shown that in this case clogging of the medium was a recurring problem that was difficult to solve. This problem reduced the applicability of the anaerobic filter for non-soluble influents and today there are few large anaerobic filters in operation or being constructed.

In Brazil, anaerobic filters have often been used in small-scale treatment systems as a unit placed after a septic tank for the treatment of the soluble fraction of domestic sewage. The success of the combination of septic tank and anaerobic filter (ST–AF) is partly due to the fact that the two together form an efficient treatment system which compensates for the weakness that each component has individually: septic tanks retain coarse solids but have low efficiency of soluble organic material removal, while anaerobic filters are efficient but clog if there are many suspended solids present. One disadvantage of the combined ST–AF system is that it is quite large and occupies a large area in residential areas, where building lots are small.

In the present work a single-family UASB reactor will be described that can replace the ST–AF combination with many benefits (Chapter 4). Although ST–AF systems are used predominantly for individual households, they are also applied in small urban or rural communities (200–5000 inhabitants), where the service of a central sewer network and treatment plant is lacking.

To evaluate the quality of the anaerobic filter as a sewage treatment unit, performance data collected by several authors in the period up to 1995 are presented in Figure 3.5. The COD removal efficiency was plotted as a function of the retention time, using a logarithmic scale for both coordinates. There is a clear influence of the retention time on the treatment efficiency. Based on the references in Figure 3.5 the following empirical expression was derived:

$$\eta = 1 - 0.87 \cdot R_h^{-0.50} \tag{3.3}$$

where

R_h = retention time in the AF, *excluding* the volume occupied by the medium
η = COD removal efficiency

3.3.3 Fluidized and expanded bed (FB, EB)

In the fluidized-bed treatment system, introduced by Jeris (1983), the biological sludge is fixed on a granular medium, the media granules being fluidized in an upward flow of the liquid phase. This upward liquid flow is maintained by introducing the influent and by recirculating the effluent by means of a recirculation pump. The biological sludge film develops on the surface of the granules and as these are in constant movement, contact with the substrate is guaranteed. The expanded bed differs from the fluidized bed only by the application of a lower upward velocity, so that the bed only expands without fluidizing all granules in suspension. The bed is expanded by 10–20% (Jewell, 1982). Sand was initially used as the granular medium, but later it was replaced by materials with a lower density such as anthracite and plastic materials, to reduce pumping costs to keep the granules in suspension. Materials with higher roughness were adopted to improve the attachment of the sludge films to the granules and improve the specific surface area available for colonization. In practical applications great difficulties were encountered in controlling the size and density of the particles and thus guaranteeing a stable performance. As biofilms grows on the granule, the density is reduced resulting in increased biomass losses with colonized carriers. A further recurring problem in the case of sewage treatment is the formation of mud balls like those often encountered in rapid sand filters for water treatment. Figure 3.6 shows experimental treatment results, collected up to 1995, of fluidized and expanded beds treating sewage. The data suggest the following empirical equation:

$$\eta = 1 - 0.56 \cdot R_h^{-0.60} \tag{3.4}$$

3.3.4 The upflow anaerobic sludge blanket reactor (UASB)

Another modern system of anaerobic treatment was developed by Prof. Gatze Lettinga and his group of researchers in the 1970s at the University of Wageningen in the Netherlands (Lettinga *et al.*, 1980). It has become the most applied system in practice: today more than 90% of large anaerobic wastewater treatment reactors are of the UASB type or one of its variants.

Figure 3.7 shows a diagram of the UASB reactor. The most characteristic device of the UASB reactor is the phase separator, which is placed in the top part of the reactor. The separator divides the reactor between a lower anaerobic digestion zone filled with biomass and an upper settling zone, having an

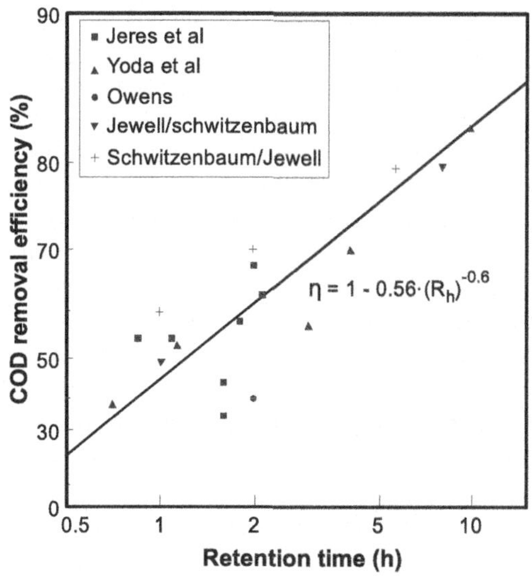

Figure 3.6 COD removal efficiency as function of the retention time for sewage treatment in FB systems.

intermediate transition zone. The influent is distributed evenly over the base of the reactor and flows upstream sequentially passing through the digestion, transition and settling zones to the top of the reactor where the effluent is collected and discharged.

The digestion zone contains the biological sludge mass that transforms organic material present in the influent into biogas. The biogas desorbs from the liquid phase and is captured by the phase separator elements placed above the digestion zone. The liquid phase passes to the settling zone through the openings of the

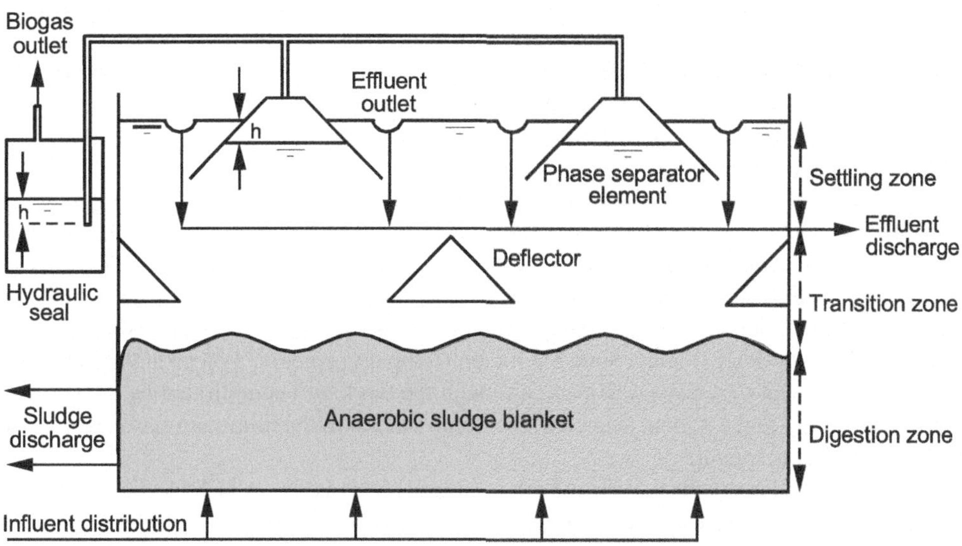

Figure 3.7 Schematic representation of the physical layout of a conventional UASB reactor.

separator. The area of this zone becomes larger when the liquid continues its upward trajectory to the top of the reactor, so that the upward liquid velocity decreases as it approaches the effluent collection system. Consequently, sludge particles carried over from the digestion zone may still settle and thus be retained.

To avoid biogas bubbles hindering the settling process, a deflector is placed under the openings of the phase separator. Thus, a tranquil zone is created where the flocs with sufficient settling velocity can settle on the deflector and separator surfaces, if their settling velocity exceeds the upward velocity of the liquid phase and they intercept with the plate surface. These surfaces are inclined and with time the accumulated solids will tend to slide back into the digestion zone. In this way an internal settler is created for retention and recirculation of sludge, while an effluent essentially free of suspended solids is discharged.

In practice, a biogas bubble may not always desorb that easily from the sludge floc in which it was generated. Instead the bubble carries the floc with it as it rises to the phase separator. Thus, a scum layer – i.e. sludge mixed with biogas bubbles – is created, floating on the surface of the liquid inside the separator elements. In order to avoid blocking of the biogas outlet, a gas chamber is created under the phase separator by the operation of an external hydraulic seal, as indicated in Figure 3.7. In this layer solids as well as oils and greases (lighter than water) that escape the digestion zone without being properly treated can accumulate. This is more likely to happen at lower temperatures, when grease removal is more problematic.

Because of anaerobic digestion there will be growth of sludge in the digestion zone and eventually there will be a need to discharge the excess sludge produced. This discharge can be continuous or periodic. Continuous discharge is typically applied when there are mechanical processes installed for the dewatering of the produced sludge (belt filter or centrifuge), while periodic discharge is used when drying beds are applied.

The produced biogas may vary in composition according to the concentration and composition of the digested organic material and the pH that is maintained in the digester. Invariably the main component will be methane, but there are always other mixed gases present: carbon dioxide, nitrogen, water vapour and sometimes hydrogen sulphide. The latter component is undesirable but cannot be avoided when the wastewater to be treated contains sulphate or organic sulphur. In Chapter 6 the production, composition and treatment of biogas is discussed in more detail.

The potential of the UASB reactor for sewage treatment was soon recognized by Lettinga and his group of researchers, but the conditions under which they developed the first experimental reactors in Holland (low temperature, dilute sewage) led to poor results (De Man, 1990; Van der Last, 1991). Only when they applied the concept of the UASB reactor in tropical regions, the performance of the UASB reactor was improved: in Cali-Colombia, the first UASB reactor was successfully applied on a demonstration scale (Schellinkhout *et al.*, 1985). Soon thereafter, the first full-scale system (3 reactors of 3.300 m^3 each) was built in Bucaramanga in Colombia.

Today, thousands of UASB reactors for sewage treatment are found in hot climate regions, with the highest number of applications in Brazil and India. In Brazil the potential of anaerobic digestion for sewage treatment was recognized early (Gomes, 1985).

Soon after the success of the first demonstration scale system in Cali, UASB reactors were built by CETESB in São Paulo (Vieira & Souza, 1986) and by the Municipality of Campina Grande (Van Haandel & Lettinga, 1994). SANEPAR (Paraná State Sanitation Company) embarked on an ambitious program to apply anaerobic treatment to sewage treatment, although the work was complicated by the relatively low temperatures prevailing in the region, located in the South of Brazil: the minimum sewage temperature is only 16°C in Curitiba, its capital.

Since then numerous large units have been built in several states as shown in Table 3.1. In Brazil, UASB reactors are operated in every state; the largest number of anaerobic systems is in the State of Paraná, but other states are following and Belo Horizonte, São Paulo and the Federal District already have larger systems installed than those in Paraná.

Table 3.1 Several recent full-scale UASB systems in Brazil. The installed post-treatment technology is also indicated (AS = activated sludge; TF = trickling filter; AS = activated sludge system; DAF = dissolved air flotation).

Location	Company	Year	Volume (m^3)	Flow Rate ($m^3 \cdot s^{-1}$)	Post-treatment
Brasília	CAESB	2005	55.000	1.5	AS
Belo Horizonte	COMPASA	2007	48.000		TF
S. José do Rio Preto	CABESP			1.34	AS
Campinas Anhumas	SANASA	2007		1.20	AS
Curitiba Atuba Sul	SANEPAR	2002	32.000	1.12	DAF
Londrina (Sul/Norte)	SANEPAR			1.07/0.85	TF/TF
Campinas-Piçarrão	SANASA	2006		0.56	AS/DAF
Ipatinga MG	COMPASA	2000	12.000		TF

Figure 3.8 shows experimental results of full-scale UASB systems treating sewage, with the efficiency of COD removal plotted as a function of the retention time with logarithmic scale on the two coordinates. Note that this is data collected up to 1995, and that most of the initial UASB were operated by research teams. In part, the large dispersion of the experimental data can be explained because it is not always clear what type of effluent is considered: sometimes the raw effluent is reported, sometimes the settled or filtered effluent. Other differentiating aspects are that the temperature is an important factor in the removal efficiency and this variable depends on the region where the systems are installed. A third factor of importance is the quality of the phase separator which also has a large influence on the performance. However, considering all data presented the relationship can be described as:

$$\eta = 1 - 0.68 \cdot R_h^{-0.68} \tag{3.5}$$

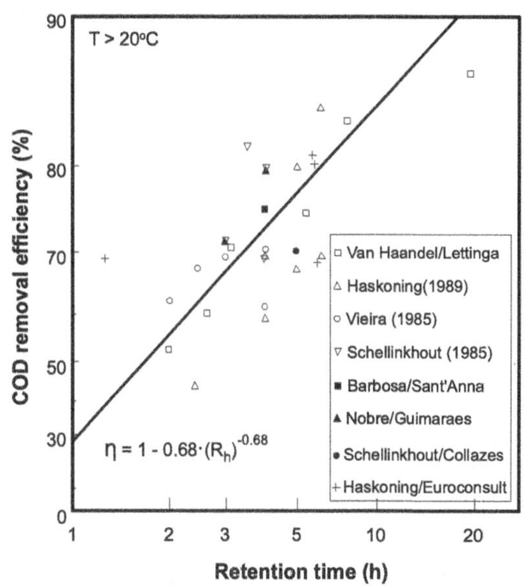

Figure 3.8 COD removal efficiency as a function of the retention time in UASB reactors.

3.3.5 The anaerobic reactor with a fluidized sludge bed: (RALF)

In the state of Paraná, Brazil, dozens of small-scale systems were designed and built by SANEPAR Sanitation Company, which use anaerobic digestion in upflow reactors, but their design is different from the conventional UASB reactor in one important aspect: they are equipped with a different device for sludge retention in the system. Instead of the three-phase separator, there is a lateral settler. Figure 3.10 schematically shows the reactor developed by SANEPAR and which was called an Anaerobic Fluidized Bed Reactor, RALF, although the name is not very applicable as strictly speaking there is no sludge bed (i.e. granules) but rather a sludge blanket (flocculent biomass) and owing to the low upflow velocities that are applied it is certainly not fluidized. It is one of three units built in Londrina and operated since 1997. The phase separator in the form of a settler was adopted in an attempt to simplify the construction of the units and reduce costs. Generally, the sewage arrives by gravity or is pumped to a central well above the treatment unit, from where it overflows levelled triangular weirs to a large number of boxes. In the boxes there are pipes through which the sewage flows to different points at the bottom of the reactor and thus the influent is distributed automatically on the lower surface of the unit.

Figure 3.9 shows some results of RALF reactors treating sewage in the state of Paraná, showing the COD removal efficiency as a function of the retention time of the liquid. In interpreting the data, it is important to note two aspects that tend to impair the reported efficiency of RALF reactors: (1) the COD of the raw effluent is generally used to calculate the removal efficiency, but since sludge discharges are not frequent in most stations, there will be sludge in the raw effluent; and (2) the sewage temperature is low, usually in the range of 17–21°C. Taking into account all systems reported, the following empirical expression is derived:

$$\eta = 1 - 1.53 \cdot R_h^{-0.64} \tag{3.6}$$

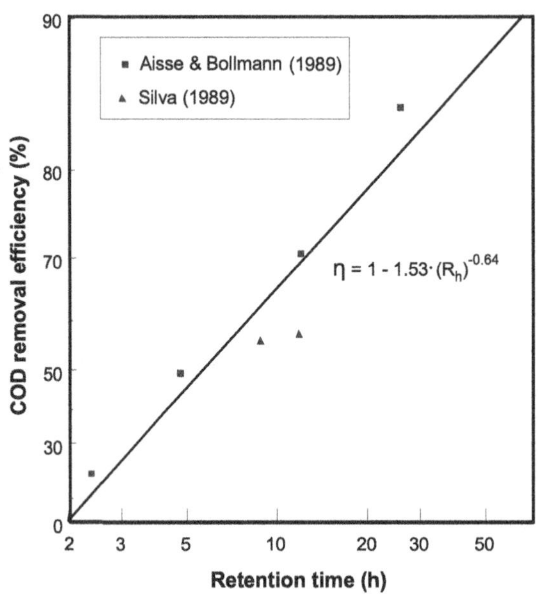

Figure 3.9 COD removal efficiency as function of the retention time in RALF reactors.

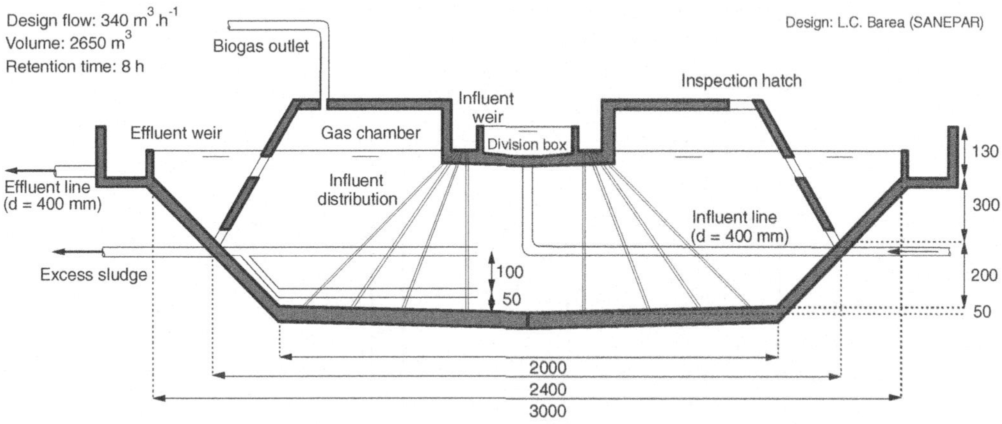

Figure 3.10 RALF – anaerobic upflow reactor with a simplified structure of the phase separator.

3.3.6 Comparison of treatment systems

The most important parameters for comparing the performance of anaerobic treatment systems are the removal efficiency of organic material, and construction and operating costs. In practice, the concentration of the organic material is usually expressed in terms of COD. Since anaerobic treatment systems are simple, with few or no mechanized parts, the investment costs are largely determined by the volume, which in turn is determined by the flow rate to be treated and the retention time. Thus, in order to compare the performance of different anaerobic systems, it is important to evaluate COD removal efficiency as a function of the retention time in the system.

In the previous Sections (3.2.2–3.3.5) experimental full-scale data was presented of the COD removal efficiency as function of the retention time (log–log scale) for different types of anaerobic systems. As demonstrated, the COD removal efficiency could be approximated with empirical expressions of the type $\eta = 1 - C_1 \cdot R_h^{C2}$. Table 3.2 summarizes the empirical expressions that were developed for the different

Table 3.2 Empirical expressions of the COD removal efficiency as a function of the hydraulic retention time in anaerobic sewage treatment systems[(2)].

Anaerobic System	Empirical Relationship	Eq. No.
Anaerobic pond[(1)]	$\eta = 1 - 2.40 \cdot R_h^{-0.50}$	Eq. (3.2)
Anaerobic filter	$\eta = 1 - 0.87 \cdot R_h^{-0.50}$	Eq. (3.3)
Fluidized/expanded bed	$\eta = 1 - 0.56 \cdot R_h^{-0.60}$	Eq. (3.4)
UASB	$\eta = 1 - 0.68 \cdot R_h^{-0.68}$	Eq. (3.5)
RALF	$\eta = 1 - 1.53 \cdot R_h^{-0.64}$	Eq. (3.6)

Note: (1) The organic removal efficiency of the anaerobic lagoon is reported as BOD removal efficiency, which is typically at least 10–15% higher than the COD removal efficiency; (2) It is always difficult to compare data published by different authors, using different wastewaters in different designs and under varying conditions.
In addition, it is not always clear whether raw or settled effluent is used for the calculations. Regardless, it does give a good indication about the performance that is to be expected and the relative efficiency of the different systems.

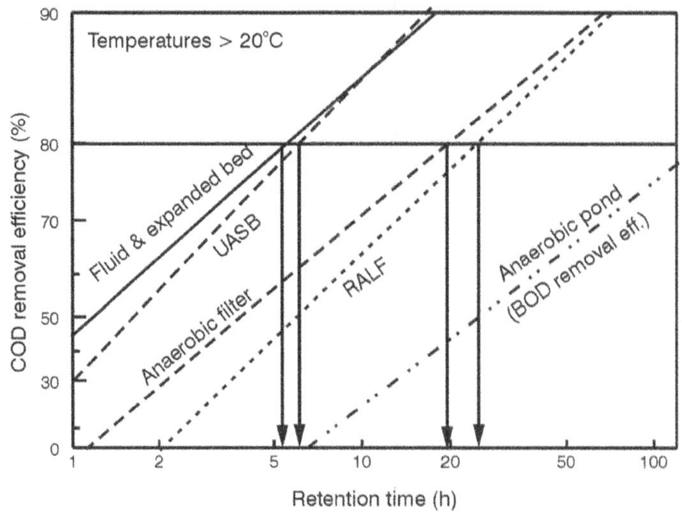

Figure 3.11 Graphical representation of the COD removal efficiency in different anaerobic treatment systems as a function of the hydraulic retention time in accordance with the expressions listed in Table 3.2.

systems, while in Figure 3.11 the expressions of Table 3.2 are represented graphically. The log–log scale was used because it results in straight lines for the different equations of Table 3.2. The empirical expressions indicate that in general terms there is a tendency to have increased efficiency with increased retention time.

From the large dispersion of the experimental data, it is clear that the data obtained in any specific anaerobic treatment unit may deviate significantly from the average value predicted by the empirical equations. However, the analysis of the results in Table 3.2 obtained with different anaerobic sewage treatment systems leads to some important general conclusions, as shown by an analysis of the equations in Table 3.2 and the graphs in Figure 3.11:

- For temperatures above 20°C, COD removal efficiencies above 80% are feasible for all systems considered in Table 3.2, but the required retention times are very dissimilar for different systems: whereas the UASB and the FB systems require a retention time of about 6 h to obtain 80% COD removal efficiency, in a RALF and an AF this increases to about 1 d and in the AP to almost a week.
- In the range of practical interest, the performance of the UASB reactor and the FB reactor tend to be similar and much better than that of AF units and RALF reactors with the same retention time. The performance of the anaerobic lagoon is much lower than those of other systems, even when the BOD instead of COD is considered (the efficiency in terms of BOD is always higher than COD, because COD also considers non-biodegradable organic material).
- Comparing the two best-performing systems (UASB and FB), the municipal UASB stands out as a simpler system with no mechanization or moving parts. For these reasons it is much preferred over other anaerobic systems. In Chapter 4 the construction elements of the UASB reactor are described in more detail.
- When the empirical equations of Table 3.2 and its graphical representation in Figure 3.11 are analysed, we arrive at the inescapable conclusion that it is no coincidence that the UASB reactor has such wide acceptance in the practice of anaerobic treatment of sewage in hot climate regions. The UASB reactor combines the advantage of a lower investment because of the relatively short retention time, with an operating cost that is also low because there are no moving parts that need

maintenance. The activities of the operators are basically limited to the removal of coarse solids and sand from the influent to prevent clogging, and excess sludge discharge and scum removal if it forms on the surface or under the separating elements.

It can be concluded from this section that the UASB reactor is with good reason the preferred system for sewage treatment in hot climates: more than 90% of large systems installed in the world are of this type. Among the reasons for this preference are not only high organic material removal efficiency, but also the simplicity of the system and its good operational stability. In this chapter some factors influencing the performance of UASB reactors treating municipal sewage are discussed in more detail. These factors include, for example, the properties of the UASB sludge, the recommended start-up procedure, the steady-state reactor performance and the effect of operational measures. Furthermore the influence of sulphate on process performance will be discussed.

Having identified the influence of these factors, it is attempted to formulate general rules for the performance that can be expected from UASB reactors treating municipal sewage. In Section 3.5 this will be evaluated as function of the hydraulic retention time, as this is an easy parameter to determine. Later, in Section 3.9 it will be demonstrated that in fact the sludge age is a much better parameter to predict treatment performance, effluent quality and the expected biomass composition.

Consequently a process model for the UASB based on sludge age is developed which will be used to develop rational design criteria for UASB reactors. In Chapter 10 this process model, together with the reactor design guidelines presented in Chapter 4, will be integrated with the most important post-treatment configurations, i.e. the activated sludge system (Chapter 8) and polishing ponds (Chapter 9), and a comprehensive design spreadsheet is presented.

3.4 PROPERTIES OF UASB SLUDGE

Due to the removal of organic material a bacterial mass is produced, which tends to attach to solids already present in the anaerobic reactor or forms macroscopic flocs. These flocs are a mixture of organic and inorganic material. The inorganic part is composed of solids present in the influent like grit and clay as well as grains of sand, although some mineral precipitation may take place as well, for example from struvite, calcium carbonate or calcium phosphate. The volatile solids are composed of three components: (1) the mass of living micro-organisms, (2) the mass of particulate non-biodegradable organic material, and (3) the mass of (mostly particulate) biodegradable organic material that has not yet been hydrolysed. The ratio of the three components depends on the characteristics of the influent and the operating conditions, notably the temperature and the applied sludge age. There is constant erosion of the sludge flocs by shear stress introduced in the digestion zone by liquid and gas flows. The floc debris that is released tends to be discharged along with the effluent, unless there is new flocculation in the settling zone, causing smaller particles to conglomerate into larger ones with higher settling velocity, so that they return to the digestion zone.

3.4.1 Sludge composition

The requirements for growth of the micro-organisms that develop in an anaerobic system are twofold: (1) organic material, from which the cells are synthesized in the anabolic process and (2) nutrients that micro-organisms incorporate into their cell structure. The main nutrients (the so-called macro-nutrients) are nitrogen and phosphorus, which contribute approximately 10% and 2.5% of the volatile sludge mass. Table 3.3, which is valid for both aerobic and anaerobic sludge, summarizes the typical fractions of nutrients present in cell mass.

In addition to the macro-nutrients there is a long series of micro-nutrients whose presence is vital for the growth of sludge and therefore for the performance of the anaerobic treatment. Among these potassium,

Table 3.3 Average composition of biological sludge.

Parameter	Formula	Anaerobic Biomass	Aerobic Biomass	Unit
Volatile fraction	$f_v = X_v/X_t$	0.55–0.70	0.75–0.80	$g\,VSS \cdot g^{-1}\,TSS$
Nitrogen fraction	$f_n = TKN/X_v$	0.08–0.10	0.08–0.10	$g\,N \cdot g^{-1} VSS$
Phosphorus fraction	$f_p = P/X_v$	0.02–0.025	0.02–0.025	$g\,P \cdot g^{-1}\,VSS$

Table 3.4 Specific anaerobic sludge production and composition in terms of sludge fractions, for both high- and low-loaded anaerobic systems at 25°C.

Sludge Fraction	High Load ($R_s \approx$ 30 d)	Low Load ($R_s \approx$ 100 d)	Unit
Total sludge	0.20	0.15	$g\,TSS \cdot g^{-1}\,COD$
Volatile sludge	0.12	0.10	$g\,VSS \cdot g^{-1}\,COD$
Biological sludge	0.02	0.03	$g\,VSS \cdot g^{-1}\,COD$
Inert sludge	0.04	0.05	$g\,VSS \cdot g^{-1}\,COD$
Flocculated biodegradable organic material	0.06	0.02	$g\,VSS \cdot g^{-1}\,COD$

sulphur and iron are the most important elements, but there are many others that must be present, even if the demand is small. Usually in sewage there is abundant presence of all macro- and micro-nutrients, but in the case of many industrial wastewaters the anaerobic treatment only proves feasible when nutrients are added.

Many metal micro-nutrients will form precipitates with hydroxides and especially sulphides, so that the availability of the free ion, the form which is available to the micro-organisms, is reduced even though the concentration of the micro-nutrient as a compound may be high.

The major source of uncertainty in the prediction of the rate of excess sludge production is that not all biodegradable particulate material is in fact hydrolysed in the digester. A part is adsorbed onto the sludge and is discharged as excess sludge without being metabolized. The fraction of biodegradable organic material remaining in the excess sludge depends basically on two factors: (1) the sludge age – the average period the sludge is retained in the digester, and (2) the temperature, which has a strong influence on the hydrolysis rate. When both the sludge age and the temperature are low, the fraction of non-hydrolysed material in the sludge may constitute a large part of the excess sludge. Table 3.4

EXAMPLE 3.2

Estimate the minimum COD/TKN/P proportions in a wastewater to ensure that the concentration of macro-nutrients is sufficient for sludge production in a low-loaded UASB.

Solution

From Table 3.4 it can be seen that volatile sludge production is probably not more than 0.10 mg $VSS \cdot mg^{-1}$ COD, while Table 3.1 shows that the mass fractions of nitrogen and phosphorus in volatile sludge are 0.10 mg $N \cdot mg^{-1}$ VSS and 0.025 mg $P \cdot mg^{-1}$ VSS respectively, so that $0.10 \times 0.10 = 0.01$ mg $N \cdot mg^{-1}$ COD and $0.10 \times 0.025 = 0.0025$ mg $P \cdot mg^{-1}$ COD are required. Hence a minimum ratio of 1:0.01:0.0025 or 400:4:1 mg COD/mg N/mg P is required.

indicates typical specific excess sludge production rate of the various sludge fractions that make up the anaerobic sludge in sewage treatment.

3.4.2 Biological sludge properties

Among the most important biological sludge properties are the specific methanogenic activity (SMA) and the sludge stability. SMA indicates the maximum rate of methane production per mass of anaerobic sludge (in g CH_4-COD $\cdot g^{-1}$ VSS $\cdot d^{-1}$) under optimized environmental conditions. The SMA of a biological sludge is of fundamental importance in evaluating if the organic load applied in the treatment system is compatible with the capacity of the sludge mass to transform this load into biogas.

The specific methanogenic activity of the sludge generated from anaerobic sewage digestion at 35°C is in the range 0.1–0.2 g CH_4-COD $\cdot g^{-1}$ VSS $\cdot d^{-1}$, which is much lower than the 0.4–1.2 g CH_4-COD $\cdot g^{-1}$ VSS $\cdot d^{-1}$ typically found for granular biomass from industrial treatment systems. Therefore only a small fraction of the sludge is composed of methanogenic biomass.

The second important parameter, sludge stability, reflects the fraction of biodegradable organic material remaining in the biological sludge that can still be transformed into methane. The larger this fraction is, the less stable is the sludge, and therefore this parameter is particularly important when evaluating the composition of excess sludge in terms of its suitability for disposal. A poorly stabilized sludge will continue digestion after removal from the reactor and thus generate both biogas and odours (H_2S).

To determine whether the anaerobic digestion of sewage was efficient, according to the US EPA (2003) criterion, the methane yield from a sample of sludge taken from the UASB reactor should be less than 100 ml $CH_4 \cdot g^{-1}$ VSS, which indicates a stable sludge. At the typical COD/VSS ratio of 1.48 for biomass, 100 ml of methane production is equivalent to 100/350/1.5 = 0.193 g VSS $\cdot g^{-1}$ VSS. In other words, when no more than 20% of the solids present in a sludge sample can be digested, the sludge is deemed stabilized. Sludge from overloaded reactors or from reactors operated at short sludge age may be unstable. In Appendix A1 the tests for SMA and sludge stability are described, while in Section 3.9.2 a theoretical approach will be presented to estimate the value of this fraction as function of sludge age, temperature and sewage characteristics.

3.4.3 Sludge settleability

In systems that contain suspended sludge and are fed in upflow mode, such as UASB reactors, settleability is a very important sludge characteristic. If the sludge has a concentration of more than approximately 1 g $\cdot L^{-1}$, the sludge particles are all interconnected and do not settle individually. Instead, when the sample is left undisturbed, a clear interface forms below which all particles tend to settle at the same rate, so that an essentially solid-free supernatant separates from a phase with the settling suspension. The velocity of the interface displacement is called the zone settling velocity (ZSV). In the UASB reactor a balance is established between the settling velocity of the solids and the upward velocity of the liquid phase, resulting in a stationary layer of sludge, where the concentration depends on the sludge settleability and the upward liquid velocity. The sludge flocs tend to settle more slowly at higher concentration. The zone settling velocity can be described by an empirical equation, proposed by Vesilind (1968):

$$v = v_0 \cdot \exp(-k \cdot X_t) \tag{3.40}$$

where

v = zone settling velocity
X_t = sludge concentration
v_0, k = Vesilind constants.

Table 3.5 Typical values of Vesilind constants for different sludges including aerobic systems (activated sludge, AS) in both continuous-flow and sequencing-batch mode) and anaerobic systems with flocculent and granular sludge generated in UASB reactors.

Type of Sludge	Value of k ($l \cdot g^{-1}$ TSS)			Value of v_0 ($m \cdot h^{-1}$)			X_t ($g \cdot L^{-1}$) for $v_u = 1$ $m \cdot h^{-1}$		
	min.	avg.	max.	min.	avg.	max.	min.	avg.	max.
AS – continuous	0.3	0.4	0.5	6	9	12	4	5	8
AS – SBR	0.1	0.2	0.3	10	15	20	8	14	30
UASB – flocculent	0.08	0.12	0.15	10	15	20	15	23	37
UASB – granular	0.02	0.035	0.05	20	15	40	46	77	150

The constant v_0 represents the extrapolated value of the ZSV when the sludge concentration tends to zero, the so-called undisturbed sludge settling velocity. The constant k denotes the compressibility, that is, how easily the sludge can increase its concentration by settling.

The Vesilind equation shows that the ZSV increases at higher values of v_0 and lower values of k. For this reason it is of great importance that an experimental method is available to determine the value of these experimental constants. Appendix A2 shows how the ZSV can be determined experimentally, using two different methods: the static test and the dynamic test.

The Vesilind equation was originally presented as an empirical description of settling in activated sludge systems. Several observers from UFCG (Consentino *et al.*, 2002; Leitão, 2004; Sales, 2013) showed that the validity of the Vesilind equation applies beyond aerobic sludge and is also valid for anaerobic sludges of flocculent nature, such as in UASB reactors treating sewage. Table 3.5 shows typical values of settling constants obtained from different sludges. It also shows the calculated concentration that can be retained in the reactor at an upflow velocity of $1.0\ m \cdot h^{-1}$, which is the maximum value that can applied in UASB reactors treating sewage, although typically the selected average reactor upflow velocity is lower at $0.5–0.7\ m \cdot h^{-1}$. The results indicate that the sludge settleability has a very large influence on the sludge concentration that can be maintained in a specific type of treatment system. As the volume of the treatment system is inversely proportional to the sludge concentration, the value of the Vesilind constants have a great influence on the minimum volume and therefore on the cost of the treatment system. It can be concluded that the Vesilind equation adequately describes the settling velocity of anaerobic sludge generated in anaerobic sewage treatment. The settleability constants are in the range of $0.08–0.15\ L \cdot g^{-1}$ TSS for the compressibility constant (k) and $10–20\ m \cdot h^{-1}$ for the unhindered settling constant (v_0).

Provided that the design recommendations with respect to the applicable upflow velocities are adhered to, as described in Section 4.3.1.2, the settling characteristics of anaerobic sludge are in general so good that it should be possible to retain all the sludge produced in the UASB reactor, as long as the maximum sludge hold-up capacity is not exceeded. Even so, part of the sludge production is found in the effluent, which can be attributed to floc erosion.

Rarely, the sludge layer in UASB reactors can extend over the whole digestion zone to the bottom of the deflector, from which it is concluded that settling is not the only process that determines the mass of sludge: erosion of flocs of sludge with detachment of small particles also plays an important role.

3.5 START-UP OF UASB REACTORS TREATING SEWAGE

For most wastewaters, the start-up of a UASB reactor without the use of seed biomass is a long and difficult process, because a large but slow-growing bacterial mass needs to be developed, adapted to the characteristics of the wastewater to be treated. During start-up there is always a risk of organic overload and, if this occurs, acid fermentation may become predominant over methanogenic fermentation, possibly resulting in souring of the reactor contents. However, the characteristics of sewage are such that the start-up of a UASB treating domestic sewage is less problematic.

Sewage differs from most industrial wastewaters in that it already contains the bacterial populations necessary for anaerobic digestion. For this reason the UASB can – if needed – be started without inoculation: feeding the empty reactor with sewage, the populations for acidogenic and methanogenic fermentations will develop spontaneously. Furthermore, the buffering capacity of the sewage is generally enough to avoid souring, even during the start-up period, when the methanogenic activity is still too low relative to that of acid fermentation. Effectively, the start-up of full-scale reactors has been accomplished without great difficulties in many new plants without inoculum. The start-up period in these cases was typically between 12 and 20 weeks.

3.5.1 Start-up without seed sludge

Figure 3.12 shows typical trends of several main variables during primary start-up of a UASB reactor treating sewage, when no seed sludge is used (Bezerra, 1997; Oliveira, 1990), namely:

(a) development of the sludge mass in the reactor
(b) removal efficiency of the organic material

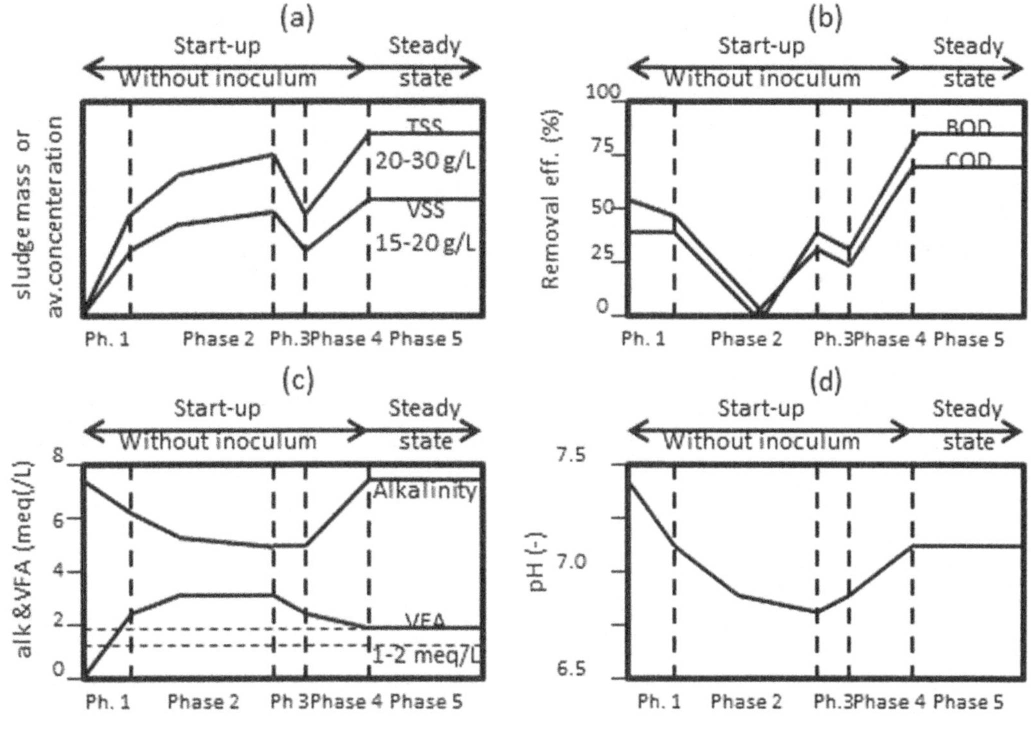

Figure 3.12 Variation of different variables during the start-up period of an municipal sewage UASB reactor without inoculum (seed sludge): (a) sludge mass or average sludge concentration; (b) COD and BOD removal efficiencies; (c) alkalinity and VFA concentration and (d) pH. The different phases in the start-up process are indicated as well.

(c) alkalinity and the volatile fatty acids concentration
(d) pH

If there is no seed sludge added at the start of the operation, it will develop during the start-up period, owing to the accumulation of settled solids and growth of the micro-organisms responsible for converting organic material into methane. As indicated in Figure 3.12 the starting period is typically divided into five phases, characterized as follows.

3.5.1.1 Phase 1: Solids settling (duration: 1–3 weeks)

In phase 1 the UASB reactor basically functions as a primary settler and retains settleable solids. At this stage the removal of settleable solids is almost complete, but that of the suspended solids is only partial, since colloids are not yet efficiently retained. Consequently, the sludge mass present in the reactor increases quickly, but is almost completely composed of settleable solids from the influent and not yet of anaerobic biomass. The organic material removal efficiency is low, comparable with that of primary treatment systems such as the septic tank or Imhoff tank: around 40% of COD and 50% of BOD is removed.

The sludge blanket grows rapidly, but there is little biogas production. The pH and alkalinity tend to remain stable or decrease slightly, because some acid fermentation already starts to occur from the first day of operation onwards.

3.5.1.2 Phase 2: Solids hydrolysis and acidification (duration: 6–8 weeks)

In this phase bacterial mass starts to develop and acid fermentation will take place. At this stage the removal efficiencies of settleable and suspended solids remain essentially constant, but the overall removal efficiency of organic material tends to decrease, owing to the solubilization of hydrolysed solids. The growth rate of the sludge blanket decreases because of the accelerated solubilization of settled material. If the rate of solids hydrolysis in the sludge blanket is higher than the rate of influent solids removal by settling, the concentration of COD and BOD in the effluent will be temporarily higher than that of the influent, i.e. in terms of organic material removal the UASB reactor may actually have a negative efficiency during this period.

At this stage a considerable concentration of volatile fatty acids (VFA) is generated, which tends to reduce alkalinity and consequently lower the pH value in the reactor. As shown in Chapter 5, it is theoretically not impossible for the pH to acquire a value below the optimal range for methanogenesis, but in practice no souring of the contents of the UASB reactor during start-up has been reported.

In early publications on the anaerobic treatment of sewage in UASB reactors (Haskoning, 1989; Schellinkhout & Collazos, 1992), it was recommended that the reactor feed should be reduced or interrupted whenever the UASB reactor showed reduced COD removal efficiency, VFA increase or pH decrease. However, it is our experience (Bezerra, 1997; Oliveira, 1990) that this is not necessary and in fact counterproductive. The UASB reactor will not sour and reduction of the feed only tends to prolong the duration of the start-up.

3.5.1.3 Phase 3: Development of methanogens (duration: 2–4 weeks)

In this phase methanogenic fermentation starts to develop. The increase in VFA concentration in the previous phase creates conditions for accelerated growth of methanogenic bacteria, which in turn results in a higher biogas production rate. The turbulence introduced by the biogas will cause the resuspension of some of the material present in the layer of settled solids, especially the lighter material. For this reason, there will be a significant (albeit temporary) increase in the concentration of settleable solids in the effluent. In parallel, the height of the sludge layer decreases due to the loss of sludge from the UASB reactor. At this stage the removal efficiency of the organic material again may be negative due to the abundance of solids in the effluent.

3.5.1.4 Phase 4: Moving to equilibrium (duration: 6–8 weeks)

This phase is characterized by full development of the required methanogenic bacterial mass. At this stage a decrease in the concentration of settleable and suspended solids in the effluent is observed, while there is also a significant increase in COD and BOD removal efficiencies. The sludge layer again increases gradually, this time with sludge of better settleability. The concentration of VFA in the effluent decreases, because the rate of methanogenesis becomes sufficiently high to convert the VFA produced. For the same reason the alkalinity increases, and the pH rises to a higher value near the neutral point.

At this stage the mass of the sludge in the UASB reactor reaches its maximum value, that is, the reactor becomes "full" in the sense that under the applied operating conditions there is no more room for a further increase of the sludge mass. Therefore, the reactor will start discharging sludge with the effluent. The rate of sludge discharge will be equal to its production rate in the reactor, so that the sludge mass tends to remain constant.

3.5.1.5 Phase 5: Establishment of steady-state conditions

After the sequential development of the different bacterial groups acting in the anaerobic digestion process is complete, the UASB reactor reaches its maximum treatment performance. The period required to reach steady-state operation varies, depending on several factors such as the composition of the influent organic material, the temperature, the geometry of the UASB reactor, the efficiency of solids retention by the phase separator and the applied retention time. However, in general the primary start-up, i.e. start-up without any inoculum, should require no more than 4–5 months at temperatures >20°C.

When the maximum sludge holdup capacity of the UASB reactor is exceeded, there are basically two ways to deal with further sludge production in the system. The first option is to apply periodic sludge discharges to avoid the presence of sludge in the effluent as much as possible. The second option is to operate the UASB reactor at the maximum sludge mass and to accept the fact that the excess sludge will be discharged in the form of settleable solids together with the effluent.

In this case it can be decided to separate the sludge from the UASB effluent, for example by the application of a final settler, or to leave the sludge in the effluent and to effect removal in the post-treatment system, for example in a polishing pond or an activated sludge system. It is important to note that although the COD associated with the settleable solids that are expelled from the reactor forms only a relatively small part of the COD content of the raw sewage (around 10–15%), it does constitute a rather large fraction of the COD content of the UASB effluent (around 30–50%).

3.5.2 Start-up procedures with seed sludge

The start-up of UASB reactors treating municipal sewage as described above – without the use of seed sludge from other anaerobic reactors – is widely applied today and almost invariably successfully. The following additional comments can be found in the literature:

- Materials such as cattle manure (Haskoning, 1989; Haskoning & Euroconsult, 1990) are sometimes used as inoculum. In this case the duration of the start-up period is reduced, but care must be taken with the seed material, which may contain many macroscopic particles (grass!) which can float and over time cause clogging inside the UASB reactor.

- In general, feed interruption is used whenever there are slight signs of souring of the reactor contents (even when weak or imagined), although practice shows that souring never materializes in UASB reactors treating sewage. The practice of periodic interruption of sewage feed does not seem effective in reducing start-up time.

With regards to the use of seed sludge for industrial UASB systems (and their clones) the situation is completely different. As industrial wastewaters are mostly soluble in nature (if required after pre-treatment for solids removal) this allows the application of anaerobic granular sludge at volumetric loading rates of up to 10 times higher than that applied in sewage UASB (15–25 g $COD \cdot L^{-1} \cdot d^{-1}$). Cultivation of granules starting from a non-granular seed biomass (or no seed at all) is possible but is a painstakingly slow process taking many months. Also the volume of an industrial anaerobic reactor is typically not very large (100–5000 m^3), where the largest size is equivalent to only two modular standard reactors of 20 × 24 × 5 m. Hence, in industry it is common practice to use granular biomass from existing anaerobic reactors as a seed material, and to start-up with a sizeable fraction of the treatment capacity already present (20–50%), depending on the application and on the contractual obligations with regards to the allowed duration of the start-up period.

3.6 STEADY-STATE BEHAVIOUR

After steady-state conditions have been established, the performance of an UASB reactor depends on the characteristics of the sewage and the operating conditions. The most important sewage characteristics are the concentration and composition of the organic material and the sewage temperature. In Section 3.6 it will be shown that the sludge age is the fundamental design variable that dictates the performance of the UASB reactor, but even for a reactor operated under steady-state conditions, it is not that easy to determine it.

In contrast, determination of the (hydraulic or liquid) retention time is extremely easy. Hence for this reason in many cases the retention time is taken as the main operational variable. The liquid retention time directly influences several aspects of the behaviour of the UASB reactor treating municipal sewage. This section attempts to establish this influence quantitatively. The following aspects are discussed:

- Effect of feed COD concentration
- Removal efficiency of organic material and suspended solids;
- Volumetric loading rate
- Profile of the concentration- and mass of the sludge in the UASB reactor;
- Specific organic load as an operational variable;
- Production, activity and stability of sludge;
- Floating solids.

3.6.1 Effect of the feed COD concentration

A factor that has already been discussed in Chapter 2 with respect to the kinetics of the removal of organic material is the concentration of organic material in the wastewater. In the case of sewage this material is predominantly of a particulate nature and to be converted to biogas it must pass sequentially through all steps from hydrolysis to final methanogenesis.

In each of these steps the conversion process will never be fully complete, leaving some residual COD, which significantly reduces the efficiency of the anaerobic digestion process as a whole.

Table 3.6 Experimental results of sewage treatment in UASB reactors with a retention time of 6 hours for different influent COD concentrations ($T = 25°C$).

Parameter	UoM	Influent COD ($mg \cdot L^{-1}$)				
		R_{816}	R_{555}	R_{298}	R_{195}	R_{92}
VLR	$kg \cdot m^{-3} \cdot d^{-1}$	3.3	2.2	1.2	0.8	0.4
Influent COD	$mg \cdot L^{-1}$	816	555	298	195	92
Effluent COD	$mg \cdot L^{-1}$	343	213	104	89	43
– SCOD	$mg \cdot L^{-1}$	177	117	70	63	30
– PCOD	$mg \cdot L^{-1}$	166	96	34	26	13
Digested COD	$mg \cdot L^{-1}$	472	347	201	107	48
Efficiency ($1 - S_{ti}/S_{te}$)	%	58%	62%	65%	54%	53%
Effluent VFA	$mg \cdot L^{-1}$	49	25	27	12	21
Sludge mass	g VSS	2104	1950	1366	1162	1377
Sludge concentration	$g\ VSS \cdot L^{-1}$	17.5	16.3	11.4	9.7	11.5
Sludge age	days	40	63	126	140	258

The sewage COD concentration is heavily influenced by the socio-economic habits of the contributing population and the type and quality of the sewage collection system. In combined sewer and drainage systems the COD concentration may have very low values under rainy conditions. The influence of the influent COD concentration on UASB reactor performance was investigated by Leitão *et al.* (2004). Table 3.6 show results from an experiment where the COD removal efficiency from sewage was evaluated for different influent COD concentrations at a fixed retention time of 6 hours in five identical UASB reactors (120 L).

Regarding the efficiency of COD removal, the data show that for the same applied retention time the efficiency of UASB reactors tends to decrease, when the COD concentration is less than 300 $mg\ COD \cdot L^{-1}$. However, with values higher than 300 $mg\ COD \cdot L^{-1}$, the reactors reached their maximum efficiency of about 58–65% of the COD of the influent. Variance analysis ($\alpha = 0.05$) showed that there were no significant differences in this respect between reactors R_{816}, R_{555} and R_{298}. In fact the Monod model does not predict efficiency variation since in this model the effluent COD concentration is independent of the concentration of the substrate in the influent.

These data corroborate the findings at Bucaramanga, where a full-scale UASB plant fed with sewage at about 300 $mg \cdot L^{-1}$ COD operated successfully, with a COD removal efficiency of 75%. However, at lower COD feed concentrations the residual VFA concentration remaining will have a stronger effect on COD removal efficiency than at higher COD feed concentrations. For example, if 20 $mg \cdot L^{-1}$ VFA-COD remains in the effluent after treatment, this would constitute only 2.5% of a feed COD of 800 $mg \cdot L^{-1}$ but no less than 20% of a feed COD of 100 $mg \cdot L^{-1}$.

It is interesting to note that sludge age increased significantly at reducing feed COD concentration: from as low as 40 days for the highest-loaded reactor (R_{816}) to more than 100 days – a typical value recommended for UASB reactors – for the R_{298} and R_{195} reactors, and even much higher for the UASB receiving the lowest COD concentration (R_{92}). The increase in SRT is what allowed the VFA concentration in the effluent to be reduced to such low values.

Tests showed that the SMA increased slightly with decreasing COD concentrations in the influent, with values of 0.18, 0.23 and 0.22 $kg\ COD \cdot kg^{-1}\ VSS \cdot d^{-1}$ at 30°C for influent COD concentrations of 816, 195 and 92 mg·L^{-1} respectively. This is also due to the increase in SRT, as particulate degradable material is degraded more slowly, which increases the fraction of methanogenic organisms in the sludge.

3.6.2 Performance in terms of organic material removal

Figure 3.13 shows the removal efficiencies of COD, BOD and TSS at the Pedregal plant in Campina Grande, which was operated at an average temperature of ±25°C. The 160 m^3 reactor was operated without intentional sludge discharge and the BOD, COD and TSS concentrations were determined for both raw and settled effluent. At the Pedregal plant, the retention time was varied from 2.1 to 17 hours. For each selected value of the retention time, the reactor was operated for a period long enough to establish a steady-state situation, with a sludge mass in the reactor compatible with the applied COD load for the applied retention time. The removal efficiencies were calculated from series of daily measurement, using raw and settled effluent samples. Settling was done in an Imhoff cone for 30 minutes. The data clearly show that the removal efficiency of all three variables depends on the applied retention time.

The influence of the retention time on the performance of the UASB reactor was also evaluated in Cali (Columbia), São Paulo (Brazil), Kanpur (India) and Bucaramanga (Columbia) as reported by Van Haandel and Lettinga (1994). The data presented in that text, and comparison with the results of other UASB reactors operating in tropical regions at more or less the same sewage temperature, allows the following conclusions:

- The data show that a high removal efficiency of COD, BOD and TSS was possible even at a relatively short retention time. However, even after settling in many cases the BOD and TSS concentrations in the effluent exceeded 50 mg·L^{-1}, which is not compatible with the standards established by the sanitary authorities in most countries. Therefore, some form of additional post-treatment after the digestion of sewage in a UASB reactor will usually be necessary;
- Although there was a reduction in the removal efficiencies of organic matter and solids when the retention time decreased, in none of the cases evaluated did the methanogenic sludge activity collapse due to an excessively short retention time. It is concluded that in this respect the UASB reactor is a very stable system.

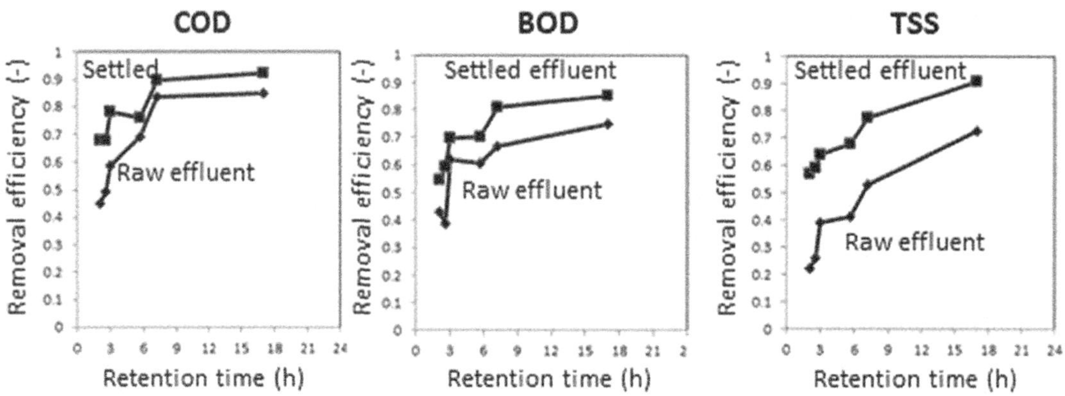

Figure 3.13 COD, BOD and TSS removal efficiencies as functions of the hydraulic retention time for UASB operating at steady state at a temperature of 25 ± 1°C.

- There are large variations in the observed values of the removal efficiency of organic material and suspended solids in systems that have similar hydraulic retention time but are treating different sewage in different systems; this leads to the conclusion that the physical configuration of the reactor, and the source and composition of the sewage are the major influences on the performance of the UASB reactor.
- The removal efficiency at lower temperatures as reported by de Man (1990) and Van der Last (1991) indicate that at a temperature below 18°C (and more markedly below 15°C) both the rate and efficiency of the organic material removal decrease considerably. This aspect of temperature dependency will be discussed further in Section 3.9.3.

3.6.3 Volumetric loading rate

The data of Figure 3.13 can be used to calculate reactor performance in terms of volumetric loading rates (VLR). This parameter indicates the daily loads of organic material (BOD or COD) per unit reactor volume of the reactor, hence typically expressed in $kg \cdot m^{-3} \cdot d^{-1}$. In this context, the term "daily organic load" is rather ambiguous and can have different meanings depending on the context. Taking COD as an example:

(1) The applied daily volumetric COD load, i.e. the load present in the influent:

$$L_a = MS_{ti}/V_r = Q_i \cdot S_{ti}/V_r = S_{ti}/R_h. \tag{3.7a}$$

(2) The removed volumetric COD load, i.e. the load that is converted into either sludge or methane and therefore is removed from the liquid phase, is given by the difference between the daily COD loads in the effluent and the settled COD load in the effluent (i.e. without sludge):

$$L_r = MS_r/V_r = Q_i \cdot (S_{ti} - S_{se})/V_r = (S_{ti} - S_{se})/R_h. \tag{3.7b}$$

(3) The digested volumetric COD load is the load that is effectively converted, predominantly to methane but also to other gaseous products like H_2S, and therefore no longer present in either the liquid or solid phases. If no sludge is discharged this load is equal to the difference between the daily COD load in the influent and the COD load in the raw (non-settled) effluent:

$$L_d = MS_d/V_r = Q_i \cdot (S_{ti} - S_{te})/V_r = (S_{ti} - S_{te})/R_h \tag{3.7c}$$

where

MS_{ti} = daily applied COD load
MS_r = daily removed COD load
MS_d = daily digested (converted) COD load
S_{te} = total effluent COD
S_{se} = settled effluent COD.

Figure 3.14 is a graphical representation of the behaviour of the volumetric loading rates defined above (i.e. applied, converted and digested COD and BOD loads) as function of the retention time, based on the data presented for Pedregal in Figure 3.13.

The following conclusions can be drawn from Figure 3.14.

- Depending on the retention time, the applied organic volumetric load MS_{ti} was quite high, and may even surpass the loading rate that is applied in aerobic treatment systems for secondary treatment (i.e. removal of organic material), which is typically around 2–3 kg $COD \cdot m^{-3} \cdot d^{-1}$ or 1–1.5 kg $BOD \cdot m^{-3} \cdot d^{-1}$.

- The removal efficiency tends to decrease at lower retention time, whereas the opposite is true for the removal rates. In other words, at lower retention times the VLR increases, but more so for the influent load than for the removed (or converted) load. Hence both removal and conversion efficiency decrease.
- At low retention (<3–4 hrs) times the digestion efficiency starts to collapse: the reactor is overloaded and the digested load decreases, even if the removed COD load still increases. At this point the hydrolysis of solids becomes the limiting process: the biodegradable particulate material present in the influent is incorporated into the sludge mass, but the hydrolysis rate is insufficient so that non-hydrolysed material accumulates and is eventually removed as excess sludge.
- The profile of the VLR as a function of the retention time clearly shows that a retention time of 2.1 hrs is excessively short: not only does the COD removal efficiency decrease but also the removal rate, despite a higher rate of substrate supply per unit volume.
- If the treatment objective is to maximize the digestion rate (i.e. to remove organic material at a maximum rate through conversion into biogas), then a retention time of 3 hrs could be chosen. However, in general the main purpose of the treatment system is to produce an effluent with a low organic material concentration, and this occurs at a longer retention time.
- When it is intended to use the UASB reactor as the only or main sewage treatment unit, the retention time should be long enough to ensure a high removal efficiency of organic material, i.e. >6 hours at 25°C.
- If the UASB reactor is designed as a pre-treatment unit, then a lower removal efficiency is sometimes permissible and, in some cases, even recommended. For example, when the UASB reactor is to be used as the pre-treatment unit of a polishing pond system, the retention time required in the pond for the removal of the thermotolerant bacteria will be longer than the retention time required for the removal of residual concentrations of organic material and suspended solids. Therefore in this case a UASB reactor could be operated with a short retention time and at a relatively low removal efficiency (see Chapter 9).

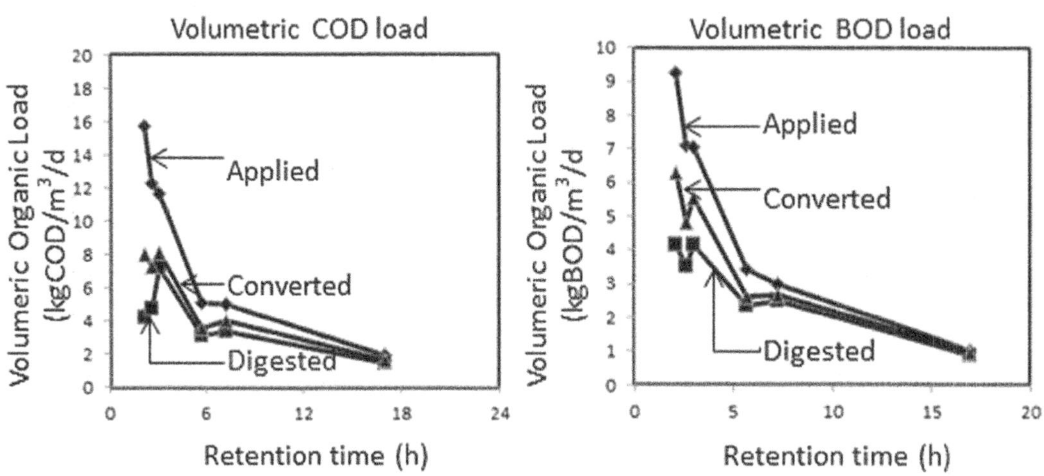

Figure 3.14 Volumetric COD and BOD loading rates of the UASB reactor at Pedregal as function of the retention time.

3.6.4 Sludge concentration profile

At Pedregal, the sludge mass retained in the UASB reactor was determined as a function of the retention time. The profiles corresponding to the steady state of the Pedregal UASB are plotted in Figure 3.15. It can be observed that this parameter strongly influences the sludge concentration profile: in the case of long residence times ($R_h = 5.6$–17 hours, with corresponding low upflow velocities of 0.24–0.71 $m \cdot h^{-1}$) a high sludge concentration was observed at the bottom of the reactor. It was reduced precipitously to very small values at a depth well below the phase separator, as shown in the right hand figure. In contrast, profiles for shorter retention times ($R_h < 3$ hrs, hence at high upflow velocities of 1–1.5 $m \cdot h^{-1}$), exhibited a gradual decrease in concentration along the liquid height in the reactor.

Table 3.7 shows aggregate data on sludge concentration and composition. The upflow velocity is indicated as well. In general there is a positive correlation between the average sludge concentration and the HRT: the sludge concentrations are highest at around 20 g $TSS \cdot L^{-1}$ at HRT values of 5.6–7.2 hrs, with corresponding upflow velocities of 0.56–0.71 $m \cdot h^{-1}$, which is in the range typically applied for municipal sewage UASB. Perhaps contrary to expectations, the average sludge concentration does not (yet) collapse at the lower range of the HRT values applied (2.6–3 hrs), which corresponds to high reactor upflow velocities

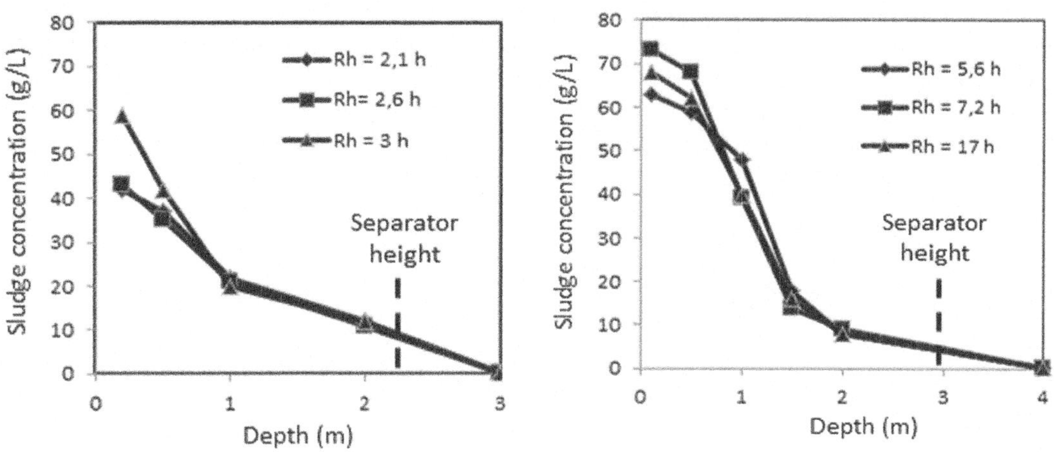

Figure 3.15 Profiles of the sludge concentration as a function of the height in UASB reactors for different values of the retention time.

Table 3.7 Aggregate data on sludge concentration and composition for HRT and upflow reactor velocities in the typical applied range of interest.

Parameter	Units	Applied HRT (hrs)			
		2.6	3.0	5.6	7.2
Upflow reactor velocity	$m \cdot h^{-1}$	1.15	1.00	0.71	0.56
Average sludge concentration	g $TSS \cdot L^{-1}$	15.6	17.8	19.6	20.2
Organic fraction	g $VSS \cdot g^{-1}$ TSS	0.7	0.68	0.55	0.56
Sludge age	days	10	15	38	46

(1–1.5 m·h^{-1}). This is probably due to a process of natural selection towards better settling sludge when higher upflow velocities are applied. However, as could also be observed in the previous section, the equilibrium between applied VLR and available methanogenic capacity is gradually being disturbed at lower HRT as a higher organic load is applied for equal (or lower) sludge mass available. The organic fraction of the biomass slowly increases at lower HRT (and at higher upflow velocity). This may seem counter-intuitive at first as one would expect sludge with a higher inert fraction, which is heavier, to be retained with preference at higher upflow velocities.

However, this behaviour can be explained when the value of the sludge age is considered as well. At a very low value of the anaerobic sludge age of 10–15 days, there is only very partial conversion of the slowly degradable particulate material present in the wastewater, which consequently accumulates in the UASB reactor and thus increases the organic fraction. To obtain a properly stabilized sludge an SRT of 10–15 days is much too low.

3.6.5 Specific sludge production

In a system operating under steady-state conditions, the average excess sludge production is equal to the sum of the discharged sludge mass and the mass of settleable solids present in the effluent. The specific sludge production is expressed as the mass of solids produced per kg of feed COD applied. The specific sludge production is an important variable as it indicates the organic load that was removed from the liquid phase, but not converted into biogas. The sludge production depends clearly on the retention time as the results from the Pedregal plant presented in Figure 3.16 clearly show. It may be noted that the specific production is influenced by the retention time. For long retention times (>6–8 hrs) there is a very gradual decrease of specific sludge production with increasing retention times.

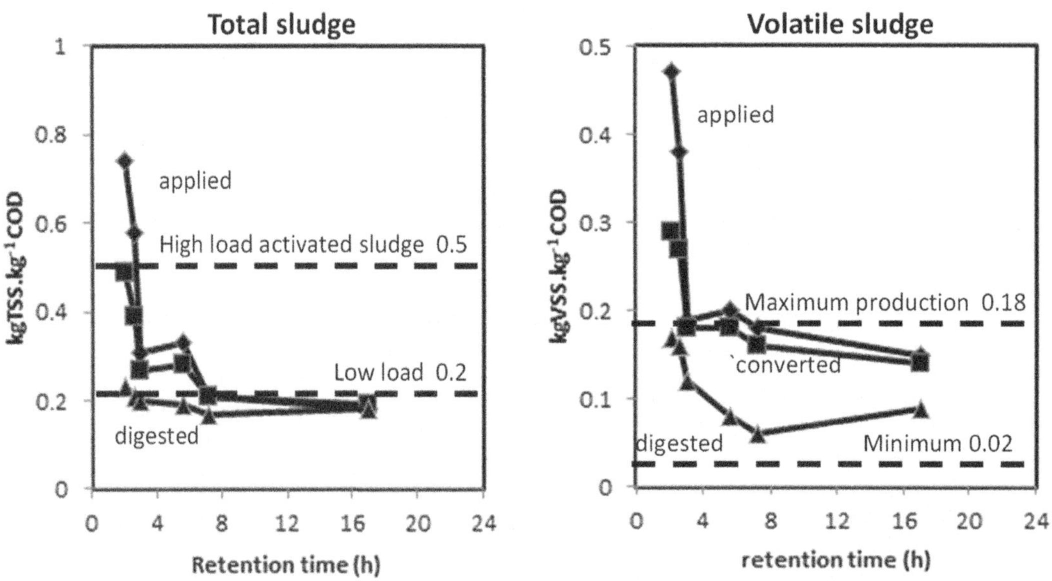

Figure 3.16 Specific excess sludge production in the UASB reactor in Pedregal as function of the hydraulic retention time.

At shorter retention times (<3–4 hrs), the specific sludge production rate increases steeply, which is explained by the flocculation of suspended solids in the influent and entrapment into the sludge blanket without the complementary biochemical processes for anaerobic digestion, because there is insufficient time for hydrolysis to develop. For comparison, Figure 3.16 also presents the theoretical minimum (methanogenic fermentation only) and maximum values (acidogenic plus methanogenic fermentation) of the specific organic sludge production in the anaerobic environment.

It is clear that the specific production of sludge at short residence times is distinctly higher than could be expected solely based on biological sludge generation. Thus, it must be assumed there is also non-biological sludge generation, due to the flocculation of the solids in the influent. Part of these solids are entrapped in the sludge blanket, while another part passes through the reactor without being metabolized. While these solids are considered to form part of the biological sludge, in fact it is biodegradable particulate material of the influent that is flocculated.

At higher retention times (>6–8 hrs), the specific organic sludge production rates are comparable with the theoretical value, so only a minor fraction of the sludge was composed of non-biodegradable solids. In Figure 3.16 the total specific sludge production rate is compared with typical values found in high-rate (0.5 kg TSS $\cdot$ kg^{-1} COD) and low-rate (0.2 kg TSS $\cdot$ kg^{-1} COD) activated sludge systems.

For relatively long residence times (>6–8 hrs), the specific sludge production in an anaerobic system tends to be smaller than the minimum value achieved for low-loaded activated sludge systems, which in practice is a major advantage of the UASB: the cost of treatment and final disposal of excess sludge will be much lower in the anaerobic system. For short retention times (<4 hrs at 25°C) the specific sludge production tends to be similar to that in high-rate activated sludge systems, again indicating that much of the sludge is composed of flocculated influent suspended solids.

3.6.6 Specific organic sludge loading rate

When both the organic load and the sludge mass present in the system are known, the specific organic sludge loading rate can be calculated as the ratio between the two, thus expressing the daily organic load per unit mass of sludge:

$$\mathrm{SOL} = \mathrm{M}S/\mathrm{M}X = L/X \ (\mathrm{kg\ COD \cdot kg^{-1} \cdot d^{-1}}) \tag{3.8}$$

where

$\mathrm{M}S$ = organic load (g COD $\cdot$ d^{-1})
$\mathrm{M}X$ = sludge mass in the system (kg X), either total or volatile solids
L = volumetric loading rate (kg COD $\cdot$ m^{-3} $\cdot$ d^{-1})
X = sludge concentration (kg $\cdot$ m^{-3})

In this context the daily organic load is defined again as the applied load ($\mathrm{SOL_a}$), the removed load ($\mathrm{SOL_r}$) or the digested load ($\mathrm{SOL_d}$).

Figure 3.17 shows the values of specific organic sludge load for the data of Figure 3.14 and the average sludge concentration calculated from Figure 3.15. Only the converted organic load (in kg COD $\cdot$ m^{-3}) is shown. It can be observed that the daily mass of organic material converted per unit mass of sludge tends to increase at decreasing retention time, with a maximum value at 3 hrs. This means that the sludge was underloaded at long retention times. However, for a very short retention time (<3 hrs), the specific COD removal rate decreases again. The explanation for this is that for very short residence times there is accumulation of non-hydrolysed organic material in the solid phase, reducing the biological sludge mass and consequently the treatment capacity per unit mass.

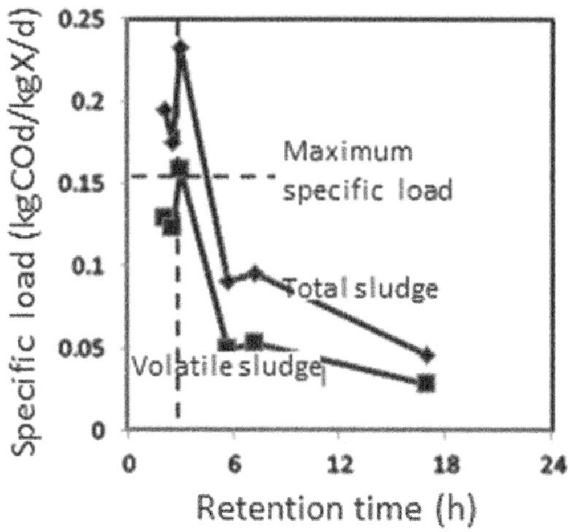

Figure 3.17 Specific organic sludge load (=digested) as a function of the retention time for the Pedregal UASB reactor.

The specific converted or digested sludge loading equals the mean specific methanogenic activity of the sludge under the operating conditions in the treatment system. The specific converted organic sludge loading rate can be conveniently linked to the methane production rate:

$$L_c = MS_d/MX_v = \sum_0^{24} Q_g \cdot f_{CH4}/0.35 \tag{3.9}$$

where

Q_g = total biogas flow (Nm^3)
f_{CH4} = methane fraction

The value of this variable, which reflects the methanogenic activity *in situ*, can be compared with the maximum specific methanogenic activity of sludge that develops in an environment without limitation of methanogenic substrate (i.e. activity test), as described in Appendix A1 In Figure 3.17 the maximum specific load that can be removed per unit mass of volatile sludge is 0.15 g $COD \cdot g^{-1}$ $VSS \cdot d^{-1}$ for a retention time of 3 hrs. However, for this retention time the removal efficiency of organic material was relatively low as shown in Figure 3.14. In practice a longer retention time is used, because the main objective of the anaerobic treatment step is not the maximum production rate of biogas, but rather a high removal efficiency of organic material.

3.6.7 Floating solids

In most sewage UASB reactors a scum layer is present on the surface of the overflow area. When the effluent collection is subsurface, by means of submerged effluent pipes, as in the Pedregal case, the floating material is not discharged: a sludge layer with a thickness of 4–6 cm was present during almost the entire year, occasionally being broken by heavy rains. When the abstraction of the effluent is superficial, by means of overflow weirs, the discharge of the surface solids can be reduced by introducing a scum baffle.

In Kanpur an investigation was carried out (Draaijer *et al.*, 1992) to test the effect of scum retention on treatment performance: one of the three reactors at the site was equipped with scum baffles and the others were not. The same retention time was applied to the three reactors and a slightly higher performance was observed in the reactor with the baffle, but the difference was not significant. Therefore, it is concluded that there was no great advantage in introducing the baffles.

In several cases (at Cali and Kanpur) a thick layer of scum developed at the liquid–gas interface under the gas hoods (see also Chapter 4). It is very likely that the rate of accumulation depends on the characteristics of the influent sewage. While estimating the accumulation rate at 125 $mm \cdot year^{-1}$ in Cali (recommending a clean-up every two years), in Kanpur a layer with a thickness of 35 cm developed over a period of only 6 months. It was assumed that the discharge of animal remains into the wastewater from the numerous tanneries in the town of Kanpur contributed to this high rate of scum accumulation.

EXAMPLE 3.3

In a UASB reactor of 5 m height, operated at a mean retention time of 6 hrs and equipped with surface scum baffles, a scum accumulation rate of 1 cm per week is observed, containing about 20% wt. solids (200 $kg \cdot m^{-3}$) with a volatile fraction of 75%. What would have been the increase in the COD concentration of the effluent if the baffles had not been installed?

Solution

The mass of volatile solids accumulating in the scum layer per week and per m^2 of reactor surface is equal to $0.01 \times 200 \times 75\% = 1.5$ kg $VSS \cdot m^{-2} \cdot week^{-1} = 9$ g $VSS \cdot m^{-2} \cdot h^{-1}$. Considering that the retention time is 6 hrs and the height of the reactor is 5 m, the upward velocity is $5/6 = 0.83$ $m \cdot h^{-1}$ and the flow rate is 0.83 $m^3 \cdot h^{-1}$ per m^2 of reactor. As 9 g VSS from 0.83 m^3 of wastewater accumulate in the scum the concentration is $9/0.83 = 11$ mg $VSS \cdot L^{-1}$. On removing the baffles these solids will go to the effluent, increasing COD concentration by $1.5 \times 11 \approx 16$ $mg \cdot L^{-1}$. Therefore, the increase in COD concentration in the effluent would be very small.

Even when the rate of scum formation is not high, it is important to take it into account by designing the phase separators in such a way that scum removal is not impossible, as discussed in Section 4.3.7.

The above example shows that the provision of baffles has a doubtful utility: the small COD reduction of the reactor effluent should be compared with the inconvenience of having to dispose of semi-solid, unstable scum with a strong odour. If there is post-treatment installed, then in general it is not worth having the baffles.

The problem of accumulating scum under the GLS separators is more complex. At SANEPAR an attempt was made to break up this scum with jets of water, but apparently without success, as the operation was abandoned. Others have provided access hatches to the gas hoods to allow for cleaning, but this remains very difficult in a long, narrow gas box. It also proves difficult to keep these access hatches gas-tight. Another strategy was proposed by Chernicharo *et al.* (2009) by installing scum extraction channels in the gas hoods, as shown in Figure 4.36. For this procedure to work it is essential to apply it regularly while the scum is still liquid: if the operator does not perform the scum removal procedure for a couple of weeks the scum solidifies, and removal will become impossible. Another possibility is to abandon the GLS phase separator altogether and place a dedicated solids settler after the UASB reactor (this will be discussed in Section 4.5).

3.7 PRESENCE OF SULPHATE IN MUNICIPAL SEWAGE

Until recently the presence of sulphate in municipal sewage was considered as a nuisance only, due to the smell and corrosion issues caused by the sulphides (H_2S, HS^-) produced under anaerobic conditions. Whereas sulphate concentrations in industrial wastewater, e.g. from the pulp and paper industry, may be very high (up to several hundred milligrams), typically in municipal wastewater the concentration is much lower. Values between 40 and 70 mg $SO_4 \cdot L^{-1}$ were reported in some of the earlier work on anaerobic sewage treatment, which took place in the Netherlands and in South America (Brazil, Colombia), where high concentrations of sulphate are not common, perhaps due to the moderate to high precipitation rates in these countries. However, in other parts of the world several recently constructed municipal UASB systems have encountered serious problems due to the presence of (very) high sulphate levels in the raw sewage. For instance, sulphate concentrations of 100–150 mg$\cdot L^{-1}$ are found in Egypt, Honduras and India, and up to 400–600 mg $SO_4^{2-} \cdot L^{-1}$ on the Arab peninsula.

When the source of domestic drinking water contains a high level of sulphate, this may eventually end up in the municipal sewage as well. An exception is the use of seawater for potable water production, as the RO brine can be discharged back into the sea. Another factor that may contribute to a high sulphate concentration in the sewage is infiltration of salty or brackish groundwater into the sewer system, a problem that might occur in coastal areas.

When sulphate is present in the feed to an anaerobic reactor then sulphide will be formed by sulphate-reducing bacteria, as discussed in Chapter 2. These sulphate-reducing bacteria compete with methanogenic bacteria for the conversion of available easily biodegradable COD. In particular, the hydrogen-oxidising sulphate reducers have kinetic advantages (higher growth rate and higher substrate affinity) over the equivalent methanogenic bacteria.

Stoichiometrically 0.67 mg of biodegradable COD per mg of sulphate is required, so in many cases potentially all sulphate present in the wastewater can be converted into sulphide. As the sulphate-reducing bacteria outcompete methane producers under all relevant process conditions, sulphate reduction tends to be complete or almost complete in anaerobic sewage treatment systems. Therefore it is prudent to assume complete sulphate reduction as a worst case scenario. Sulphate reduction in anaerobic–aerobic treatment systems has two detrimental effects on process performance: (1) in the anaerobic part the organic material used for sulphate reduction is no longer available for methane production, and (2) in the aerobic post-treatment the produced sulphide (which contains COD) is oxidized back to sulphate, consuming oxygen in the process.

An additional problem of sulphate reduction is that H_2S is known to be very toxic to nitrifying bacteria, significantly reducing their growth rate and requiring a significant increase in sludge age in order to meet ammonium effluent limits (if these apply).

Therefore, in general, it is recommended to install a pre-aeration step upstream of the main aerobic system in order to reduce sulphide toxicity. Such a pre-aeration reactor is basically a once-through tank that is operated without biomass retention at a hydraulic residence time of 20–60 minutes. In this reactor either biological oxidation or a biologically mediated chemical oxidation of H_2S/HS^- will occur at a very fast rate, in contrast to the slow reaction of the purely chemical oxidation process that is applied in treatment of refinery wastewaters, even in the presence of a catalyst such as $FeCl_3$. In addition to sulphate, sulphide may also be present in the raw sewage if the residence time in the sewer system is very long. Under anaerobic conditions a significant portion of the SO_4^{2-} present in the sewer will be reduced to H_2S, which subsequently reacts with iron to form ferric sulphide (FeS), which is responsible for the black colour of many sewages. This ferric sulphide represents a fraction of the COD demand in both anaerobic influent and effluent and will not be removed in the UASB.

EXAMPLE 3.4

A municipal UASB treats a wastewater containing 600 mg COD·L^{-1} and 50 mg sulphate·L^{-1}. The COD conversion efficiency is 65%.

Estimate the fraction of the COD demand in the anaerobic effluent that is due to the presence of sulphides and indicate the effect on COD digestion efficiency.

Solution

The initial COD/SO_4 ratio is 600/50 = 12 and it may thus be expected that enough biodegradable COD is available for complete reduction of the sulphate. Therefore, 100% sulphate reduction is assumed, although in practice the conversion may be less. Knowing that the molecular weight of sulphate is 96 mg·L^{-1}, with a sulphur content of 32/96 = 0.33 g S·g^{-1} SO_4^{2-}, the effluent sulphide concentration is equal to 0.33 × 50 = 17 mg S·L^{-1}. From stoichiometry, 1 mg of hydrogen sulphide-S is equivalent to two mg of O_2. Thus, the oxidized COD concentration = 16.67 × 2 = 33.3 mg COD·L^{-1}. This represents a fraction of 33.3/600 = 5.5% of the influent COD and reduces the methane production by 33.3/(0.65 × 600) = 8%.

3.8 INFLUENCE OF UASB OPERATION

The influence of operational procedures on the performance of the UASB reactor treating sewage is discussed in this section. Experiments were conducted on the following subjects: (1) comparison of parallel and serial operation of UASB reactors, (2) excess sludge discharge strategies and (3) influence of the return of settleable solids from the effluent. The effects of the return of aerobic sludge from post-treatment units to the UASB reactor will be discussed in Chapter 8.

3.8.1 Parallel versus series operation of UASB reactors

In most applications of the UASB reactor there are several reactors in a single treatment system. When there are two or more reactors, it is at least in principle possible to operate these either in parallel or in series configurations. While parallel operation is the more common approach, there are several examples in wastewater treatment systems where in series reactors are applied. For example, this is the case when one has first-order processes such as the decay of thermotolerant bacteria in polishing ponds (refer to Chapter 9).

There are two basic approaches to the operation of in series reactors, as shown in Figure 3.18a: (1) maintain the same height as in the unit reactor, which implies that the upflow velocity will be doubled or (2) maintain the same velocity but reduce the height of the reactor by half. It is not *a priori* clear which of the three options in Figure 3.18a results in higher removal efficiency of organic material. For this reason an experimental study was carried out in which the efficiency of the three system configurations was determined as a function of the residence time (Araújo, 2000).

Pilot reactors with a height of 4 m (systems A and B) and 2 m (system C) operated on raw sewage with retention times ranging from 2 to 10 hours. The efficiency of organic material removal in each of the reactors was determined and the performance of the single UASB reactor (A) was compared with the system of two reactors in series. All systems had equal total volume and were operated without intentional sludge discharge.

Figure 3.18b shows the COD removal during the research study. In Figure 3.18c the average sludge concentration in the reactors is shown. It can be observed that the settled effluent COD of the second reactors (B_2 and C_2) is almost the same as in the first reactor (B_1 and C_1), so that it is concluded that the second reactors removed very little COD. From the diagrams in Figures 3.18b and 3.18c the following can be concluded:

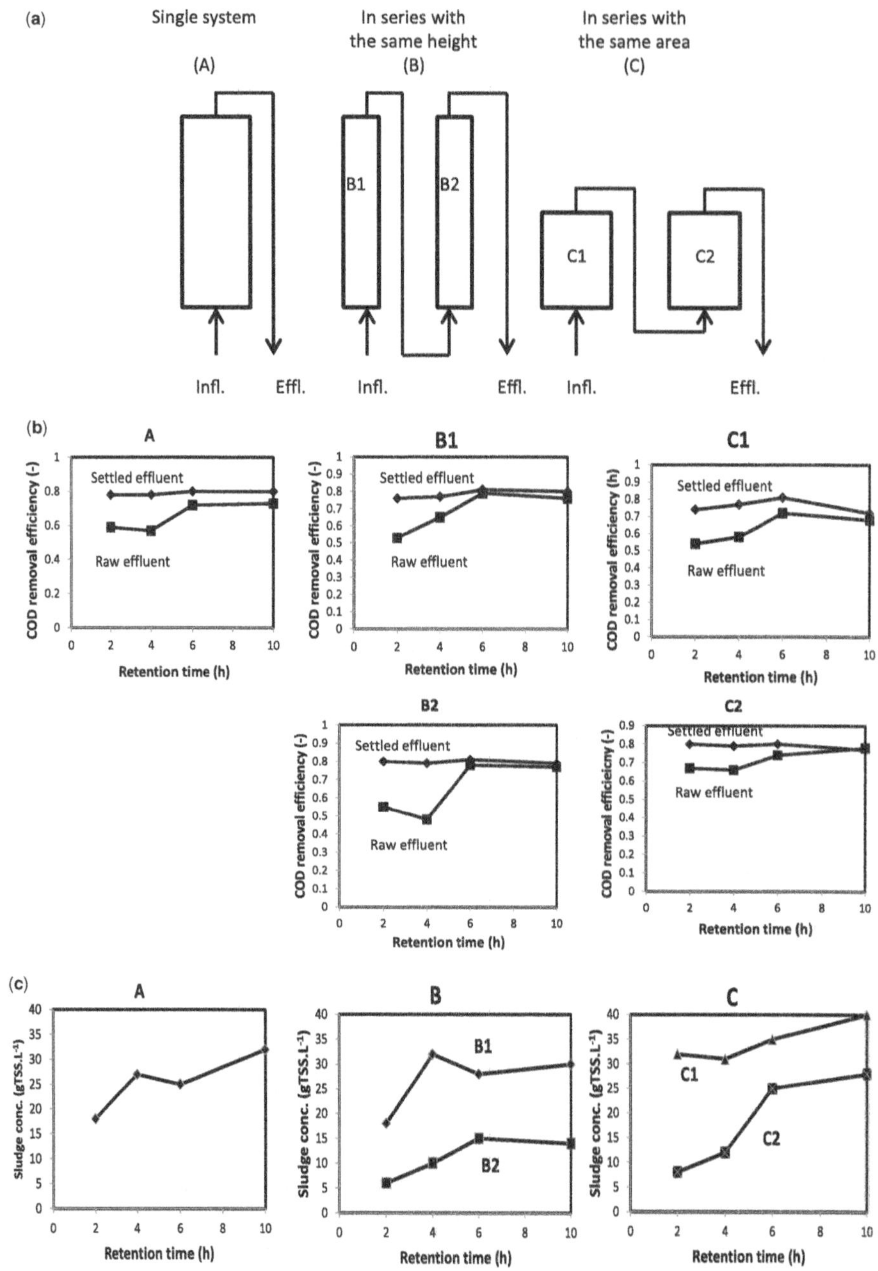

Figure 3.18 (a) Schematic representation the systems used in the study: the single-reactor system (A), compared with two reactors in series: (B) with the same height and (C) with the same area. All systems have equal total volume. (b) COD removal efficiency as a function of the retention time for the three systems of Figure 3.18a. The efficiency in the first reactors of the series systems is also indicated (B1 and C1). (c) Average sludge concentration as a function of the retention time in the five reactors of the three experimental systems used in the research study.

(1) All three systems exhibited a high COD removal efficiency, even for very short residence times; approximately 80% was removed from the final settled effluent in the three investigated systems.
(2) Comparing the performance of the three systems for equivalent retention times, there was no significant difference in the efficiency of organic material removal.
(3) The difference between the raw and settled effluent COD increased with decreasing retention time, indicating a greater presence of settleable solids in the effluent, that is, the production of sludge increased at decreasing retention time, presumably because the shorter retention time led to less efficient particulate organic material hydrolysis.
(4) Comparing the removal in the first reactors (i.e. B_1 and C_1) of the two series systems (i.e. $B_1 + B_2$ and $C_1 + C_2$), it is observed that almost all the removal of the organic material occurs in the first reactor; the second reactor has a very low efficiency. It is concluded that under the conditions prevalent in the research study, in fact, little was gained in the operation of two UASB reactors in series: the removable COD was almost completely eliminated in the first reactor, even with an extremely short retention time of only 2 hours.
(5) It can be observed that the first reactors of the series have a comparable sludge concentration and its value does not change considerably with the retention time. By contrast the average concentration in the second reactors (B_2 and C_2) is much smaller even though the reactors were operated for more than a year and these reactors received the sludge expelled from the first reactors (B_1 and C_1). It is concluded that these reactors produced very little sludge of their own and could not even retain the sludge discharged from the first reactors.

It is concluded that little was gained in the operation of two UASB reactors in series: the removable COD was almost entirely removed in the first reactor, even with an extremely short retention time of only 2 hours (corresponding to an upflow velocity of 2 $m \cdot h^{-1}$!). From the experiment it can be concluded that parallel operation is the preferred configuration for anaerobic sewage treatment.

3.8.2 Excess sludge discharge strategy

An experimental study was carried out to establish the maximum possible discharge of sludge that could be applied without affecting the effluent quality of the UASB reactor in a significant way (Cavalcanti *et al.*, 1999). It was established that for retention times of 4 and 8 hours, a discharge of up to 50–60% of the sludge mass (when the mass in the reactor was at its maximum) did not diminish the performance of the UASB reactor significantly. The experimental study was performed at a pilot scale using identical UASB reactors with volumes of 126 L each, fed with raw sewage. Table 3.8 summarizes the operational conditions of the experiments carried out. The study had the following objectives: (1) to determine the maximum feasible discharge volume of excess sludge without appreciably reducing reactor performance; and (2) to determine the rate of sludge production during steady-state conditions.

Figure 3.19 shows the minimum COD removal efficiency as a function of the percentage of the sludge mass discharged as excess sludge from the "full" reactor, for retention times of 4 and 8 h respectively. This value was determined immediately after the discharge took place. The diagrams indicate that there was no significant decrease in COD removal efficiency in the period subsequent to a sludge discharge if the percentage of discharged sludge did not exceed more than 50% of the sludge mass present before discharge.

The sludge mass in the reactor was calculated from weekly sludge concentration profiles. The effluent settleable solids were also measured, which allowed calculation of the rate of sludge loss in the effluent (involuntary sludge discharge). The sludge age was calculated as the ratio of total daily sludge discharge and the mass of sludge in the reactor. Figure 3.20 (centre) shows an example of the sludge mass growth in the reactor as a function of time, which was observed after a discharge of 54% of the sludge from a

Table 3.8 Average operational conditions and process performance during the two phases of the experimental study during periods without sludge discharges (steady state).

Variable	Unit	$R_h = 4$ h	$R_h = 8$ h
Volume	L	126	126
Height	m	4	4
Flow	$L \cdot d^{-1}$	750	375
Influent COD	$mg \cdot L^{-1}$	589	622
Specific volumetric load	$kg \cdot m^{-3} \cdot d^{-1}$	3.5	1.8
Upflow velocity	$m \cdot h^{-1}$	1.0	0.5
Temperature	°C	26	25
Total sludge mass	kg TSS	2.95	3.35
Volatile sludge mass	kg VSS	1.62	1.76
Total sludge production	$g\ TSS \cdot d^{-1}$	66	33
Volatile sludge production	$g\ VSS \cdot d^{-1}$	36	17
Sludge age	days	45	101

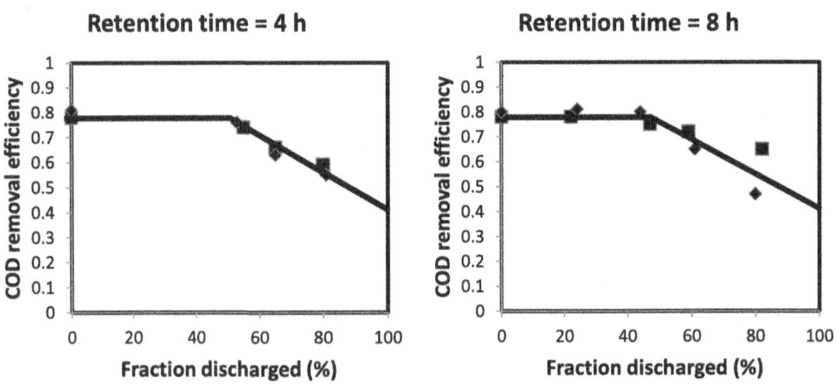

Figure 3.19 Value of the *minimum* COD removal efficiency remaining after waste sludge discharge, as function of the fraction of the reactor sludge mass that was discharged, for $R_h = 4$ hours and $R_h = 8$ hours.

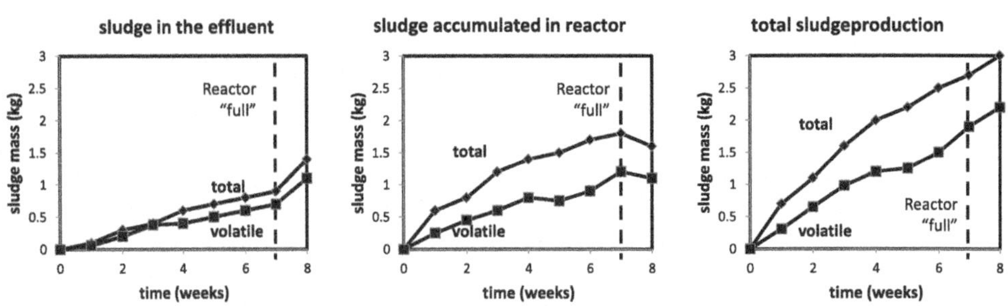

Figure 3.20 Values of the total and volatile sludge mass discharged in the effluent (left), accumulated in the reactor (middle) and the total combined sludge production (right) as a function of time after a sludge discharge of 54% of the total sludge mass ($R_h = 8$).

reactor operated with a $R_h = 8$ h. Directly after sludge discharge, the reactor was restarted with 740 g VSS of sludge and after 7 weeks this had increased to a maximum value of 1952 g VSS, at which value it essentially stabilized. Hence it was concluded that after 7 weeks of operation the reactor had reached its maximum sludge hold-up capacity.

The sludge mass discharged in the effluent as a function of time shown in Figure 3.20 (left) also leads to the conclusion that the maximum interval between sludge discharges is seven weeks: to good approximation, the accumulated mass of sludge in the effluent increased linearly with time (indicated by a constant settleable solids concentration in the effluent) until the seventh week. In the eighth week, the discharge of sedimented solids suddenly increased, presumably because the reactor was full. The experimental data highlight the following behaviour:

- before discharge, the sludge mass in the "full" UASB reactor remains essentially constant and all additional sludge produced is discharged with the effluent;
- after excess sludge discharge, the sludge concentration (settleable solids) in the effluent decreases, remaining at a constant value while the sludge mass in the reactor builds up until the maximum sludge holdup capacity of the reactor is again exceeded.

It was concluded that phase separation in the reactor was only partial: while there was 1953 g of VSS accumulation of solids in the reactor in the seven weeks after discharge, simultaneously there was also a discharge of 545 g of VSS with the effluent. Therefore the total sludge production was 1953 + 545 = 2498 g VSS, but only 1953/2498 = 82% was retained in the reactor, while 18% was discharged. When the experiment was repeated in the same reactor, but at a shorter retention time (4 hrs) and consequently with a larger sludge production, it was found that the sludge discharge with

EXAMPLE 3.5

Calculate the value of the specific organic sludge load and the sludge age based for the experimental system based on the data shown in Figure 3.20. Use the following additional data: COD of the influent = 750 mg·L^{-1} and volume of the reactor = 120 L.

Solution

The volatile sludge mass increased by 2498 g VSS in seven weeks or 51 g VSS·d^{-1}. The sludge mass after sludge discharge was 740 g VSS and the involuntary discharge was linear with time at a rate of 545 g VSS in seven weeks or 9.1 g VSS·d^{-1}. Therefore the initial sludge age directly after sludge discharge can be calculated as $R_s = 740/9.1 = 80$ days.

The sludge age gradually increased and at the end of the seven-week period, with a mass of 1952 g VSS in the reactor and still only a discharge of 9.1 g VSS·d^{-1} in the effluent, the sludge age would have been, if this situation persisted, at a maximum value of 1952/9.1 = 214 days. However, the reactor sludge mass stabilized at 1952 g VSS, but the daily solids discharge in the effluent increased to 51 g·d^{-1} (as the reactor was apparently full), so that the sludge age stabilized at 1952/51 = 38 days. In order to calculate the specific sludge load, one needs to define the organic load which is equal to L = 0.75 g·L^{-1}/0.33 d = 2.25 g·L^{-1}·d^{-1}.

Therefore the specific sludge load varies between 2.25/(740/120) = 0.38 g COD·g VSS^{-1}·d^{-1} directly after the sludge discharge to 2.25/(1952/120) = 0.14 g COD·g VSS^{-1}·d^{-1} when the reactor is full.

The specific sludge load value of 0.38 g COD·g^{-1} VSS·d^{-1} just after a discharge is probably higher than the SMA of the sludge. However, from the fact that no significant effluent COD increase or pH decrease was observed, it was concluded that the remaining sludge in the reactor could metabolize the VFA load fed to the reactor with the influent or produced in the reactor. This may be attributed to the slow rate of hydrolysis that leads to a delayed production of VFA. The negative side of this delayed conversion is that part of the biodegradable particulate influent COD leaves the reactor not converted.

the effluent was equal to 32% of the sludge production, as long as the maximum holdup capacity of the reactor was not exceeded. The upflow velocity of the reactors with a height of 4 m was 1 $m \cdot h^{-1}$ for $R_h = 4$ h and 0.5 $m \cdot h^{-1}$ for $R_h = 8$ h. The experimental data show that an appreciable part of the sludge produced could not be retained in the reactor, even when the reactor was not operating at its maximum sludge hold-up capacity. It is concluded that even when a very efficient phase separator was used, as was the case in the study, the fraction of sludge that could not be retained in the reactor was of the order of 1/6th for a retention time of 8 hrs to 1/3th for 4 hrs. This important conclusion was later corroborated with reactors operating at different retention times and a general expression could be derived for the fraction of sludge that is discharged even when the reactor is not operating at its maximum sludge hold-up.

$$f_{di} = 1.33/R_h \tag{3.10}$$

where

f_{di} = fraction of the produced sludge that will be discharged with the effluent, independent of whether the reactor operates at its maximum sludge hold-up capacity or not
R_h = retention time (in hours).

The fraction f_{di} was determined at 25°C and at other temperatures may have a different value. In addition it is likely that the phase separator design influences the value of f_{di}. At full-scale plants an empirical expression was developed for the sludge concentration in the effluent in systems with intentional sludge discharges:

$$X_{te} = 70 \cdot v_u + 10 \text{ (for } 0.4 < v_u < 1.0 \text{ m} \cdot \text{h}^{-1}\text{)} \tag{3.11}$$

These values have been corroborated with full-scale data collected at the Ajman plant (UAE), which is very well operated (Heffernan *et al.*, 2010a). The results of Eqs. (3.10) and (3.11) lead to very different results as can be observed in Figure 3.21, where the sludge concentration in the effluent is plotted against the retention time for a total sludge production of 150 mg $TSS \cdot L^{-1}$. It would thus appear that the low effluent solids concentration in pilot systems cannot be obtained in full-scale systems, even when these are well operated.

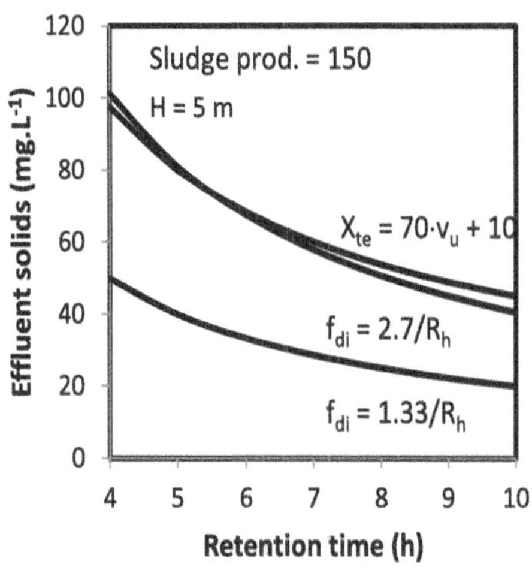

Figure 3.21 Effluent sludge concentration calculated from Eqs. (3.10) and (3.11).

Therefore it is recommended that the expression is checked and eventually corrected to express the experimental involuntary discharge of sludge with the effluent in a specific UASB reactor. If the expression of Eq. (3.10) is changed to $f_{di} = 2.7/R_h$, an almost perfect fit between Eqs. (3.10) and (3.11) is obtained, as can be observed in Figure 3.21. The value of the effluent sludge concentration is important, not so much for the design of the UASB reactor itself, but rather for the design of post-treatment units, especially in the case of aerobic post-treatment: the settleable COD in the effluent (which in reality forms part of the anaerobic sludge) makes up an important fraction of the effluent.

3.8.3 Settleable solids return from the anaerobic effluent into the reactor

In the previous section it was established that there is always sludge being discharged, independent of whether the reactor is "full" of sludge or not. It was hypothesised that if this sludge could be separated from the effluent and returned to the reactor, it could not only improve the effluent quality, but at least potentially increase the sludge mass in the reactor and therefore the sludge age, and hence its performance. In this section a study is described to establish if return of settled sludge improves the performance of the UASB reactor (Silva, 2016).

The predominant mechanism responsible for the loss of sludge from the UASB reactor is that the flocs tend to break up (floc erosion), and that the small particles that are produced do not have enough settling velocity, so that they are carried upward with the liquid phase and end up being discharged with the effluent. This occurs even without the reactor being full of sludge as was discussed in the previous section, i.e. under stationary operating conditions the sludge blanket in the UASB reactor generally does not extend to the phase separator. Therefore it would in principle be possible to accumulate more sludge. From the fact that this does not occur, it is concluded that the mechanism of unintentional discharge of sludge is not that the flocs themselves have insufficient settling velocity, but rather that it must be attributed to floc erosion. In the reactor a balance will be established between the mechanisms of floc erosion, producing flocs that cannot be retained on one side and sludge production due to bacterial growth and flocculation of particulate matter from the influent on the other side. The separation of solids from the UASB effluent and its return to the reactor serves two purposes: (1) the quality of the effluent improves when the settleable solids are removed; and (2) an increase of the sludge mass in the reactor increases the reactor sludge age, which in itself leads to increased treatment efficiency. During the experimental study the settleability of three different sludge types was determined: (1) the sludge inside the UASB reactor; (2) the sludge ejected from the reactor; and (3) the sludge in the settler of the UASB effluent.

Figure 3.22 shows the experimental set-up. One UASB reactor was operated as a reference (without settler) while the other three each had an additional settler to separate and retain the solids in the effluent. The three settlers had different diameters and liquid upflow velocities, leading to different separation efficiencies of the settleable solids. Table 3.9 shows the dimensions of settlers as well as the resulting upflow velocity of the liquid. As shown in Table 3.9 the upflow velocity in the settlers was well above the upflow velocity in the UASB reactor, which was only $0.2\ m \cdot h^{-1}$. Thus, at least in principle, solids that were not retained in the settler of the UASB reactor should also not be retained in the settlers. The fact that there was settling of solids in the settlers indicated that there must have been flocculation of the discharged flocs, resulting in larger flocs with higher settling velocity.

Therefore if this sludge is returned there is the possibility that at least part of it will be retained in the UASB reactor. It is however also possible that the returned sludge will disintegrate again in the UASB reactor and thus be ejected again without leading to an increase in the mass of retained sludge.

The efficiency of the settlers in the retention of solids was measured as the difference in the concentration of suspended solids and of organic material at the effluent of the reference UASB and in the effluent from

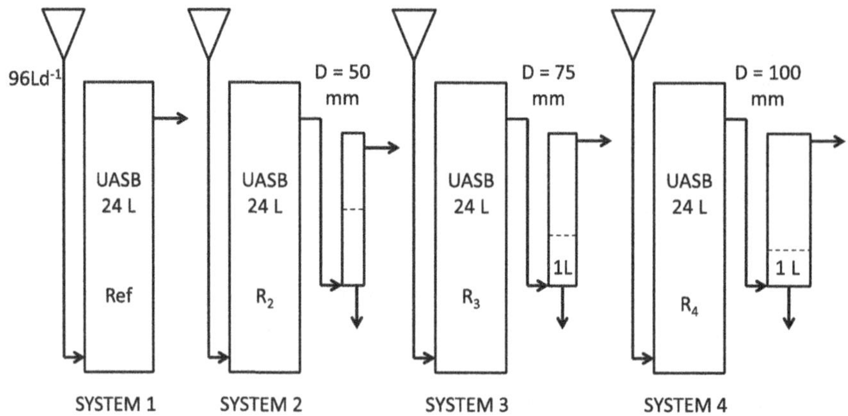

Figure 3.22 Layout of the experimental system operated to establish the feasibility of increasing the sludge mass in the UASB reactor by settling and returning the sludge present in the effluent.

Table 3.9 Area, settling velocity and volume of the settlers as a function of the diameter for an influent flow of 96 $L \cdot d^{-1}$ (4 $L \cdot h^{-1}$ or HRT = 6 hrs) in a 24 L reactor.

Unit	Diameter (mm)	Area (cm^2)	Overflow Velocity ($m \cdot h^{-1}$)	Settler Volume (*L*) for *H* = 1 m
UASB	150	88	0.2	
Settler 1	50	20	2.0	2.0
Settler 2	75	44	0.90	4.4
Settler 3	100	78	0.51	7.8

the settlers, as shown in Table 3.10. The results indicate that, although the settlers were operated at a much higher upflow velocity than the UASB settlers, sludge retention did occur. This indicates that flocculation occurred in the additional settler, so that flocs with a greater settling velocity were formed. The data in Table 3.10 clearly show that the average sludge concentration and therefore the mass in the UASB reactors were not measurably affected by the return of settled sludge. It is concluded that the solids expelled from the UASB reactor, even when retained in external settlers and returned to the UASB reactor, did not result in an increase in their sludge mass. This leads to the conclusion that the returned solids disintegrated in the

Table 3.10 Evaluation of the retention of organic material and suspended solids in the settlers and its effect on the average sludge concentration in the UASB reactors.

Variable	Reference	D50	D75	D100
COD removal in the settler ($mg \cdot L^{-1}$)		15	16	18
TSS removal in the settler ($mg \cdot L^{-1}$)	–	28	36	43
TSS concentration in the reactors ($g \cdot L^{-1}$)	39.8	40.0	43.9	40.8
VSS concentration ($g \cdot L^{-1}$)	24.2	25.6	28.6	25.5
Volatile fraction (g VSS $\cdot g^{-1}$ TSS)	60	60	59	62

reactors, so that they were discharged again with the effluent. It is concluded that it is innocuous to attempt to increase the sludge mass in the UASB reactors through the return of settleable solids in the effluent.

3.9 KEY PROCESS DESIGN PARAMETERS FOR UASB REACTOR PERFORMANCE

For aerobic systems it has been convincingly demonstrated both mathematically and in practice that it is not the hydraulic retention time, but rather the sludge age that determines system performance in terms of organic material removal efficiency, oxygen demand and excess sludge production (Van Haandel & Van der Lubbe, 2012). The hydraulic retention time is only important to the extent that it determines the sludge concentration as it defines the reactor volume for a given sewage flow. Thus the sludge age is the key design parameter for aerobic treatment. In anaerobic treatment systems it is much harder to reliably measure the sludge concentration, hence the sludge age is usually difficult to control. Furthermore, contrary to aerobic systems, anaerobic systems in general are operated at the maximum possible sludge age for the available reactor volume, with the objective of maximising the digestion efficiency or producing a sufficiently stabilized excess sludge. Operation at a high sludge age does not incur a "penalty" like excessive aeration costs for aerobic systems, although the required reactor volume will increase at higher anaerobic sludge age. In the next sections it is shown that the sludge age is a fundamental variable to describe the performance of UASB reactors and an empirical model is presented. It is also demonstrated that the sludge age that can be applied is influenced by the efficiency of the GLS separator in sludge retention. Finally, an important variable that defines the performance of the UASB reactor is the sewage temperature. This aspect is addressed in Section 3.9.3.

3.9.1 Influence of the GLS separator design quality on UASB performance

In this section it will be demonstrated that as in aerobic systems, the sludge age is also the key design parameter in anaerobic systems. Santos *et al.* (2016) treated raw sewage with two UASB reactors of equal size (1.2 m^3), but with a different GLS separator design, as schematically shown in Figure 3.23. One reactor was equipped with a conventional GLS separator like the one presented in Section 3.3.4, while the other reactor was additionally equipped with parallel plates, placed above the conventional separator, effectively transforming it into a high-rate settler.

Construction details of the configuration of the reactor with the enhanced settler are shown in Figure 3.24. The digestion zone of each reactor consisted of two concrete rings with a height of 1 m each and a diameter of 0.80 m. The conventional separator, made of fiberglass, was placed in the square masonry section above the rings, which formed the settling zone. In the other reactor parallel fiberglass plates were added in the part above the conventional phase separator. The plates had a width of 0.5 m and a depth of 0.35 m and were placed at a 45° angle with a spacing of 0.07 m. Both systems were operated without intentional excess sludge discharge, so that each reactor was operated with the maximum sludge mass that could be retained for the applied organic and hydraulic loads.

After steady-state conditions had been established, the two systems, although operated at the same hydraulic retention time, showed a very different performance as the retained sludge mass, and thus also the operational sludge age in the systems, was not the same.

The maximum sludge mass that could be obtained was influenced by the efficiency of the phase separation. As discussed in Section 3.6, the addition of parallel plates in the enhanced GLS design allows for a much more efficient sludge retention.

In Section 4.5, it will be demonstrated that there is a large difference between the capture efficiency of flocs in a conventional UASB settler and that in the settler with added parallel plates. Hence, it is possible

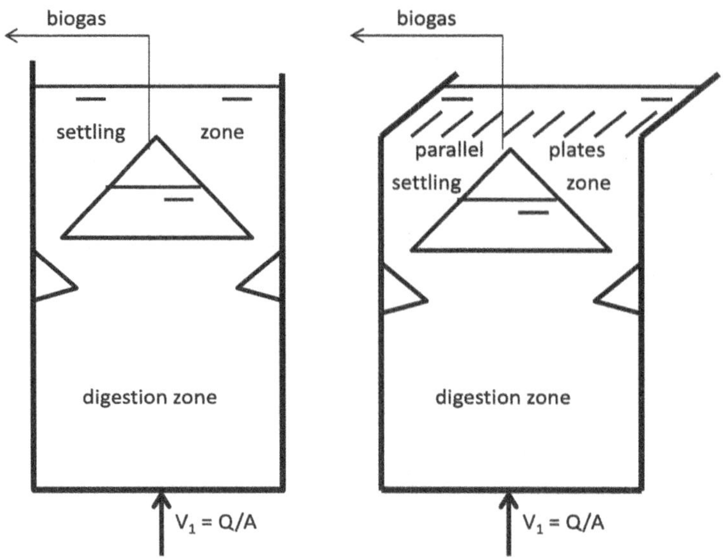

Figure 3.23 Schematic designs of two separators for the UASB reactors: conventional (left) and with additional parallel plates (right).

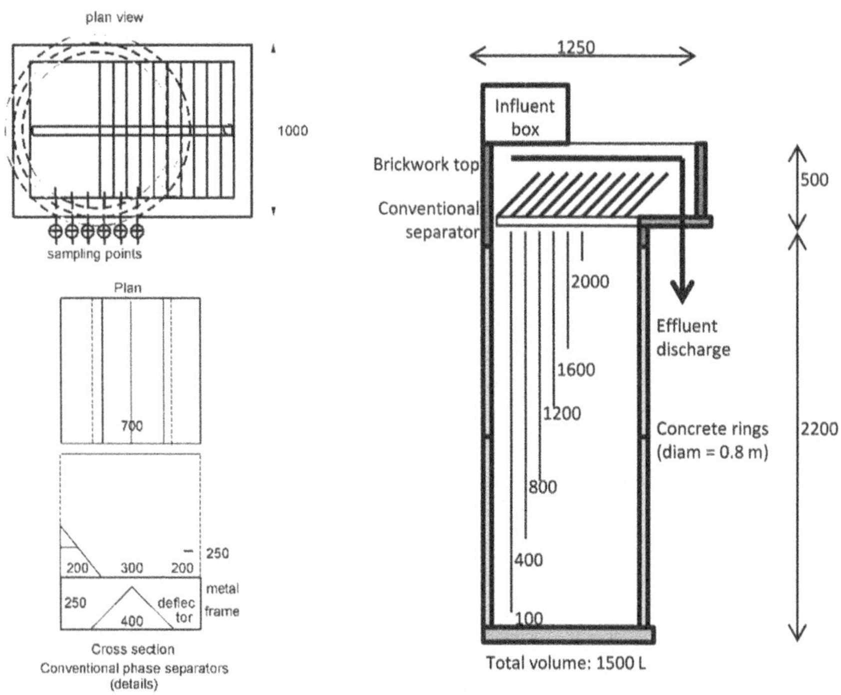

Figure 3.24 Detailed representation of the physical layout of the reactor with plate pack separator used in the experimental study (all dimensions are in mm).

to increase the hydraulic load of a reactor equipped with parallel plates and still maintain the same floc retention efficiency as in a similar reactor with only a conventional separator. It is however not possible to increase the sewage flow proportional to the ratio s_c/s'_c, because then also the organic load (and hence the sludge production) would increase.

The two UASB reactors were operated under identical operating conditions and without intentional discharge of excess sludge, so that after the maximum sludge mass was established, excess sludge was being discharged in the effluent at the same rate as it was produced in the reactor. The reactors were fed with raw sewage at constant flow rates and operated at hydraulic retention times that were varied from 12 to 1.5 hours. After imposing a particular hydraulic retention time, the reactors were operated under steady-state conditions for a period of at least two months prior to obtaining the experimental data, so that a representative sludge was established.

The evaluated parameters were related to: (1) operational stability; (2) removal efficiency of organic matter; and (3) production and composition of the excess sludge. Regarding the operational stability of the reactor, the following parameters were measured in the effluent: pH, total alkalinity (Alk_t) and volatile fatty acids (VFA).

All suspended solids that settled in an Imhoff cone within 30 minutes were considered part of the excess sludge and the residual COD of the supernatant was regarded as the true "soluble" effluent COD. For this reason, both raw (unsettled) and settled effluent COD concentrations were determined. The difference between these two COD values was used to estimate the COD concentration of the sludge in the effluent and hence its concentration in terms of VSS. Knowing that the COD content of a unit mass of organic sludge (f_{cv}) is approximately 1.5 g COD·g^{-1} VSS (Marais & Ekama, 1976), the organic sludge concentration in the effluent was calculated as $X_{ve} = (S_{te} - S_{se})/1.5$, where X_{ve} = VSS concentration in the effluent; S_{te} = raw or total COD concentration in the effluent (unsettled) and S_{se} = settled COD concentration in the effluent.

Table 3.11 shows the average total (S_{te}) and settled (S_{se}) effluent COD concentrations as well as the VFA concentrations for the two reactors, as a function of the retention time (R_h), while Table 3.12 presents the sludge mass and its composition in terms of TSS concentration and the organic fraction. The sludge mass (total and organic) was calculated from the linearized sludge concentration profiles, using the sludge concentrations determined at the sampling points (Figure 3.15). The production of organic sludge was

Table 3.11 COD concentration of the influent (S_{ti}), raw (S_{te}) and settled effluent (S_{se}) and VFA in the effluent (all in mg·L^{-1}) for different values of the hydraulic retention time in the conventional UASB reactor (A) and in the reactor with improved GLS separator (B). The temperature is 25°C.

R_h (hrs)	COD Concentrations (mg·L^{-1}) and (mg HAc·L^{-1})						
	Influent (S_{ti})	Reactor A (Without Plates)			Reactor B (With Plates)		
		Raw Effl. (S_{te})	Sett. Effl. (S_{se})	VFA	Raw Effl. (S_{te})	Sett. Effl. (S_{se})	VFA
12	587	157	88	18	155	86	26
10	492	143	78	22	139	84	20
8	554	189	108	18	163	80	23
6	480	186	102	24	172	92	28
4	526	252	133	38	166	85	45
3	619	360	195	73	236	134	87
2	561	454	236	97	304	167	78
1.5	613	–	–	–	386	215	69

Table 3.12 Average reactor sludge concentration (X_t), volatile fraction (f_v) and sludge age (R_s) for different values of the hydraulic retention time R_h for reactors A and B. The temperature is 25°C.

R_h (hrs)	Reactor A (Without Plates)			Reactor B (With Plates)		
	Xt (g TSS·L−1)	fv (g VSS·g−1 TSS)	Rs (days)	Xt (g TSS·L−1)	fv (g VSS·g−1 TSS)	Rs (days)
12	20.6	0.54	122	36.6	0.56	205
10	18.0	0.57	98	29.7	0.58	155
8	16.1	0.58	58	27.0	0.57	120
6	16.0	0.61	44	19.7	0.57	64
4	17.5	0.65	21	28.2	0.61	47
3	16.9	0.67	13	25.4	0.61	28
2	14.6	0.68	6	23.7	0.63	17
1.5	–	–	–	28.8	0.68	11

estimated from the difference between the COD concentrations of the raw and settled effluent as explained above.

The data presented in Tables 3.11 and 3.12 can be used to calculate the fractions of the influent COD transformed into effluent COD, excess sludge COD and digested COD:

$$f_s = S_{se}/S_{ti} \tag{3.12}$$

$$f_x = (S_{te} - S_{se})/S_{ti} \tag{3.13}$$

$$f_d = 1 - f_s - f_x \tag{3.14}$$

where

f_s = COD fraction in the settled effluent
f_x = COD fraction converted into organic sludge
f_d = digested COD fraction
S_{ti} = influent COD
S_{te} = total effluent COD (non-settled)
S_{se} = settled effluent COD

Figure 3.25 shows the graphs corresponding to three COD fractions defined above (f_s, f_x and f_d), as function of the liquid retention time for both the conventional UASB reactor (left) and the reactor with improved separator (right).

In this experimental study, the digested fraction was not actually determined from the methane production, but rather calculated by Eq. (3.14). The measured methane production was smaller than the value based on the mass balance, even if the presence of dissolved methane in the effluent was considered. In the study it proved to be impossible to avoid undetected losses of methane.

The data from Figure 3.25 are presented in a different format in Figure 3.26, which shows the division of the influent COD into the three COD fractions. At lower values of the retention time R_h the proportion of both f_s and f_x increases at the cost of a decrease in f_d. It can be observed that the reactor with the improved separator demonstrated a much better performance, as at all retention times the fraction of digested COD was much larger than in the conventional reactor. This is a clear indication that the hydraulic retention

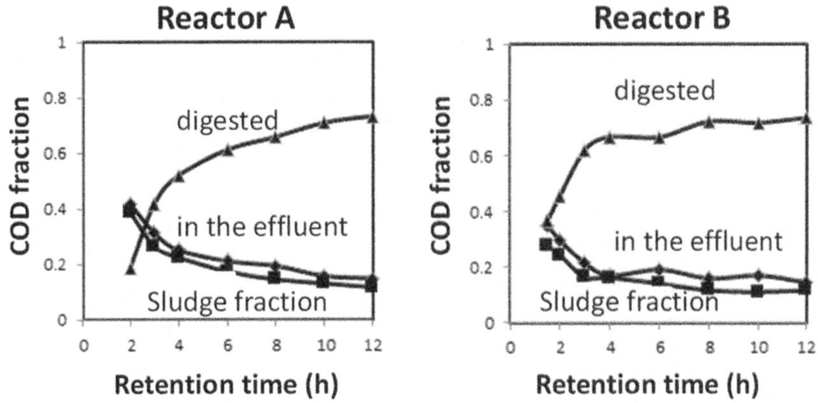

Figure 3.25 COD fractions in the effluent, discharged as sludge and digested as functions of the retention time in reactors A and B, operated at $T = 25°C$.

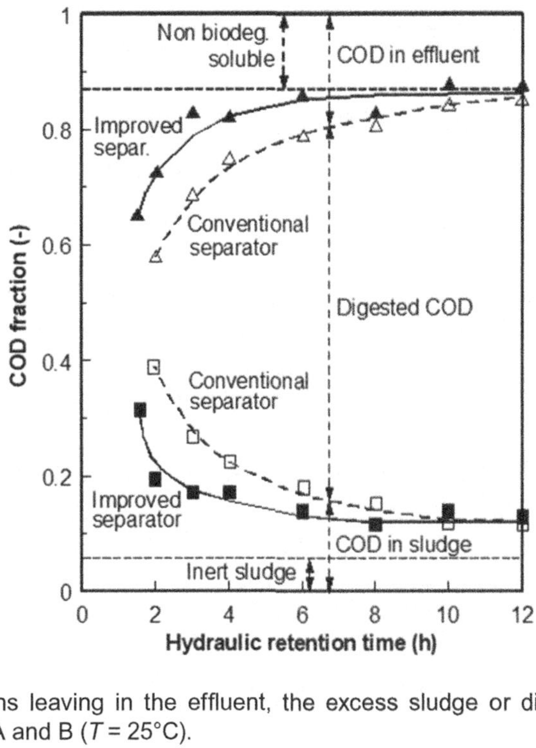

Figure 3.26 COD fractions leaving in the effluent, the excess sludge or digested, as function of the retention time in reactors A and B ($T = 25°C$).

time alone is not a suitable indicator of the expected performance of a UASB reactor: the quality of the phase separator design, which influences sludge retention and hence the sludge age, is important as well.

The data from Tables 3.11 and 3.12 can also be used to plot the three COD fractions (f_s, f_x and f_d) as function of the sludge age, as shown in Figure 3.27 (log scale on the abscissa). If the sludge age is used as an independent variable, the fractions f_s and f_x of reactors A and B, in good approximation, can be

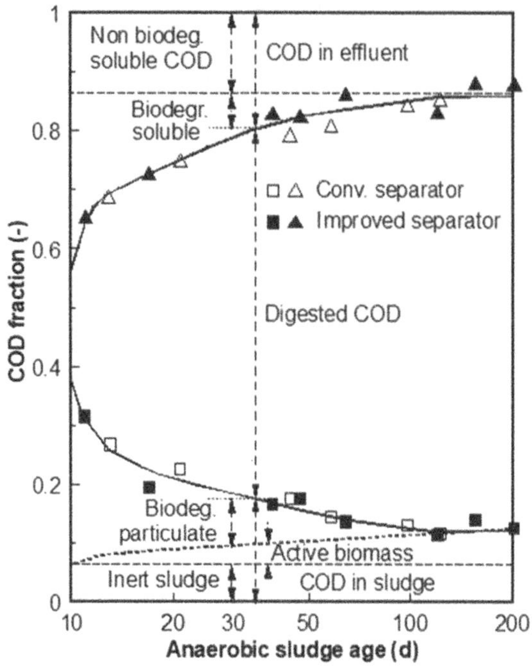

Figure 3.27 COD fractions leaving in the effluent, the excess sludge and digested, as function of the sludge age in reactors A and B ($T = 25°C$).

described with a single curve. Now, as can be observed, the data for the two GLS configurations turns out to be very comparable. This means that, for a given sludge age, the fractions of the influent COD ending in the effluent or converted to sludge (and hence also the digested fraction) are the same regardless of the design of the phase separator or the liquid retention time that is applied.

3.9.2 Empirical model for UASB performance based on sludge age

From the experimental data, presented in Table 3.11, Table 3.12 and Figure 3.27, empirical expressions can be derived to estimate the values of f_s, f_x and f_d in Eqs. (3.15) to (3.17) for raw sewage treatment to predict the fraction of non-settleable COD in the UASB effluent (f_s) and the fraction of COD used for excess sludge production (f_x), as well as the digested COD fraction (f_d) as functions of the anaerobic sludge age R_s, valid for a temperature of 25°C:

$$f_s = 0.14 + 0.25\exp[-0.04\,(R_s - 6)] \tag{3.15}$$

$$f_x = 0.12 + 0.20\exp[-0.04\,(R_s - 6)] \tag{3.16}$$

$$f_d = 1 - f_s - f_x = 0.74 - 0.45\exp[-0.04\,(R_s - 6)] \tag{3.17}$$

where

f_s = influent COD fraction discharged as non-settleable material in the effluent

f_x = influent COD fraction transformed into settleable material in the effluent or discharged as excess sludge
f_d = influent COD fraction that is transformed into methane.

In Eq. (3.15) the term 0.14 is identified as the fraction of the non-biodegradable soluble material in the influent, f_{ns}, although two considerations must be made here:

(1) If sulphate is present in the influent, it will for the most part be converted into sulphide by sulphate-reducing bacteria (SRB) at a rate of 2 mg COD per mg SO_4-S, in the process oxidising organic material that will therefore no longer be available for methane production. Hence, in that case part of the effluent COD is in fact sulphide COD, which will exert an oxygen demand in subsequent oxidation. This however does not affect UASB performance, but it should be considered when designing an aerobic post-treatment.
(2) Monod kinetics shows that even at a very long sludge age the biological processes of acidogenesis, acetogenesis and methanogenesis are not complete, so that there is a residual concentration of their products: primary and secondary metabolites and volatile fatty acids. However it is considered that these concentrations are so low at long sludge ages ($R_s > 100$ days) that these can be neglected. This is also apparent from the results in Figure 3.27, where the soluble effluent COD is virtually constant in the range $100 < R_s < 200$ d. In the next section it is shown that for temperatures between 12 and 25°C this is also valid.

In Eq. (3.16) the term 0.12 that is obtained for a long sludge age depicts both the non-biodegradable influent COD fraction and the COD fraction that is transformed in active biomass and a provision must be made to express the biological sludge production. This can be done in the following manner: if a fraction f_d of the influent COD is transformed into methane (catabolized), the COD fraction that is metabolized is equal to:

$$mS_{mb} = f_d/(1 - f_{cv}Y_{an}) \tag{3.18}$$

Thus the COD fraction transformed into active (live) microbial biomass for anaerobic digestion becomes:

$$mS_{xa} = f_{cv}Y_{an}f_d/(1 - f_{cv}Y_{an})/(1 + b_{an}R_{su}) \tag{3.19a}$$

Now, considering that the decay rate is very low (<0.01 d^{-1}), Eq. (3.19a) can be simplified to:

$$mS_{xa} = f_{cv}Y_{an}f_d/(1 - f_{cv}Y_{an}). \tag{3.19b}$$

while owing to the low decay rate of anaerobic bacteria the COD fraction transformed into endogenous sludge can be neglected:

$$mS_{xe} \approx 0.0.$$

The COD fraction accumulating as inert volatile sludge is equal to the non-biodegradable particulate influent fraction, f_{np}:

$$mS_{xi} = f_{np} \tag{3.20}$$

The maximum fraction of influent COD transformed is obtained when all biodegradable COD is digested:

$$f_{dm} = 1 - f_{ns} - f_{np} \tag{3.21}$$

Thus the maximum fraction of biological sludge can be expressed as:

$$
\begin{aligned}
f_{bx} &= mS_{xam} + mS_{xi} = f_{cv}Y_{an}f_{dm}/(1 - f_{cv}Y_{an}) + f_{np} \\
&= f_{cv}Y_{an}(1 - f_{ns} - f_{np})/(1 - f_{cv}Y_{an}) + f_{np}
\end{aligned}
\tag{3.22}
$$

where

f_{ns} = non-biodegradable soluble fraction of the influent COD
f_{np} = non-biodegradable particulate fraction of the influent COD
mS_{mb} = metabolized biodegradable COD fraction
mS_{xa} = influent COD fraction converted into active sludge
mS_{xe} = influent COD fraction converted into endogenous residue
mS_{xi} = influent COD fraction converted into inert sludge
f_{bx} = COD fraction transformed into biological sludge
f_{dm} = maximum influent COD fraction that can be digested (for $R_s > 100$ days)
Y_{an} = yield coefficient of the anaerobic micro-organisms
R_{su} = sludge age in the UASB reactor
f_{cv} = ratio COD/VSS of sludge (biomass) = 1.5 mg COD·mg^{-1} VSS.

Equation (3.22) still does not permit the calculation of the active sludge fraction, because the yield coefficient for the mixed anaerobic biomass population is unknown.

An informed estimate can be made considering that the value of Y_{an} must at least be higher than the yield coefficient for pure methanogenic biomass: $Y = 0.03$ mg COD·mg^{-1} COD. On the other hand Y_{an} must also at least be lower than the yield that would be observed if all the biodegradable organic material in the feed were subjected to the full anaerobic cycle from hydrolysis, acidogenesis, acetogenesis to methanogenesis ($Y = 0.18$ mg COD·mg^{-1} COD). In Figure 3.28 the values of mX_a and mX_i have been plotted as function of the yield coefficient. A value of $Y_{an} = 0.05$ mg COD·mg^{-1} COD would appear to be reasonable.

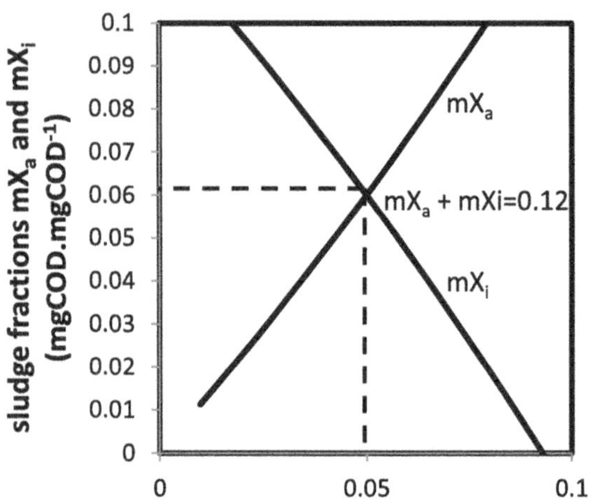

Figure 3.28 Active and inert sludge fractions as functions of the yield coefficient.

A smaller value is improbable, since much of the biodegradable COD is particulate and hence must be hydrolysed. A greater value would imply a very small fraction of non-biodegradable particulate influent COD. For $Y_{an} = 0.05$ mg COD · mg^{-1} COD, the calculated values are: $mS_{xam} = mS_{xi} = 0.06$ mg COD · mg^{-1} COD. Hence the non-biodegradable and particulate COD fraction is estimated at $f_{np} = 0.06$ and now the active sludge fraction mS_{xa} can be calculated for any sludge age from Eq. (3.19b).

Equations (3.15) and (3.16) clearly show that it is in fact the anaerobic sludge age and not the hydraulic retention time that is the fundamental design and operating parameter of anaerobic treatment systems. It can be concluded that the performance of different types of anaerobic reactors treating the same wastewater will be comparable, if the operational sludge ages are equal. In other words, the division of the influent COD into fractions that end up in the solid, liquid and gaseous phases tends to be identical in different anaerobic treatment systems treating similar wastewater, if these are operated at the same sludge age. In this respect, anaerobic treatment systems are thus very similar to aerobic systems.

In practice it is difficult to predict the sludge age of an anaerobic system, which makes it difficult to use it as a design parameter. Therefore, the hydraulic retention time is often used instead, although, as explained above, different systems with the same hydraulic retention time may exhibit significantly different behaviour. In the next section empirical equations will be presented that allow an estimate of the sludge mass and composition that will develop as functions of the influent COD load and composition and the applied anaerobic sludge age. This in turn facilitates the use of the anaerobic sludge age for the process design of UASB reactors.

Apart from highlighting the influence of the sludge age as the key determinant to the performance of an anaerobic treatment system, the experimental curves of Figure 3.27 reveal some additional important characteristics. In addition to anaerobic biomass and inert sludge, depending on the operational conditions of the anaerobic reactor (i.e. sludge age and temperature), there may be a significant fraction of biodegradable particulate organic material present in the sludge mass as well, resulting from the incomplete hydrolysis of organic material. Therefore, the active biomass fraction in the organic excess sludge from an anaerobic reactor will be small. At a low sludge age, the particulate biodegradable organic material will constitute the largest fraction in the organic sludge mass, so the excess sludge will be unstable and may even be putrescible. In fact in many cases the minimum required sludge age in UASB reactors is set by the requirement to obtain a stabilized sludge rather than achieving the maximum COD removal efficiency.

As the sludge age decreases, the COD fraction that ends up in the effluent will increase due to a reduction in the removal of soluble biodegradable material in the anaerobic reactor, although this is not due to insufficient methanogenic capacity to metabolize the VFA produced: in Table 3.11, the VFA concentration remains low, even at very short retention times. It is concluded that the increased biodegradable COD concentration in the effluent is due rather to primary and secondary metabolites that are not converted in the acidogenic and acetogenic processes. There is a minimum sludge age below which the anaerobic reactor fails altogether, i.e. the biological activity ceases because the bacterial populations are not able to develop. The reactor will then act as a primary settler. Hence for both reactors used in the study there are minimum values for R_h and R_s below which methanogenesis will not occur and consequently all organic material will leave the reactor either in the effluent or as flocculated material in the excess sludge.

The minimum value of R_h is *different* for both reactors and depends on the efficiency of the phase separator. On the other hand, the minimum value of R_s is *the same* for both reactors and depends on the maximum growth rate of methanogens, which in turn depends on temperature, among other factors. The data presented in this section lead to the following important conclusions:

- The sludge age is the fundamental variable to describe the performance of the UASB reactor for the treatment of wastewater, particularly the division of the influent COD into three fractions: (1) discharged into the effluent, (2) converted into sludge, and (3) digested into methane. The shorter the sludge age that is applied, the larger will be the fractions of the influent COD that are discharged either in the effluent or in the excess sludge.
- The sludge age that can be maintained in a UASB reactor strongly depends on the design quality of the GLS phase separator. The addition of parallel plates to a conventional phase separator to form a high-rate settler proved to be very effective in increasing sludge retention and can easily double the treatment capacity.

The selection of the optimal anaerobic sludge age depends on the comparison between the benefits of increased digestion efficiency, better effluent quality and reduced and more stabilized excess sludge production versus the higher investment costs of the UASB reactor required to retain the large sludge mass that will develop at a long sludge age. For efficient treatment the required sludge age will be much higher than in the case of aerobic treatment. On the other hand, it is usually possible to accommodate several times more sludge per unit of volume in anaerobic systems, which allows the application of a high sludge age in a relatively compact system.

EXAMPLE 3.6

Predict the process performance of UASB reactors operated at a sludge age of 60 days at a temperature of 25°C and treating 50,000 $m^3 \cdot d^{-1}$ of raw sewage with a COD concentration of 600 $mg \cdot L^{-1}$, assuming the empirical equations Eqs. (3.15) and (3.17) are applicable. What is the expected COD removal efficiency? What will be the effect on COD removal efficiency and excess sludge production from an increase of the sludge age to 100 days?

Solution

The soluble and settleable COD fractions in the effluent of a UASB reactor operated at 25°C can be expressed as:

$$\begin{aligned} f_s &= f_{ns} + 0.25\exp[-0.04(R_s - 6)] \\ &= 0.14 + 0.25\exp[-0.04(60 - 6)] = 0.169 \end{aligned}$$

$$\begin{aligned} f_x &= f_{np} + f_{bx} + 0.20\exp[-0.04(R_s - 6)] \\ &= 0.12 + 0.20\exp[-0.04(60 - 6)] = 0.143 \end{aligned}$$

Therefore: $f_d = 1 - f_s - f_x = 1 - 0.169 - 0.143 = 0.688 = 68.8\%$

Assuming an *ideal* situation, i.e. when no sulphate is present and methane and sludge losses with the effluent are ignored, then for the influent COD load of 30,000 kg $COD \cdot d^{-1}$, the daily mass flows are:

$$MS_e = f_s MS_{ti} = 0.169 \cdot 30{,}000 = 5065 \text{ kg COD} \cdot d^{-1}$$
$$MS_x = f_x MS_{ti} = 0.143 \cdot 30{,}000 = 4292 \text{ kg COD} \cdot d^{-1}$$
$$MS_d = MS_{ti} - MS_e - MS_x = 30{,}000 - 5065 - 4292 = 20{,}643 \text{ kg COD} \cdot d^{-1} \ (=5161 \text{ kg } CH_4 \cdot d^{-1})$$

The expected COD removal efficiency in the ideal situation – when it is assumed all sludge is retained in the reactor – is equal to $1 - f_s = 83.1\%$. To assess the effect of the increase of the sludge age to 100 days, the values for f_s and f_x need to be recalculated: $f_s = 0.146$ and $f_x = 0.125$

The COD removal increases from 83.1% to $1 - f_s = 85.4\%$, which is only a very modest increase.

On the other hand the sludge production would decrease from a fraction of $f_x = 0.143$ to $f_x = 0.125$, i.e. a reduction of $0.018 \times 30.000 = 552$ kg $COD \cdot d^{-1} = 368$ kg $VSS \cdot d^{-1}$.

EXAMPLE 3.7

For the wastewater of Example 3.6, it is known that the sulphate concentration is 36 mg·L^{-1}. Establish the values of the inert soluble COD fraction (f_{ns}) and the inert particulate COD fraction (f_{np}). For the latter, assume that Y_{an} = 0.05 mg COD·mg^{-1} COD.

Solution

To start with f_{np}, at a sludge age of R_s = 100 days and at 25°C, the digested fraction f_d is maximum and can be calculated considering that the exponential terms in Eqs. (3.15 and 3.16) can be neglected, i.e. $f_{dm} = 1 - 0.14 - 0.12$. The COD fraction converted into active sludge can be calculated from Eq. (3.19b):

$$mS_{xa} = f_{cv}Y_{an}f_{dm}/(1 - f_{cv}Y_{an}) = 1.5 \times 0.05 \times 0.76/0.925 = 0.06$$

Using the calculated value of mS_{xa}, it follows that $mS_{xi} = f_{np} = f_{x\,(100d)} - mS_{xa} = 0.12 - 0.06 = 0.06$.

As to the value of f_{ns}, the COD to SO_4 ratio of the wastewater is 600/36 = 16.7, which is very high so that, especially at high sludge age, it is perfectly justified to assume that the conversion of sulphate to sulphide will be complete. This conversion will yield 36/3 × 2 = 24 mg COD·L^{-1} in the form of (inorganic) sulphide, which amounts to a COD fraction of 40/600 = 0.04. Hence, the true value of f_{ns} = 0.14 – 0.04 = 0.10. Thus the non-biodegradable COD fractions f_{ns} = 0.10 and f_{np} = 0.06 have been established.

EXAMPLE 3.8

Continuing with Example 3.6, if the maximum average sludge concentration of the reactor is 22.5 g TSS·L^{-1} with an organic fraction f_v = 0.67 g VSS·g^{-1} TSS, what would be the liquid retention time of the reactor at a sludge age of 60 days? What would be the required increase of the volume to operate the reactor at a sludge age of R_s = 100 d?

At a sludge age of 60 days the COD load used for sludge production was calculated as 4292 kg·d^{-1} so that the volatile sludge mass is 4292 × 60/1.5 = 171,670 kg VSS and for an average sludge concentration of 0.67 × 22.5 = 15 g VSS·L^{-1}, the volume is calculated at 171,670/15 = 11,445 m^3 and the retention time is $R_h = V_r/Q_i$ 1 = 1445/50,000 × 24 = 5.5 h.

At a long sludge age (100 days), the fraction f_x = 0.125, so the sludge production rate is $ME_v = f_x \cdot MS_{ti}/f_{cv}$ = 0.125 × 30,000/1.5 = 2493 kg VSS·d^{-1}. Therefore the sludge mass is R_sME_v = 249,310 kg VSS. If the organic sludge concentration is 15 g VSS·L^{-1}, the volume is 16,620 m^3 and the retention time R_h increases to 16,620/(50,000/24) = 8.0 h, an increase of 45% in comparison with that of a sludge age of 60 days.

In Example 3.8 it was tacitly assumed that the organic fraction of the anaerobic sludge for both sludge ages was equal. However, this assumption is not completely true. As in aerobic systems, it can be observed that the organic fraction of the sludge tends to decrease at higher sludge age. Municipal sewage contains a certain concentration of mineral solids, and when it is assumed that this fraction, due to its higher density, will be preferentially retained in the reactor, then the inert sludge mass in the reactor will increase proportionally with the sludge age:

$$MX_m = R_s \cdot f_m \cdot MS_{ti} \tag{3.43a}$$

where f_m = mineral fraction in sewage (=S_m/S_{ti}), in mg FSS·mg^{-1} COD.

If f_m and f_x are both known, the organic fraction f_v can also be directly calculated as:

$$f_v = (f_x/f_{cv})/(f_x/f_{cv} + f_m). \tag{3.43b}$$

Interestingly, organic biomass, even when grown fully on organic substrate like acetate (but with all micro and macro nutrients present), will always develop a certain innate inorganic fraction, typically around 2–4% of the cell mass, owing to take-up of inorganic minerals for growth. Lastly, and quantitatively much more important, there may also be extensive precipitation of poorly soluble salts like $CaCO_3$, FeS, $CaHPO_4$ or struvite. For example, in the recycled paper industry the calcium concentration in the wastewater can easily reach values up to 800–1000 $mg \cdot L^{-1}$, and it is not uncommon – at least when no preventive or curative measures are taken – to encounter completely calcified sludge in the anaerobic reactor, with TSS concentrations >500 g $TSS \cdot L^{-1}$ and with an organic fraction of less than 20%. Such precipitation is rarely encountered in most municipal sewage, and the organic fraction of the UASB sludge usually in the ranges between 55–70% (Table 3.3).

EXAMPLE 3.9

Assume that for the wastewater in Example 3.6 the mineral solids concentration was equal to 28.5 mg $FSS \cdot L^{-1}$. What according to Eq. (3.43) is the expected organic fraction of the sludge at a sludge age of 60 respectively 100 days?

Solution

The mineral fraction can be calculated as $f_m = 28.5/600 = 0.0475$ mg $FSS \cdot mg^{-1}$ COD, so that the mineral sludge mass at a sludge age of 60 days can be estimated as:

$$MX_m = R_s \cdot f_m \cdot MS_{ti} = 60 \times 0.0475 \times 30{,}000 = 85{,}500 \text{ kg FSS}$$

Using the organic sludge mass calculated in Example 3.8, the organic fraction is calculated as $f_v = MX_v/(MX_v + MX_m) = 171{,}670/257{,}180 = 66.8\%$, as adopted in Example 3.8. At a longer sludge age the organic fraction tends to decrease slightly. For $R_s = 100$ days one has:

$$MX_m = R_s \cdot f_m \cdot MS_{ti} = 100 \times 0.0475 \times 30{,}000 = 142{,}500 \text{ kg FSS}$$

From Example 3.8, $f_v = 249{,}300/391{,}800 = 63.6\%$, which is indeed lower than the organic fraction at 60 days. This is not a large difference, which can be explained by the fact that the difference in specific organic sludge production at 60 and 100 days is also very small.

Equation (3.43b) could also have been used for a direct calculation of the organic fraction of the biomass: for example for $R_s = 60$ d:

$$f_v = (f_x/f_{cv})/(f_x/f_{cv} + f_m) = (0.143/1.5)/((0.143/1.5) + 0.0475) = 66.8\%.$$

EXAMPLE 3.10

For the conditions of Example 3.8, estimate the fraction of the anaerobic sludge production that is lost with the effluent, for a reactor liquid height of 5.0 m, at sludge ages of 60 and 100 days respectively. Use Eqs. (3.10) and (3.11) and compare the results obtained.

Solution

Using Eq. (3.10) and considering retention times of 5.5 and 8.0 h for sludge ages of 60 and 100 d respectively, the fraction of sludge that ends up in the effluent is estimated at $1.33/R_h = 24.2\%$ and 16.6% respectively, representing 0.242 × 0.143 = 3.5% and 0.166 × 0.125 = 2.1% of the influent COD. Hence for an influent COD of 600 mg·L^{-1}, as in Examples 3.6 to 3.9, the solids in the effluent represent 21 mg COD·L^{-1} at 60 days and 13 mg COD·L^{-1} at 100 days sludge age, which is very low.

For a liquid reactor height of 5.0 m, the upflow velocity is $v_u = H_{liq}/R_h$ as 0.91 m·h^{-1} for a sludge age of 60 d, so that the expected effluent sludge concentration can be calculated with the aid of Eq. (3.11) as $X_{te} = 70v_u + 10 = 70 \times 0.91 + 10 = 74$ mg·L^{-1}. Hence for a flow of 50,000 m^3·d^{-1}, the sludge mass present in the effluent is 3685 kg TSS·d^{-1}. Knowing from Example 3.9 that the daily excess sludge production $ME_t = 257{,}180/60 = 4286$ kg TSS·d^{-1}, this represents a fraction of 1 − 3685/4286 = 86%.

For a sludge age of 100 d, the upflow velocity is 0.63 m·h^{-1}, which results in 54 mg·L^{-1} of TSS in the effluent and a solids loss that is considerably lower (2700 kg TSS·d^{-1}), i.e. the sludge fraction that leaves the reactors with the effluent decreases to 2700/3920 = 69%

While these values may seem high, it should be considered that the UASB reactor used in the experimental study demonstrated excellent performance, owing to the enhanced settler. In Example 3.6 a COD removal efficiency of 83.1% and a digestion efficiency of 68.1% were calculated, both of which are at least 10% higher than observed in full-scale systems (Heffernan *et al.*, 2010a). In other words, in these full-scale systems the expected excess sludge production would be significantly higher, and the proportion lost with the effluent would be concomitantly reduced.

In any case, from this example it is apparent that Eqs. (3.10) and (3.11), give very different results, indicating that depending on the quality of the construction and the operation of the UASB reactor the results can be very different.

Returning to Figure 3.27, it shows that, depending on the sludge age, the fraction of the influent COD that accumulates as biodegradable particulate solids in the UASB reactor may be significant and even predominant, especially when the temperature is low. These solids are undesirable, first because the objective of the treatment is to transform biodegradable material into biogas and second because the sludge quality deteriorates as the biodegradable fraction increases and the sludge tends to become unstable and even putrescible. Therefore it is convenient to establish a limit for the anaerobic sludge age of a UASB reactor treating sewage, linked to the sludge composition and more specifically to the fraction of biodegradable material in the sludge. In other words, along with of having a minimum sludge age for methanogenesis, as usually is suggested, there would also be a minimum sludge age for hydrolysis, recognizing that this process is limiting for the entire anaerobic digestion process and eventually leads to a deterioration of the effluent quality (unintentional discharge of sludge) and the sludge production. As discussed previously in this section, the fractions of the feed COD ending up as volatile sludge in the anaerobic reactor is composed of three components:

(1) inert sludge from flocculation of non-biodegradable particulate COD present in the influent (f_{np});
(2) biological sludge generated by anabolic activity (mS_{xa}), which can be calculated with Eq. (3.19b);
(3) biodegradable (particulate) material that is not yet hydrolysed (f_{bp}).

Thus the fraction of the COD present as residual biodegradable material in the sludge can be expressed as $f_{bp} = f_x - mS_{xa} - f_{np}$, and subsequently the biodegradable fraction in the anaerobic sludge can be calculated as $f_{bp}/(f_{bp} + mS_{xa} + f_{np})$.

In Section 3.4.2 the EPA guideline for a stable sludge was already discussed: to qualify as stable, the residual digestion of solids from the sludge should not be more than 20% of the initial organic sludge mass. This value can be compared with that of a similar parameter used to qualify the performance of the aerobic digestion of primary and/or secondary sludge f_{ae}, which is defined as the active sludge fraction remaining after aerobic digestion (Van Haandel & Van der Lubbe, 2012). The maximum limit of f_{ae} is 0.1 g VSS·g^{-1} VSS for a well stabilized sludge. While the value of the EPA guideline is higher, it should be considered that the organic fraction of the UASB excess sludge will be lower than that of the excess sludge of a comparable

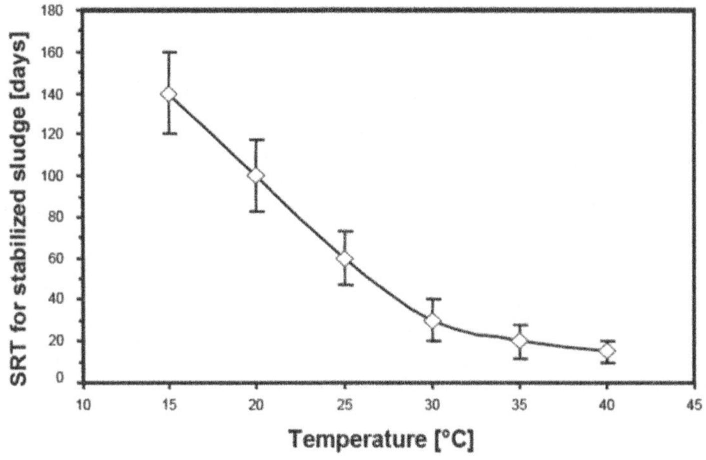

Figure 3.29 Guidelines for the minimum anaerobic sludge age to obtain a stabilized UASB sludge as function of the reactor temperature (adapted from Van Lier *et al.*, 2010).

activated sludge system. Also, aerobic active sludge decays more rapidly than anaerobic sludge and is in itself the source of instability, as in aerobic sludge the residual biodegradable influent COD concentration is normally very low. In contrast, in anaerobic systems the active sludge decays very slowly and the instability is mainly due to the presence of residual biodegradable particulate material. In our experiments at 25°C and at sludge ages above 60 days, the excess sludge indeed appeared to be stable (no floating solids nor generation of bad smells), when the calculated biodegradable fraction in the sludge is 20% or less. Under the conditions of Figure 3.27, the minimum required sludge age to obtain a stabilized sludge is about 65 days at 25°C.

The value of the minimum required sludge age will increase with decreasing temperature, as the hydrolysis process becomes increasingly slow. Figure 3.29 presents the guidelines recommended by Van Lier *et al.* (2010) with respect to the minimum sludge age required to produce a stabilized sludge, as function of the temperature. It can be observed that at 25°C the recommended sludge age is 60 days, which corresponds very well indeed to the value that was deduced from Figure 3.27.

3.9.3 Temperature influence on the empirical model

It has already been noted that the temperature is the dominant factor that determines the feasibility of the application of anaerobic digestion, resulting in limited possibilities outside the tropical and subtropical regions. In this section an experimental study is presented, describing the influence of temperature on the performance of UASB reactors by treating sewage in the range of 12–25°C (Santos *et al.*, 2016). From the data, empirical equations of the COD removal efficiency and the sludge production as a function of both temperature and sludge age are presented.

The experimental study on the temperature effect was carried out at pilot scale with four UASB-type reactors with a useful volume of 25 L, fed with raw sewage. The four reactors were operated at different temperatures: 25°C (ambient temperature), 20°C, 15°C and 12°C. Figure 3.31 shows a schematic representation of the process set-up of the reactors. Three reactors were placed in water tanks with temperature control, while the fourth reactor was operated at the environmental temperature of the lab. The influent of the reactors was fed with dosing pumps to maintain the desired COD load and retention time. The reactors were operated without intentional sludge discharge and the effluent was collected in

EXAMPLE 3.11

For the conditions of Examples 3.6 to 3.10, estimate the fraction of biodegradable organic material in the effluent and in the produced sludge production for sludge ages of 60 and 100 d.

Solution

In Figure 3.30 the composition of organic material in both the effluent and in the sludge are shown as functions of the sludge age, using the equations of Example 3.6. The ratio between the biodegradable and the total organic material in the effluent (non-settleable) is also indicated ($=f_{bs}/(f_{bs}+f_{ns})$). At a sludge age of 60 d the biodegradable fraction in the effluent is small, about 10% but at shorter sludge ages it gradually increases. In so far as the sludge composition is concerned, the biodegradable fraction (i.e. the fraction of flocculated biodegradable particulate influent material) in the anaerobic sludge is considerable, even at long sludge ages: at 60 days a biodegradable fraction f_{bp} of 20% is calculated. Therefore in this case that would be the minimum sludge age that should be applied.

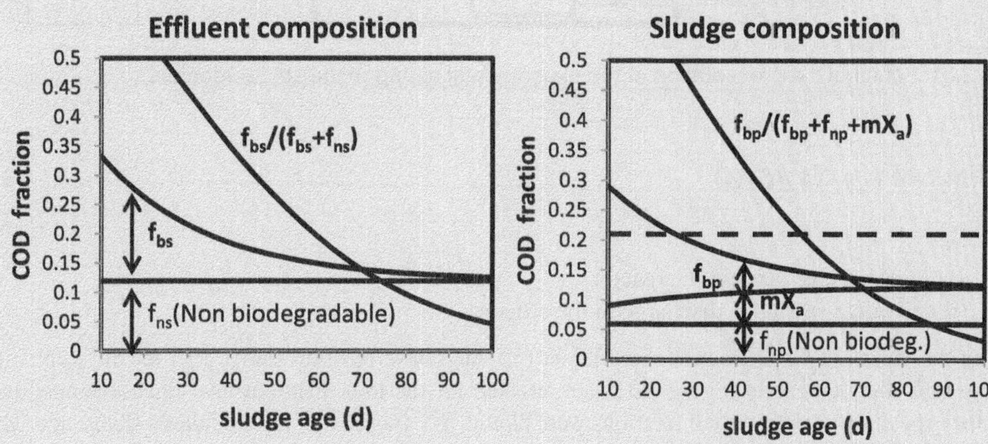

Figure 3.30 Residual biodegradable fractions in the effluent (left) and in the produced anaerobic sludge (right) as functions of the sludge age.

tanks for raw and settled COD analyses, allowing the determination of the non-settleable and settleable COD fractions in the effluent. Periodically the contents of the reactors were discharged and mixed after which the sludge characteristics were determined: average TSS concentration, organic fraction as well as biological (SMA and stability) and mechanical (settleability) properties. These tests were performed both on the sludge kept inside the reactor as well as the sludge separated from the effluent. Biogas production was determined in Mariotti flasks.

For each of the temperatures a range of experimental retention times was defined and at each retention time the reactors were operated at the required flow rate until steady-state performance was obtained, without any additional sludge discharge from the reactor. When a steady state was reached, the collected COD data was used to calculate the fractions of feed COD present in the settled effluent (f_s), transformed into sludge (f_x), and digested (f_d). The sludge age was estimated from the organic sludge discharge in the effluent and the organic sludge mass in the reactors:

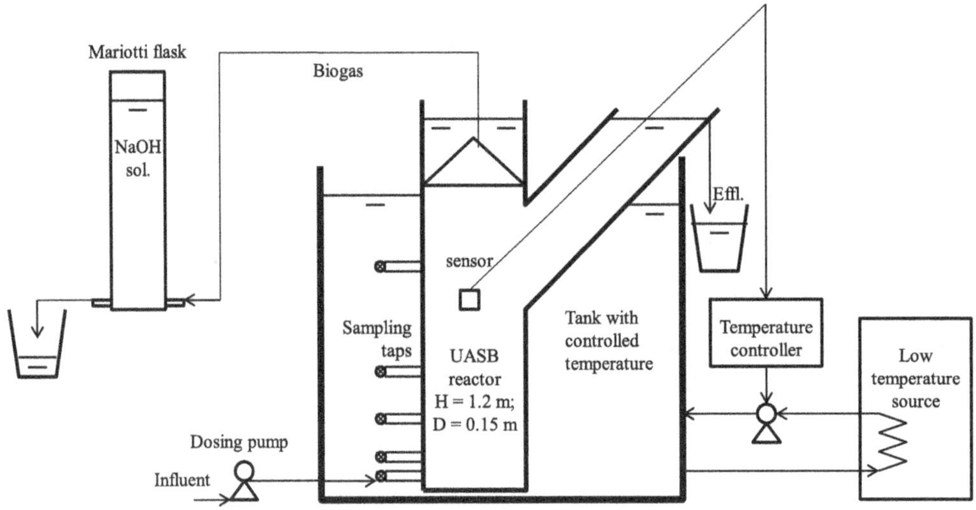

Figure 3.31 Schematic representation of the experimental set-up of the UASB reactors.

$$R_s = MX_v/MX_{ve} = X_v V_r/(Q_i X_{ve}) \quad (3.23)$$

where

MX_v = organic sludge mass in the reactor
MX_{ve} = daily organic sludge discharge in the effluent

Using the values of COD fractions in effluent (f_s) and converted into sludge (f_x) for different values of sludge age, diagrams as in Figure 3.32 were created for the four different investigated temperatures, where the experimental determined fractions were plotted as a function of the calculated sludge age. These diagrams were used to establish empirical expressions to describe the f_s and f_x fractions as a function of the two variables: temperature and sludge age. The following expressions were able to describe the fractions f_s and f_x well as functions of the sludge age and for different temperatures (based at a reference temperature of 35°C):

$$f_s = f_{ns} + f_{bs} = f_{ns35} \times 1.037^{(35-T)} + 0.20 \times 1.011^{(35-T)} \exp[-0.04\,(R_s - 3.8 \times 1.06^{(35-T)})] \quad (3.24)$$

$$f_x = f_{np} + f_{bp} = f_{np35} \times 1.015^{(35-T)} + 0.16 \times 1.014^{(35-T)} \exp[-0.04\,(R_s - 3.8 \times 1.06^{(35-T)})] \quad (3.25)$$

where

f_{ns} = non-biodegradable soluble COD fraction in influent
f_{bs} = soluble degradable COD fraction in the anaerobic effluent
f_{np} = non-biodegradable particulate COD fraction
f_{bp} = particulate degradable COD fraction in the anaerobic effluent (the sludge)

In Figure 3.32 and in Eqs. (3.24) and (3.25) the non-biodegradable COD fractions depend on the temperature, whereas the residual degradable COD fractions depend on both the temperature and the

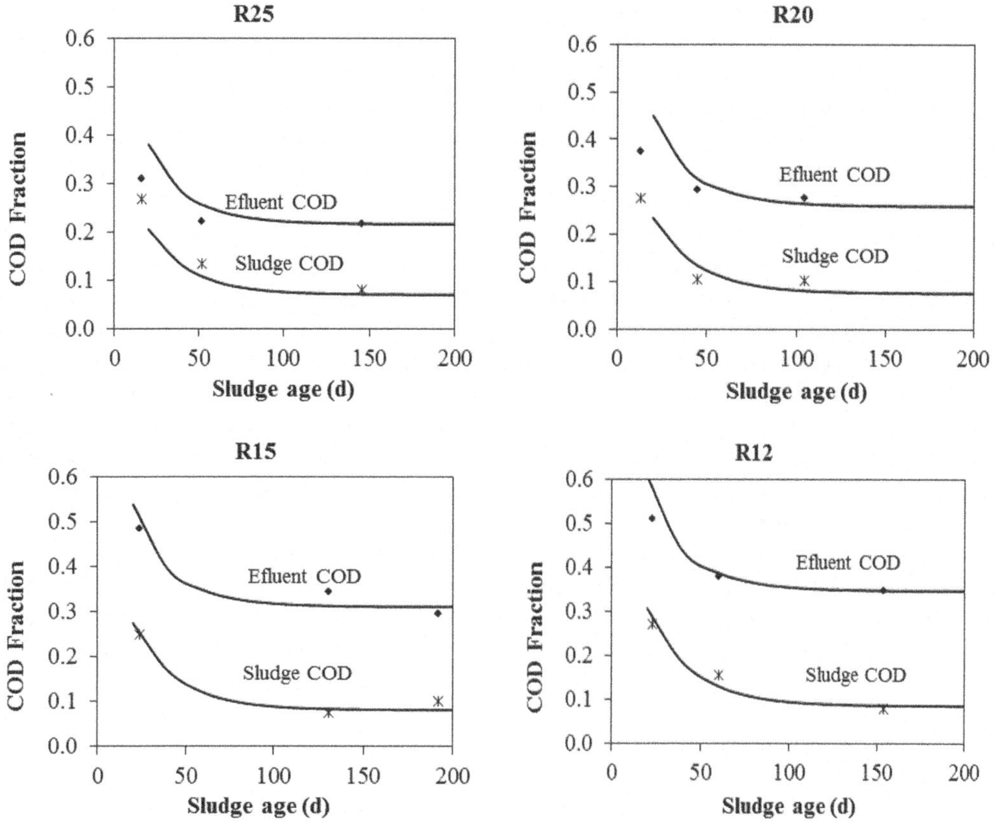

Figure 3.32 Experimental values of f_s and f_x as functions of the sludge age 25°C and 20°C (top left and right) and for 15°C and 12°C (bottom left and right).

sludge age. The temperature dependency of the non-biodegradable fractions has been observed by several authors (for example O'Rourke (1968), see Figure 2.13).

For a long sludge age the equations Eqs. (3.24) and (3.25) simplify to:

$$f_{ns} = f_{ns35} \times 1.037^{(35-T)} \quad (3.26)$$

$$f_{np} = f_{np35} \times 1.015^{(35-T)} \quad (3.27)$$

As can be observed in Figure 3.32, Eqs. (3.24) and (3.25) describe reasonably well the COD fractions in the effluent and in the sludge for the range of temperatures and sludge ages that were investigated and therefore will also describe the digested COD fraction. It is interesting to note that the non-biodegradable fractions determined in this research study (with $f_{ns} = 0.23$ and $f_{np} = 0.08$) are very different from the results obtained earlier, discussed in Section 3.8.1 (with $f_{ns} = 0.14$ and $f_{np} = 0.12$), even though in both studies the same wastewater was used, i.e. the municipal sewage of the city of Campina Grande, Brazil. This can be explained by the fact that in the period between the two studies (four years), there was a serious deterioration of the sewer network resulting in the loss of a large fraction of the sewage (more than 80%) to the rainwater drainage network. As a result the sewage flow rate became very low and extensive digestion already occurred in the sewer network. This had significant impact on the sewage characteristics: the COD concentration decreased from a previous average value of 800 $mg \cdot L^{-1}$ when the network functioned well to an actual average value of about 400 $mg \cdot L^{-1}$. For the same reason the values of the non-degradable COD fractions changed: the inert soluble COD fraction increased (almost doubled), because a large fraction of the biodegradable COD was removed, while the non-biodegradable soluble COD concentration remained constant. On the other hand, the non-biodegradable particulate fraction decreased, presumably because of retention (settling) in the sewer network due to the low flow.

The fact that the state of maintenance of the sewage system has an impact on important characteristics of the wastewater leads to the conclusion that the model to describe anaerobic digestion of sewage in a UASB reactor will only predict the performance correctly if the sewage characteristics are known, particularly the numerical values of the non-biodegradable COD fractions in the influent. Therefore, ideally the sewage composition should be established before building the sewage treatment system. However, in practice one is generally very far away from ideality: the sewage treatment plant often will be designed in a situation where the sewage to be treated does not even exist. In those cases it is necessary to resort to information from treatment systems that operate under comparable conditions. If no information is available at all, a conservative estimate would be $f_{ns35} = f_{np35} = 0.10$.

In the experiment, the digested COD fraction measured in the Mariotti flasks was smaller than the fraction calculated from Eq. (3.17). The difference could not be explained by the fact that part of the methane remained dissolved in the effluent, although the dissolved fraction is considerable for low temperatures and high flow rates (i.e. at low retention times). It is concluded that there must have been methane leakage from the system. In Figure 3.33 the calculated values of the digested fraction f_d are plotted against the sludge age for different temperatures.

The values of f_d are almost constant for sludge ages above 100 days, but are consistently lower than this maximum value at lower temperatures. The reason for this is that at such long sludge age the digestion efficiency is near its maximum for all temperatures. The maximum digestion efficiency is heavily dependent on the reactor temperature: at 25°C, the maximum digestion efficiency was about 70%, while at 12°C it was only marginally above 50%. This is consistent with earlier findings by O'Rourke as shown in Figure 2.13.

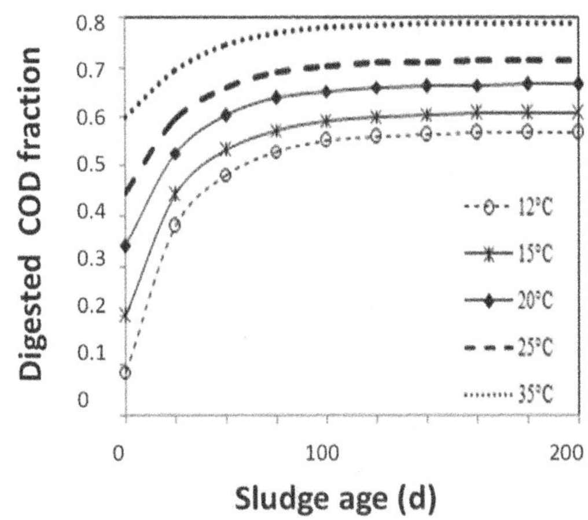

Figure 3.33 COD fraction converted into methane as function of the sludge age for different temperatures.

Even under favourable conditions (long sludge age, high temperature), the methane production potential that can be expected from municipal sewage is small and its use as fuel is only feasible for large systems, for example with more than 500.000 inhabitant equivalents.

The average reactor sludge concentration during the experiment was $X_v = 17$ g VSS·L^{-1} with a total organic sludge mass MX_v of 423 g VSS. These values were little influenced by either temperature or sludge age. Therefore the production of organic (and total) excess sludge can be expressed by Eqs. (3.28a) and (3.28b):

$$ME_v = f_x MS_{ti}/f_{cv} = f_x Q_i S_{ti}/f_{cv} \quad (3.28a)$$

$$ME_t = f_x MS_{ti}/(f_{cv} f_v) = f_x Q_i S_{ti}/(f_{cv} f_v) \quad (3.28b)$$

where ME_v, ME_t are organic and total excess sludge production (g VSS·d^{-1} and g TSS·d^{-1}).

The sludge age can be expressed as the ratio between the mass of sludge and the rate of production of sludge:

$$R_s = MX_v/ME_v \ (=MX_t/ME_t) \text{ or} \quad (3.29a)$$

$$MX_t = R_s ME_t \ (\text{and } MX_v = R_s ME_v) \quad (3.29b)$$

Figure 3.33 also demonstrates that it is not necessary to operate a UASB reactor at a sludge age higher than 100 days, because at such a sludge age the efficiency of the digestion is already close to the maximum for all temperatures. In order to compare our results, which are based on sludge age, with the Brazilian guidelines, which are based on hydraulics, the following steps must be followed. First, for the maximum performance of the anaerobic digestion process, the organic sludge mass that develops can be calculated. For a sludge age of 100 days this yields:

$$MX_v = V_r X_v = R_s ME_v = 100ME_{v100} \quad (3.30)$$

This can be rewritten to define the required liquid reactor volume as function of selected sludge age and the predicted excess sludge production.

$$V_r = R_s ME_v/X_v = 100ME_{v100} \quad (3.31a)$$

$$= R_s ME_t/X_t = 100ME_{t100} \quad (3.31b)$$

Finally, the hydraulic retention time required for essentially complete digestion, at a sludge age of 100 days, is defined as:

$$R_h = V_r/Q_i = R_s f_x S_{ti}/(f_{cv} X_v) = 100 f_{x100} S_{ti}/(1.5X_v) \quad (3.32a)$$

$$= R_s f_x S_{ti}/(f_{cv} f_v X_t) = 100 f_{x100} S_{ti}/(1.5 f_v X_t) \quad (3.32b)$$

where f_{x100} = fraction of feed COD converted into sludge at a sludge age of 100 days.

Figure 3.34 shows the hydraulic retention time required as a function of temperature for maximum digestion efficiency and based on the results obtained in the research study, i.e. $R_s = 100$ days, for values of $S_{ti} = 0.8$ g·L^{-1} and $X_v = 17$ g·L^{-1}. The values of the COD and sludge concentrations used were the average values obtained during the experimental study and may be different for other conditions. Particularly, variation of the non-biodegradable fractions of the raw sewage can have a significant impact on the performance of the reactors. The calculated value of the hydraulic retention time required for maximum efficiency of the UASB reactor can be compared with the value recommended by the Brazilian Standard, NBR 12.209 (ABNT, 2011).

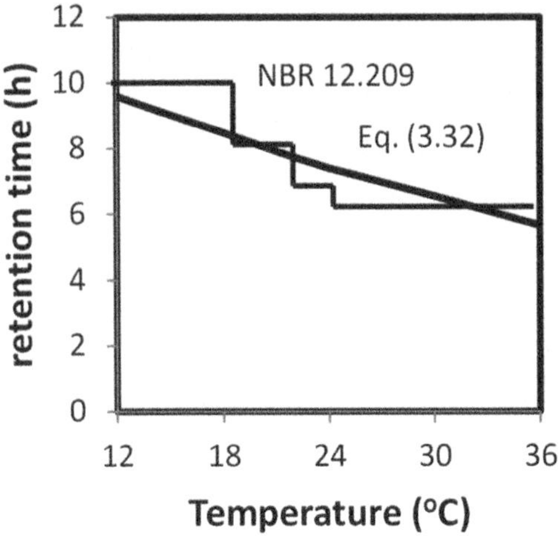

Figure 3.34 Required retention time as per Eq. (3.32) for maximum digestion efficiency (at a sludge age of 100 days) as a function of temperature, compared to NBR 12.209/2011.

This standard specifies the required retention time as a function of the temperature and was established based on the observed organic material removal efficiency in full-scale UASB systems operating in Brazil. It is concluded that there is a good correlation between the required retention time for maximum digestion efficiency calculated from the data presented in this section and the values recommended by NBR 12.209/2011. The Brazilian standard does not explicitly consider the stability of the produced anaerobic excess sludge. However, it can be observed in Figure 3.29 that a sludge age of 100 days is sufficiently high for sewage temperatures ≥20°C. Only in the south of Brazil will the lowest monthly sewage temperature be lower (15–18°C), but this only occurs a few months per year.

In Figure 3.35 the maximum digestion efficiency, calculated from Equation (3.17), is plotted as function of the temperature at a sludge age of 100 days. Thus, to analyse the feasibility of anaerobic treatment of

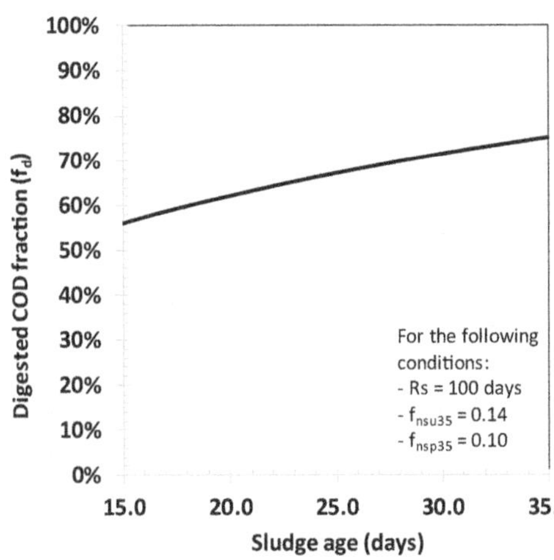

Figure 3.35 Maximum digestion efficiency as a function of the temperature for a sludge age of 100 days.

sewage in a specific design situation, it must be assessed whether it is economically feasible to construct a UASB reactor with enough volume to maintain a sludge age that allows the maximum digestion efficiency (at $R_s = 100$ days) to be reached for the temperature of the influent in question. For this, the digested influent COD fraction as well as the fractions that will be discharged in the effluent or converted into sludge (Eqs. (3.24) and (3.25)) must be calculated. Consideration should be given to the need for additional (aerobic) treatment, especially if the temperature is low.

In the experimental study the settleability and specific methanogenic activity (SMA) were also determined, both for the reactor sludge and the solids discharged with the effluent. As to sludge settleability Santos *et al.* (2016) concluded that:

- There is selective discharge of the sludge from the UASB reactor: the settleability of the discharged sludge is much lower than that of the sludge remaining in the reactors.
- The sludge settleability tends to improve slightly with temperature.
- There is no clear trend of the value of the sludge settleability as a function of the retention time or sludge age.

As to the specific methanogenic activity, the data showed that:

- At lower temperature the value of the SMA decreased.
- Both the retention time and sludge age have no clear effect on the SMA value.
- The SMA of the sludge in the reactors is approximately equal to that of the sludge discharged.

It is important to note that in practice UASB reactors are not operated as in the experiment described above, where no intentional sludge discharge was applied. Usually periodic sludge discharges will be applied to reduce the discharge of solids in the effluent. Depending on the liquid retention time, the settleable COD can make up an important part of the total effluent COD as discussed in Section 3.9.1, but its presence cannot be avoided unless an external settler is applied to treat the effluent from the UASB reactor.

3.10 PROCESS DESIGN PROCEDURE USING THE ANAEROBIC SLUDGE AGE

Equations (3.15) to (3.17) and their temperature-dependent counterparts Eqs. (3.24) and (3.25) clearly show that the efficiency of anaerobic digestion of organic material in sewage increases with the sludge age. Therefore, at least in principle, it is advantageous to operate the UASB reactor at the maximum sludge age. Unfortunately, knowledge about UASB reactor designs is still insufficient to make an *a priori* estimate of the exact maximum sludge age that can be maintained in a UASB reactor for a specific phase separator design and for the applied operating conditions. The retained sludge mass depends not only on the design of the phase separator, but also on the mechanical properties of the sludge that develops, particularly its settling characteristics. These properties depend on the operating conditions in the reactor, as well as the influent. On a practical level however, based on experience from full-scale reactors, which will be discussed in Section 4.3.1, it is certainly possible to provide an initial estimate of the maximum sludge concentration that can be maintained in a UASB reactor. A typical value of 20–25 g $TSS \cdot L^{-1}$ may be assumed, under the condition that the digestion zone comprises at least 60% of the liquid reactor volume (refer to Section 4.3.1.3). Hence, once the required anaerobic sludge mass is known, the required reactor volume can be calculated according to a relatively simple iterative procedure:

(1) Select an initial design value for the anaerobic sludge age, based on either the requirements for a certain COD removal efficiency or to produce a stabilized sludge (for example using Figure 3.34 to obtain an initial estimate for the required value of the sludge age).

(2) Calculate the value of f_s with Eq. (3.24), which allows calculation of the soluble COD load in the effluent.

$$f_s = f_{ns} + f_{bs} = f_{ns35} 1.037^{(35-T)} + 0.20 \times 1.011^{(35-T)} \exp[-0.04(R_s - 3.8 \times 1.06^{(35-T)})]$$

The non-biodegradable soluble fraction f_{ns35} ideally must be determined experimentally, which is very laborious work. If wastewater is not available, use typical values from other projects, or as a last resort use a conservative estimate of $f_{ns35} = 0.1$. The design temperature is the lowest monthly average sewage temperature in the region.

(3) Calculate the value of f_x with Eq. (3.25):

$$f_x = f_{np} + f_{bp} = f_{np35}\, 1.015^{(35-T)} + 0.16 \times 1.014^{(35-T)} \exp[-0.04(R_s - 3.8 \times 1.06^{(35-T)})]$$

The fraction f_{np} is composed of two parts: the inert particulate COD fraction which only depends on the temperature, but not on the sludge age, and the active sludge fraction, which is dependent on both.

(4) Calculate the value of f_d with Eq. (3.17):

$$f_d = 1 - f_s - f_x$$

(5) Calculate the maximum active fraction from Eq. (3.22) assuming complete removal of biodegradable material ($R_s > 100$ d) with $Y_{an} = 0.05$ mg VSS·mg^{-1} COD as a typical value. Proceed to calculate the inert fraction (Eq. 3.20), which is valid for any sludge age and finally the biodegradable sludge fraction:

$$mS_{xa} = f_{cv}\, Y_{an} f_d/(1 - f_{cv}\, Y_{an}) \quad (3.19b)$$

$$mS_{xi} = f_{np} \quad (3.20)$$

$$f_{dm} = 1 - f_{ns} - f_{np} \quad (3.21)$$

$$\begin{aligned} f_{bx} &= mS_{xam} + mS_{xi} = f_{cv}Y_{an} f_{dm}/(1 - f_{cv}Y_{an}) + f_{np} \\ &= f_{cv}Y_{an}(1 - f_{ns} - f_{np})/(1 - f_{cv}Y_{an}) + f_{np} \end{aligned} \quad (3.22)$$

(6) Calculate the fraction of the COD that will end up as residual biodegradable material in the sludge as $f_{bp} = f_x - mS_{xa} - f_{np}$. Subsequently the biodegradable fraction in the anaerobic sludge can be calculated as: $f_{bp}/(f_{bp} + mS_{xa} + f_{np})$. This value should be limited to 20%. If it is higher, it is recommended to increase the sludge age.

(7) Use Eq. (3.32a) or (3.32b) to calculate the retention time and the reactor volume

$$R_h = V_r/Q_i = R_s f_x\, S_{ti}/(f_{cv}\, X_v) = R_s f_x\, S_{ti}/(f_{cv} f_v\, X_t) \quad (3.32a)$$

To calculate the retention time an estimate of the average reactor sludge concentration is indispensable. This concentration is typically around 20–25 g TSS·L^{-1}, with a typical organic sludge fraction of $f_v = 0.6$–0.7.

Equation (3.43b) shows that the organic fraction can also be calculated as a function of the sludge age if the ratio between mineral solids and influent COD (f_m) is known:

$$f_v = (f_x/f_{cv})/(f_x/f_{vc} + f_m) \quad (3.43b)$$

(8) Use either Eq. (3.10) or Eq. (3.11) to estimate the sludge concentration in the effluent and the sludge mass to be discharged as excess sludge:

$$\begin{aligned} f_{di} &= 1.33/R_h \text{ for GLS separators with plate-pack} \\ &= 2.7/R_h \text{ for GLS separators without plate-pack} \end{aligned} \tag{3.10}$$

$$X_{te} = 70v_u + 10 \text{ (for } 0.4 < v_u < 1.0 \text{ m}\cdot\text{h}^{-1}) \tag{3.11}$$

The latter case requires selection of a design value for the upflow velocity v_u. As the hydraulic retention time has already been calculated in step (7), the reactor liquid height is calculated as $R_h v_u$. The liquid reactor height should be between 4–6 m. If it is outside this range, adapt the selected upflow velocity

(9) Based on the value for f_{di} or X_{te} from step (8), calculate the expected particulate COD concentration that will end up in the effluent as per Example 3.10

$f_{xe} = f_x f_{di}$ or

$f_{xe} = Q_i f_{cv} f_v X_{te}/MS_{ti}$

(10) Calculate the COD removal efficiency. Do not only consider the fraction f_s, but also f_{xe} to account for the unintentional sludge discharge in the effluent:

$\eta_{COD} = 1 - f_s - f_{xe}$

If the COD removal efficiency is too low, it can be considered to increase the sludge age. However, above a sludge age of 100 days there will be little benefit from additional digestion, and this applies to all temperatures as demonstrated in Figure 3.35.

(11) By trial and error decide the optimal sludge age, pondering the advantages and disadvantages:
   (a) the main advantages of a long sludge age are the following: low effluent BOD and COD concentrations, high biogas production, low sludge production, stabilized sludge, good process stability;
   (b) the disadvantages of a long sludge age are mainly: high cost (large reactor volume) and large area requirement.

(12) With the value of f_x and the influent COD load, MS_{ti}, the daily total excess sludge production ME_t can be calculated, using $f_{cv} = 1.5$ g COD$\cdot$g^{-1} VSS) and the value for f_v calculated or adopted in step (7):

$$ME_t = f_x MS_{ti}/(f_{cv} f_v) \tag{3.28b}$$

(13) Calculate the total sludge mass that will need to be accommodated in the UASB:

$$MX_t = R_s ME_t \tag{3.29b}$$

The value of MX_t, the total anaerobic sludge mass, is a very important input in the reactor design procedure that will be discussed in Sections 4.3 and 4.4, as it defines the volume of the digestion zone

(14) Finalize the process design calculations (calculate methane production, effluent quality, required sludge disposal, etc.)

EXAMPLE 3.12

Develop the design for optimal anaerobic digestion for 50,000 $m^3 \cdot d^{-1}$ of sewage containing 800 mg $COD \cdot L^{-1}$. The monthly minimum and maximum sewage are 15 and 25°C respectively. The envisaged post-treatment system consists of polishing ponds, so that the BOD and settleable solids concentrations in the effluent of the UASB reactor are not important, as they will be easily removed in the post-treatment (see Chapter 9).

Adopt the following values for the non-biodegradable fractions: $f_{ns35} = f_{np35} = 0.08$. Hence, using the temperature dependency presented in Eqs. (3.26) and (3.27) the following adjusted values for f_{ns} and f_{np} are calculated: $f_{ns25} = 0.12$, $f_{np25} = 0.10$; $f_{ns15} = 0.17$ and $f_{np15} = 0.12$.

Solution

Equations (3.23) to (3.25) can be used to calculate the effluent, sludge and digested COD fractions f_s, f_x and f_d for both temperatures as functions of the sludge age. The results are shown in Figure 3.36, where it can be observed that the influence of temperature is not very pronounced, but the sludge production is somewhat greater at 15°C, especially at shorter sludge ages. To achieve a minimum COD digestion efficiency of 50%, the sludge age must be at least 35 d at 15°C, but the digestion efficiency gradually increases at higher sludge. At a digestion efficiency of 50%, the methane production is $0.5 \times 40{,}000/4 = 5000$ kg $CH_4 \cdot d^{-1}$ with a maximum electric energy production potential of about $0.2 \times 5000 = 1000$ kW or 1 MW, if all methane produced is collected (which will never be the case). Depending on the required investment costs, the local price of energy and relevant taxes and incentives, it may be attractive to use the methane.

However, when the composition of the sludge is considered (shown in Figure 3.37), then clearly a sludge age of less than about 70 days is inadequate: for shorter sludge ages the biodegradable fraction (non-hydrolysed particulate biodegradable influent COD) constitutes more than 20% of the anaerobic sludge. This biodegradable COD fraction in the sludge is calculated by considering that the sludge is made up of three fractions:

$$f_x = f_{bp} + mS_{xa} + mS_{xi}$$

where

f_{bp} = non-hydrolysed biodegradable particulate influent COD fraction

mS_{xa} = influent COD fraction transformed into bacterial mass ($Y = 0.05$ mg $VSS \cdot mg^{-1}$ COD)

mS_{xi} = influent non-degradable particulate COD fraction ($= f_{np}$)

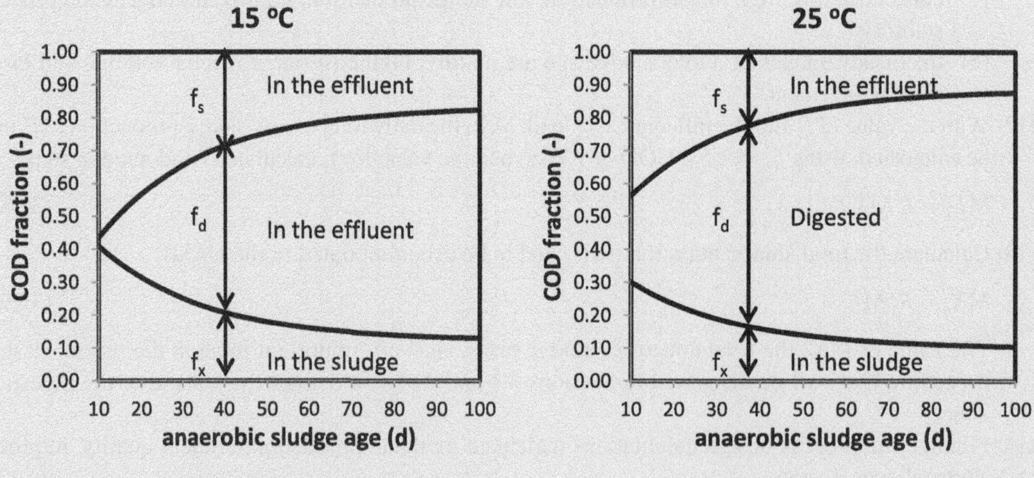

Figure 3.36 Influent COD fraction ending up in the effluent, in the sludge and digested as function of the sludge age at temperatures of 15 and 25°C.

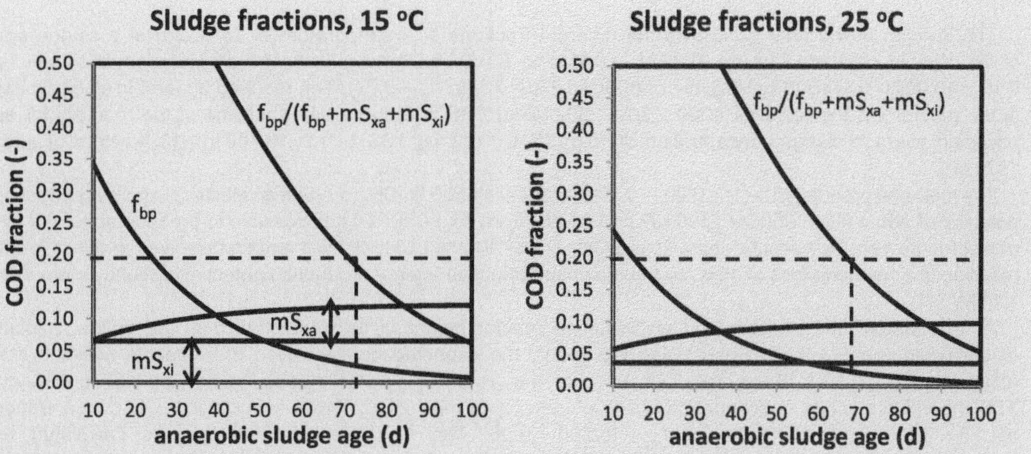

Figure 3.37 Anaerobic sludge composition as function of the sludge age at 15 and 25°C.

Figure 3.37 also shows the maximum allowable value of 20% for the biodegradable fraction in the sludge, which allows the sludge to be classified as "stable".

The next step is to evaluate what the effect of the sludge age is on the main factor in the investment costs: the reactor volume. In Figure 3.38 the hydraulic retention time has been plotted as a function of the anaerobic sludge age for an assumed average sludge concentration in the reactor of 15 g VSS $\cdot$ L^{-1} and for temperatures of 15 and 25°C. At the minimum sludge age for stabilized sludge production (70 d), the required retention time is estimated at almost 9 hours for the minimum design temperature of 15°C. If the system is operated at a sludge age of 100 d, in order to obtain maximum digestion efficiency, the retention time increases to almost 11 hours.

Now, taking into consideration the information presented in Figures 3.36 to 3.38, the hydraulic retention time that would probably be selected corresponds to the minimum sludge age required to produce a stabilized sludge, at the minimum design temperature of 15°C, which is about 70 days (Figure 3.37).

As can be observed in Figure 3.36, the biodegradable COD concentration in the effluent is only a small fraction of the total COD concentration in the raw sewage and poses no difficulty for treatment in polishing ponds.

If a sludge age of 70 days is selected for operation at 15°C, the higher rates of the biological reactions permit a longer sludge age to be applied at 25°C. As can be seen in Figure 3.38, at 25°C the sludge age for a retention time of 9 hrs is about 100 days, so that in fact it becomes possible to attain the maximum digestion efficiency for this temperature.

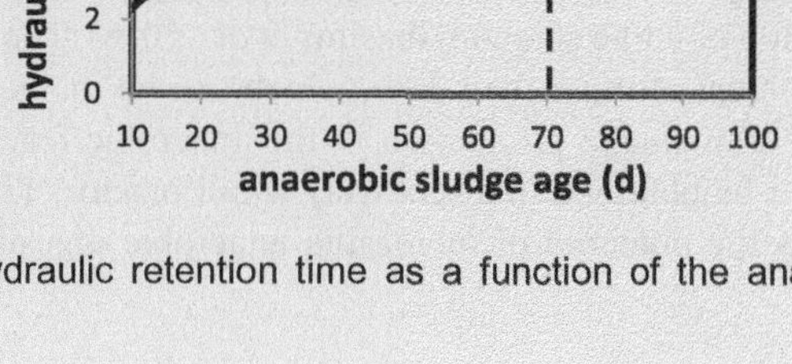

Figure 3.38 Required hydraulic retention time as a function of the anaerobic sludge age at 15 and 25°C.

The division of the feed COD over the different fractions at a temperature of 15°C and at a sludge age of 70 days is calculated as $f_s = 0.20$, $f_x = 0.15$ and $f_d = 0.65$. For $f_x = 0.15$, the sludge production will be $0.15 \times 40{,}000/1.5 = 4000$ kg VSS·d^{-1} or 6000 kg TSS·d^{-1} (for $f_v = 0.67$). The required anaerobic sludge mass in the reactor will therefore be $4000 \times 70 = 280{,}000$ kg VSS and the required volume of the reactors for an assumed average sludge concentration of 15 gVSS·L^{-1} (22.5 g TSS·L^{-1}) is 18,700 m^3 (10 reactors of each 1870 m^3).

The methane production is $(1 - 0.20 - 0.15)\,40{,}000/4 = 6500$ kg CH_4·d^{-1} with an electrical energy production potential of about $0.2 \times 6500 = 1300$ kW or 1.3 MW (from 1 kg CH_4·d^{-1} it is possible to produce about 0.2 kW of electric power). However, at least 20 mg CH_4·L^{-1} or 1000 kg CH_4·d^{-1} will remain dissolved in the effluent, resulting in a methane loss of 15%, so the maximum effective energy production potential is slightly more than 1.1 MW.

At a retention time of 9 hrs and an assumed reactor height of the reactor of 5 m, the upflow velocity of the liquid is $5/9 = 0.56$ m·h^{-1}. Using Eq. (3.11), the expected concentration of settleable solids in the effluent is $70 \times 0.56 + 10 = 49$ mg TSS·L^{-1}. For the assumed volatile sludge fraction of 0.67 g VSS·g^{-1} TSS the volatile solids concentration in the effluent is 32 mg VSS·L^{-1} with associated COD concentration of $1.5 \times 32 = 48$ mg COD·L^{-1}, or $f_{xe} = 48/800 = 0.06$. The COD removal efficiency is calculated as $\eta_{COD} = 1 - f_s - f_{xe} = 1 - 0.20 - 0.06 = 0.74 = 74\%$.

The calculation can be repeated for the more favourable conditions of 25°C, when the reactor can be operated at a sludge age of 100 days, leading to a better quality effluent, a lower sludge production and a higher potential for energy production.

3.11 PRODUCTIVE USE OF THE HEATING POTENTIAL FROM THE BIOGAS

In the previous section the influence of temperature on the efficiency of the anaerobic treatment was quantitatively described: both the soluble non-biodegradable and biodegradable fraction increased at lower temperature, resulting in an increase in the COD fraction present in the effluent. The COD fraction ending up in the sludge also increases, mainly because the hydrolysis rate at low temperature is low. Therefore, the digested COD fraction decreases at lower temperatures. This decrease can be partly compensated by an increase of retention time, which results in an increase in the sludge age. However, as the reactor volume needs to be increased, this measure has a cost, which may become prohibitive. In any case Figure 3.33 shows that there is no point in increasing the sludge age beyond 100 days, as the maximum digestion efficiency has been reached. In this section another option is explored, which can be applied in the case of a low sewage temperature (<18°C): application of the heat that is generated from the combustion of the generated biogas. This energy is not enough to significantly heat the entire sewage flow, but it can be used to increase the temperature of certain part of the total flow. For example, in sewage treatment primary settlers are often used which separate suspended solids from the liquid phase. The settled solids (the primary sludge) can be concentrated by gravity to a volume of less than 1% of the initial sewage volume. In this case, when applying the heat from the combustion of the biogas, a very significant temperature increase of this primary sludge can be achieved.

Instead of using gravity for phase separation of the solids in the influent, a more efficient process may be used such as application of a fine screen (Ghasimi *et al.*, 2016). The produced primary sludge is concentrated with a solids content of more than 20%, which can be incinerated. The screened sewage contains very little solids, so that sludge production in the anaerobic reactor is significantly reduced, and a sludge age of 100 d can be obtained in a relatively small reactor. This concept has not yet been applied in practice, but it has the potential of increasing anaerobic digestion well into regions with a temperate climate.

EXAMPLE 3.13

Estimate the maximum temperature increase that can be achieved using the combustion heat of methane, for an anaerobic digestion efficiency of 50% and an influent COD concentration of 0.8 g·L^{-1}. The combustion heat of methane is 12 kCal·g^{-1} CH_4.

Solution

The concentration of the digested COD is 0.5 × 0.8 = 0.4 g COD·L^{-1} or 0.1 g CH_4·L^{-1}. The solubility of methane in water is around 20 mg CH_4·L^{-1}, so the collected biogas contains 0.1 − 0.02 = 0.08 g CH_4·L^{-1}, which can generate 0.08 × 12 = 1 kCal·L^{-1}. As per definition 1 kCal can increase the temperature of 1 L of water by 1°C, the maximum increase in temperature will be 1°C. In practice the temperature increase will be even smaller because the efficiency of heat transfer is never 100%.

EXAMPLE 3.14

For the same conditions as in Example 3.13, estimate the required heat transfer efficiency to increase the primary sludge temperature from 14 to 54°C (thermophilic conditions), knowing that typically the primary sludge flow is only 1% of the sewage flow.

Solution

The heat generated by anaerobic digestion was estimated in the above example at 1 kCal per L of raw sewage or 1/0.01 = 100 kCal per litre of primary sludge. To increase the temperature of 1 L of primary sludge from 14 to 54°C, the required heat is 54 − 14 = 30 kCal. Therefore under these conditions the required heat transfer efficiency would only have to be 30/100 = 0.3 or 30%. In practice such an efficiency can be easily achieved, as most boilers can reach up to 85% combustion efficiency.

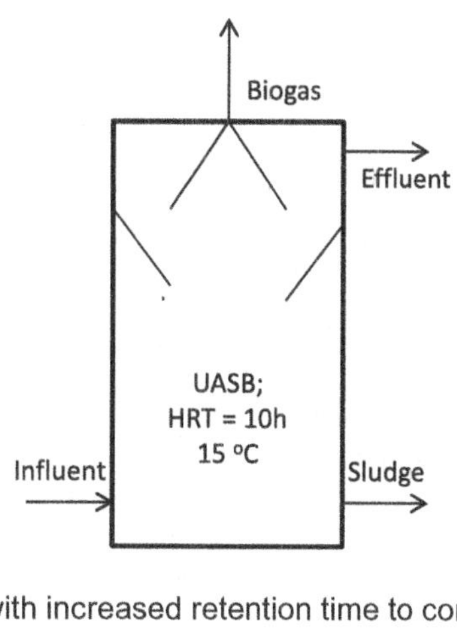

Figure 3.39 Single UASB reactor with increased retention time to compensate for a low temperature.

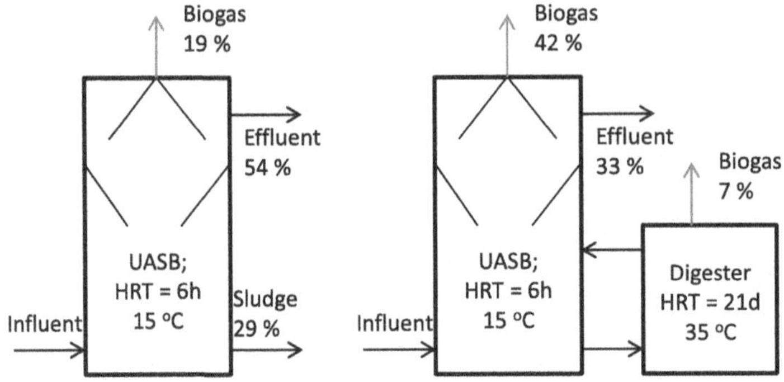

Figure 3.40 UASB reactor with a sludge digester coupled to the reactor (Mahmoud *et al.*, 2004).

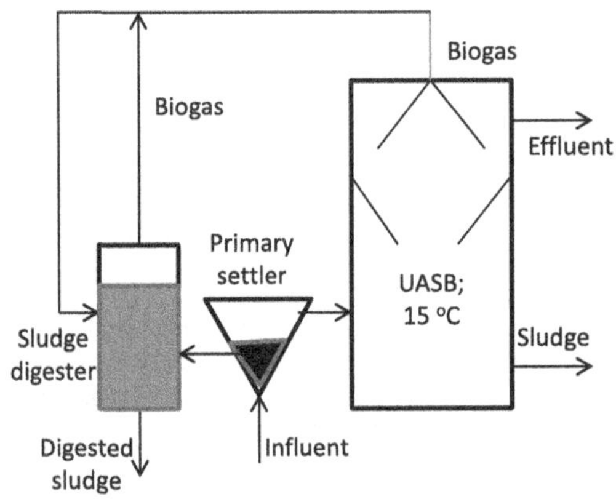

Figure 3.41 Primary settler with sludge digester as pre-treatment unit before a UASB reactor.

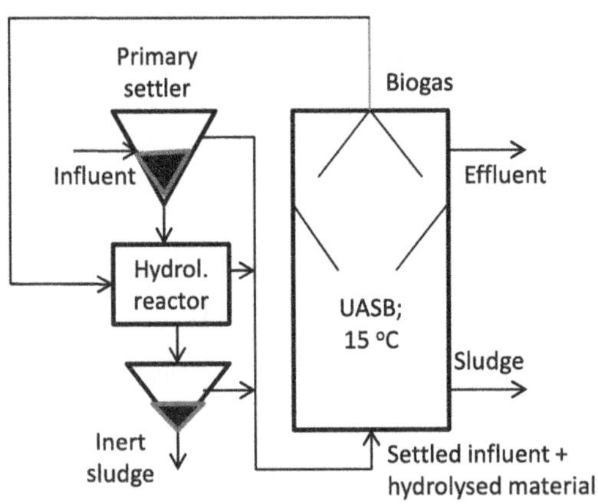

Figure 3.42 Primary settler with hydrolytic reactor as a pre-treatment before a UASB reactor.

In practice various configurations may be used to increase the efficiency of the anaerobic digestion process. Most solutions focus on the acceleration of the hydrolysis of the biodegradable particulate material, which is usually the rate-limiting process in the overall digestion process. In Figure 3.40 to Figure 3.42 some alternatives can be observed for an assumed temperature of 15°C. These alternatives should be compared with the base alternative of simply increasing the reactor volume, shown in Figure 3.39. In the previous section it has been shown that for a sludge age of 100 days the required retention time is about 8–10 hrs to reach a maximum efficiency of 55% at 15°C (Figure 3.36).

Figure 3.40 shows a configuration studied by Mahmoud *et al.* (2004) in Palestine. The performance of a UASB reactor (at 15°C and a retention time of 6 hrs) was compared with the performance of an identical UASB reactor, coupled to a sludge digester (35°C and 21 days SRT), that was fed with the waste sludge extracted from the UASB reactor. The digested sludge was either returned to the UASB reactor to boost treatment capacity (if needed) or discarded. Figure 3.40 shows that the introduction of the digester, taking advantage of the energy in the generated biogas, allowed also for a very significant increase in treatment efficiency, not only of the complete system, but also of the UASB reactor, which implies that the addition of digested sludge to the reactor had a significant impact Figure 3.41 and 3.42 show additional possibilities to use the combustion heat of the produced methane to increase the digestion efficiency at low temperatures.

Chapter 4

UASB reactor design guidelines

4.1 INTRODUCTION

In Chapter 3 it was demonstrated that the UASB reactor is the most suitable treatment system for anaerobic treatment of sewage. Furthermore a process model was introduced, allowing for anaerobic process design based on sludge age instead of the more commonly applied hydraulic retention time. In this chapter the guidelines for the design for the different construction elements that comprise the UASB will be discussed and a rational reactor design procedure is presented. In Chapter 10 a design spreadsheet will be presented, based on the theory developed in Chapters 3 to 9, which allows for rapid evaluation of UASB design alternatives, including the post-treatment section.

At the end of this chapter, different modifications of the standard UASB design will be discussed: where Section 4.5 highlights the impact on performance by introducing a high-rate external settler, Section 4.6 presents several small-scale UASB configurations that are very suitable for *in situ* (decentralized) treatment, for example on the level of individual households.

4.2 CONSTRUCTION ASPECTS AFFECTING UASB PERFORMANCE

The potential for anaerobic treatment in regions with a warm climate is obvious, considering the attainable COD removal efficiency of 70–80% at a relatively short hydraulic retention time, as reported in many of the early pilot and full-scale UASB studies (Section 3.3). However, when the performance of some of the recently constructed large full-scale UASBs is considered, the results are less favourable (Chernicharo, 2007). The difference can be partially explained by considering the operational conditions such as temperature, but mainly may be attributed to imperfections of the reactor operation, such as a lower than expected efficiency of the pre-treatment and insufficient sludge removal capacity which resulted in the presence of sludge particles in the effluent.

In Figure 4.1 the COD removal efficiency according to Equation (3.5) is plotted (for $S_{ti} = 600\ mg \cdot L^{-1}$), considering that part of the sludge production will be discharged in the effluent. It can be concluded that the

 Anaerobic Sewage Treatment: Optimization of Process and Physical Design of Anaerobic and Complementary Processes
Editor(s): Adrianus van Haandel and Jeroen van der Lubbe
doi: 10.2166/9781780409627_0133

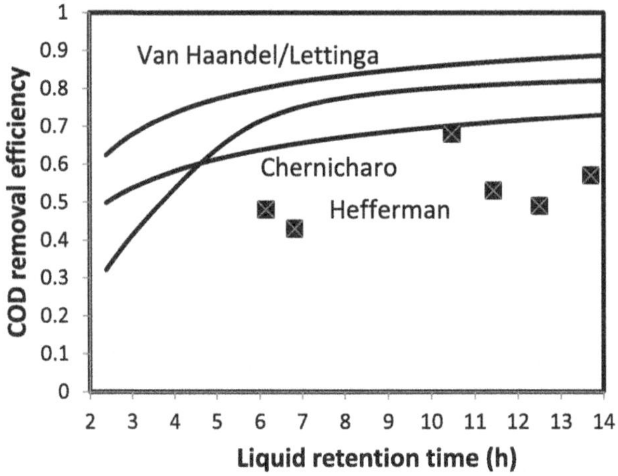

Figure 4.1 Predicted values of COD removal efficiency according to Van Haandel and Lettinga (1994), Chernicharo (2007) and calculated using the process model presented in Chapter 3 (Eqs. 3.24–3.26). Experimental data reported by Heffernan *et al.* (2011) from full-scale UASB plants are also indicated.

differences between the three models can mainly be attributed to the imperfect separation of sludge solids that will end up in the effluent. Figure 4.1 also shows the findings from an extensive field survey conducted in 2009 (Heffernan *et al.,* 2010a), which included several large full-scale UASBs located in Brazil, India and the United Arab Emirates, with installed capacities ranging from 30,000 to 160,000 $m^3 \cdot d^{-1}$. The observed COD removal efficiencies are shown for different operating retention times. In all cases the COD removal efficiency was well below any of the models of Figure 4.1.

An extensive report of the findings with respect to plant design, operation and performance can be found in Heffernan *et al.* (2010a); Van der Lubbe *et al.* (2010a,b) and in Van Lier *et al.* (2010). The authors concluded that the disappointing treatment results were not caused by any inherent flaw in the concept of UASB treatment. On the contrary, for each of the UASBs visited, the lower than expected COD removal efficiency could be fully explained by errors in design and/or engineering, lack of qualified supervision and insufficient maintenance. As to the quality of operation, at the time the UASB technology was introduced (by the end of the 1980s), there was often a significant involvement by universities and consultancy companies, with PhD and MSc students on site. Obviously, this may have had its effect on the treatment results reported in for example Van Haandel and Lettinga (1994).

Another finding from the review study is that in the last two decades the progress regarding the design of municipal anaerobic sewage treatment plants has been rather slow. Notable exceptions are the work of Chernicharo (2007), who contributed on reactor design and engineering, construction and operational procedures, and the research from the team of Professor Van Haandel about process design and system behaviour, which has been discussed extensively in the previous chapter. Notwithstanding these individual endeavours, the basic design of the UASB is still pretty much the same as it was in the early 1990s, summarized by Van Haandel and Lettinga (1994).

A possible reason for the relatively slow development of UASB technology might have been that originally the UASB was meant as a low-cost, low-tech system to provide at least partial sewage treatment in developing countries.

To minimize power consumption, maintenance requirements and considering the unavailability of skilled labour in the poorer countries, the design was based on one-time pumping at the head of the STP with gravity flow through the rest of the installation. Therefore, the current design of the UASB reactor tends to be rather complicated. For example, the influent distribution system is mounted on the top of the reactor while flow is upward, which results in practical difficulties during design and construction.

It should also be considered that without a suitable infrastructure in terms of technical and institutional capabilities, even the low-tech UASB will ultimately fail. Reactors will fill up, effluent weirs are misaligned, and influent systems will block. Implementation of UASB reactors in countries without an appropriate infrastructure is therefore not easy. On the other hand, the UASB will continue to provide at least partial treatment for a long time, even when poorly operated and maintained. Fortunately, a significant part of the formerly underdeveloped world is now experiencing rapid development, for example countries like Brazil, India and China, resulting in increased technical and financial capabilities. Simultaneously stricter effluent limits often require application of more complex aerobic post-treatment. So, there is no need to maintain the rock-bottom low-cost approach towards the UASB, and significant improvements in municipal UASB design (or perhaps more generally in anaerobic sewage treatment) are possible. Lessons can be learnt from the more advanced designs that are now commonplace in the anaerobic treatment of industrial wastewaters.

4.2.1 Design and engineering issues

In this section several key issues observed in the review of existing full-scale UASBs are summarized (Heffernan *et al.,* 2010a; Van der Lubbe *et al.,* 2010a,b):

- Inadequate process design;
- Improper pre-treatment;
- Incorrect flow division and distribution;
- Poorly constructed effluent collection system;
- No provisions for scum removal and/or poor access for cleaning;
- Poor means of sludge sampling and excess sludge withdrawal;
- Poor quality of gas collection and utilization devices;
- Improper material selection and poor quality of construction.

4.2.1.1 Inadequate process design

The first requirement for a well performing sewage treatment plant is that the process design is adequate. The following faults are often observed:

- Failure to consider the detrimental effect of sulphate on anaerobic process performance. Prior to starting a new project, the sulphate concentration in the raw sewage should be determined or estimated (for instance from data of drinking water production companies).
- Use of the UASB for thickening and stabilization of aerobic excess sludge has often been cited as an important advantage. However, the aerobic sludge will only be partly digested under the sub-optimal conditions prevailing in the UASB (low temperature, poor mixing). The non-degraded aerobic waste sludge tends to wash out in the UASB effluent. Hence, such an operational mode is only feasible if precautions are taken to separate the solids in the UASB effluent from the liquid phase, for example through the addition of a settling step, which entails higher investment costs. Otherwise these solids will accumulate in the combined system because it is operated without an excess sludge outlet.
- Return of back flush water from filter units (e.g. sand filters or submerged aerated filters) to the UASB creates a similar problem. Apart from the presence of poorly degradable suspended solids which decrease the anaerobic sludge age, the hydraulic load to the UASB is increased as well. The subject of returning waste solids streams to the UASB was discussed in greater detail in Section 3.8.3.

4.2.1.2 *Improper pre-treatment*

Three constituents need to be removed from the raw sewage prior to treatment in a UASB:

- Sand and grit, which otherwise accumulate in the digestion zone of the UASB, reducing the volume available for anaerobic biomass and potentially blocking the inlet pipes. In general, a mechanized grit chamber with bottom scraper is enough. For small systems, the combination of Parshall flume and sand trap may be used (Van Haandel & Lettinga, 1994).
- Large and fine debris, such as fibres and hairs, which cause blockages throughout the whole influent distribution system, most notably in the inlet pipes and at the overflow weirs, as well as also tending to promote scum formation. In general, a 40 mm bar screen (Figure 4.2) followed by a 6 mm fine screen (preferably a step screen) is considered the most cost-effective combination to remove most of the debris. Raking mechanisms for screen cleaning should be avoided, as the rakes tend to push the debris through the bars.
- Free fat, oil and grease fat (FOG). Given enough time and provided that the mixing intensity is sufficient, these components are perfectly biodegradable. However, in the UASB the conditions for degradation of this material are less favourable and formation of a scum layer is almost unavoidable. Apart from being a nuisance, scum can block the overflow weirs and fill up the gas hoods. An aerated sand trap can significantly reduce the rate of scum formation as at least part of the FOG will be removed. However, these systems are currently seldom installed in the pre-treatment section of an UASB.
- Improper construction of the screens (as shown in Figure 4.4) requires high levels of operator attendance for cleaning and unblocking of the distribution system; if blockages are left unchecked, severe short-circuiting in the UASB may result.
- Thus, the minimum required to ensure proper pre-treatment for UASB-based sewage treatment is composed of 40 mm coarse screening followed by a lifting pump station (centrifugal pumps or, for larger treatment plants, Archimedian screw pumps – Figure 4.3), 6 mm fine screens (step screens, rotating drums or equivalent) and a sand/grit removal step (preferably aerated for simultaneous FOG removal).

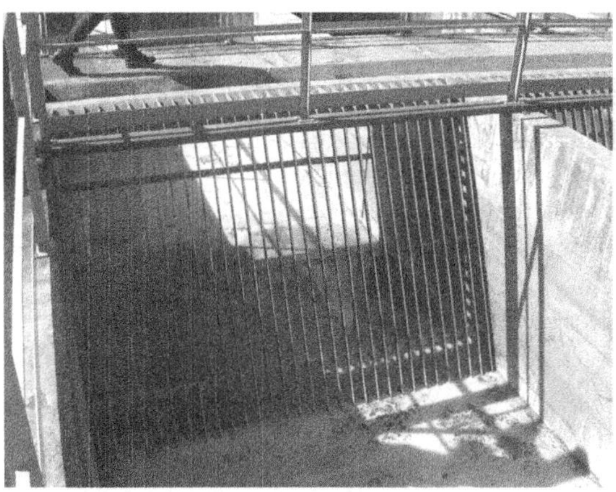

Figure 4.2 A coarse 60 mm bar rake screen as first step in the required pre-treatment process.

Figure 4.3 Archimedean screw pumps lift the raw sewage to the head of the sewage treatment plant so further passage can be by gravity.

Figure 4.4 A chain screen with clearance on the side allowing debris to pass the device.

Figure 4.5 Debris floating on top of the UASB overflow section, clearly indicating insufficient quality of the pre-treatment step.

4.2.1.3 Incorrect flow division and distribution

Proper distribution of the inlet flow over the reactor surface area is important for proper UASB performance. As each inlet pipe typically covers 2–4 m^2 of reactor surface area, the inlet flow must be divided into many equal parts. In systems with more than one reactor, a series of distribution chambers is used, to divide the flow over the individual reactors. At this level the final division over the individual inlet pipes is

Figure 4.6 Conventional circular grit/sand trap with bottom raking device.

Figure 4.7 Detail of the bottom grit raking device.

Figure 4.8 A poorly designed final distribution box (3 out of 6 overflows not visible) resulting in uneven flow distribution (left). Covering these boxes resulted in the formation of scum, with a further impact on distribution (right).

achieved through one or more final distribution boxes or final turndishes. At every distribution chamber or distribution box, it is important that a free-fall overflow is ensured, even at peak flow. If the overflows are correctly aligned (often they are not!), then equal flow division is ensured for any flow rate.

If the height of the free-fall overflow is insufficient and flooding of one or several outlets occurs, then differences in the pressure drop over the outlets will determine the flow through each pipe and imbalances may occur. To facilitate even distribution, it is also recommended to construct the distribution box as symmetrically as possible. Although all of this seems obvious, in many plants the influent distribution system was poorly designed and constructed, as can be observed in Figure 4.8. Examples of a well designed final distribution box layout are shown in Figure 4.9, where the circular design is to be preferred.

Another design issue is the installation of pipes with multiple outlets (Figure 4.10). While this saves on material costs, blockages of individual openings will be difficult to detect, as to a certain extent the flow will be divided over the other openings. Even if eventually the blockage is detected, cleaning of

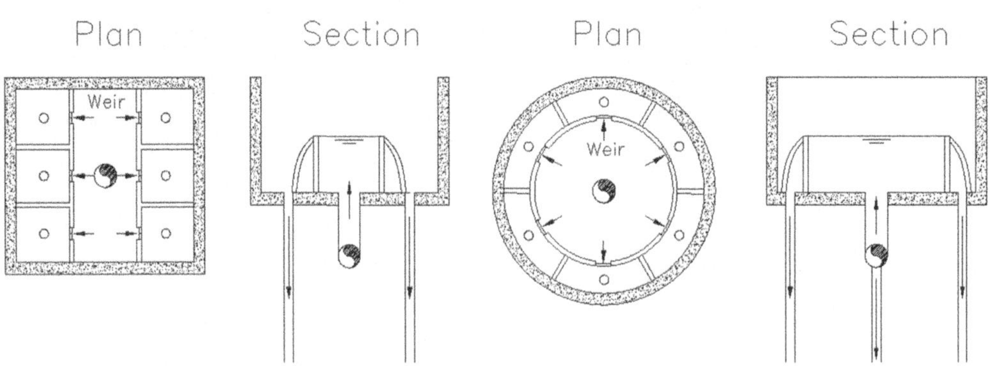

Figure 4.9 A schematic representation (plan and section) of a properly designed rectangular (square) and circular final distribution box.

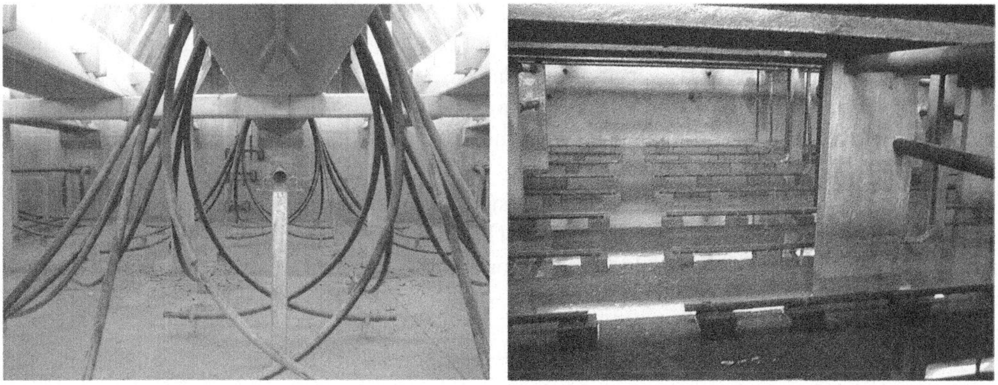

Figure 4.10 Recommended inlet distribution system, where each inlet pipe discharges at a single outlet point (left) and the alternative configuration where each inlet pipe has several discharge points (right).

the plugged opening is by no means guaranteed. In contrast, in a one- pipe–one-outlet configuration the blockage of a pipe will be immediately detected, as the liquid level in the corresponding inlet box will increase. Unblocking the pipe means unblocking the opening: the success of cleaning is indicated by the resulting decrease in liquid level in the inlet box.

4.2.1.4 Poorly constructed effluent collection system

Effluent withdrawal should be balanced over the surface of the reactor, in order to prevent short-circuiting between influent and effluent, resulting in localized high overflow velocities and the occurrence of dead (stagnant) zones in the reactor. In practice, two collection systems are used: effluent gutters fitted with overflow weirs and submerged perforated pipes.

If effluent gutters are installed, it is recommended to equip the overflow weirs with v-notches (triangular overflows) as this allows for easier alignment. However, frequently too many v-notches are used, i.e. more than the 1–2 per m^2 reactor area as recommended by Van Haandel and Lettinga (1994). The resulting flow rate per v-notch will then be too low and the liquid level above the base of the v-notch will be insufficient to facilitate weir alignment. Furthermore, at the low liquid height resulting from an excessive number of v-notches, a small misalignment has proportionally a much larger effect on the overflow rate than at a higher liquid height. This can be demonstrated when one considers the following equation, which calculates the flow rate in $m^3 \cdot s^{-1}$ as function of the liquid height above the base of the v-notch (h, in meters) and the base angle α:

$$Q = 1.465h^{2.5}\tan(0.5\alpha) \tag{4.1}$$

For example, consider the effect of a misalignment of only 2 mm for a 90° v-notch weir. At low liquid height, the flow per v-notch increases from 0.053 $m^3 \cdot h^{-1}$ at 10 mm above the base of the v-notch to 0.083 $m^3 \cdot h^{-1}$ at 12 mm, which represents an increase in flow rate of 58%. The same misalignment error of 2 mm, but now at increased liquid height above the base of the v-notch has a much smaller effect: for example from 0.52 $m^3 \cdot h^{-1}$ at 25 mm to 0.63 $m^3 \cdot h^{-1}$ at 27 mm, an increase of 21%.

Yet another detrimental effect of an excessive number of v-notches is that it will make the weir susceptible to blockages by scum. Due to the low flow rate the v-notches are no longer self-cleaning: the scouring intensity is insufficient to remove the blockage. This even applies when scum baffles are used.

Once one v-notch is blocked, the floating scum layer starts to accumulate, and soon adjacent v-notches will be blocked as well, resulting in non-uniform effluent collection. At several full-scale plants the quality of construction of the effluent weir was very poor, in one case with up to 5 cm difference in weir level over the length of the reactor. Obviously, the difference in local overflow rates must have been enormous. The effluent gutter of another reactor did not have any v-notches at all and consequently whole sections of the effluent collection channel received no flow at all.

If submerged pipes are used then alignment is easier, while the effect of a remaining misalignment on reactor performance will be significantly reduced. The flow rate through a circular perforation is much less sensitive to variations in liquid height than the flow over a v-notch. The flow rate through a submerged perforation can be described as:

$$Q = \omega A(2gh)^{0.5} \tag{4.2}$$

where

ω = contraction coefficient (0.66 for a circular opening)
g = the gravitational constant (9.81 $m \cdot s^{-2}$)
h = is the liquid level above the perforation (m).

In this case it can be calculated that an increase in liquid level above the openings from 10 to 12 mm yields an increase in flow rate from 0.57 to 0.63 $m^3 \cdot h^{-1}$ (+12%), while an increase from 25 to 27 mm results in an even lower flow increase from 0.85 to 0.89 $m^3 \cdot h^{-1}$ (+5%). Thus, the submerged pipe is indeed less vulnerable to misalignment during construction, with consequent differences in liquid height above the perforations.

Another advantage of submerged pipes is that less turbulence is created on the surface area, thus minimizing the emissions of CH_4 and H_2S to the atmosphere, as corroborated by the results of Souza *et al.* (2011), discussed in Chapter 6. This offers the possibility of localized vent gas collection in the effluent collection system. Finally, the risk that a perforation will be blocked by scum is low, provided that the velocity through the opening is sufficiently high.

4.2.1.5 No provisions for scum removal or poor access for cleaning

Scum problems are a real nuisance for UASB reactors at two locations: (1) the surface of the settling compartment, and (2) the gas–liquid interface under the gas hoods. As discussed previously, scum formation is almost inevitable due to the presence in the raw sewage of fats, oil and grease as well as surfactants and larger floating debris. This is partly a dietary issue, as the problems observed are much more serious in Brazil where meat consumption per capita is high, compared with predominantly vegetarian India. In Brazil scum formation is much higher in the south than in the north, partially because the temperature is lower and partly because southerners are known as more voracious grease eaters.

Proper pre-treatment will significantly reduce the rate of scum formation and has been discussed previously. In principle, scum formation on the reactor surface does not have to be a problem, provided scum baffles are not used, as the scum is then discharged together with the effluent over the effluent weir. However, as discussed above, the flow rate over the v-notches is often so low that they are easily blocked, as demonstrated in Figure 4.11. Cleaning v-notches is easy, but the scum problem is compounded by the recent trend to cover UASB reactors in order to prevent odour problems. Apart from the question whether covering the entire reactor is really necessary, as most of the odour is released at the overflows, the use of fixed concrete covers makes inspection and cleaning of the v-notch weirs very difficult, even if inspection hatches are installed. Should covers be required, then lightweight removable covers are recommended.

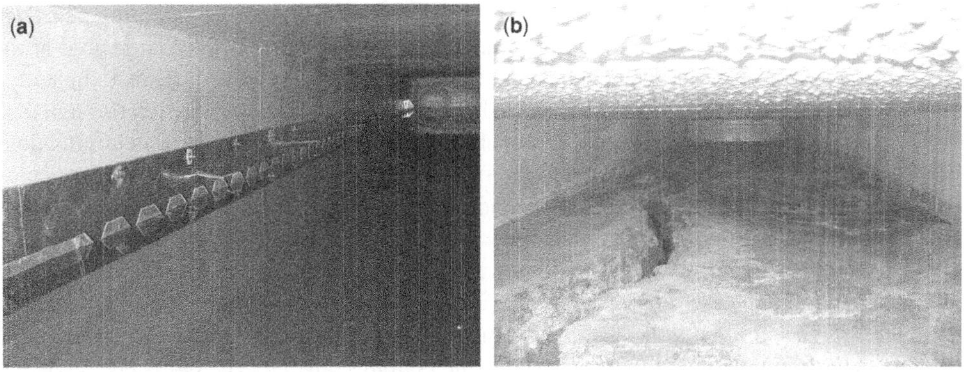

Figure 4.11 Scum accumulation in the settler section (no scum baffle installed). The effluent gutter is clearly visible during commissioning (a), but after 1 year a scum layer of approximately 30 cm in thickness has developed (b).

Scum formation under the gas hoods is a more serious problem, as access is difficult. Release of gas from the liquid phase will be much more difficult when a hard and thick scum layer is present; this tends to form if the removal frequency is low. When release of gas is erratic and violent (as it commonly is), solids (scum) can be entrained in the gas piping and create obstructions. The resulting increase in gas pressure might damage the gas hoods. Furthermore, the gas hoods may fill up completely with scum at which time erratic and uncontrolled discharges of gas into the settling zone may occur over the length of the gas hood. At this point there may be no alternative but to drain the reactor and clean it.

Removal of scum from the gas hood is difficult. Some plants have resorted to the installation of access hatches on top of the gas box. However, the construction of an access hatch that remains gas tight after repeated opening and closing has proven difficult, even when the size of the hatch is minimized. Insufficient gas-tightness of the gas hoods may again result in odour problems while it also complicates operation at the gas pressure required to transport the biogas to its point of treatment and utilization. Furthermore, as gas boxes are long and narrow, cleaning it from the restricted access provided by the hatch is a daunting task indeed. Different approaches will be suggested in Section 4.2.6.

4.2.1.6 Poor means of sludge sampling and excess sludge withdrawal

In principle the treatment objectives of a UASB will be to maximize COD removal and/or to produce a stabilized sludge. Therefore, the UASB should be operated at the highest sludge age possible, i.e. with the highest biomass inventory possible for the given reactor volume. This requires operation at the highest sludge blanket level possible and at maximum biomass concentration. A first condition for proper UASB solids inventory management is that it should be possible to determine the height of the sludge blanket accurately.

Generally, sludge sampling connections are installed in the reactor wall for this purpose. However, the number of sample connections is often inadequate. Take for instance a UASB with only three sampling points installed over the height of a 2.5 m digestion zone (at 0.3, 0.9 and 2.0 m). When the top of the sludge blanket is more than 2 m above the reactor floor, it can no longer be measured. The only tell-tale sign that the maximum sludge blanket height is exceeded is when the concentration of suspended solids in the effluent starts to increase. This can be detected by observing the volume of settleable solids in an Imhoff cone. A second problem associated to this method of sludge sampling is that the sampling pipes are often mounted close to the reactor wall. It is not altogether certain if the concentration measured at the wall represents the true average reactor sludge concentration at that height, as mixing tends to be less intense near the walls. A much better (and simpler) solution is to install a sampling connection through the top of the gas hood and to sample the sludge blanket from above. This not only allows exact determination of the sludge blanket height but also for sample collection from any desired level.

A proper system of sludge withdrawal is also important. In general, it is recommended that at least two levels in the sludge blanket are available for sludge withdrawal. The reason for this is that a sludge profile will develop in the digestion zone with the denser and better settling sludge (40–80 g TSS $\cdot$ l^{-1}) at the bottom of the reactor and the less-dense sludge (15–25 g TSS $\cdot$ l^{-1}) at the top of the sludge blanket. Thus, the active biomass concentration will be higher in the bottom sludge. Unless there is a specific reason to do so (e.g. excessive accumulation of inert material in the bottom sludge) it is recommended to waste from the top of the sludge blanket. This way the denser, better settling and more active sludge is maintained in the reactor preferentially. Several UASBs with only a bottom withdrawal system in place exhibited poor biomass retention, characterized by a shallow sludge blanket and a low sludge concentration. Finally, the sludge abstraction points should be distributed evenly over the reactor surface area. In general, one abstraction point per 50–60 m^2 is considered as a good compromise between optimal performance and required investment costs.

4.2.1.7 Improper material selection and poor quality of construction

Due to the presence of sulphides in the anaerobic effluent the potential for corrosion in a UASB is high, especially at those locations where a gas–liquid interface exists. This is especially true for covered reactors where the presence of condensed water, oxygen and H_2S accelerates the formation of H_2SO_4, which is very corrosive towards concrete. It is therefore recommended to apply lining (or coating, provided it is applied well, which is a difficult task) on all concrete parts upwards from 0.5 m below the waterline. Use of lining inside the gas box has a second advantage in that it is impermeable to gas and therefore allows the gas boxes to be pressurized. The biogas can then be transported to the gas treatment equipment without the use of a gas compressor. Furthermore, all piping and valves on top of the reactors should be made of corrosion-resistant (and solar-radiation-resistant) material, or a suitable coating is required. This also applies to other metal parts such as hand railings and ladders.

4.2.2 Operational and maintenance issues

The main issues observed are the following (Van der Lubbe *et al.*, 2010a,b):

- Insufficient operator attendance and maintenance
- Insufficient process knowledge.

In most treatment plants, many examples of insufficient operator attendance were observed during the site visits. In UASB reactors most frequently this led to blocked inlet pipes and accumulated scum on the surface of the settler blocking off whole sections of the effluent weir, which obviously should have been cleaned. In defence of the operators often no adequate means of unplugging the inlet pipes are provided. However, the solution can be simple, as observed at a Brazilian plant where a cleaning pipe with rubber seal is used. The pipe is placed on top of a blocked inlet pipe and flushed with pressurized service water (effluent). This solution is shown in Figure 4.12. In this particular case, although the inlet pipes were constructed with many bends and connections, problems with blocked inlet piping were absent.

Figure 4.12 Effective method for unblocking clogged inlet pipes with pressurized water using a simple cleaning tube that seals off the inlet of the pipe.

As to the issue of insufficient maintenance, one area where this is particularly obvious is in gas collection and treatment. Working gas flow meters are rare and, even worse, some of the flares are out of operation as well, resulting in continuous venting of methane to the atmosphere. Fine screens sometimes are in a poor condition, especially the raking mechanisms. The same applies to the grit classifiers that are used to transport the deposited sand/grit out of the sand trap.

Finally, to the issue of insufficient process knowledge: one cannot expect an operator to know all the fine detail of biological treatment, but a plant or operations engineer should at least have some grasp of the basic concepts involved. For instance, if a UASB is fed with wastewater, the result will be that sludge is produced. This sludge needs to be wasted as otherwise it will inevitably end up in the effluent.

Several UASBs discussed in the study by Van der Lubbe *et al.* (2010a,b) either did not withdraw excess sludge at all or the solids control strategy consisted simply of "opening a valve for a couple of turns for one minute per day". A simple yet effective sludge management strategy is to discharge a sludge batch with a fixed volume whenever the height of the sludge blanket exceeds a certain upper set point value. However, this presupposes that the operator has the means to determine the sludge blanket level with a certain degree of accuracy, which often is not the case, as discussed in the previous section. An alternative measurement that can be used is to take effluent samples and determine the sludge volume in an Imhoff cone after 30 minutes of settling: if the volume of settled sludge is more than 1–2 $ml \cdot L^{-1}$, it is time to discharge sludge. A more detailed strategy for excess sludge discharge was presented in Section 3.8.2.

4.2.3 Undeliverable expectations of UASB performance

Whereas there are doubtless many advantages associated with anaerobic treatment, these benefits have sometimes been exaggerated. Consequently, when performance is less than expected, the UASB system is blamed for it. Some common examples of undeliverable expectations of UASB performance are presented below.

4.2.3.1 Overestimated methane production

This is perhaps one of the most important items where appropriate management of expectations is due. The theoretical methane production is equal to 0.35 $Nm^3 \cdot kg^{-1}$ COD *converted*. However, sometimes this is taken as per kg COD *applied*. Whereas for the industrial application of anaerobic treatment, often receiving wastewater with a high soluble biodegradable COD fraction, this approximation may be justified, for municipal sewage it is certainly not, since a considerable COD fraction is non-degradable. The digestion efficiency of influent COD into methane for municipal sewage is often in the range of 55–75% (if sulphate concentrations are low), depending on the wastewater composition, the applied sludge age and the temperature. The rest of the COD will end up either in the excess sludge or in the effluent. This digestion efficiency is much lower than for industrial applications, where digestion efficiencies of up to 90% are not uncommon. A factor that reduces the methane yield in municipal sewage treatment even further is the diluted nature of the wastewater, resulting in poor methane recovery. Even though methane is not very soluble in water (18–20 mg · l^{-1} at a 80–90% methane fraction in the biogas), at atmospheric pressure and for temperatures between 15–30°C (see Example 6.2), the associated COD demand of 70–80 mg COD · l^{-1} may easily represent up to 20–30% of the digested influent COD fraction.

Finally, there may be a significant sulphate concentration present in the raw sewage, resulting in a competition between sulphate-reducing and methanogenic bacteria for available biodegradable COD. As a result of the factors discussed here, the actual methane production is typically much lower than the theoretical maximum of 0.35 Nm^3 · kg^{-1} COD.

4.2.3.2 Less reduction in sludge production than expected

The reduced methane production can be attributed mainly to incomplete hydrolysis of particulate biodegradable COD. The accumulation of non-biodegradable or non-degraded particulate material may constitute the main contribution to excess solids production in a municipal UASB. For a municipal sewage UASB an apparent yield of 0.10–0.15 mg TSS · mg^{-1} COD may be considered as a typical value. However, this figure reflects only the suspended solids discharge with the excess sludge and does not consider the suspended solids content in the UASB effluent. At a concentration of 40–100 mg TSS · l^{-1} this discharge into the effluent is considerably higher than in activated sludge systems. This "effluent excess sludge" is only partly degradable and accumulates in the aerobic post-treatment system, where it will contribute to the secondary excess sludge production.

The total apparent sludge yield (i.e. excess sludge plus the effluent suspended solids) for a municipal UASB depends on the applied sludge age and the efficiency of the GLS separator but will in general be around 0.15–0.20 mg TSS · mg^{-1} COD, of which up to 40–50% may end up in the effluent. If an aerobic post-treatment step is added for effluent polishing, then the total *discharged* specific excess sludge production (from aerobic plus anaerobic reactors) will be around 0.2–0.25 mg TSS · mg^{-1} COD.

In comparison, the organic sludge production in a conventional activated sludge system treating raw sewage at a sludge age of 5 days and at 20°C typically has a value around 0.4–0.45 kg TSS · kg^{-1} COD applied and slightly higher (up to 0.50–0.6 kg TSS · kg^{-1} COD) if a primary settler is installed. However, when this sludge is digested, a high fraction is converted into biogas and the actual final sludge production will be of the same order of magnitude as in anaerobic systems.

4.3 DESIGN ASPECTS OF THE UASB REACTOR

In the previous chapter the UASB reactor was identified as the most suitable anaerobic system for sewage treatment. Figure 3.7 showed a generic sketch of a UASB reactor with the main elements of which the unit consists. In this section guidelines will be presented that can be used for design of the UASB reactor with regards to the following aspects:

(1) Reactor dimensions: reactor volume, surface area and height
(2) GLS separator and gas box

(3) Deflector
(4) Feed flow division
(5) Influent distribution system
(6) Effluent collection
(7) Sludge sampling and discharge
(8) Scum removal
(9) Biogas collection
(10) Reactor cover and ventilation.

Finally, in Section 4.4 all presented design guidelines will be integrated into an extensive UASB reactor design example.

4.3.1 Reactor dimensions

Often the required UASB reactor volume is calculated as the product of the design feed flow and the selected hydraulic retention time. However in Chapter 3 it was shown that the determining variable for the design of the UASB reactor is the organic rather than the hydraulic load. Consequently the sludge age and not the liquid retention time is the key design parameter, which not only determines the reactor volume but also system performance in terms of digestion efficiency, COD removal and excess sludge production.

4.3.1.1 Required reactor volume

A process design procedure based on the use of the anaerobic sludge age has been presented in Section 3.10; based on set objectives with respect to either desired COD removal efficiency, degree of excess sludge stability or both, an initial design value for the sludge age is selected. Using Eqs. (3.24, 3.25 and 3.29b) the mass of sludge is calculated that needs to be accommodated in the reactor, or more specifically in the digestion zone, as the volume occupied by the GLS separator should not contain any sludge. The fraction of the reactor volume allocated to the digestion zone is not fixed and depends on design decisions made with respect to the GLS separator, applied upflow velocities and so on; this will be highlighted in Example 4.1.

A reactor with a proportionally larger GLS separator will have a smaller digestion zone and thus a lower average sludge concentration. For this reason it is better to calculate the volume of the digestion zone separately. Normally an iterative calculation is required to determine the optimal value of the sludge age, for which a spreadsheet is very convenient. The use of the spreadsheet supplied with this text will be explained in Section 10.

When it is decided to use the hydraulic retention instead as a reactor design criterion, the required reactor volume can readily be estimated using the specifications in the Brazilian Standard for anaerobic treatment facilities, NBR 12.209. This standard specifies the liquid retention time as a function of the sewage temperature as shown in Table 4.1. It has already been demonstrated in Section 3.8.3 that the reactor retention time calculated on the basis of the sludge age agrees well with the recommended retention time in the standard. Table 4.1 also shows the calculated reactor upflow velocity, as function of the selected reactor height, which is typically between 4 and 6 m, as will be discussed later.

It is important to emphasize that the design of wastewater treatment plants should be based on the *minimum* monthly sewage temperature, when process performance is worst, and not on the *air* temperature which may be quite different. This fact is demonstrated in Table 4.2, where the maximum and minimum ambient temperature are shown compared with the corresponding sewage temperatures for several regional capitals in Brazil. With respect to the Brazilian Standard 12.209 it is important to note that this standard was developed by a committee of scientists and engineers, based on the considerable experience gained in

Table 4.1 Recommended values for the design retention time in UASB reactors for different temperatures in accordance to the Brazilian Standard 12.209. Calculated values for the average reactor upflow velocity for different reactor heights are presented as well.

Temperature (°C)	Retention Time (h)	Upflow Velocity at Different Reactor Heights		
		4 m	5 m	6 m
<18	10	0.40	0.50	0.60
18–21	8	0.50	0.63	0.75
21–24	7	0.57	0.71	0.86
24–26	6	0.67	0.83	1.00

Brazil with full-scale UASB reactors treating raw sewage. However, it must be remembered that the range of retention times (or hydraulic loads) in the Standard reflects the average composition of Brazilian sewage and that a different composition will have a different result, because it is not the hydraulic but the organic load that is the governing variable for UASB reactors.

Thus, for example, for certain dry locations in the Middle East the COD values may reach higher values (sometimes as high as to 1500 mg $\cdot$ L^{-1}), also because of the habit of using tankers to truck in concentrated sewage from rural areas without connection to the sewer grid. For a comparable HRT the organic loading rate will be higher than that for Brazilian sewage, with an average COD of 500–800 mg $\cdot$ L^{-1}. When the design procedure based on sludge age is used, these kinds of issues are prevented.

4.3.1.2 Reactor surface

Once the reactor volume is defined, the design engineer will have to decide on the height and surface area of the treatment unit. The surface area is dependent on the selected reactor upflow velocity. Although the upflow velocity of liquid in a UASB is usually regarded as a pivotal variable in UASB design, it is within certain boundaries really only of minor interest. This was demonstrated in a research study conducted by Leitão *et al.* (2004) on this subject. In this study several identical UASB reactors were fed with the same

Table 4.2 Maximum and minimum temperatures in air and sewage in several regional capitals of Brazil. The latitude (degrees to the north (+) or south (–) of the equator) is indicated as well.

City	Latitude	Ambient Air (Monthly Average)		In Sewage (Monthly Average)	
		Maximum	Minimum	Maximum	Minimum
Manaus	+3°	32	22	30	30
Fortaleza	–4°	30	21	28	28
Campina Grande	–7°	20	20	25	25
Salvador	–13°	30	20	24	26
Belo Horizonte	–22°	29	15	21	23
São Paulo	–23°	28	12	19	22
Curitiba	–26°	23	8	16	21

sewage but at different dilutions. Hence, the volumetric organic loading in all reactors was equal, but the applied hydraulic retention times and the upflow velocities were not. It was established that for the same volumetric organic loading rate the performance of the reactors was independent of the applied liquid upflow velocity in the range of 0.7–1.8 $m \cdot h^{-1}$. Hence, at least at pilot scale it was demonstrated that if a high upflow velocity is continuously applied, a sludge with good settleability will develop which will be retained in the reactor, although part of the influent solids might be washed out.

Although the liquid upflow velocity in itself may not be the most important variable, there are certain practical limits to consider. It is generally accepted that the maximum value of the upflow velocity should not exceed a value of 1.0 $m \cdot h^{-1}$ for the average flow rate. In Table 4.1 it can be seen that if the Brazilian Standard 12.209 is accepted, the upflow velocity of the liquid only reaches this value if both the retention time is low (6 h) and the UASB reactor is tall (a height of 6 m). Even so, the results of Leitão *et al.* (2005) show that good performance of the reactor is possible at much higher velocities.

There is also a minimum allowed value for the design upflow velocity: the VSS/TSS ratio of the sludge is markedly influenced by the upward velocity of the liquid and for upflow velocities lower than 0.3 $m \cdot h^{-1}$, a high degree of accumulation of inorganic solids in the bottom section of the reactors is observed (Souza, 2000). This in practice leads to serious operational problems, since the heavy inorganic sludge accumulates in the reactor, is difficult to remove and eventually will block the inlets of the feed lines.

An important question that has not received its due attention to this point is on which flow rate to base the reactor design: for example the average or the peak flow rate. In practice the design of the UASB is often based on the *daily maximum* flow rate, where large rainfall events are excluded. Within the day there are typically marked variations in flow rate and organic load due to the cyclic nature of human activities, with distinct peaks in both flow and load in the morning and early evening. Under the condition that the sludge blanket height is sufficiently high, then typically both the particulate and soluble COD loads can be absorbed. If the ratio between the peak flow and load and the design flow and load is very high or sustained for more than a couple of hours, this may result in wash-out of part of the sludge blanket with the reactor effluent. In this design case it might be better to calculate the upflow velocity for this maximum flow rate. In the case of heavy rainfall events, then typically after the first flush has been received the COD concentration in the feed rapidly drops, sometimes even to below the UASB effluent concentration. At this point it may be very advantageous to bypass part of the peak flow directly to the effluent. For this purpose, bypass weirs can be provided in the common feed channel, which overflow when the liquid height in the feed channel exceeds a certain level, which corresponds to a maximum acceptable feed flow rate. If necessary, the bypassed flow can be subjected to some form of partial treatment, for example with high-rate lamella settlers, prior to disposal.

As the first step in the reactor design process the minimum reactor surface area is determined, which is equal to the surface area required to comply with the recommended upflow velocity under design flow rate conditions:

$$A_{min} = Q_i/(24 v_l) \tag{4.3}$$

Note that this is a *minimum* surface area, and it may be decided at a later stage in the design procedure to select a larger surface area if this is convenient. At this stage in the design procedure only the upflow velocity in the digestion zone is considered; later when the GLS separator has been designed the velocity in the apertures (v_a) and overflow area (v_o) should be checked as well. The reduction in area available for liquid upflow, especially at the height of the apertures between the gas hoods, will result in a significant increase in local liquid upflow velocity. The guidelines following in Table 4.3 are recommended for the three velocities.

Whereas for small flow-rates (<250 $m^3 \cdot h^{-1}$) the reactor size might be customized, for larger applications a standardized modular design is often used. Generally, the reactors are square, though a rectangular shape can also be used. Circular designs are not recommended when more than one reactor is constructed, as

Table 4.3 Recommended values for the design upflow velocity (in m · h^{-1}) at different heights in the UASB reactor (digestion zone, apertures and overflow level) for municipal sewage.

Parameter	Recommended	Minimum	Maximum
v_l – upflow velocity in the digestion zone	0.5–0.8	>0.3	<1.0
v_a – upflow velocity in the apertures	<1.5	N/A	<2.5
v_o – upflow velocity at the overflow level	<1.0	N/A	<1.2

sharing of reactor walls will not be possible, which would reduce the usage of concrete. Typically, the length (L_u) and width (W_u) of a modular UASB reactor vary between 15–30 m. Once the surface area is determined and the values of L_u and W_u have been specified, this will yield the number of UASB reactors required (N). If more than one reactor is required, normally these are constructed as completely independent units that can be taken out of service for maintenance, without disturbing the other reactors.

In Brazilian practice an unofficial standard is respected in that the surface area should not be larger than 400–500 m^2, in order to avoid a very costly foundation and reactor bottom. The reason for is that the concrete must be resistant to the development of cracks from variations of sewage temperature that naturally occur. These temperature variations lead to dilatation and contractions of the concrete, so that its structure must be reinforced, especially when the surface area is large.

4.3.1.3 Reactor height

The volume of the UASB reactor is defined by the wastewater flow rate and the retention time to be applied. In Table 4.3 recommendations were given with respect to the upflow velocities at certain key levels in the UASB (digestion zone; aperture area and overflow level). In the previous section it was pointed out that based on the selected reactor upflow velocity there is a minimum surface area, which leads to a maximum height that can be applied. The total reactor height is dependent on the following factors:

- The GLS design and most specific the total height of gas box + gas hood + deflectors;
- The required minimum digestion zone volume and height; and
- Construction costs, as higher reactors do not only require higher walls but also thicker walls and stronger foundations.

The area to depth ratio – for a given total volume – affects both the flow regime of the liquid and the expansion of the sludge bed in the UASB reactor. In taller reactors with reduced area, the upward velocity is high, the sludge blanket is more expanded and as a result contact between the organic material from the influent with the biomass contained in the sludge blanket will be more intense than in shallower reactors. Furthermore the turbulence introduced by the biogas bubbles will increase for taller reactors: not only is the distance to the GLS device longer, but also the biogas production per unit area of reactor will be higher. The establishment of good contact between influent organic material and sludge particles in the digestion zone is advantageous for the performance and the stability of the reactor. The disadvantage is that owing to the higher upward velocity and turbulence there is also a possibility of increased sludge loss, especially in peak flow periods. An upward velocity in the digestion zone of higher than 1.0 m · h^{-1} for design (i.e. daily maximum) flow rates is normally not applied. Knowing that the retention time in a UASB reactor in the tropical zones is typically around 6 h, it is concluded that a UASB digester will thus rarely have a liquid height of more than 6 m.

On the other hand, if – again for a given total volume – the reactor is very shallow, then the sludge blanket layer will be relatively thin, so the contact between the organic material from the influent and the biomass is more restricted, which is further accentuated as the biogas production rate per unit area will be lower as well. Moreover, insofar as the reactor is shallower, there is more possibility of hydraulic short circuiting, through flow of sewage in preferred channels.

The upper and lower limits for the reactor liquid height can be derived as follows: from the results of Souza (2000) it was concluded that a minimum rising velocity of 0.3 $m \cdot h^{-1}$ is required to prevent extensive settling of inorganic solids in the reactor.

At the minimum required HRT of 6 hours this would mean that the minimum depth is $0.3 \times 6 = 1.8$ m. However such a shallow reactor is not at all practical, because it would be impossible to construct a GLS separator and still leave enough space for the digestion zone.

As will be demonstrated in Section 4.3.2, the GLS separator (including the deflector) typically has a minimum height of at least 1.5–2 m. With a minimum digestion zone height of 2–2.5 m, the liquid height required for an UASB reactor is then at least 3.5–4 m. In practice a liquid height or depth of 4 m is indeed the minimum value applied in full-scale UASB reactors. On the other hand the maximum depth is not clearly defined from the experimental data and depends rather on economic considerations and the availability of land for the treatment plant. In practice a depth of more than 6 m is very exceptional, also because the construction costs will rapidly increase above this height.

Once the surface area has been determined as per the guidelines in Section 4.3.1.2, and the GLS separator has been designed taking into account the considerations in Sections 4.3.1 and 4.3.3, which will supply the height of the GLS separator (H_{gls}), the only additional element required to calculate the total liquid height of the UASB reactor is the height of the digestion zone (H_{dig}):

$$H_{liq} = H_{dig} + H_{gls} \tag{4.4}$$

For the selected surface area, the digestion zone height follows directly from the volume that is needed for the digestion zone. The digestion zone volume is calculated as the total required anaerobic sludge mass divided by the average biomass concentration in the digestion zone. The required anaerobic sludge mass is the product of the selected anaerobic sludge age and the predicted anaerobic excess sludge production under steady-state design conditions. The equations that can be used to estimate the excess sludge production as function of the sludge age have been discussed in the process design procedure presented in Section 3.10.

The sludge concentration in the digestion zone typically varies from 40–80 g $TSS \cdot l^{-1}$ at the bottom of the reactor to 15–25 $g \cdot l^{-1}$ at the top of the sludge blanket (Van der Lubbe *et al.*, 2010a,b), as can be observed in Figure 4.13, which shows the sludge profiles of two large well designed and operated UASB plants.

The average digestion zone concentration is around 30–40 g $TSS \cdot L^{-1}$. Hence, knowing that the digestion zone comprises typically around 60% of the total liquid reactor volume, it should be feasible to obtain an average sludge concentration of 20–25 g $TSS \cdot l^{-1}$ in well designed and operated UASB reactors.

In practice the top of the sludge blanket will have to be maintained at some distance (typically at least 0.25 m) from the widest point of the deflector, in order to prevent the sludge blanket from spilling over into the settling zone. The total height of the digestion zone is thus calculated as:

$$H_{dig} = MX_t/(N \cdot X_t \cdot A) + 0.25 \text{ (m)} \tag{4.5}$$

where

MX_t = total required anaerobic sludge mass (kg TSS) = $R_s \cdot ME_t$
R_s = selected anaerobic sludge age (d)
ME_t = anaerobic excess sludge production ($kg\ TSS \cdot d^{-1}$)
N = number of UASB reactors
X_t = average biomass concentration in the sludge blanket ($kg\ TSS \cdot m^{-3}$)
A_u = UASB reactor surface area (m^2), which is $\geq A_{min}$

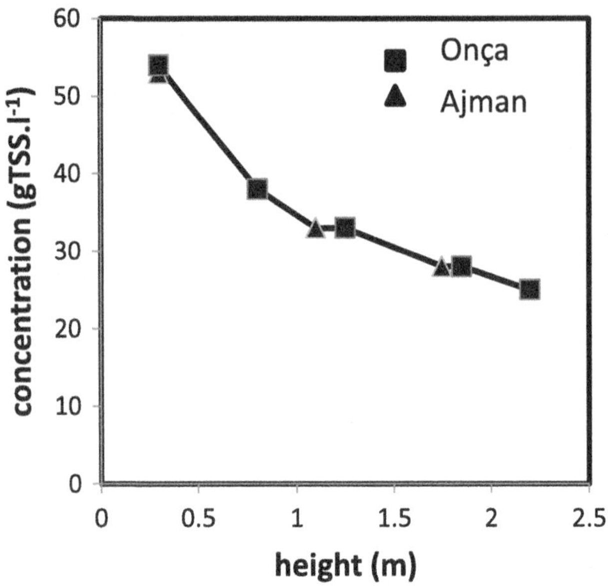

Figure 4.13 Observed sludge concentration profiles in the digestion zone of two large, well designed and operated full-scale UASB sewage treatment plants (Ajman, United Arab Emirates; Onça, Belo Horizonte, Brazil).

In practice there may be a maximum value to the height to which the sludge blanket can extend. As a result of the applied shear force from mixing due to upward liquid and gas flows, floc erosion will occur. In practice sludge blanket heights up to 2.5–3 m have been observed for municipal UASB, so it seems reasonable to have a maximum value for the digestion zone of 3.5 m, which includes the 0.25 m of minimum distance to the deflector.

If the liquid height is outside the recommended interval of 4–6 m, then either the reactor surface area should be adapted or else the GLS will require a redesign. For example, increasing the number of GLS separators will lead to a reduction in required GLS height, although at the expense of increased overflow velocity and increased construction costs.

Operation at a liquid height of less than 4 m is not recommended and in practice most municipal UASB reactors have liquid heights between 4 and 5 m, but apart from the construction costs there is no fundamental reason not to go higher, for example to 6 m, which is a typical (minimum) height of concrete UASB reactors treating industrial wastewaters.

4.3.2 Gas–Liquid–Solid (GLS) separator

The phase or GLS separator of the UASB reactor is certainly its most characteristic device. It is called the GLS separator as it separates the upward flowing mixture of biomass, sewage and gas into separate gas, liquid and solids fractions. For this reason, the GLS separator is composed of two parts: an upper part that contains the gas collection elements (the gas collection plates and the gas box, together forming the gas hoods) and a lower part composed of the gas deflectors. These deflectors are installed directly under the apertures between the gas hoods.

The GLS separator performs four main functions:

(1) Collection of the biogas that develops in the digestion zone under the GLS separator.
(2) Return of sludge to the digestion zone from the interface under the separator after the biogas has been separated.
(3) Retention of the suspended solids that enter into the settling zone, the upper part of the reactor above the GLS separator, and transfer of these solids back to the digestion zone; this allows a large sludge mass to be maintained in the digestion zone while achieving a low concentration of settleable solids in the effluent.
(4) The settling zone above the separator can be used for sludge storage during periods of hydraulic overload, without necessarily resulting in loss of sludge.

Figure 4.14 shows a schematic representation of the components of the GLS separator, where it can be observed that the GLS separator is constructed of three main parts:

(1) The deflector, which prevents the biogas from entering the settling compartment;
(2) The gas collection plates, which direct the produced biogas into the gas box and simultaneously create a quiescent top settling compartment;
(3) The gas box, in which the biogas produced in the reactor is collected and from where it is discharged to the biogas treatment and disposal units.

Either the GLS separator may be completely submerged or alternatively the gas box extends above the liquid phase, which is by far the more usual arrangement. The advantage of the submerged GLS separator is that the entire surface reactor is available for settling, thus having a minimum overflow velocity, increasing the likelihood that even slow-settling particles may be retained. On the other hand the loss of these particles, having a low settling velocity, is not really a great loss. Also, due to being submerged, corrosion will be less of a problem as the GLS separators are not exposed to the oxygen present in the air, especially where oxygen and water are in contact with metal. The disadvantages of the submerged unit are: (1) that it becomes extremely

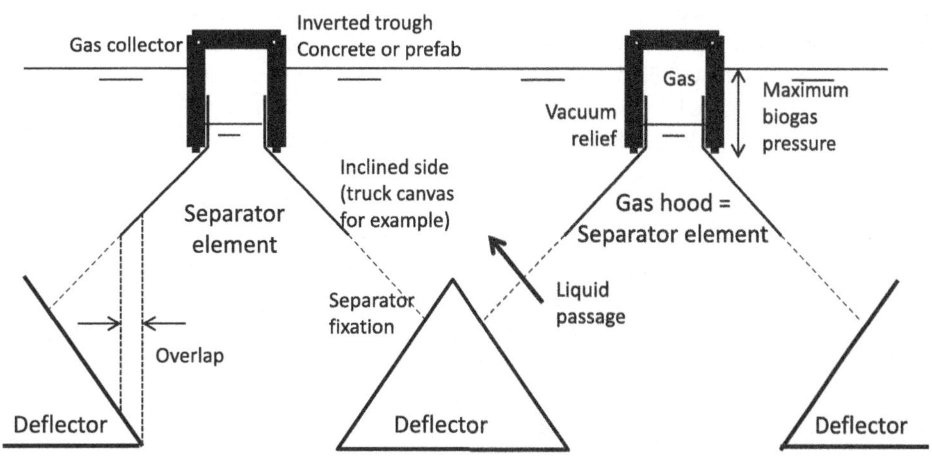

Figure 4.14 GLS separator composed of its elements the gas box, the collection plates (which together form the gas hood) and the deflectors.

difficult (and painstaking, since it is mostly manual labour) to remove scum that may have accumulated under the separator; and (2) that the separator should be very well anchored in the UASB structure to be able to resist the upward force due to buoyancy from accumulating biogas if the gas outlet is blocked.

In practice, the vast majority of GLS separators are constructed with part of the gas boxes extending above the liquid surface, which also allows access to remove scum that is formed on the surface of the overflow zone, even though it may still not be easy to remove it. In Section 4.3.7 several design measures that allow for easier scum removal are discussed in greater detail.

Knowing that gas bubbles follow a slightly oscillating trajectory when they rise in water, it is necessary that gas hood and deflectors overlap as indicated in Figure 4.14, so that any gas bubbles leaving from under the deflector are securely captured by the gas hoods. Practice has shown that a 10–15 cm overlap between deflector and gas hood is enough to prevent biogas escape into the settling zone (Heffernan *et al.,* 2010b). A larger overlap than 15 cm is counterproductive because it results in an unnecessary reduction of the aperture area and consequently in an increase in the upward velocity of the liquid entering the aperture, increasing the risk of entraining both gas bubbles and biomass.

The upward velocity varies with the height in the UASB and is at the maximum value where the area available for liquid upflow is minimum. This is normally at the level of the apertures between the gas collection plates.

Hence it is of particular importance in sewage treatment to maximize the surface area of the apertures between the gas hoods, because this ensures that the liquid velocity remains low and the escape of sludge particles to the settling zone is reduced. In this text it will be shown that to obtain the lowest possible upflow velocity in the aperture area between the gas hoods – and thus ensure the maximum return of flocs to the digestion zone – it is necessary to select an equal width for the base (or widest point) of the deflector and the (projected) base of the gas hoods. Then, if for example a 10% overlap between the aperture and the deflector is adopted, the total combined width of all apertures will be equal to 45% of the reactor width. Therefore, both the combined widths of the gas hoods and that of the deflectors is equal to 55% of the reactor width (see also Example 4.1).

Above the level of the aperture, the area available for upward flow increases gradually and the liquid velocity decreases, reaching a minimum at the level of the effluent collection or overflow. In principle, only those flocs with a settling velocity lower than the maximum liquid upflow velocity will enter the settling zone, while flocs with a settling velocity higher than the upward velocity of the liquid at the overflow level can be retained. These solids will be deposited on the gas collection plates until their combined weight is enough to overcome the surface friction, allowing the solids to slide back into the digestion zone below the gas hoods. Particles with a settling velocity lower than the upflow liquid velocity at the overflow level are in principle not retained and are discharged together with the effluent. However, given enough time even these particles may – owing to adsorption or flocculation – conglomerate with other particles into larger, better settling particles with a settling velocity high enough to overcome the drag force of the liquid phase.

The top of the deflectors must not extend into the aperture above the base of the gas hoods, because this will occupy part of the aperture area, which consequently is not available for liquid flow, so that the upflow velocity is correspondingly higher. Sometimes the top of the deflector is even lowered somewhat below the aperture (by up to 20 cm) to ensure the maximum area is available and to make the passage from the transition zone to the settling zone as smooth as possible.

The physical dimensions of the GLS separator are to a certain extent determined by the construction material as indicated in Figure 4.15. In the first UASB designs, the entire GLS separator was made of concrete and to obtain enough structural strength a minimum width is required: typically at least 150 mm. Later composite GLS separators were introduced with the gas boxes and gas collection plates made of different materials.

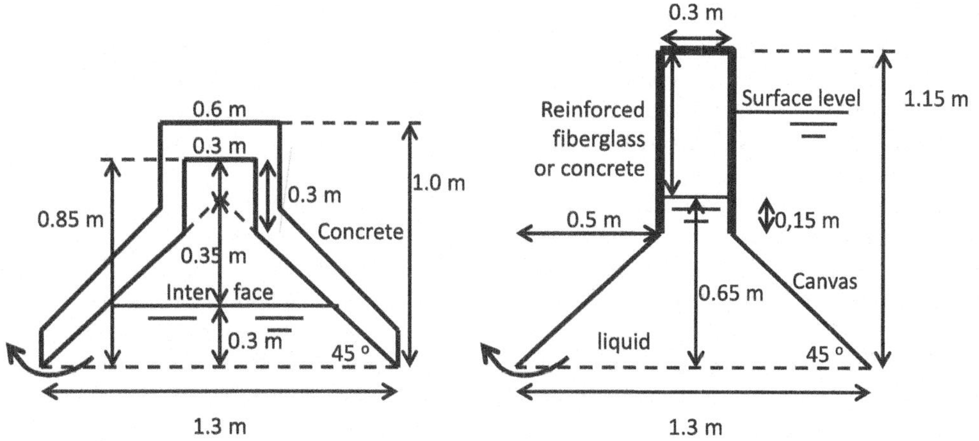

Figure 4.15 Examples of GLS separator elements (gas box + collection plates) made of concrete (left) and of lighter material like glass reinforced plastic (GRP) for the gas box and truck canvas for the plates (right).

At the surface of the gas–liquid interface inside the gas hood, a layer of scum may form which in time can slowly build up to a thick, semi-solid layer. At some point these solids may be released under the side of the gas collection plate into the overflow zone. To reduce this risk of unintentional solids discharge, the minimum required liquid level in the gas hood should be 0.3 m above the bottom of the gas hood, as indicated in the left-hand side of Figure 4.14. In the case of composite GLS settlers (right-hand side of Figure 4.14) the gas–liquid interface is typically located at least 0.2 m above the bottom of the gas box (and thus much further away from the bottom of the gas hood).

In several large full-scale UASB reactors, for example in the large 160,000 $m^3 \cdot d^{-1}$ Onça treatment plant located in Belo Horizonte, truck canvas has been applied with very good results as a construction material for the gas collection plates while concrete (lined or coated to ensure gas-tightness) was used for the gas box. However, use of these materials only is possible if the sole function of the gas collection plates is to convey the biogas to the gas box, as in this case the maximum force it must withstand is only its wet weight in the air. The (rigid) gas box, which can be made of GRP or concrete (the latter material should be coated or lined to make it gas-tight!) should be of sufficient height to comply with the following conditions: (1) to maintain sufficient distance between the gas–liquid interface and the biogas outlet to prevent clogging (see below); and (2) to prevent the escape of the gas through the relatively weak connection between flexible plates and rigid gas box.

In the gas box the biogas is pressurized by either an external hydraulic seal or a pressure-regulating valve. In principle the surface area required for gas desorption from the liquid phase, i.e. the combined surface area of all the gas boxes in a reactor, is determined by the maximum allowable gas desorption rate expressed in $m^3 \cdot m^2 \cdot h^{-1}$ or $m \cdot h^{-1}$. Souza (1986) recommended a maximum gas desorption rate in the range 3–5 $m \cdot h^{-1}$, but owing to the low rate of gas production in municipal UASB reactors, this will in practice never be a limiting design factor and the size of the gas boxes is determined by more practical considerations.

To prevent the build-up of biogas pressure under the flexible plates an emergency exit for the biogas at the level of the gas box must be installed, as shown in Figure 4.14. The connection of the gas collection plates to the gas box can constructed in such a way that there is an opening below the normal level of the liquid–gas interface, creating an automatic "safety valve" that is activated when clogging occurs in the gas piping and the produced biogas cannot be removed through the normal exit. In that case, the gas will accumulate, and the interface level will decrease until it reaches the opening. Any additional gas that is collected will then be expelled through this opening. As an additional benefit, the escaping gas bubbles will be visible on the surface of the reactor if it is not covered and will serve to alert the operator to the fact that the gas line is clogged.

As mentioned above, it is important to prevent the solids present at the gas–liquid interface from clogging the gas outlet. Actually, this is more likely to happen than one may think: the gas production rate in UASB reactors can be highly irregular as biogas is not continuously released but rather accumulates in the sludge blanket in the digestion zone. When enough gas builds up, suddenly large biogas bubbles can be released from the sludge blanket to the gas–liquid interface where they burst rather violently into the gas chamber, provoking high turbulence (this may even be audible from the outside). Thus it is necessary that a minimum distance is maintained between the gas–liquid interface and the outlet of the biogas pipe; in other words, a minimum height of the gas box has to be maintained.

The same turbulence induced by biogas release may also cause the release of solids at the bottom of the gas hood. In the gas hood there is a built-up of solids composed of a mixture of scum and sludge with biogas bubble inclusions. If the distance between the bottom of the gas hood and the interface is too small these solids may be released to the settling compartment and will tend to be discharged with the effluent, because this material is lighter than the liquid phase. Therefore a minimum distance of 0.3 m is adopted. Thus the minimum height occupied by the gas chamber in the gas hoods is 0.5 m and that by the liquid layer is 0.3 m, resulting in a total minimum height of 0.8 m for the gas hoods.

In practice, the liquid (submerged) height or sidewall depth of the gas box (H_{gb}) is typically at least 0.5 m (Chernicharo, 2007), which also allows for installation of effluent collection channels with overflow weirs or submerged effluent collection pipes. Including a 0.3 m freeboard height, the minimum total height of the gas box is then equal to 0.8 m. In practice the gas–liquid interface inside the gas box is displaced about 25–30 cm below the level of the effluent collection device. This is defined by the minimum biogas pressure maintained in the headspace, typically a gauge pressure of 25 mbar, which is required to operate the biogas flare or electricity generator. Taking this into account this means that the gas–liquid interface will be located approximately 0.2–0.3 m above the lowest point of the gas box. A typical example of gas hood design which respects the guidelines presented above is indicated in Figure 4.17.

The width of a gas box (W_{gb}) is usually determined by the requirement to provide safe and easy access for operators to the top of the reactor, as it doubles as a walkway. The minimum recommended (outer) gas box width is therefore 0.6–0.7 m. It is counterproductive to make the boxes wider as this will reduce the overflow area available for settling at the top of the reactor and thus increases the overflow rate at this critical level leading to undesired sludge discharge. When a minimum wall thickness of 150 mm is assumed, the internal gas box width that remains is between 0.3–0.4 m. Smaller gas boxes are prone to clogging and are only accessible and operated with difficulty.

The main factors that define the height of the GLS separator are the number of gas boxes, the height of the separator elements (as discussed above), and the inclination of the gas collection plates and the deflector. The latter typically ranges between 45° and 60°. The benefit of a high inclination of the gas collection plate is that it results in an increased likelihood of interception of sludge particles in the overflow section. On the other hand, it also results in an increased height of the gas hood, thus increasing the total height of the reactor because the height of the digestion zone is already determined by the criterion of the minimum required sludge mass. An inclination of 45° is often taken as the minimum value required ensuring that the settled solids can slide back into the digestion zone. However, a higher value might be beneficial, although this will increase the height of the GLS separator. Often the gas deflector has the same inclination as the gas collection plate, though this is not strictly necessary: a lower inclination is possible. In the absence of conclusive data with respect to operational performance with different plate inclinations, it is recommended to apply a plate and deflector inclination between 45° and 52°.

Gas boxes usually span the full length of the reactor, and the number of gas boxes installed per reactor depends on the available width of the reactor and on the requirement to minimize the total height of the GLS separator. For example, if a single gas box is selected, then in order to cover the full reactor the (projected) height of the gas collection plates will be very large. This reduces the height and volume available for the digestion zone. On the other hand, a large number of gas boxes will result in a low projected height of the gas collection plates and thus allows for a small total height, but with little surface area remaining for settling in the overflow zone. Therefore, a sensible compromise must be selected, ideally by calculation of the GLS layout as function of the number of gas boxes installed. It is recommended to base the selection of the installed number of gas boxes per reactor (N_{gb}) on the following two criteria:

- A minimum required height for quiescent settling of 0.8 m. This height is basically equal to the liquid height of the gas hood, i.e. the liquid sidewall depth of the gas box plus the projected height of the collection plates. If the deflector is positioned below the aperture, then the vertical distance between the base of the gas hood and the top of the deflector can be considered as part of the settling zone as well.
- A minimum fraction of the reactor volume must be available to accommodate the sludge mass in the digestion zone (at least 60%), to treat the applied sewage load.

Figure 4.16 shows the installation of a modular gas box while Figure 4.18 shows the completed GLS separator including submerged effluent pipes in a large full-scale UASB reactor. An important advantage

Figure 4.16 Photo of the installation of a prefab solid concrete gas box.

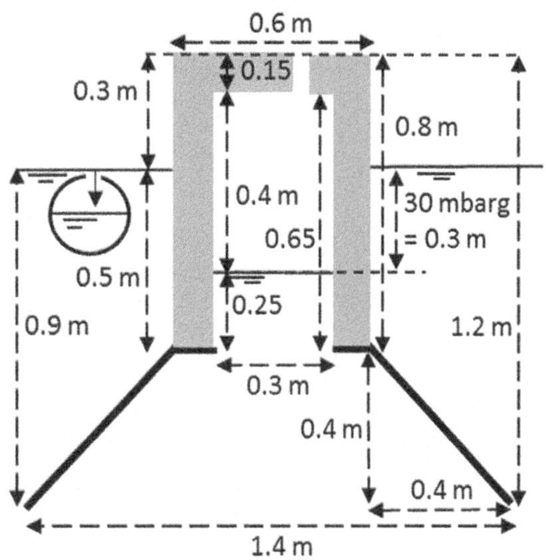

Figure 4.17 GLS design with concrete gas box complying to all design criteria.

Figure 4.18 Construction of a GLS phase separator consisting of a concrete gas box and canvas gas collection plates (sheets) to direct the gas production: (left) seen from above and (right) from the bottom of the reactor. The concrete deflector can also be seen and is used to attach the canvas gas collection plates. (Courtesy of CMS).

of the modular gas box design presented in Figure 4.16 is the relatively easy access it allows to the gas–liquid interface. There is a tendency for a layer of floating scum material to accumulate at the surface of this interface, which can turn into a hard cake layer especially in the coldest time of the year. As the gas box typically is heavy enough to remain in place under its own weight, it does not need to be permanently fixed to the reactor support structure. Therefore, if required it can be lifted for cleaning, and then put back in place. To allow this, provisions must be made so that the gas collection plates can be easily disconnected from the collector. Maintenance of gas hoods that are built as an integrated part of the reactor (particularly of the submerged GLS type) is much more complicated.

Example 4.1

Design the GLS separator of a 2000 m^3 UASB reactor (20 × 20 × 5 m) to be operated at a hydraulic retention time (HRT) of 8 hrs with different numbers of gas boxes per reactor (4–12), and evaluate the effect on the total height of the GLS separator and the digestion zone and the overflow velocity at the reactor outlet.

Calculate the fraction of the reactor volume occupied by the digestion, transition and settling zones and check the upflow velocity that develops at different levels in the reactor (digestion zone, aperture between the gas hoods and overflow area) with the recommended ranges specified in Section 4.3.1.2.

The following additional information is given:

- Select an equal width for the base of the gas hood and the deflector, while the overlap between the deflectors and gas collection plates is 10%;
- Use a 45° inclination for both the gas collection plates and the deflector;
- The gas box has a minimum total height of 0.8 m, a sidewall depth of 0.5 m and a width of 0.3 m; ignore the thickness of the gas box material.

Solution

The gas hood (consisting of gas collection plates + gas box) and the deflector have an equal width and thus both cover both at least 50% of the reactor surface to prevent gas bubbles from entering the settling zone. When it is considered that both elements have an overlap of 10% this increases to 55% of the reactor area for each. The total width of the reactor occupied by gas hoods will thus be 55% of 20 m = 11 m. Now the width of the individual gas hoods can be calculated as a function of the number of gas hoods installed. For example, for 12 gas boxes (=12 gas hoods) the width of each individual gas hood (which in this example is the same as the deflector width) is equal to 11/12 = 0.92 m. Similarly the width of the apertures is equal to (20 − 11)/12 = 0.75 m.

As specified above, the gas box width is 0.3 m, leaving (0.92 − 0.3)/2 = 0.31 m to be covered by each of the two collection plates included in the gas hood: the projected width of the collection plate. At a plate inclination of 45° the projected height of the collection plate is by definition equal to the projected width, but the general applicable calculation is 0.31 · tan(45°) = 0.31 m. The liquid height of the gas hood is calculated by adding the gas box sidewall depth of 0.5 m, which increases the total height of the GLS separator to 0.81 m, which complies with the first design criterion that the minimum height available for settling should be more than 0.8 m. The deflector being 0.92 m wide will have a height of 0.92/2 = 0.46 m, increasing the liquid height of the GLS separator to 0.81 + 0.46 = 1.27 m. The total height of the reactor was given as 5.0 m with a freeboard of 0.3 m, so that the reactor liquid height is 4.7 m. Hence the height of the digestion zone is 4.7–1.27 = 3.43 m, which will occupy 3.43/4.7 = 73% of the liquid reactor volume available. Hence this solution also complies with the second criterion that the digestion zone should be >60%.

The results of the main calculation steps or different numbers of gas hoods are summarized in Table 4.4, while the schematic reactor design for 12 gas boxes is shown in Figure 4.19. In Table 4.4 it is can be seen that the configuration for 6 gas hoods has insufficient volume available for the digestion zone (54% < 60%) while for 14 gas hoods the height of the separator units is insufficient (0.74 m < 0.80 m). Note that if the deflectors are moved down by, for example, 0.1 or 0.2 m, the solution with 14 gas hoods would also have been feasible. It can be concluded that between 8 and 12 gas boxes per reactor is adequate.

The GLS separator design now can be checked with regards to recommendations in Table 4.3 for the upflow velocities of the liquid at different levels in the reactor, for the design HRT of 8 hours. Naturally for all solutions

Table 4.4 Summary of the main parameters of the GLS separator design and the reactor configuration as function of the selected number of gas hoods in Example 4.1.

Parameter	UoM	16	14	12	10	8	6	4
Width of deflector/gas hood	m	0.69	0.79	0.92	1.10	1.38	1.83	2.75
Aperture width	m	0.56	0.64	0.75	0.90	1.13	1.50	2.25
Projected width/height plates	m	0.19	0.24	0.31	0.40	0.54	0.77	1.23
Height separator unit = available height for settling[1]	m	0.69	**0.74**	0.81	0.90	1.04	1.27	1.73
Height deflector	m	0.34	0.39	0.46	0.55	0.69	0.92	1.38
Liquid height GLS	m	1.04	1.14	1.27	1.45	1.73	2.18	3.10
Height digestion zone	m	3.66	3.56	3.43	3.25	2.98	2.52	1.60
Fraction digestion zone	%	78%	76%	73%	69%	63%	**54%**	34%
Fraction transition zone	%	7%	8%	10%	12%	15%	20%	29%
Fraction settling zone	%	15%	16%	17%	19%	22%	27%	37%

[1]In this design example the two are equal, as the top of the deflector is at the level of the apertures. Should the deflector be displaced vertically, this distance together with the height of the separator unit is then available for settling.

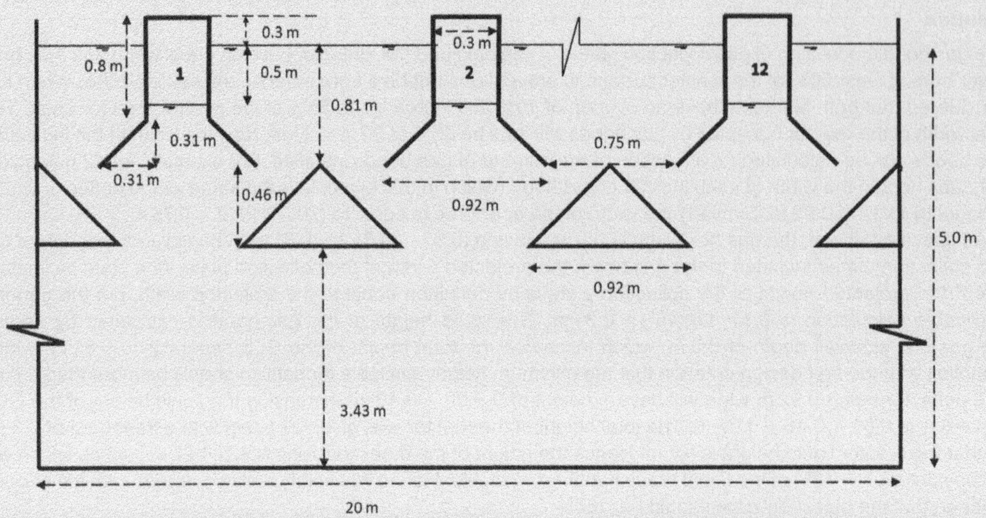

Figure 4.19 Schematic representation of the reactor GLS design for 12 gas hoods for the conditions of Example 4.1 (NTS). Only the first two and the last of the GLS separators are shown.

the upflow velocity v_u has the same value and is calculated as H_{liq}/HRT = 4.7/8 = 0.59 m · h^{-1}. The total area available for the apertures is also equal for all configurations, and the aperture velocity v_a can be calculated proportional to the total width available for the apertures, i.e. $v_a = 20/(20 - 11) \cdot 0.59 = 1.31$ m · h^{-1}.

The only velocity that actually depends on the number of installed gas boxes is the overflow velocity at the effluent discharge level: $v_o = W_u - N_{gb} \cdot W_{gb}$. Two values are indicated in Table 4.5: the first for the design HRT of 8 hrs, the second for the minimum HRT of 6 h that might be applied in practice, in which case the overflow velocity would increase by (8 − 6)/6 = 33%. It can be observed that all velocities are such that proper performance of the reactor is to be expected for both values of the HRT.

When a higher number of gas boxes is selected, then the overflow velocity at the discharge level will increase. On the other hand the volume fraction available for digestion also increases, as the height of the GLS decreases, which may compensate for the effect of the higher overflow velocity. Since the performance will be affected only marginally, the selection of the optimal number of elements is decided mainly by economic considerations and the ease of construction. In general it should be expected that a higher number of gas boxes will result in increased construction costs.

Table 4.5 Summary of the design velocities in the reactor for an HRT of 8 hrs as a function of the number of gas hoods installed in Example 4.1 – only valid solutions are considered. The recommended and limit values from Table 4.3 are indicated as well.

Velocities (all in m · h^{-1})	Symbol	Recommended	Limit	12	10	8
Reactor upflow velocity	v_u	0.5–0.8	0.3–1.0	0.59	0.59	0.59
Upflow velocity in apertures	v_a	<1.5	<2.5	1.31	1.31	1.31
Overflow velocity (for HRT = 8 hrs)	v_o	<1.0	<1.2	0.72	0.69	0.67
Overflow velocity (for HRT of 6 hrs)	v_o	<1.0	<1.2	0.96	0.92	0.89

In Chapter 10 the Excel sheet for UASB design is introduced, which can be used to calculate the optimal configuration of the GLS separator as a function of the width of the reactor, the number of gas boxes and the specified angles of the gas collection plates and the deflector.

4.3.3 Biogas deflector

The deflector should be located right beneath the apertures between the gas hoods. As shown in Figure 4.14, the top of the deflector is typically placed a little below the base of the gas hoods, to make sure that the combined distance between deflector and the base of the each gas hood is not less than the aperture width. However, if the plate and the deflector inclination are both equal to 45°, this is not required, as shown in Figure 4.19. Furthermore, the aperture upflow velocity should preferably be below the recommended value of 1.5 m · h^{-1} as per Section 4.3.1.2.

As for the design of the deflector, both open and closed units are used in practice. Open deflectors are typically constructed as inverted triangles, whereas the cross section of a solid unit can be triangular, square or even quadrilateral as shown in Figure 4.20(d–f). The main benefit of an open deflector is that it requires less material in construction. On the other hand, biogas that is captured by open deflectors must be diverted to the gas hoods, which is normally achieved with gas deflection tubes as shown in Figure 4.20(a–c).

An important design decision is the level of the gas–liquid interface with respect to the exit level of the biogas. Usually a scum layer will form on the surface of the interface in such a way that the biogas exit can easily become clogged.

One preventive measure is to maintain only a very small interface under the deflector as shown in Figure 4.20(b), so that scum formation is reduced and thus clogging may be avoided to a great extent. Another option is to use a curved gas deflection pipe so that the scum layer will be located below the deflection pipe as shown in Figure 4.20(c). If the deflector is solid, then formation of scum is not an issue. Triangular units may be applied as shown in Figure 4.20(d), but care must be taken that the bottom is completely flat

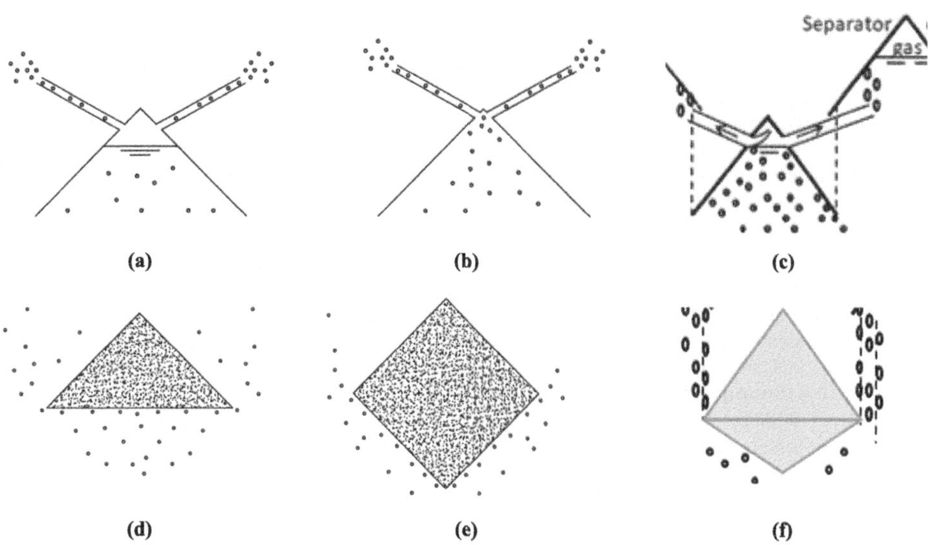

Figure 4.20 Open (a–c) and closed (d–f) biogas deflectors in different configurations.

Figure 4.21 Examples of assembling pre-fabricated solid concrete deflectors for a large UASB reactor (Courtesy of Companhia Matonense de Saneamento).

and horizontally well aligned to avoid accumulation of biogas under the deflector which, when enough volume is built up may suddenly escape over the side of the deflector and cause massive turbulence in its upward path, carrying sludge upward through the deflector. Therefore, it is better to design the bottom of the deflector not flat in order to actively deflect gas bubbles towards the gas hoods. The deflector need not necessarily be symmetrical as in Figure 4.20(e), although this facilitates construction, also a more artistic approach as in Figure 4.20(f) may be applied.

Example 4.2

The velocity of rising gas bubbles in water is typically around $0.2\ m \cdot s^{-1}$. Hence if the biogas production in a UASB reactor is such that the velocity in the biogas deflection pipes from the deflectors is more than $0.2\ m \cdot s^{-1}$, the biogas will no longer rise in the form of independent bubbles but rather as a jet flow. If the biogas production is $0.2\ m^3$ of gas per m^3 of influent, what should be the surface density of 50 mm gas deflection pipes to maintain a gas velocity of at least $0.2\ m \cdot s^{-1}$ in the deflection pipes? Assume a hydraulic retention time of 6 hours and a reactor height of 4 m.

Solution

The retention time of 6 hours and the height of 4 m together define the upward liquid velocity as $4/6\ m \cdot h^{-1}$ which is $0.67\ m^3 \cdot m^{-2} \cdot h^{-1}$, or, to put it another way, $0.67\ m^3$ of sewage is fed to the UASB per m^2 per hour. Hence the biogas production is equal to $0.2 \times 0.67 = 0.13\ m^3 \cdot m^{-2} \cdot h^{-1}$.

The cross-sectional area of a 50 mm tube is $0.002\ m^2$. To sustain a velocity in the tube of $0.2\ m \cdot s^{-1}$ or $0.2\ m^3 \cdot m^{-2} \cdot s^{-1}$ the flow of gas must be $0.2\ m^3 \cdot s^{-1}$ per m^2. Thus for 1 tube or $0.002\ m^2$ the biogas production must be $0.2 \times 0.002 = 0.0004\ m^3 \cdot s^{-1}$ or $1.4\ m^3 \cdot h^{-1}$. Hence per pipe this requires a reactor area of $1.4/0.13 \approx 10\ m^2$ for it to blow rather than bubble biogas out.

In Example 4.1, when 12 gas boxes were selected the width of the deflector was 0.91 m, so that the surface area of a single deflector is $0.91 \times 20 = 18.2\ m^2$. In this case only enough biogas is collected for one effluent pipe. If the diameter is reduced to 40 mm the required area is reduced by $(40/50)^2 \approx 2/3$ so that the required area is $7\ m^2$ and thus two pipes could be installed per deflector.

4.3.4 Influent distribution system

The purpose of the influent distribution system is to divide the influent flow as uniformly as possible over the bottom of the UASB reactor, so that good contact with the biomass is ensured when the sewage flows upwards through the sludge blanket. Except for single-family units, a distribution system is needed that divides the often large influent flow (between 200 and 500 $m^3 \cdot h^{-1}$ for large full-scale reactors) into many smaller flows, each of equal size. Furthermore, if more than one UASB reactor is installed, then even before the wastewater reaches the UASB reactor, it has already passed through a flow distribution network, where the wastewater is divided over the individual reactors.

4.3.4.1 Wastewater flow division

A large sewage treatment plant, where flows can reach as high as $100–150 \times 10^3\ m^3 \cdot d^{-1}$, requires an elaborate flow distribution network, often relying on a series of distribution chambers. At the reactor level, the flow is typically received in a division box called the reactor distribution chamber, from where flow is divided over several final distribution boxes which, if they are circular, are named final turndishes.

A full-scale design layout is shown in Figure 4.22, where the flow for twenty modular UASB reactors is divided into five equal parts in the central distribution chamber, each part directed to one of six second distribution chambers. One of these six distribution chambers is a spare unit for a potential future expansion with another four UASB reactors. This system was designed to treat the sewage of more than 1 million inhabitants.

In the second distribution chamber the flow is divided again, this time into four equal parts for each of the four UASB reactors this chamber serves. From the reactor distribution chamber the flow is distributed over the six final turndishes. Hence, before arriving at the final turndish the flow has already been divided

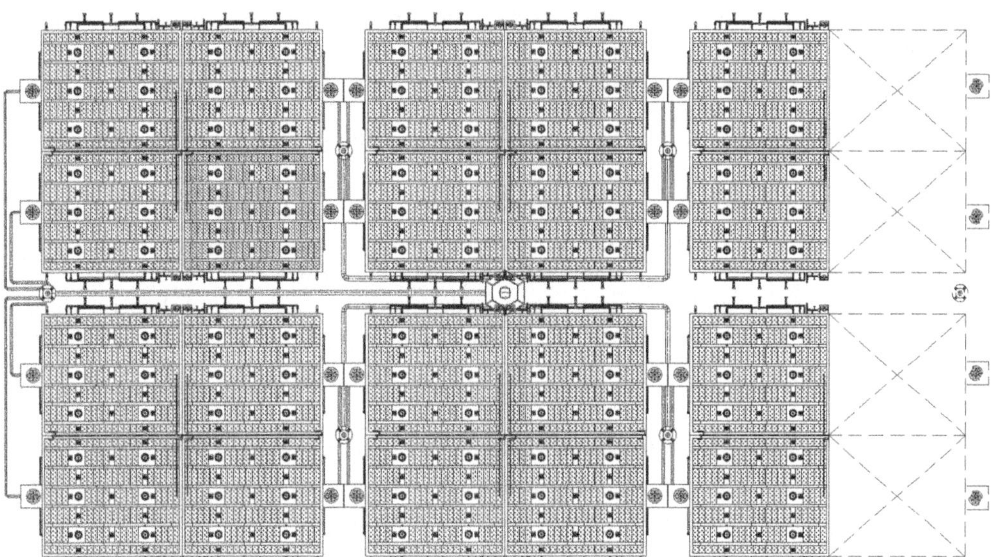

Figure 4.22 Design layout for a large-scale sewage treatment plant consisting of 20 modular UASB reactors of each 20 × 24 × 5 m, with room for expansion with another four reactors.

Figure 4.23 Reactor distribution chamber under construction at the Al Fileyah sewage treatment plant in the United Arab Emirates, dividing the reactor feed flow over eight final turndishes. Provisions for placing removable covers are in place.

$6 \times 4 \times 6 = 144$ times. It is clearly not feasible to install flow meters in even part of these lines or channels – typically only the incoming flow at the inlet of the sewage treatment plant is monitored and gravity is used for the division of the flow.

Some sensible guidelines for the design of the feed flow distribution system are summarized below:

- Ensure that all overflows are free-fall overflows, so that the pressure loss in downstream channels of pipes does not affect the liquid level at the overflow weir and hence the flow distribution.
- Make sure that even under peak flow conditions the division system is not flooded, i.e. ensure the height between the different steps in the hydraulic profile is sufficient.
- The distribution chambers and final turndishes must be easily accessible by operators, so that they can be cleaned and maintained if necessary.
- It is recommended to cover the feed flow distribution system – especially at the overflows – to prevent smell issues (H_2S) and connect the covers to a vent gas treatment system, such as a compost filter. Use removable covers to allow for inspection and maintenance.
- Provision must be made to be able to block individual outlets at each step of the flow division, not only to be able to isolate an individual reactor but also to distribute the excess flow uniformly over the remaining units.

4.3.4.2 Final influent distribution system

The design of the final influent distribution system (IDS) must be such that: (1) it is verifiable that the flow rate for each final discharge point is the correct fraction of the total UASB feed flow; (2) any clogging in the influent system is easily detectable; and (3) the obstruction that causes the clogging can easily be removed.

It is crucial to design the UASB influent distribution system correctly in order to allow for good performance of a UASB reactor, without short circuiting and local overloading. The maximum area

Table 4.6 Suggested final inlet distribution point densities for municipal sewage UASB reactors (adapted from Chernicharo, 2007).

Sludge Concentration in Digestion Zone (kg TSS · m^{-3})	Volumetric Loading Rate (kg COD · m^{-3} · d^{-1})	Suggested Density (m^2 Per Inlet Point)
>40	<1.0	0.5–1.0
	1.0–2.0	1.0–2.0
	>2.0	2.0–3.0
20–40	<2.0	1.0–2.0
	>2.0	2.0–5.0

covered by a single-feed entry point has been evaluated by several researchers. In Cali, Columbia, no difference in UASB performance was detected for feed inlet densities ranging from 1 to 4 m^2 reactor area per feed inlet (Haskoning, 1989). This range was later confirmed by several other authors (van Haandel & Lettinga, 1994). In recent practice, feed inlet densities ranging from a maximum of one inlet per 2 m^2 to a minimum of one per 4.5 m^2 have been observed (Heffernan *et al.*, 2010a), again without any clear and distinguishing effect on treatment performance. Table 4.6 shows the guidelines suggested by Chernicharo (2007), which are based on the expected sludge concentration in the digestion zone and the applied volumetric loading rate. At lower volumetric loading rates the recommended inlet point density increases, due to the reduction in mixing intensity that will result from the decrease in both influent flow rate and biogas production.

When a high density of inlet points is selected, this influences the construction costs, bearing in mind that for a typical full-scale UASB reactor (20×20 m = 400 m^2) and for a selected density of 2 m^2 surface area per pipe no less than 200 inlet pipes are required, if each inlet pipe will have only one outlet. The result will be a veritable "spaghetti" of pipes that all must be led down from the final distribution boxes through the limited space available of the aperture between gas hood and deflector and from there to the designated discharge location on the bottom of the reactor.

There are basically two approaches to distribution of the sewage over the reactor bottom: (1) to pump the influent directly into the reactor using feed pipes with outlet holes spaced at regular distances; or (2) to pump the influent into a distribution box (final turndish) and use gravity to distribute the influent flow further over the inlet points at the bottom. While the first alternative is often used for industrial wastewater treatment, it is not recommended for municipal sewage treatment: unless exceptionally efficient pre-treatment for coarse solids retention is applied, the distribution holes will eventually become clogged and the feed distribution will already be non-uniform after a short period of operation.

A widely used distribution system using gravity-based flow is indicated in Figure 4.24, with either a central feed well or a feed channel overflowing into feed boxes connected by pipes or tubes to corresponding feed inlet or distribution points at the bottom of the reactor. In general the circular arrangement is preferred, as it allows much easier hydraulic balancing of the individual overflow rates, because the water level in principle has the same height over the full perimeter of the device. In a channel however, as influent is diverted to the final inlet points, the flow will decrease and with it the pressure drop, so that the liquid level at the first overflow weirs will be appreciatively higher than that at the last overflow. This is especially problematic when several reactors are sequentially fed from one long common channel.

A removable cover can be placed over the influent distribution system to avoid bad odours and fly proliferation. In practice, however, the odours in anaerobic treatment plants can often be attributed to volatile influent material rather than to the anaerobic processes themselves. As the pressure in the boxes is atmospheric, they can be open (or be opened) for visual inspection. If there is clogging in the tubing of

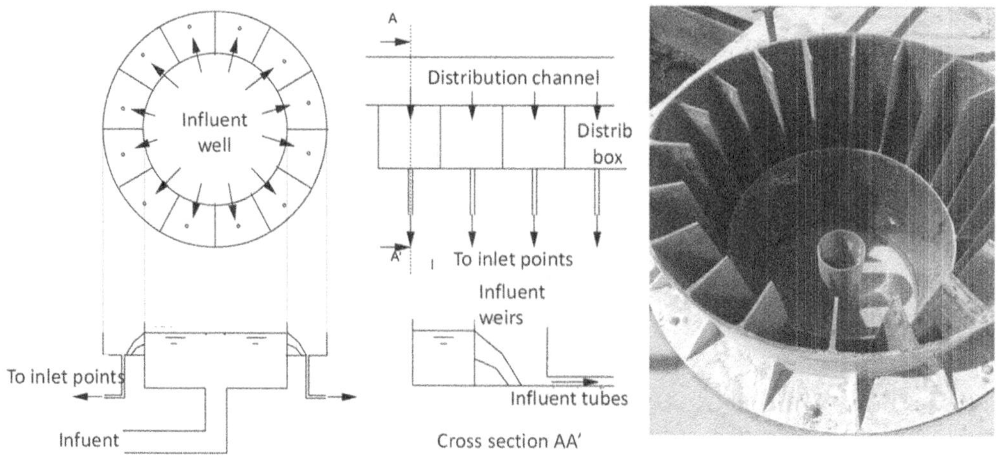

Figure 4.24 Schematic representation of final influent distribution box with both circular and rectangular design. On the right-hand side an installed circular final turndish is shown.

one of the boxes, its level will be equal to that of the inlet channel and higher than in the other boxes, so that this clogging is easily detectable.

In Figure 4.25 each inlet tube has several feeding points, whereas in Figure 4.26 each tube feeds one inlet. The latter option requires more feed piping to be installed but it ensures that pipes can be unblocked, which is not the case if multiple outlet points (or holes!) are used. As the final turndish is installed above the liquid surface in the reactor, it means that the feed pipes leading to the bottom of the reactor either will either have to pass through the apertures between the gas hoods and then bend around the deflector

Figure 4.25 Influent distribution system with multiple influent inlet points per feeding tube.

Figure 4.26 Influent distribution system with a single feed inlet point per feeding tube.

Figure 4.27 Different final turndish designs: on the left several smaller turndishes are installed that each fit in the overflow section between two gas boxes, on the right one central turndish is installed.

or, alternatively, they can pass directly through the gas collection plates or the deflector. The latter is only possible if the gas collection plates are constructed from a rigid material. If the tube passes through the GLS separator below the liquid–gas interface, then the perforation of the separator required for passage of the influent tube will not cause many problems provided the hole is reasonably tight, though there might be the occasional loss of biogas bubbles passing near the hole. On the contrary, it may be an advantage to have the orifice, as it can function as an emergency biogas relief valve in case of gas pipe clogging.

Example 4.3

Estimate the diameter and height of a circular influent distribution box with 100 feed points (and thus also 100 v-notch overflows), assuming there is one feed inlet point per 4 m^2, for a reactor height of 5 m and a retention time of 6 hrs. Allow for a peak flow at 3 times the average feed flow rate. A triangular weir is used with a bottom angle of 90°. The following formula gives the weir flow rate in as function of the liquid height above the base of the weir:

$$Q = 8/15 \cdot \sqrt{(2 \times 9.81)} \cdot C_e \cdot \tan(\alpha/2) \cdot h^{5/2} \ (m^3 \cdot s^{-1}) \qquad (4.6)$$

where

C_e = contraction coefficient, which is around 0.58
α = angle of the base of the v-notch
h = weir height in m

Solution

For the retention time of 6 h and a height of 5 m, the upward velocity of the liquid is 5/6 = 0.83 m · h^{-1}. For the given feed inlet distribution density of one per 4 m^2, the discharge per feed inlet point is 4 × 0.83 = 3.33 m^3 · h^{-1} and at peak flow rate 3 × 3.33 = 10 m^3 · h^{-1} per inlet point. This results in a required water height of 54 mm above the base of the weir at average flow and 84 mm at peak flow.

Taking 100 mm as the design liquid height of the weir, the minimum width of a single weir is 200 mm (for a 45° angle) and therefore for 100 overflow weirs mounted in a circle, an inner perimeter of 100 × 200 mm = 20 m, or a diameter of 6.5 m, is required for the final turndish. Assuming the feed boxes have a length of 0.3 m this brings the total outer diameter to 7.1 m (the thickness of the box material is not considered). If this is considered to be too large there are two solutions: (1) construct several instead of one final turndish, or (2) have a double row of distribution boxes as shown schematically in Figure 4.29.

If two final turndishes are installed there are 50 distribution boxes per turndish and the perimeter of the well is only 10 m so that the diameter is reduced to 3.2 m. Again taking a length of 0.3 m for the overflow boxes the total diameter is now 3.8 m. With a double row of distribution boxes the number of distribution boxes in both the first and second row is equal to 50. The inner perimeter is again 50 × 0.2 = 10 m, so that the inner diameter is 3.2 m and the outer diameter is 3.2 + 4 × 0.3 = 4.4 m assuming the second row of distribution boxes has the same length of 0.3 m.

To feed the second row of distribution boxes influent must be provided at the outside of the second row of boxes which requires an additional 0.2, m yielding a total external diameter of 4.4 + 2 × 0.2 = 4.8 m.

Thus the following possibilities exist:

(1) A single turndish with 100 influent points and a diameter of 7.1 m;
(2) Two turndishes with 50 influent points each and a diameter of 3.8 m;
(3) A single turndish with a double row of distribution boxes and a diameter of 4.8 m.

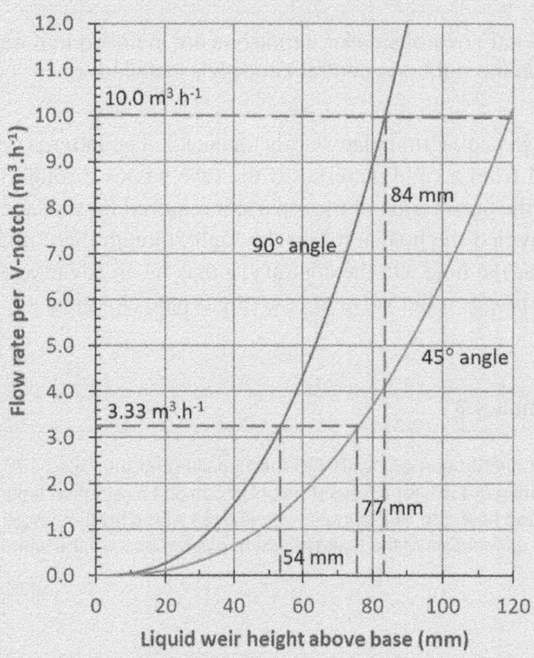

Figure 4.28 Flow rate in m^3 · h^{-1} as function of the weir liquid height in mm for a 45° and a 90° v-notch.

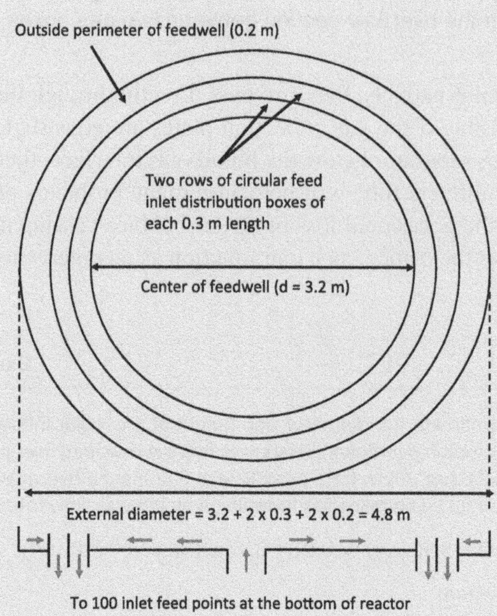

Figure 4.29 Circular turndish for 100 inlet points in a double row of feed boxes.

Whereas having only one concrete turndish is certainly attractive in terms of cost, having more turndishes (for Examples 6–8 per reactor) offers other advantages: (1) it avoids the construction problems involved in routing 100 pipes through the apertures located below a single turndish; and (2) the turndishes might fit between the gas boxes, i.e. with no interference with the construction of the GLS separators.

The relation between the overflow rate and the liquid height above the base of the weir for different base angles (of 45° and 90°) is given in Figure 4.28. The smaller angle of the weir has the advantage that at small flows there is still a significant water column over the weir, but at larger flows the height may increase beyond the height of the weir, so that control over the flow division is partially lost.

Example 4.4

Consider again a UASB reactor with a volume of 2000 m^3 and a liquid height of 5 m. The feeding point density is one per 4 m^2. Estimate the total length of feed pipe that is required if a single central turndish is used with one inlet point per tube. The connection height of the feed pipes to the final turndish is 0.5 m above the liquid surface.

Solution

The area of the digester is $2000/5 = 400\ m^2$ so that both length and width are 20 m for a square design. Hence 100 feed pipes are required for the required feed density of one per 4 m^2, the outlets of which can be arranged in 10 rows, each with 10 feed outlets. If a diameter of $6.5 + 2 \times 0.15 = 6.8$ m is adopted for the turndish, assuming the feed pipe is attached to the middle of the distribution boxes (see Example 4.3), then the length of the shortest feeding tube is about 5.5 m and the longest about 12.5 m, so that the average tube length is about 9 m long, which is 900 m tubing for all feeding points.

With a retention day of 6 hrs or a quarter of a day, the sewage flow that can be treated is $4 \times 2000 = 8000\ m^3 \cdot d^{-1}$, which for a typical per capita contribution of $100\ L \cdot d^{-1}$ is equal to the wastewater produced by 80,000 inhabitants. Hence the tubing for each inhabitant is 900/80,000 m or about 1.2 cm. It follows that the cost of having a single tube for an inlet is insignificant, while it guarantees uniform distribution of the influent, which is a big advantage.

It is important that the design of the feed pipes is such that air bubbles can escape back up the pipe. Otherwise the entrained air bubbles might conglomerate into larger air pockets that (temporary) block the inlet flow, until the upward force becomes so large that the air pocket is released back into the distribution box. This has a negative impact on the influent distribution. Other reasons why entry of these bubbles in the reactor should be prevented are: (1) the oxygen introduced may inhibit methanogenesis, and (2) it could result in the formation of a potentially explosive mixture with the biogas in the separator elements, though this is only a problem if the amount of air entrained is very large. According to Van Haandel and Lettinga (1994), a downflow velocity less than $0.2\ m \cdot s^{-1}$ at peak flow rate will be enough to prevent air bubbles with a diameter of 2 mm or more being entrained. On the other hand, a high velocity is required at the outlet of the pipe, typically mounted at 100–200 mm from the bottom of the reactor, in order to prevent deposition around the pipe outlet, which might result in blockages. At the maximum flow rate an outlet velocity of around $0.6\ m \cdot s^{-1}$, which is twice the design settling velocity for a grit channel, is considered to be enough. The material of the feed pipe may be rigid PVC or other plastic or flexible tubing, so that connection of the feed boxes to the discharge points is easy. It is important to note that the costs of the inlet pipes constitute a negligible fraction of the construction costs: in Example 4.4 it was calculated that the required length of the inlet pipes is not more than a few centimetres per inhabitant equivalent.

Example 4.5

Calculate the minimum diameter of the feed tubes in Example 4.4 to avoid that air is entrained. Assume that an air bubble has a rising velocity in water of 0.2 m · s^{-1}.

Solution

For the flow rate of 3.33 m^3 · h^{-1} per feed line and a maximum allowed velocity of 0.2 m · s^{-1} in the influent feed tubes, it is calculated that the diameter must be at least 78 mm, but since there is no such diameter in practice, a diameter of 100 mm will be selected. This diameter will not prevent air from entering the reactor when the flow is very large. For a 100 mm diameter feed tube a velocity of 0.35 m · s^{-1} is reached for the expected peak flow of 10 m^3 · h^{-1}. However when the flow is large a liquid layer be present at the bottom of the distribution box, largely impeding the entrance of air into the tube.

4.3.5 Effluent collection system

The purpose of the effluent collection system is to uniformly collect and discharge the treated wastewater at the top of the UASB reactor. If part of the effluent collection system takes much more water than another part, short circuiting can occur in the reactor, consequently with part of the GLS taking a much higher liquid loading rate than foreseen. As discussed elsewhere, it is recommended to use submerged effluent pipes instead of effluent gutters as the effect of misalignment on the flow distribution in the reactor is less severe and, perhaps more importantly, as emissions of sulphide and methane to the atmosphere are also minimized. Thus, the effluent pipes, especially when operated semi-flooded, can in principle be integrated in an effluent methane recovery scheme (refer to Section 4.3.9).

Most UASB reactors however have a collection device like that used in the final effluent settlers of activated sludge systems: horizontal gutters with triangular or v-notch overflow weirs placed at regular distances as indicated in Figure 4.30. The use of these gutters in open air causes many corrosion problems, due to the oxidation of dissolved H_2S to sulphuric acid, when they are made of some metallic material, even when using more resistant materials such as stainless steel. Preferably, fibre glass or a similar material should be used. Sometimes the effluent gutters are equipped with scum retainers or scum baffles (see Figure 4.30). The function of these scum baffles is to prevent the floating sludge from being discharged along with the effluent. Part of the floating material is sludge attached to biogas bubbles. When these bubbles escape into the atmosphere, there is a possibility for these solids (with a higher density than water) to return to the digestion zone. In Kanpur, India the performance of a UASB reactor with scum baffles was compared with that of an equivalent reactor without these scum baffles. It was observed that

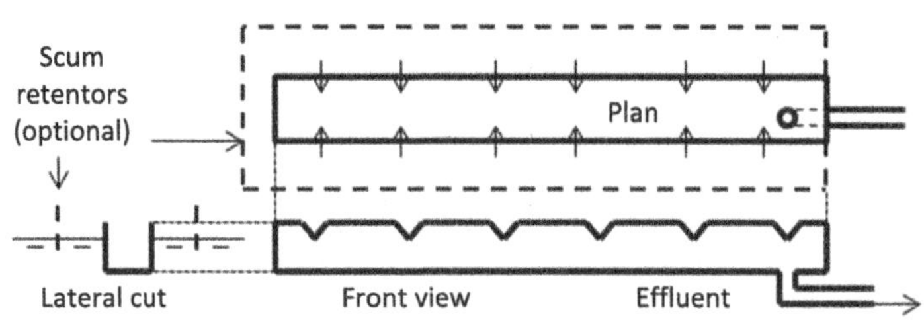

Figure 4.30 Triangular or v-notch overflow weirs with scum retention baffles for effluent collection and discharge, installed at the surface of the UASB reactor.

the scum baffles only marginally improved the quality of the effluent, which is according to expectation: the amount of floating material that is retained is very small and can be neglected, especially if there is a post-treatment system installed.

A frequent problem with the v-notch overflow weirs is that part of the gutter is blocked due to floating solids, even when scum baffles are applied. As a result, there is no longer a uniform collection of effluent. In order to have an overflow weir relatively free from operational problems it is recommended that the number of effluent v-notches does not exceed the number of feed inlet points. This way at least a reasonable flow rate can be applied per v-notch, which has a considerable self-cleaning effect.

Another problem with the above effluent collection system is that the effluent gutters need to be very well aligned during construction, which in practice can be quite difficult due to poor construction and workmanship. If for example there is a 5 mm difference in height of the gutter over a length of 10 m, something that can easily happen in practice, the difference between a water height of 40 and 45 mm in the weirs represents a difference of 34% in flow rate through the v-notch, so that the effluent collection would definitely not be uniform.

To reduce the effects of misalignment, at least 25 mm of head above the base of the v-notch should be available under conditions of average flow. For most municipal applications, this corresponds to 1 or 2 v-notches per m^2 of surface area, as calculated by van Haandel and Lettinga (1994).

Example 4.6

In the preceding examples what is the water level for rectangular effluent weirs if the same density is used as for influent points: i.e. 1 per 4 m^2 reactor surface area?

Solution

If each v-notch weir collects the effluent of 4 m^2 there must be 100 v-notches for the reactor area of 400 m^2. For the sewage flow of 8000 $m^3 \cdot d^{-1}$ or 0.093 $m^3 \cdot s^{-1}$ and each weir receiving 1% of this flow, it is calculated using the formula presented in Example 4.3 that the water level in the effluent weirs will be 54 mm above the base of the v-notch when the angle of the weir is 90°. For 45° the liquid height is 77 mm. It might be advisable to choose the latter option since the higher water level would tend to be more effective to avoid blocking of the weir by floating solids.

An alternative design for the effluent collection system is shown in Figure 4.31. In this case there are no gutters: the effluent collection device consists of a perforated PVC pipe that discharges the collected effluent below the water level into a common effluent channel. The perforations in the pipe are on the top, so that they are easily accessible for cleaning, providing there is no reactor cover, or alternatively that the covers are removable. It can be observed that during normal operation the level of the liquid in the reactor is the same as in the effluent channel and is set by the level of the final effluent weir as indicated in Figure 4.31.

Fouling by biomass growth and solids deposition is expected to occur both inside the effluent pipe and in the effluent channel. Both can be cleaned periodically by removing the final effluent weir: when this weir is opened (or removed), the sudden reduction of the water level in the effluent channel will result in large flush of effluent through the effluent pipes, which will thus be cleaned. At the end of the cleaning the final effluent weir is replaced. Care must be taken in the design that it is not possible to lower the water level in the reactor too much as it may cause the GLS separators to collapse due to the vacuum that is created in the gas chamber. For this reason vacuum reliefs should be provided in the biogas header or the gas boxes.

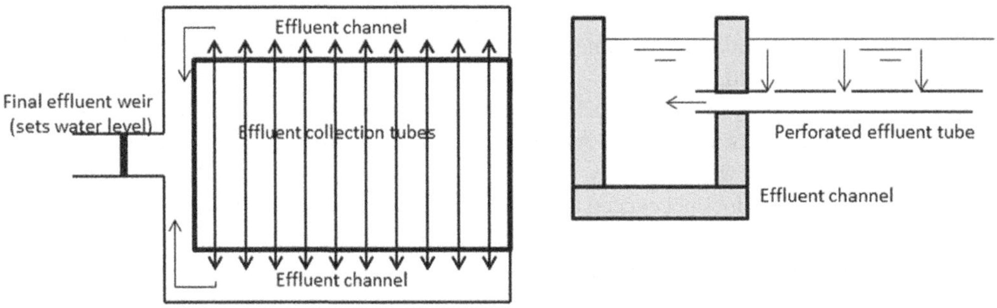

Figure 4.31 Effluent collection device with subsurface collection to guarantee uniform effluent withdrawal at each collection point (see also Figure 4.18 for an installation of submerged pipes at a full-scale UASB).

A disadvantage of the particular variation of the submerged pipe configuration shown in Figure 4.31 is that slightly unbalanced effluent withdrawal may develop. The reason for this is that the effluent lines are operated full and therefore a pressure drop will develop in the effluent pipes. As a result, the flow through the perforations at the end of the pipe furthest away from the outlet into the channel will be lower. This can be compensated by selecting a small diameter of the perforation, such that the pressure drop over it is much higher than the pressure drop over the pipe length. In that manner an essentially uniform discharge of the effluent can be guaranteed. In the design spreadsheet discussed in Chapter 10 the diameter of the orifice can be chosen by trial and error.

A variation on this design is to use free-fall overflows from individual effluent pipes into a common effluent channel. This way, provided the perforations in the effluent are properly sized (i.e. not too large), the flow through the perforations will be limiting and therefore the effluent pipes will be only partly filled with liquid. This guarantees uniform effluent withdrawal over the full reactor surface. As an additional benefit this allows the effluent pipes to be used in a vent gas extraction system, as will be discussed in more detail in Section 4.3.9.

4.3.6 Sludge sampling and discharge

Profiles of the concentration and quality (activity, stability, settleability) of sludge over the reactor height are important in evaluating the performance of the UASB reactor and in deciding on the need to discharge excess sludge.

To determine when a new sludge batch discharge is due, it is important to be able to establish the exact height of the sludge blanket. Often sampling connections are provided at the wall of the reactor, as shown in the left-hand side of Figure 4.32 and in Figure 4.33. If this method of sludge sampling is selected, then it is recommended to install a minimum of five sampling connections over the height of the digestion zone, at intervals not exceeding 50 cm. However, even then it is not possible to use the reactor volume between the top sampling point and the widest point of the deflector for digestion, as there is no information available about the sludge blanket level and concentration above the highest sludge sampling point. Furthermore, it is important that the sampling points extend at least 50 cm into the reactor as hydraulic effects can result in locally increased sludge concentrations at the wall. The diameter of the sampling pipes should be at least 50 mm.

Alternatively – and in our view preferably – samples can also conveniently be taken through openings in the phase separator as shown in in the right-hand side of Figure 4.32: sludge samples are taken from the

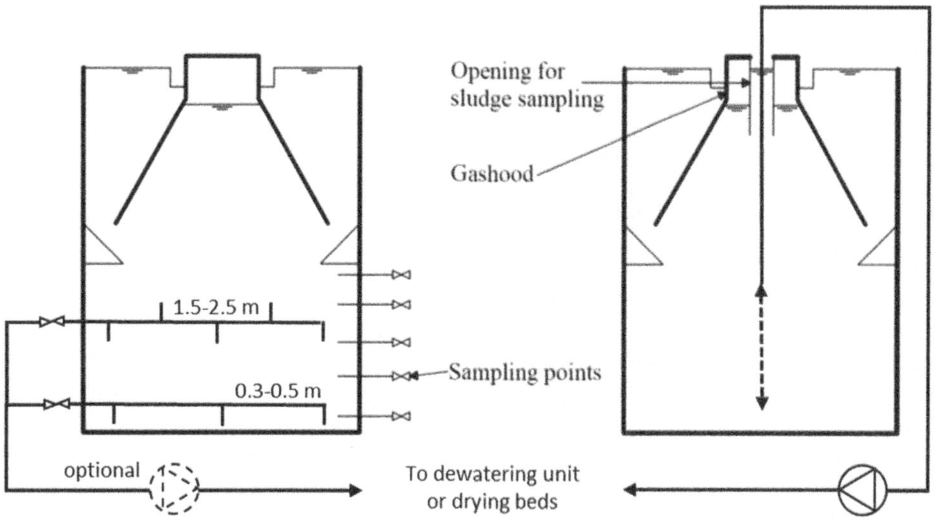

Figure 4.32 Different methods of sludge blanket sampling and excess sludge discharge: (left) sampling and discharge connections at fixed levels in the wall of the UASB, and (right) flexible height sampling and discharge through the gas box.

Figure 4.33 Typical sludge sampling manifold (left) and arrangement of the sampling lines in the interior of the reactor. Note that the distance from the wall is at least 0.5 m.

contents of the reactor at different heights with the help of a sampler and the biomass concentration profile is determined. The sampling tube is also useful for two other reasons:

(1) If the biogas outlet is blocked, the biogas production will start to accumulate, first in the gas box and when full also in the gas hood. The sampling tube can then relieve the gas pressure and avoid the development of an excessive gas pressure in the GLS separator, which could damage the equipment.
(2) If for some reason (for example a rate of excess sludge discharge that is too high) a vacuum is created in the gas box, this may lead to an implosion of the gas hood. The sampling tube will allow air to be sucked into the gas box and thus avoid an excessively low pressure.

The quality of the effluent tends to improve with increasing amounts of biomass in the reactor, and is maximum when the digestion zone is operated as full as possible and with the highest concentration of active biomass. However, the solids retention capacity in the UASB is not unlimited, and when it is depleted any additional amount of sludge produced in the reactor will inevitably lead to an equivalent unintentional discharge of excess sludge with the effluent. The presence of these solids reduces effluent quality: the presence of sludge particles results in increased concentrations of BOD, COD and TSS. For this reason, part of the sludge should be discharged periodically when its mass (and volume) reaches a certain pre-set value. This procedure is particularly recommended when there are no post-treatment units and the quality of the effluent in terms of suspended solids is important.

In general the bottom part of the digestion zone will contain the densest sludge (up to 60–80 g TSS $\cdot$ l^{-1}), while this gradually decreases to 10–20 g TSS $\cdot$ l^{-1} at the top of the sludge blanket. Thus in principle it is good practice to install two systems for excess sludge discharge: a top one from which the less dense sludge is discharged by preference, and a bottom one that is mostly used to control the inorganic fraction in the biomass: i.e. to prevent the build-up of heavy inert material such as sand, silt and clay.

In principle, there are two sludge extraction setups: (a) using a sludge discharge manifold installed at fixed height inside the reactor as shown in the left-hand side of Figure 4.32, and (b) flexible discharge from the top of the reactor through openings in the gas separator as shown on the right-hand side of Figure 4.32.

In the first case there are typically bottom sludge extraction pipes installed at 0.3–0.5 m height to remove the heavy inert sludge and a top sludge extraction manifold installed at 1.5–2.5 m height (this depends on the height of the digestion zone) to remove the less dense and inactive sludge. An example of these extraction setups is shown in Figure 4.34. The diameter of the bottom excess sludge discharge pipes should be at least 150 mm to prevent clogging, according to Chernicharo (2007). Furthermore, the bottom sludge pipes typically have only one inlet as this allows for unblocking should this be required. The diameter of the second extraction pipe can be larger (e.g. 300–400 mm) and this pipe has several lateral suction pipes, each of 150 mm diameter.

In the second approach to sludge discharge, using weights a flexible hose can be lowered through the sample opening in the gas hood to the height from which sludge discharge should take place. This hose should be reinforced to prevent collapse when a vacuum is pulled by the sludge discharge pump. As discussed earlier, the same openings can also be used for sludge sampling. As the opening extends to below the level of the gas–liquid interface, there is no risk of biogas escape.

Several combined sampling/sludge extraction points should be provided, distributed over the surface area of the reactor, allowing the sludge to be withdrawn from different locations in the reactor. Thus, by applying a rotary sludge discharge scheme, the accumulation of mineral solids in the bottom of the reactor can be avoided or at least reduced.

The maximum sludge mass that can be maintained in the UASB reactor depends on the settling characteristics of the sludge, which will depend on various factors such as the composition of the influent, which in turn varies from one location to another. It is also necessary to experimentally establish the

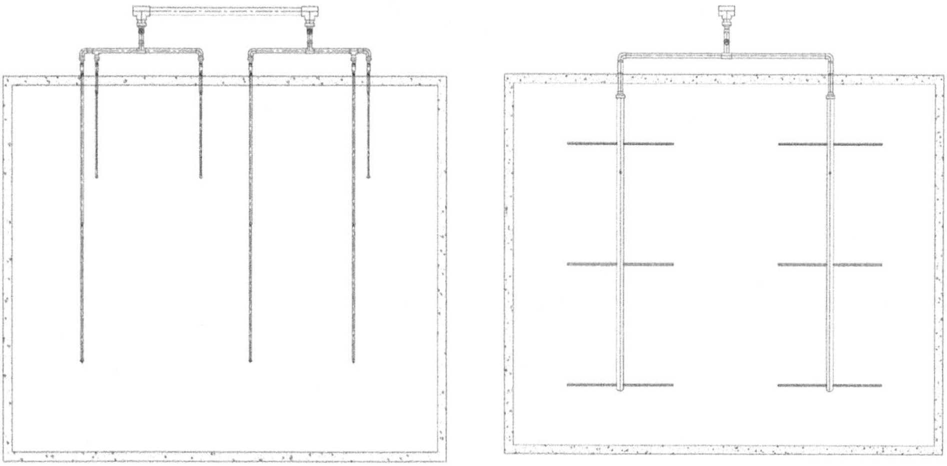

Figure 4.34 Example of sludge removal setup with a bottom sludge extraction system (left) consisting of two headers with each three sludge abstraction points and a top header (right) with two larger diameter central collection headers with each six small sludge inlet points.

maximum mass of excess sludge that can be discharged at one time, without severely affecting the quality of the effluent. The frequency of sludge discharge depends mainly on the method for dewatering and drying used. Mechanical dewatering devices such as decanter centrifuges and belt presses will operate normally in continuous mode but only during working hours. Therefore in this case the UASB must have a (semi-) continuous controlled discharge rate in order not to exceed the capacity of the dewatering equipment. Alternatively, a batch of excess sludge can be received in a sludge collection sump for gradual transfer to the dewatering equipment. The sump needs to be equipped with a mixer to prevent settling of heavy solids, and ventilation to prevent accumulation of explosive or toxic gases.

If the excess sludge is discharged to sludge drying beds, then it is important to fill up the beds entirely in one discharge: i.e. once a sludge bed has received a sludge batch it is no longer allowed to send subsequent batches to the same drying bed. Hence, in the case of sludge drying beds, sludge discharge will tend to have a pre-established routine in that at regular intervals (e.g. weekly or monthly) a certain volume of sludge (e.g. 20–50% of the sludge mass of a UASB reactor) is discharged. In both cases, the sludge discharge can be carried out on a more solid basis when it is preceded by a determination of the sludge concentration profile. A strategy for sludge discharge to drying beds is discussed in detail in Chapter 7.

4.3.7 Scum removal

The formation of scum is perceived by many operators as an important problem, especially in subtropical regions with a low temperature, where fat degradation proceeds at a lower rate. However, many UASB reactors treating raw sewage do not have specific devices to remove the scum that forms under the phase separator or on the liquid surface of the UASB reactor. At present, most attempts to reduce the formation of this scum layer are by means of an efficient pre-treatment, designed to remove both floating material and oils in conventional or (preferably) aerated grit channels.

With regards to the overflow area, this is especially a problem if v-notch weirs are installed and the design guidelines are not respected; thus with too many v-notches installed the self-cleaning effect is lost and blockages of individual v-notches will soon occur with scum spreading rapidly over the entire surface area. If the reactor is not covered it may be sufficient to have regular manual cleaning of the v-notches, preferably frequently enough that the scum layer does not solidify.

For covered reactors so-called "scum bowls" may be installed as shown in Figure 4.35: collection points distributed over the reactor surface area and connected to a scum discharge manifold, which are regularly opened to remove the still fluid scum and discharge it to the effluent discharge line (not to the sludge lines, as the liquid is very dilute!). Typically one or two scum bowls are provided in each overflow section between two gas boxes. Alternatively overflows can be provided in the reactor wall, connected to the effluent piping to the post-treatment units, but this is less efficient as the scum extraction points are far apart.

Preventing the build-up of a scum layer in the gas box is much more difficult because this build-up cannot be observed, as the gas box is not readily accessible during operation, not only being pressurized but also containing a potentially explosive and toxic atmosphere. In addition to introducing a more intensive pre-treatment, devices can be installed that allow the periodic removal of the scum layer from the gas box (see Figure 4.36). Two different mechanisms are proposed:

(1) Provide overflow openings at both ends of the gas box, connected to a scum removal pipe linking into the effluent line. Scum is discharged by closing the normal discharge of the effluent, which results in a liquid level increase in the gas box and hence an overflow into the scum removal pipes (upper part of Figure 4.37).
(2) Discharge into a scum collection trough placed within the gas box and then into the effluent pipe, by raising the level of the gas–liquid interface. This can be done using a sudden relief of the biogas pressure or by temporarily interrupting the discharge of effluent. Thus, periodically the contents of the gas hood are removed through a sequence of steps, as shown in the lower section of Figure 4.37.

Figure 4.35 Scum removal device or "scum bowl" installed for removal of scum from the surface of the overflow area.

Figure 4.36 Scum removal device or "scum trough" installed inside the gas box for removal of scum from the liquid-gas interface (Onça, Belo Horizonte, Brazil).

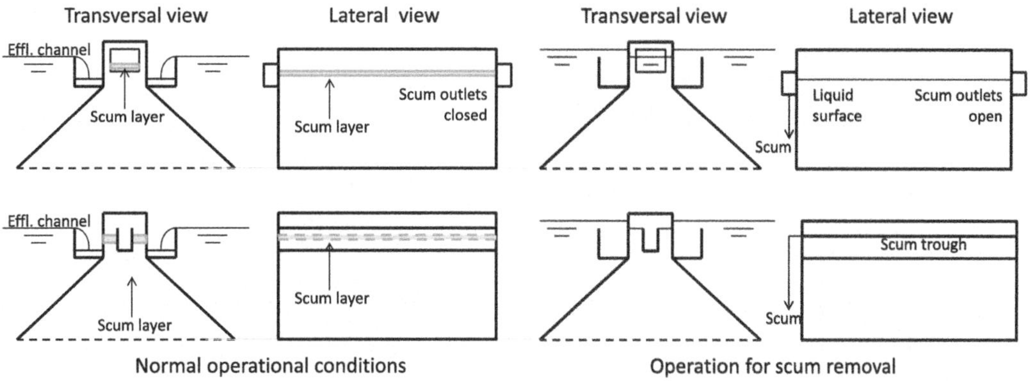

Figure 4.37 Example of a scum removal procedure from the gas hoods by means of a scum removal trough installed inside the gas box.

Note that both approaches will only work if the scum is not yet solidified – hence regular, preventive discharges are needed to remove the still soluble precursors of the scum.

Care should also be taken that biogas escape through the scum disposal lines is prevented, so a hydraulic lock should be in place. Furthermore, this procedure can only work if the gas boxes are gas-tight. As an alternative to increasing the liquid level in the gas box through a reduction of the gas pressure, it is also possible to close the effluent line. An emergency overflow should then be provided to prevent uncontrolled effluent spillage during the discharge procedure, in case the operator forgets to re-open the effluent line again.

The key factor in sustained and successful operation of a scum removal system is to prevent the scum layer from turning into a hard or viscous layer. As an example, in one of the plants visited a scum removal system had been installed, but it had not been used in the first year after start-up, owing to commissioning problems in the sludge dewatering unit. As a result, the whole scum discharge system had become completely blocked by scum. Hence, there was no way of rehabilitating the scum removal system except for a complete reactor drain. That experience illustrates how scum disposal should take place on a regular basis. It could even be considered to automate the procedure. In the case described above, it would have been possible to prevent the blockage of the scum removal system if the scum could have been discharged into the effluent line and not towards the sludge treatment units. As long as the scum discharge flow rate is high enough, the scum will be very dilute so that the COD content will be low. The decrease in UASB effluent quality will be very small while the scum will be removed in the aerobic post-treatment system anyway. Also, discharge to the effluent line decreases the hydraulic load to the sludge treatment units. Therefore, in our view, notwithstanding the problems associated with installing and operating a scum removal system in the gas hoods, it is strongly recommended to do so, as the alternative of emptying and cleaning a UASB reactor is a very labour-intensive task.

4.3.8 Biogas collection

The biogas collection system is designed to allow for removal of the biogas produced in the reactor while maintaining a constant gas box pressure, which will in turn result in a constant level of the liquid–gas interface in the gas box. It should be noted that – in contrast to most industrial UASB reactors and even to most other anaerobic reactor configurations, such as completely mixed digesters, for example – the

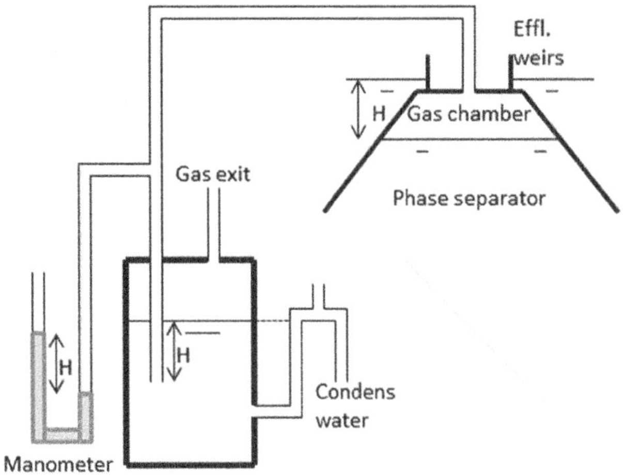

Figure 4.38 Schematic representation of a hydraulic seal connected to a GLS separator.

Figure 4.39 Hydraulic seal sized for a design production of 8000 $m^3 \cdot d^{-1}$ of biogas.

headspace of the UASB reactor itself is not contained and pressurized, but only the relatively small volume occupied by the gas box.

Biogas production rates are not very high in the case of sewage treatment, even for large treatment plants, owing to the low COD concentration in the sewage, the low digestion efficiency and the substantial loss of methane with the effluent. However, the diameter of the gas pipe cannot be very small because there may be solids (scum) present in the gas that can otherwise cause clogging.

Assuming the gas–liquid interface is located below the water level in the reactor, then the biogas must pass through a hydraulic UASB reactor seal, the height of which determines the pressure of the accumulated biogas. Alternatively a pressure-regulating valve can be used. A typical operating pressure is around 25 mbar, requiring 25 cm water column in the hydraulic seal, if the biogas is to be flared only, and up to 30–35 mbar if the biogas is treated prior to utilization (sulphide removal typically followed by biogas drying), although utilization of the biogas is rarely observed for municipal sewage in a UASB.

Owing to biogas cooling, condensate will accumulate in the hydraulic seal. Therefore, it is necessary to install an overflow for the condensate, as shown in Figure 4.38, so that a constant level and pressure can be maintained in the gas box. Furthermore, the biogas header should be installed without pockets or low points where condensate could accumulate, eventually blocking the biogas piping. Preferably the biogas piping should drain towards the hydraulic seal and/or to the condensate trap located directly upstream of the biogas flare. If pockets or low points in the biogas piping cannot be prevented, then these should be equipped with a condensate trap and drain.

4.3.9 Reactor cover and ventilation

The first UASB reactors for sewage treatment were built without reactor covers, hence with the liquid surface of the overflow area open to the atmosphere. However, owing to concerns about smell issues, the trend in more recent large-scale projects is to include reactor covers to ensure that smell problems in the vicinity of treatment systems are avoided. However, it should be considered that often the emission of more intense odours does not occur at the surface of the UASB reactor, but rather in the preliminary treatment

Figure 4.40 Example of a reactor that is completely covered by a fixed concrete roof (left), versus an example where lightweight aluminium covers are used (right).

section, i.e. at the wastewater lifting station, fine screens, grit trap or feed distribution system. Other points of higher emissions are typically the effluent collection channels, especially at the overflow weirs and at any free-fall overflow in the treated effluent discharge lines.

Hence it is recommended to cover only these high-risk areas, which is especially easy if submerged pipes are used instead of overflow weirs. The gaseous emissions from submerged collection pipes are considerably lower than those from overflow weirs into open effluent channels. Furthermore, the effluent pipes can be easily connected to – and form part of – the vent gas collection system if they are operated only partially filled. If it is desired to do so, then vent gas is extracted from the (covered) effluent channel in which the effluent pipes discharge. To prevent increased turbulence at the inlet perforations of the effluent collection pipe, a separate ventilation air inlet is provided at the far end of the pipe.

If it is decided to install reactor covers then there are two options: (1) to use removable covers made of lightweight UV- and corrosion-resistant materials such as fibreglass-reinforced plastic or aluminium; or (2) to construct a fixed concrete roof over the whole surface of the UASB reactor. The latter option requires the provision of several hatches which still provide (albeit very limited) access to the effluent channels or pipes for cleaning and maintenance. From the point of view of operational convenience there is no doubt that both alternatives to avoid undesirable gas leaks have a negative impact, and especially so the fixed cover.

The installation of a cover over the reactor liquid surface is not enough by itself to prevent odour emissions: a small vacuum should be also created in the headspace under the reactor cover to prevent odorous gases from passing into the atmosphere through small leaks and openings. Hence it is recommended to connect the reactor headspace with extraction fans operating a low differential pressure to a vent gas treatment system, for example compost filters.

Another reason why (partial) surface coverage of the UASB reactor could be applied is to prevent desorption of dissolved methane. As discussed in Chapter 6, a quite significant part of the methane produced in the UASB is lost with the effluent. In the UASB reactor the liquid phase is in equilibrium with biogas containing between 70–85% of methane, resulting in a high methane partial pressure, so that about 20 mg $\cdot$ L^{-1} of methane remains dissolved at the overflow level. When the anaerobic effluent is exposed to the atmosphere all this methane will be eventually be released. The reactor cover itself may delay, but not prevent, desorption of the dissolved methane: at some point the anaerobic effluent will encounter the atmosphere and at this moment the desorption process will begin.

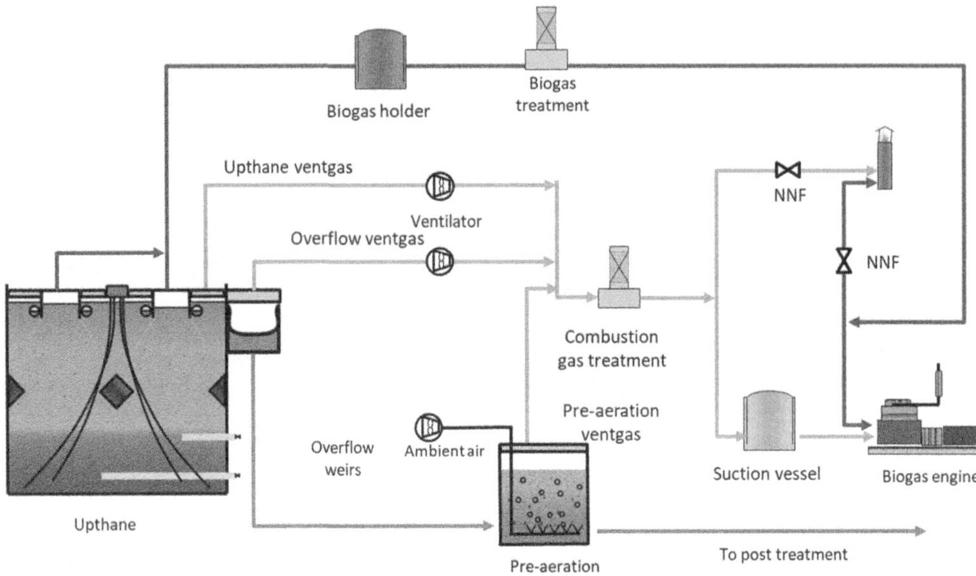

Figure 4.41 Simplified flow scheme for methane recovery from the covered overflow section of a UASB and from the pre- or flash aeration off-gas for use as combustion gas in either the biogas flare or engine. In the latter case, treatment of the combustion gas for H_2S removal might be required.

If the reactor is operated with a slight vacuum underneath the roof and with continuous vent gas extraction the methane is thus retained in the vent gas which subsequently can be used to burn the methane as combustion gas. One of the concepts is shown in Figure 4.41, which was developed by Biothane for a patent for the recovery and use of methane from their Upthane municipal UASB reactor (van der Lubbe & Heffernan, 2012).

Alternatively, the vent gas may pass through a biological filter of wetted bagasse or similar material where there may be biological oxidation of methane and other gases (H_2S), however the extent and efficiency of methane removal from the vent gas in this kind of biological filter has to our knowledge not yet been evaluated.

4.4 REACTOR DESIGN PROCEDURE

In this section an integrated reactor design example for the reactor and the GLS is presented which is based on the design considerations discussed in the previous sections. It consists of the following steps:

(1) If the process design is based on HRT, then adopt a suitable design hydraulic retention time R_h taking into consideration the minimum monthly sewage temperature (for example using the Brazilian standard BR 12.209) and possibly other area, budget or construction restraints (Section 4.2.1). Using the design sewage flow, for example, the daily maximum flow rate Q_i in $m^3 \cdot d^{-1}$, calculate the required liquid reactor volume V_u and continue with step (3):

$$V_u = Q_i/(24\,R_h) \text{ in } m^3 \tag{4.7}$$

(2) If the process design is based on the sludge age then select a suitable value for the design sludge age R_s, which satisfies the requirements with respect to COD removal and/or sludge stabilization: follow the procedure outlined in Section 3.10:

(a) Use Eqs. (3.16) or (3.25) to calculate the COD fraction that will end up as volatile sludge. From the organic load and the volatile sludge fraction calculate the sludge mass MX_t that will develop (in kg TSS). Either use a typical value for the organic fraction (0.55–0.7) or calculate this with Equation (3.43b).

(b) Assume a design value for the average sludge concentration in the *digestion* zone X_{td}, for example 30–40 g TSS · L^{-1} and calculate the required digestion zone volume:

(c) $V_{dig} = MX_t/X_{td}$ (m^3) (4.8)

(3) Determine the total required *minimum* reactor surface area based on the specified design wastewater flow rate and by adopting a maximum design upflow velocity v_l in the digestion zone, for example 0.7 m · h^{-1} (Section 4.3.1.2):

$$A_{min} = Q_i/(24v_l) \text{ in m}^2 \quad (4.9)$$

(4) Adopt a value for the area (A) for one reactor (for example 400–500 m^2) and calculate the number of reactors (N) that is required. Roundup to the next integer:

$$N = A_{min}/A \quad (4.10)$$

The total reactor area is equal to $N \cdot A$ (m^2) (4.11)

(5) Select the length (L_u) and calculate the width (W_u) of the reactor. For single reactors typically a square configuration is selected $L_u = W_u = (A_u)^{0.5}$

(6) Only for the design procedure based on sludge age: based on the selected reactor surface area A_u, calculate the height of the digestion zone:

$$H_{dig} = V_{dig}/(N \cdot A_u) + 0.25 \text{ (m)} \quad (4.5)$$

If this is outside of the recommended interval of $2.0 < H_{dig} < 3.5$ m, consider the following actions:

- Decrease the selected design reactor upflow velocity v_l, which will increase the area available to the digestion zone;
- Decrease the design value of the sludge age R_s, which will decrease the total required anaerobic sludge mass and hence the volume of the digestion zone V_{dig}.

(7) For the selected reactor surface area A_u and as function of the number of gas boxes installed in the reactor (N_{gb}) calculate the overflow velocity v_o that will result and check whether it is in the recommended range (<1.0–1.2 m · h^{-1}, Table 4.3). Assume a typical value for the external width of the gas box W_{gb} (including wall thickness) of, for example, 0.6–0.7 m.

$$v_o = Q_i/[(A_u - N_{gb} \cdot W_{gb} \cdot L_u) \cdot 24 \cdot N] \text{ in m} \cdot \text{h}^{-1} \quad (4.12)$$

If not, discard that configuration and reduce the number of gas boxes until a suitable overflow velocity is found. In principle the largest possible number of gas boxes is selected, as this will minimize the total liquid height of the GLS and thus also that of the reactor.

(8) Select a maximum value of the aperture velocity v_a, for example, 1.5 m · h^{-1} (refer to Table 4.3). For the selected value of v_a, calculate the required aperture area and from this the aperture width. Consider that there is half an aperture width on both extremes of the reactor, so that the number of apertures is equal to the number of gas boxes N_{gb}:

$$A_a = Q_i/(v_a \cdot 24 \cdot N) \text{ in m}^2 \quad (4.13)$$

$$W_a = A_a/(L_u \cdot N_{gb}) \text{ in m} \quad (4.14)$$

(9) Calculate the projected width of the collection plate. The combined projected width of all the gas collection plates is equal to the part of the reactor not occupied by either the apertures or the gas boxes. Considering that each gas box has two collection plates, the projected width a of a single gas collection plate is given by:

$$a = (W_u - \Sigma\, W_{gb} + \Sigma\, W_a)/(2 \cdot N_{gb}) = (W_u - N_{gb} \cdot (W_{gb} + W_a))/(2 \cdot N_{gb}) \text{ in m} \tag{4.15}$$

(10) Select the angle of the gas collection plate inclination α, for example 52°. For this angle, calculate the projected plate height "b":

$$b = a \cdot \tan(\alpha) \text{ in m} \tag{4.16}$$

(11) For the calculated aperture width W_a, calculate the deflector width W_{df} required to cover the aperture. Consider that the width of the deflector should overlap at least 0.1–0.15 m with each of the gas collection plates:

$$W_{df} = W_a + 2 \times 0.15 \text{ (m)} \tag{4.17}$$

(12) Calculate the height of the deflector H_{df}. Unless there is a specific reason not to do so, accept the inclination used for the gas collection plates.

$$H_{df} = (W_{df}/2) \cdot \tan(\alpha) \tag{4.18}$$

Note that only the top part of the deflector, i.e. above the widest part of the deflector, will be included in the calculation of the total liquid height of the GLS, as in principle the digestion zone can extent to just below the widest part of the deflector. The required safety distance between the sludge blanket and the widest part of the deflector is already included in the calculation of the required height of the digestion zone (Equation (4.5)).

(13) If the inclination of either gas collection plates or deflector is greater than 45°, decrease the installed level of the deflector with maximum 0.2 m (H_{dist}) to ensure that that it is the aperture area that is limiting, and not the area between the edge of the gas collection plates and the deflector.

(14) Assume typical values for the total gas box height (0.8–0.9 m) and the gas box freeboard (0.3 m), which sets the liquid height of the gas box H_{gb}. The total liquid height of the GLS separator (or more exactly the height available for undisturbed settling of entrained solids) is equal to the sum of the projected height of the gas collection plate, the deflector, the liquid gas box height and if applicable the vertical displacement of the deflector:

$$H_{gls} = b + H_{df} + H_{gb} + H_{dist} \tag{4.19}$$

(15) Check if enough settling height is available ($H_{gls} - H_{df} > 0.8$ m). If this is not the case, decrease the number of gas boxes, which will increase the projected height of the gas collection plates, and repeat the calculation of the GLS

(16) Calculate the liquid height of the reactor. For the design based on HRT this is equal to:

$$H_{liq} = V_u/A_u \tag{4.20}$$

and for the design based on sludge age:

$$H_{liq} = H_{gls} + H_{dig} \tag{4.4}$$

The liquid height of the reactor should be in the range $4.0 < H_{liq} < 6.0$ m. If the maximum liquid reactor height is exceeded, decrease the reactor upflow velocity v_l, increase the number of gas boxes or reduce the sludge age. If the minimum height is not achieved, do the opposite.

(17) For designs based on HRT, calculate the digestion zone height and check if the digestion zone occupies at least 60% of the total liquid reactor volume. This check is not strictly required in the procedure based on the sludge age, as the sludge mass is explicitly considered, but it does no harm to check.

(18) Add the freeboard to obtain the total reactor height:

$$H_u = H_{liq} + H_{fb} \tag{4.21}$$

In the next two Examples the application of the reactor design procedure presented above will be demonstrated in practice. Where Example 4.7 follows the above calculation procedure on a step-by-step basis, Example 4.8 will focus on obtaining an optimized solution for the process design made in Chapter 3, Example 3.12.

Example 4.7

Design the UASB reactors for the following design case, where the daily sewage flow and load are 26,000 $m^3 \cdot d^{-1}$ and 15.6 ton COD $\cdot d^{-1}$ respectively. For the (given) design sludge age of 70 days (at 22°C), it has been calculated that 200 tons of anaerobic sludge mass are required ($f_v = 0.6$). An average sludge concentration in the digestion zone of 40 g TSS $\cdot L^{-1}$ is assumed. The following additional data are given to complete the reactor design:

- Design upflow velocity $v_l = 0.7$ m $\cdot h^{-1}$;
- Outer width of the gas box $W_{gb} = 0.6$ m; total gas box height = 0.8 m;
- Reactor freeboard = 0.3 m;
- Standard dimensions of the reactor are 20 × 20 m or 400 m^2
- The inclination of both the gas collection plates and the deflectors (α) = 52°;
- A standard triangular deflector is used, with 0.15 m overlap on each collection plate;

Solution

First determine the required total digestion zone volume:

$$V_{dig} = MX_t/X_{td} = 200{,}000/40 = 5000 \text{ m}^3 \tag{4.8}$$

Then calculate the minimum required UASB surface area:

$$A_{min} = Q_i/(24 \cdot v_l) = 26{,}000/(24 \times 0.7) = 1548 \text{ m}^2 \tag{4.9}$$

For the standard UASB reactor size of 20 × 20 m, according to Equation (4.10) the required number of reactors = 1548/400 = 3.87 = 4, each with a surface area $A_u = L_u \times W_u = 400$ m^2. The actual upflow velocity is 26,000/(24 × 4 × 400) = 0.68 m $\cdot h^{-1}$, which is close enough to the design upflow velocity to be accepted. Now that the total reactor area is known, the height of the digestion zone can be calculated as:

$$H_{dig} = V_{dig}/(N \cdot A_u) + 0.25 \text{ (m)} = 5000/(4 \times 400) + 0.25 = 3.38 \text{ m} \tag{4.5}$$

The available overflow surface area depends on the area occupied by the gas boxes. The overflow velocity is therefore calculated as function of the number of gas boxes installed per reactor (N_{gb}). This calculation is demonstrated here for ten gas boxes and for the given external gas box width W_{gb} of 0.6 m:

$$v_o = Q_i/[(A_u - N_{gb} \cdot W_{gb} \cdot L_u) \cdot 24 \cdot N] = 26{,}000/[(400 - 10 \times 0.6 \times 20) \cdot 24 \times 4] = 0.97 \text{m} \cdot \text{h}^{-1} \tag{4.12}$$

This is almost equal to the recommended value and thus this solution is accepted. Now the aperture width needs to be calculated. A design aperture velocity of $v_a = 1.5$ m $\cdot h^{-1}$ is selected:

$$\begin{aligned} A_a &= Q_i/(v_a \cdot 24 \cdot N) \text{ in m}^2 \\ &= 26{,}000/(1.5 \times 24 \times 4) = 181 \text{m}^2 \text{ and} \end{aligned} \tag{4.13}$$

$$W_a = A_a/(L_u \cdot N_{gb}) \text{in m} = 181/(20 \times 10) = 0.9\ \text{m} \quad (4.14)$$

The projected width of the gas collection plates can be calculated as:

$$a = (W_u - \Sigma W_{gb} + \Sigma W_a)/(2N_{gb}) = (W_u - N_{gb}(W_{gb} + W_a))/(2N_{gb}) = (20 - 10(0.6 + 0.9))/(2 \times 10) = 0.25\ \text{m} \quad (4.15)$$

Which leads, for a plate inclination $\alpha = 52°$, to a projected collection plate height of:

$$b = a \cdot \tan(\alpha) = 0.25 \tan(52°) = 0.32\ \text{m}. \quad (4.16)$$

The deflector width required for the specified overlap of 0.15 m is calculated as:

$$W_{df} = W_a + 2 \times 0.15 = 0.9 + 0.3 = 1.2\ \text{m}$$

The inclination of the deflector is also specified as 52°. Hence the deflector height is:

$$H_{df} = (W_{df}/2) \cdot \tan(\alpha) = 1.2/2 \cdot \tan(52°) = 0.77\ \text{m} \quad (4.18)$$

For the gas box height of 0.8 and the freeboard of 0.3 m, the liquid gas box height H_{gb} is equal to $0.8 - 0.3 = 0.5$ m. Furthermore, as the plate inclination $\alpha > 45°$, the top of the deflector should be displaced 0.2 m below the base of the gas hoods. The resulting GLS height is:

$$H_{gls} = b + H_{df} + H_{gb} + H_{dist} = 0.32 + 0.77 + 0.5 + 0.2 = 1.79\ \text{m} \quad (4.19)$$

The height available for settling is $H_{gls} - H_{df} = 1.79 - 0.77 = 1.02 > 0.8$ m (= OK)

The total liquid height is calculated with Eq. (4.4) as $H_{liq} = H_{gls} + H_{dig} = 1.79 + 3.38 = 5.16$ m and the total reactor height (Eq. (4.21)) as $H_{liq} + H_{fb} = 5.16 + 0.3 = 5.46$ m.

The digestion zone occupies 3.38/5.16 = 65% of the liquid reactor volume, which confirms that the sludge-age-based design was correct. The total liquid reactor volume is $5.16 \times 4 \times 400 = 8260\ \text{m}^3$, so that the average sludge concentration is $200{,}000/8260 = 24.2\ \text{g TSS} \cdot \text{L}^{-1}$, which is nicely in the range recommended in Section 4.3.1. A final check on HRT: $8260/26{,}000 \times 24 = 7.62$ hrs, which confirms to the Brazilian standards in Table 4.1.

Example 4.8

Finalize the UASB reactor design for the sewage treatment plant of Chapter 3, which was designed on the basis of the anaerobic sludge age as shown in Example 3.12. The daily sewage flow is $50{,}000\ \text{m}^3 \cdot \text{d}^{-1}$ and the design organic load is $40\ \text{ton COD} \cdot \text{d}^{-1}$. The sewage temperature fluctuates between 15 and 25°C. The reactor must be able to hold 280 ton VSS or 418 ton TSS ($f_v = 0.67$), at a design sludge age of 70 days at 15°C. Assume an average *reactor* sludge concentration of $22.5\ \text{g TSS} \cdot \text{L}^{-1}$ ($15\ \text{g VSS} \cdot \text{L}^{-1}$). The following additional data are given:

- Outer width of the gas box $W_{gb} = 0.6$ m; liquid height of the gas box $H_{gb} = 0.5$ m;
- Reactor freeboard = 0.3 m;
- Maximum size of the reactor is $2000\ \text{m}^3$;
- The inclination of both the gas collection plates and the deflectors $\alpha = 45°$;
- A standard triangular deflector is used, with 10% overlap on each collection plate.

Solution

In Example 3.12 the retention time was calculated as $R_h = 9$ h for 15°C and at a sludge age of 70 d. For the average sludge concentration of $22.5\ \text{g TSS} \cdot \text{L}^{-1}$, the required volume was $18{,}700\ \text{m}^3$, suggesting 10 reactors of

1870 m^3 each. The values of the area and the height of the reactors must now be selected such that the various limitations to reactor design are all adequately addressed. These are:

(1) The upflow velocity in the digestion zone must be between 0.3 and 1.0 m · h^{-1}, while the overflow velocity must also be adequate (<1.2 m · h^{-1}) (Table 4.3);
(2) The volume of the digestion zone must be sufficiently large to accommodate the necessary sludge mass.

A convenient approach is to verify for different reactor depths whether the above criteria are obeyed. Initially this will be evaluated for a depth of 4 m; the results of the evaluation are summarized in Table 4.7. In this table the dimensions of the GLS separator elements (columns 2–6) and the fractions of the volume for the digestion, transition and settler zone (columns 7–9) have been calculated for different numbers of GLS separators (column 1) for the selected depth of 4 m.

As in Example 4.1, it has been assumed here that the total width of the separator elements and that of the deflectors is 55% of the reactor width. The reactor width is the square root of the area given as the ratio of volume to liquid depth: $A_u = 1870/4 = 479$ m^2, i.e., $L = W = 21.7$ m. It can be observed that the height of the separator elements (column 5) is adequate for solutions with 8–12 gas boxes; the height is then more than the minimum value of 0.8 m. However, the fraction of the digestion zone (column 7) is too small: it does not reach the required minimum value of 60% of the total volume. Only solutions with 14 and 16 gas boxes comply with this requirement, but the height of the separator elements is insufficient (<0.8 m in column 5). It is concluded that it is not possible to find an adequate reactor design for a liquid depth of 4 m.

When the calculations are repeated with a liquid height of 4.5 m, it can be demonstrated that this height also does not allow a design that complies with the imposed criteria (results not shown). When the calculations are repeated for a depth of 5 m (Table 4.8), it can be seen that a suitable design is possible for 9 or 10 gas boxes: the height of the separator elements and the digestion fraction of the reactor volume are both adequate.

Now it must be checked whether the velocities at the different levels in the reactor are within the constraints of Table 4.3. The results are shown in Table 4.9. The values at the overflow level are calculated considering the area occupied by the gas boxes ($N_{gb} \cdot W_{gb} \cdot L_u$). Note that the upflow velocity and the aperture velocity, which is based on a fixed value of 45% of the surface area available to the apertures, do not depend on the number of gas boxes selected, but only on the reactor surface area.

If the calculations are repeated for a height of 5.5 m, it can be shown that the design is possible for 7 or 8 separator elements, but the overflow velocity is higher than the recommended value (for $N_{gb} = 7$) or even just above (for $N_{gb} = 8$). It is concluded that under the specifications of the example only a liquid height of 5 m yields a satisfactory design for the GLS separator.

Now the last criterion must be checked: for 10 separator elements the digestion zone has the largest digestion zone at 65% of the volume of the reactor, so that this zone has a height of $0.65 \times 5 = 3.25$ m and a concentration of $5/3.25 \times 22.5 = 35$ g TSS · L^{-1}. This is in the middle of the range suggested in Section 4.3.1.3, but if it was too high, the design must be repeated all over again, this time with a longer retention time.

Table 4.7 Dimensions of the GLS separator and division of the liquid reactor volume over different reactor zones for a selected depth of 4 m.

1	2	3	4	5	6	7	8	9
# Elements	**Width of Elements**	**Width of Apertures**	**Height Plate**	**Height Separator**	**Height Deflector**	**Fraction Digestion Zone**	**Fraction Transitional Zone**	**Fraction Settling Zone**
8	1.49	1.22	0.54	**1.04**	0.74	*0.44*	0.20	0.28
9	1.32	1.08	0.46	**0.96**	0.66	*0.49*	0.18	0.26
10	1.19	0.97	0.40	**0.90**	0.60	*0.52*	0.16	0.24
12	0.99	0.81	0.30	**0.80**	0.50	*0.58*	0.13	0.22
14	0.85	0.70	0.23	*0.73*	0.43	**0.61**	0.11	0.20
16	0.74	0.61	0.17	*0.67*	0.37	**0.64**	0.10	0.18

Table 4.8 Dimensions of the GLS separator and division of the liquid reactor volume over different reactor zones for a selected depth of 5 m.

1	2	3	4	5	6	7	8	9
# Elements	**Width of Elements**	**Width of Apertures**	**Height Plate**	**Height Separator**	**Height Deflector**	**Fraction Digestion Zone**	**Fraction Transitional Zone**	**Fraction Settling Zone**
8	1.33	1.09	0.47	**0.97**	0.67	0.59	0.14	0.21
9	1.18	0.97	0.39	**0.89**	0.59	**0.62**	0.13	0.19
10	1.07	0.87	0.33	**0.83**	0.53	**0.65**	0.11	0.18
12	0.89	0.73	0.24	0.74	0.44	**0.69**	0.09	0.16
14	0.76	0.62	0.18	0.68	0.38	**0.71**	0.08	0.14
13	0.82	0.67	0.21	0.71	0.41	**0.70**	0.09	0.15

Table 4.9 Check on velocity limits and recommended values of Table 4.3 for the reactor designs for liquid heights of 5.0 and 5.5 m.

Velocities ($m \cdot h^{-1}$)	Optimum	Limit	$H = 5.0$ m	$H = 5.5$ m
Reactor upflow velocity	0.5–0.8	0.3–1.0	0.56	0.62
Upflow velocity in apertures	<1.5	<2.5	1.23	1.36
Overflow velocity	<1.0	<1.2	0.77/0.80[1]	1.19/1.24[2]

[1]For 9 or 10 gas boxes.
[2]For 7 or 8 gas boxes.

4.5 ANAEROBIC REACTOR WITH EXTERNAL SETTLER

In the conventional UASB reactor discussed in the previous sections the settler is placed above the anaerobic reactor, so that it has in fact two functions: to transform the organic influent material into biogas in the digestion zone at the bottom and to retain the sludge in the upper settling zone. The UASB reactor is a single unit with two conflicting objectives: (1) to achieve maximum digestion of the organic material present in the influent, and (2) to retain as much sludge as possible in the reactor, thus making the first objective possible. It could be debated whether these objectives would not be served better if the single reactor were substituted by two separate units, each with its own objective, so that specific designs could be made for the two parts to attain maximum efficiency. Along with a possible gain in performance this could reduce construction costs, because the specialized digester need not have GLS separators, which are a significant cost item in the conventional UASB reactor. This composite system certainly would be shallower, saving in costs for the walls. It might also be smaller and easier to construct. Thus the conventional UASB unit could with important advantages be replaced by a system composed of two parts: a digester followed by a settler. From the operational point of view having a digester with access to the entire surface has a great advantage: any scum formed could be very easily removed and even avoided by using a continuous skimming device on the surface as in secondary activated sludge settlers. The settler could be equipped with plates as in water treatment practice, which is much more efficient than the settlers in the conventional unit. A disadvantage would be that the area of the system composed of two parts would be greater, as the two parts would be constructed one alongside the other.

An alternative to replacing a three-phase separator by two biphasic separators was recently presented by Pessotto (2015) and is represented schematically in Figure 4.42. The reactor consists of a digestion zone and a gas collection cover, followed by a high-rate settler. On the reactor surface biogas bubbles are separated from the adhered sludge particles. After the bubbles are released, sludge settles back to the digestion zone by gravity. The biogas is captured by a traditional gas holder or a flexible plastic cover placed over the reactor that can be easily removed to gain access to the surface, where a layer of scum can form. It is possible to reduce or even prevent scum formation by slow mechanical stirring of the liquid surface, which also enhances the separation of biogas and the return of the sludge. The effluent from the reactor is collected from below the surface of the reactor and passes to a high-rate settler. The settled sludge is transferred back into the digestion zone of the UASB reactor, while an essentially solids-free effluent is discharged from the settler. A settler equipped with parallel plates permits retention of flocs with a much lower settling velocity than in a conventional settler.

An additional advantage of the presented alternative is that the height of the treatment unit is reduced, because the GLS separator is no longer required and in many cases it will be possible to avoid pumping of the raw sewage and apply gravity feeding, which is a very important operational advantage. The shallower unit will also reduce construction costs.

The settling theory of high-rate settlers is well established in many textbooks of water treatment and can be used to estimate the ratio between the area of the high-rate settler and that of the reactor, which is equal to that of the conventional settler. In a conventional settler the flow is upward, and a floc is not retained if its settling velocity is less than the upward velocity of the liquid, as shown in Figure 4.43 Therefore, there is a critical settling rate for floc retention:

$$s_c < v_l \tag{4.22}$$

where

s_c = critical settling velocity that allows retention of a floc ($m \cdot h^{-1}$)
v_l = upward velocity of the liquid (flow divided by area) in $m \cdot h^{-1}$

For the settler with parallel plates, the retention efficiency of the solids is determined by the critical settling velocity in the zone of the parallel plates. To calculate the critical settling velocity the following procedure can be used: Figure 4.43 shows the path of a sludge particle moving between two parallel

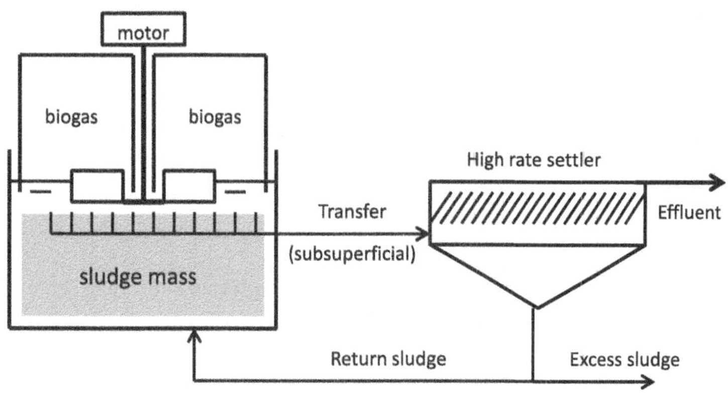

Figure 4.42 UASB reactor with a separate high-rate settler as solid-liquid phase separator.

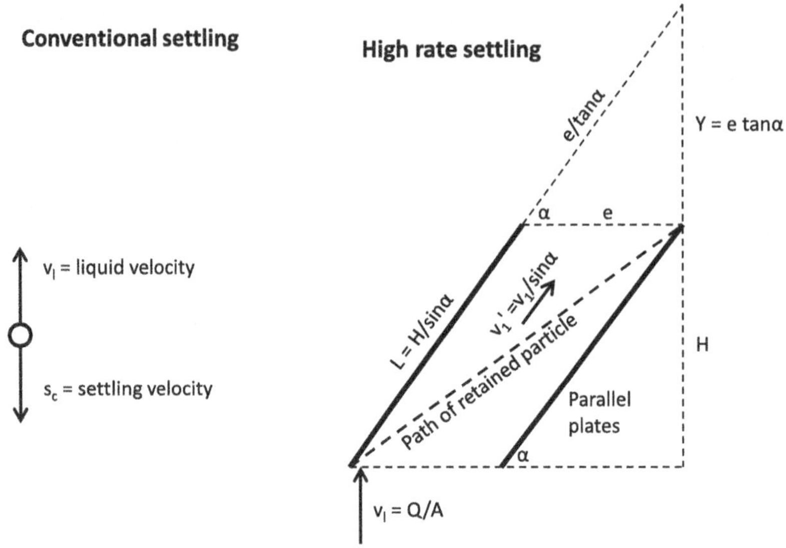

Figure 4.43 Settling in a conventional separator (left) compared to that with parallel plates (right).

plates. The particle enters the plate zone directly adjacent to the first (left-hand) plate. While the liquid flows upward between the plates, the particle will be subjected to a settling velocity and follow a different (flatter) trajectory than the liquid.

In this case the particle touches the second (right-hand) plate just before the effluent leaves the settling zone. Hence, this particle is retained and will eventually be returned to the digestion zone. While the liquid travels a distance $L + E/\cos(\alpha)$, the particle settles a distance $E\tan(\alpha)$. Ignoring the thickness of the plates, the liquid velocity between the plates can be expressed as:

$$v_l' = v_l/\sin(\alpha) \quad (4.23)$$

Hence to retain the flocs, the ratio between the critical settling velocity of a particle and the velocity of the liquid between the plates must be:

$$s_c' = v_l' = E\tan(\alpha)/(L + E/\cos(\alpha)) \quad (4.24)$$

where:

v_l' = liquid velocity between plates ($m \cdot h^{-1}$)
s_c' = critical settling velocity ($m \cdot h^{-1}$)
α = angle of the plates
L = length of the plates (m)
E = spacing between plates (m)

The critical settling velocity permitting floc retention in a high rate settler can be expressed as:

$$s_c' = v_l' \cdot E \cdot \tan(\alpha)/(L + E/\cos(\alpha)) = (v_l/\sin(\alpha)) \cdot E \cdot \tan(\alpha)/(L + E/\cos(\alpha)) \quad (4.25)$$

$$s_c'/s_c = [(v_l/\sin(\alpha)) \cdot E \cdot \tan(\alpha)/(L + E/\cos(\alpha))]/s_c = 1/[(L/E) \cdot \cos(\alpha) + 1)] \quad (4.26)$$

$$s_d/s_c = A_r/A_d = [(v_l/\sin(\alpha)) \cdot E \cdot \tan(\alpha)/(L + E/\cos(\alpha))]/s_c = 1/[(L/E) \cdot \cos(\alpha) + 1] \quad (4.27)$$

Hence:

$$(A_d/A_r) = 1/[(L/E) \cdot \cos(\alpha) + 1] \tag{4.28}$$

where

s_d = settling velocity of floc that can be retained in settler with parallel plates ($m \cdot h^{-1}$)
A_r = area of the reactor, approximately equal to that of a conventional settler (m^2)
A_d = area of the high rate settler (m^2).

Equation (4.28) shows that the ratio between the area of a high-rate settler and a conventional settler (A_d/A_r) depends on three factors: (a) distance between the plates (spacing), (b) the plate inclination α, and (c) the length L of the plates. These three factors are limited by practical considerations:

- The distance between the plates must be sufficiently large (at least > 10 cm) to avoid clogging of the plates with settled solids;
- The plate inclination must be steep enough to ensure that the settled sludge flocs do slide easily back into the digestion area (in practice between 45–60°);
- For economic reasons, the length of the plates and consequently the height of the settling zone where the parallel plates are mounted cannot be very high.

Figure 4.44 shows the ratio between the area of a high-rate settler and a conventional settler of a UASB reactor as function of the depth of the plates, for different values of plate spacing and plate angle.

It can be observed that even for a relatively small plate depth selected there is already a very considerable decrease of the required area for the high-rate settler. The spacing also has an important effect. In this case the benefit of installing a smaller plate pack area if the spacing is decreased must be balanced against the

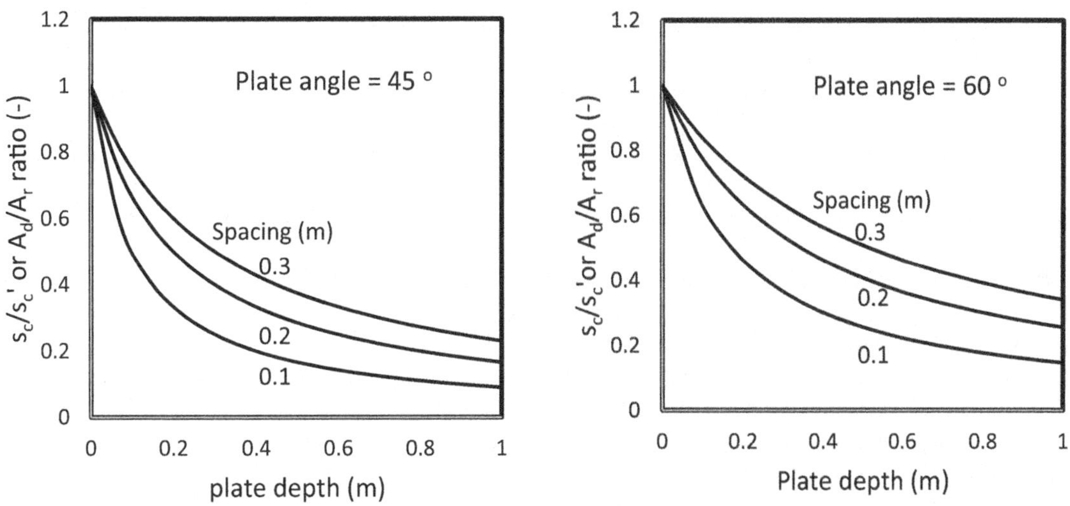

Figure 4.44 Ratio between the area required for a high-rate settler and a conventional settler and the ratio of the critical velocity for floc settling in UASB reactors with a phase separator plus parallel plates and a conventional separator as a function of the depth of the plates for different plate spacing and angles of the plates.

risk of more operational problems resulting from plate blocking. The angle of the plates also has an effect, a lower angle requires a smaller height, but it must be ensured that retained flocs will slide from the plates to be returned to the reactor.

Example 4.9

Calculate the ratio of required settling area for a high-rate settler to that of the settling zone of an UASB reactor (ignore the reduction in overflow area due to the plates). The plates have a length of 1.0 m, an angle of 45° and are spaced 0.2 m apart.

Solution

Using Eq. (4.28) one has $A_d/A_r = 1/((L/E) \cdot \cos(\alpha) + 1) = 1/((1/0.2) \cdot 0.71 + 1) = 0.22$.

Therefore the area of the settler needs to be only 22% of the area of the UASB reactor. Although there are no full-scale units with this configuration in operation at present, the significant reduction calculated for the settler area indicates that it is very likely that there will be a reduction of the construction costs of the treatment unit, which in itself is a reason to pursue this option.

4.6 SMALL-SCALE UASB DESIGNS

4.6.1 $UASB_b$ with two bi-phase separators

Perhaps surprisingly, until now the application of the UASB reactor as a single-family sewage treatment unit has been very limited, although there is an enormous potential for upgrading existing units such as the septic tank (ST) or the Imhoff tank. Coelho *et al.* (2003) reported on an experiment with a small-scale modified unit of the UASB reactor, comparing its performance with that of a conventional 1800 L septic tank when fed the same wastewater. Figure 4.45 shows the schematic design of this very simple unit, as well as a full-scale example of the reactor.

The UASB reactor had a liquid volume of 360 L (only one-fifth of the volume of the septic tank!) and was constructed from two PVC pipes connected with fibreglass reinforced polyester. The vertical tube, the actual digestion zone, had a diameter of 500 mm and a height of 1.80 m. The other pipe, placed at a 45° upward angle, had a diameter of 350 mm and a length of 450 mm (when the top of the pipe was measured) and 1000 mm (bottom of the pipe). The weight of the empty UASB reactor was only 18 kg.

Phase separation in the UASB reactor thus constructed was different from the classic design discussed in Section 4.3.2 and consisted of two parts: gas–liquid separation occurred in the upper part of the vertical pipe, while in the lateral arm liquid–solids separation took place.

The sludge that settled in the inclined lateral arm tended to slide back into the digestion zone towards the bottom of the vertical pipe, allowing a large sludge mass to be maintained. Thus the conventional tri-phase (GLS) separator was substituted by two bi-phase separators (G/L and S/L). For that reason this small scale UASB reactor is designated $UASB_b$.

Figure 4.46 compares the COD and TSS removal efficiencies of both systems in time, while Table 4.10 compares the $UASB_b$ and the septic tank in terms of performance, construction costs and operation & maintenance requirements. Clearly the $UASB_b$ reactor has important advantages: the performance in terms of COD and suspended solids removal efficiency is superior, it is much easier to implement, the

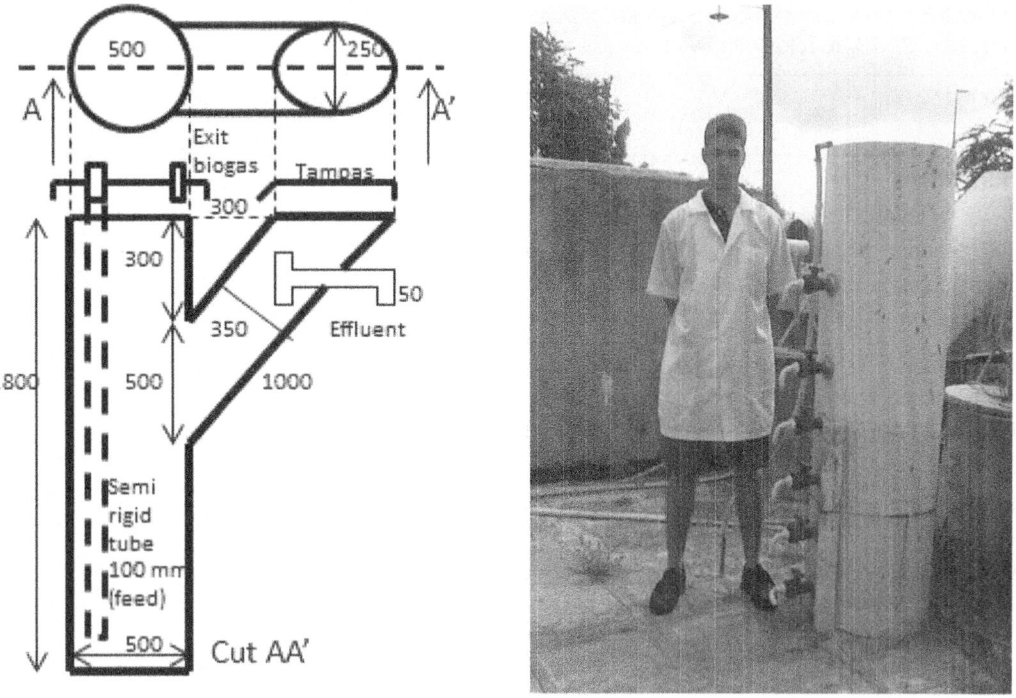

Figure 4.45 Schematic representation and picture of the UASB$_b$ reactor design for single families, used in the experimental investigation.

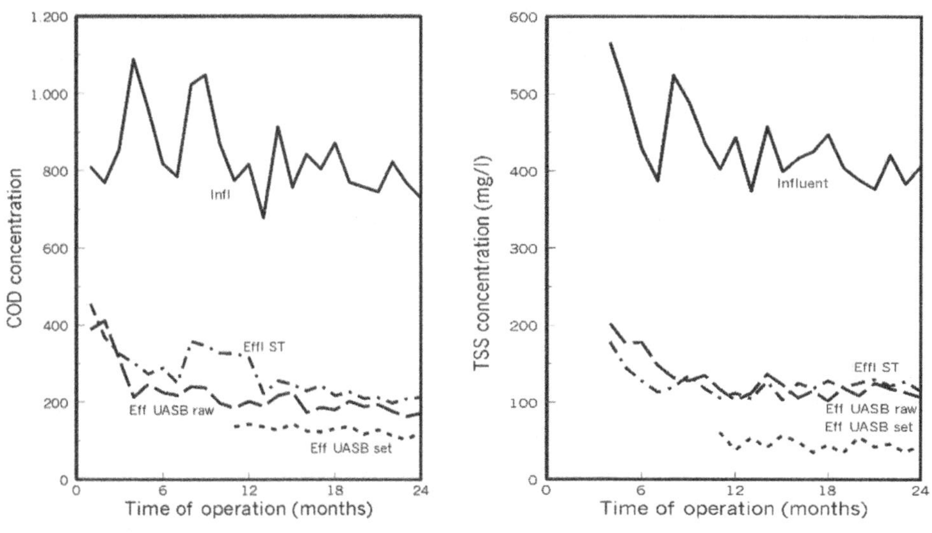

Figure 4.46 COD concentration (left) and suspended solids concentration (right) in the influent and effluent of the septic tank and the UASB$_b$ reactor as a function of time.

Table 4.10 Comparison of two sewage treatment systems treating equal flow and -load of household sewage: the $UASB_b$ reactor with two bi-phase separators and the septic tank.

Parameter	Septic Tank	$UASB_b$ Reactor	
		Raw Effluent	Settled Effluent
Performance:			
• COD removal efficiency	60–75%	75–80%[1]	80–85%
• TSS removal efficiency	55–70%	70–75%[1]	80–95%
• Helminth eggs removal efficiency	95%	50–60%[1]	–
• Post treatment (in Brazil)	Anaerobic filter	No	No
Construction:			
• Construction time	1 week	1 day	
• Required area	$1.5 \times 2.5 = 3.8\ m^2$	$1 \times 1.5 = 1.5\ m^2$	
• Retention time	2 days	0.5 d	
• Depth	1.5 m	1.8 m	
• Weight	Very heavy	18 kg	
• Construction costs[2]	US$ 500	US$ 100	
Operation & maintenance:	De-sludging every 3–5 years	Without sludge discharge	

[1]After establishing steady state conditions.
[2]For Brazilian conditions.

footprint is much smaller (important for low-cost housing with small plots!) and the construction costs are much lower.

It is not required to discharge waste sludge from the $UASB_b$ reactor, even when there are some settleable solids present in the effluent: the unit needs to be operated as full of biomass as possible, as this is vital to maintain the organic material removal efficiency of the reactor.

4.6.2 $UASB_c$ with conical three-phase separator

Recently, an experiment with small-scale UASB reactors with a different design was completed (Santos, 2016). The UASB reactor is characterized by the fact that both the digestion and settling zones are cylindrical, but the settling zone is expanded as it has a larger diameter which is attached to the digestion zone by means of a conical connection. The phase separator has also a conical shape and therefore the reactor is indicated as $UASB_c$. Figure 4.47 shows the schematic design of the UASBc reactor. Compared to a UASBb reactor, the $UASB_c$ reactor has the advantage of being compact and very simple to construct. Thus the unit could be mass-produced and sold in stores. The essence of the $UASB_c$ reactor design is that different upflow velocities can be selected in the settling and digestion zones as, unlike in the conventional UASB reactor, these diameters are not the same in the $UASB_c$ reactor. The upflow velocity in the digestion zone should be higher than about $0.3\ m \cdot h^{-1}$ in order to avoid the deposition of inorganic solids with fine sand and silt.

In the settling zone the velocity should be as low as possible, but the value is limited by cost considerations: the diameter becomes excessive at very low velocities which leads to a high cost and a large footprint. The

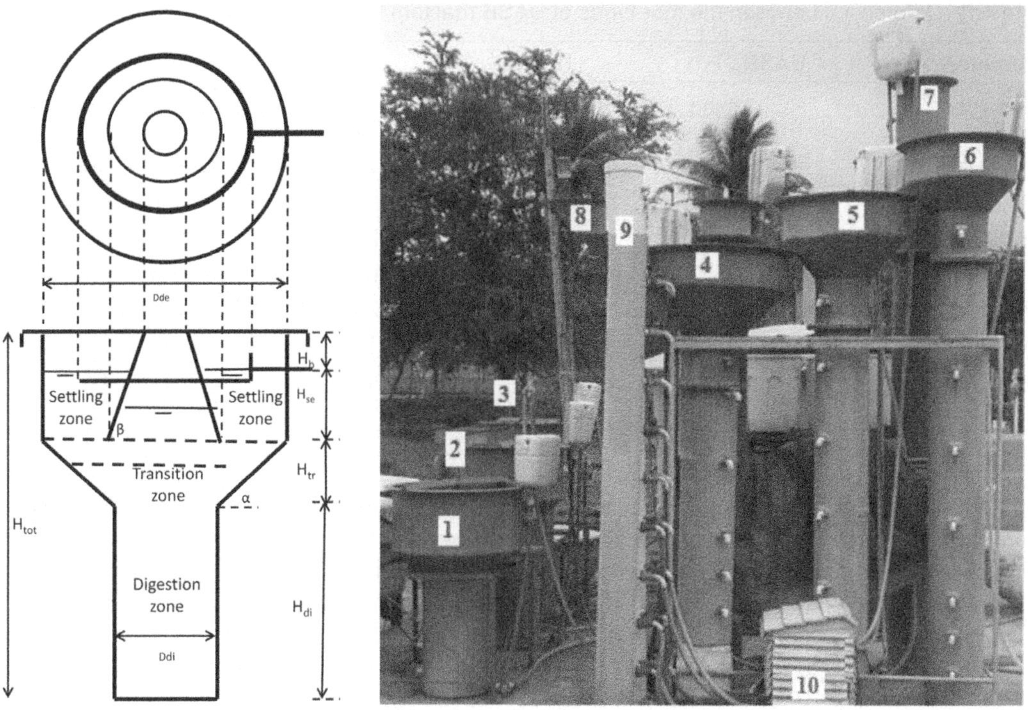

Figure 4.47 Schematic representation of a $UASB_c$ reactor (left) and a picture of the pilot reactors containing six $UASB_c$ reactors, one $UASB_b$ and one conventional reactor. Legend: 1. Reactor S_1: $D_{dig} = 0.4$ m; $D_{sed} = 0.75$ m; 2. Reactor S_2: $D_{dig} = 0.4$ m; $D_{sed} = 0.70$ m; 3. Reactor S_3: $D_{dig} = 0.4$ m; $D_{sed} = 0.65$ m; 4. Reactor T_1: $D_{dig} = 0.4$ m; $D_{sed} = 0.8$ m; 5. Reactor T_2: $D_{dig} = 0.4$ m; $D_{sed} = 0.7$ m; 6. Reactor T_3: $D_{dig} = 0.4$ m; $D_{sed} = 0.6$ m; 7. Conventional reactor: $D_{dig} = 0.3$ m; 8. Reactor $UASB_b$: $D_{dig} = 0.3$ m; 9. Feed vessel; 10. Dosing pumps.

$UASB_c$ concept can be applied to individual houses, neighbourhoods or small towns. The application for single-family treatment was tested, with different units up to about 250 litres as shown in Figure 4.47. A series of experiments was performed with units with diameters of 0.3 or 0.4 m in the digestion zone and different diameters of the settling zone.

4.6.3 Geometry of the three phase separator $UASB_r$

The $UASB_b$ and $UASB_c$ designs are simple and have a low construction cost, but as will be shown in Chapter 10 they can only be used for the treatment of small flows. For sewage treatment of larger communities the conventional trapezoidal separator, discussed in Chapter 3, can used. In Chapter 10 the influence of the physical dimensions of the separator on the design and performance of the UASB reactors will be discussed.

The differences between the reactors with the different types of separators is not limited to geometry, but also in other respects, notably the size of the reactors and the upward velocity of the liquid phase at strategic points of the reactors, which has a large impact on the retention efficiency of sludge and therefore

Table 4.11 Differences between the four types of UASB reactors for sewage treatment.

Parameter	$UASB_b$	$UASB_c$	$UASB_r$	$UASB_s$
Material reactor (separator)	PVC tubing	Glass fibre	Concrete (fibre glass)	Concrete (fibre glass)
Separator type	Two biphasic	Conical triphasic	Trapezoidal	Parallel plates
Digestion zone	Cylindrical	Cylindrical	Rectangular	Cylindrical or rectangular
Transition zone	No	Conical	Rectangular	No
Settling zone	Cylindrical (inclined)	Cylindrical	Rectangular	Rectangular
Range of volumes (m^3)	0.025–2.5	0.25–250	25–2500	25–2500
Flow rate ($m^3 \cdot d^{-1}$)	0.1–10	1–1000	100–10,000	100–10,000
Height (m)	1–5	1.5–5	4–6	2–3 (digester) 2–3 (settler)

on treatment capacity. As for the size of the units with different separators, the unit with two biphasic separators is limited by its geometry to have a small volume and therefore is destined to the treatment of waste water with small flow like the sewage of one or a few houses. Table 4.11 shows the main differences between the UASB types. The characteristics of the UASB reactor with separate digestion and settling zones ($UASB_s$), presented in the next section have also been included.

Chapter 5

Value and stability of the pH in anaerobic digesters

5.1 INTRODUCTION

In Chapter 2 it was established that anaerobic digestion develops at its maximum rate when the pH value of the reactor is in the neutral range between 6.8 and 7.2, while pH values between 6.5 and 7.5 are still considered acceptable, although methanogenic activity will decrease. In particular, pH values below 6.5 are critical as the methanogenic activity is then effectively stopped, whereas acid production continues unabated. The value and stability of pH value in wastewater treatment systems depends on the interaction of the acid–base systems that are either present in the influent or generated during the digestion processes.

In this chapter it will be shown that in the case of domestic sewage the pH will be in the ideal range without the need for addition of chemicals. The neutral pH value is established owing to the buffering capacity of the acid–base systems present in the digester, the most important being the carbonic system. This system is not only important for the ionic balance in the liquid phase, but its presence also has implications for the liquid–biogas equilibrium and it also affects directly the liquid–solid equilibrium with respect to $CaCO_3$ precipitation or dissolution. The influence of these equilibriums on the composition of biogas and the tendency of the liquid phase to be either aggressive or precipitate (scaling) will also be analysed in this chapter.

In more concentrated wastewaters generally the buffering capacity of these acid–base systems is not enough to maintain the pH in the optimal range. In this case there are basically two procedures to adjust the pH: (1) add alkalinity or (2) remove acidity, but both are rarely applied for anaerobic sewage treatment.

Addition of an alkalizing agent such as caustic soda (NaOH) has cost implications and it affects effluent quality negatively: the addition of more ions can reduce its quality for irrigation. The removal of acidity in practice means that CO_2 is removed from the digester and released into the atmosphere. This CO_2 release can be done in a desorption unit by recirculation of the alkaline effluent, which can be mixed with influent, or by recirculating of biogas and absorbing CO_2 in an alkaline solution. Effluent recirculation not only influences the value and stability of the pH, but it also reduces alkalinity requirements. Furthermore, it also increases the flow of liquid entering the reactor and thus helps to maintain an adequate upward velocity to

 Anaerobic Sewage Treatment: Optimization of Process and Physical Design of Anaerobic and Complementary Processes
Editor(s): Adrianus van Haandel and Jeroen van der Lubbe
doi: 10.2166/9781780409627_0193

optimize the conditions in the anaerobic sludge bed and stimulates the development of anaerobic granules. At the same time removal of CO_2 from the digester leads to biogas production with a higher partial pressure of methane, which will increase its value as a fuel.

5.2 WEAK ACID–BASE SYSTEMS

The relative stability of dilute wastewaters like sewage is due to the buffering capacity provided by the presence of systems of weak acids and their conjugated bases, the most important being the carbonic system. In general the dissociation reaction of an acid (HA) in an aqueous solution and the consequential formation of its conjugate base (A^-) can be expressed as:

$$HA \xleftrightarrow{K_a} H^+ + A^-$$

or

$$K_a = [H^+][A^-]/[HA] \tag{5.1}$$

where:

$[X]$ = molar concentration of compound X, i.e. H^+, HA (acid) or A^- (conjugate base)
K_a = dissociation constant of system HA/A^-.

The dissociation constant K_a characterizes the acid–base system HA/A^-. The smaller the dissociation constant, the weaker the acid. Correspondingly the conjugated base becomes stronger as the value of K_a decreases. In practice the constant K_a for many acid–base systems has a very small value and it is convenient to adopt the pK_a value, which is the negative logarithm of K_a. So basically three regions of K_a values can be distinguished:

(1) $K_a > 1$ or $pK_a < 0$: strong acid with very weak conjugated base
(2) $10^{-14} < K_a < 1$ or $1 < pK_a < 14$: weak acid and base
(3) $K_a < 10^{-14}$ or $pK_a > 14$: very weak acid and strong conjugate base.

Strictly speaking Eq. (5.1) is only valid for solutions in pure water. For water with impurities (such as sewage) a correction should be made to quantify their influence on the equilibrium. In this case the *activity* of the various compounds rather than their *concentration* should be used.

5.3 BUFFER INDEX

In biological treatment systems such as anaerobic digesters not only the current value of the pH is important, but also its stability. The suitable parameter to assess the stability of pH is the buffer index, defined as the strong acid or base concentration required producing a unit pH change:

$$B_i = dC_b/dpH = -dC_a/dpH \tag{5.2}$$

where:

B_i = buffer index
C_b = concentration of added strong base
C_a = concentration of added strong acid.

A mathematical expression for the buffer index can be derived (Butler, 1964) for any acid–base systems by using the definition of Eq. (5.2) in the equilibrium equation of Eq. (5.1):

$$B_i = 2.3C_tK_a[H^+]/(K_a + [H^+])^2 \tag{5.3}$$

In the case of so-called polyprotic acids (acids with more than one dissociation step, such as carbonic acid or phosphoric acid), with a difference of more than two units between successive pK_a values, the buffering index can be expressed as the sum of the buffer capacities associated with the different dissociation steps:

$$B_i = 2.3C_t\sum_{i=1}^{N}[K_{ai} \cdot [H^+]/(K_a + [H^+])^2] \tag{5.3a}$$

where:

B_i = buffer index of the acid–base system
C_t = total concentration of species belonging to the acid–base system
K_{ai} = dissociation constants
N = number of dissociation steps

The dissociation for water itself also generates a buffer capacity. Equation (5.4) shows the buffer index equation for the hydrolysis of water:

$$B_{H2O} = 2.3([H^+] + K_w/[H^+]) \tag{5.4}$$

Table 5.1 presents the dissociation constants of acid–base systems normally present in anaerobic sewage digesters, together with typical ranges of their concentrations in sewage.

The sum of volatile fatty acids concentrations (VFA) in mmol $\cdot$ l^{-1} are grouped together as an equivalent acetic acid solution, as the values of the dissociation constants of the individual VFA components are very similar. Figure 5.1 shows plots of the buffer capacity as a function of the pH for the acid–base systems in Table 5.1. The carbonic system is diprotic with two dissociation constants: $pK_1 = 6.3$, which is associated with the $CO_2 \leftrightarrow HCO_3^-$ equilibrium and $pK_2 = 10.3$ associated with the $HCO_3^- \leftrightarrow CO_3^{2-}$ equilibrium. The buffer index of water is also indicated. From Eqs. (5.3) and (5.4) and Figure 5.1 it can be concluded that:

Table 5.1 Reactions, dissociation constants and typical range of concentrations of acid–base systems commonly encountered in municipal sewage treatment plants.

Acid/base	Reactions	pK_a	K_a	Concentration (mmol · L^{-1})
Carbonic	$H_2CO_3 \leftrightarrow HCO_3^- + H^+$	6.35	4.5×10^{-7}	5–10
	$HCO_3^- \leftrightarrow CO_3^{2-} + H^+$	10.33	4.7×10^{-11}	
Ammonium	$NH_4^+ \leftrightarrow NH_3 + H^+$	9.26	5.5×10^{-10}	2–5
Phosphoric acid	$H_3PO_4 \leftrightarrow H_2PO_4^- + H^+$	2.23	5.9×10^{-3}	0.2–0.5
	$H_2PO_4^- \leftrightarrow HPO_4^{2-} + H^+$	7.21	6.2×10^{-7}	
	$HPO_4^{2-} \leftrightarrow PO_4^{3-} + H^+$	12.32	4.9×10^{-13}	
Hydrogen sulphide	$H_2S \leftrightarrow HS^- + H^+$	6.96	1.1×10^{-7}	0.2–0.5
	$HS^- \leftrightarrow S^{2-} + H^+$	14.00	1.0×0^{-14}	
Acetic acid	$HAc \leftrightarrow Ac^- + H^+$	4.74	1.8×10^{-5}	0.5–2.0
Propionic acid	$HPr \leftrightarrow Pr^- + H^+$	4.90	1.3×10^{-5}	<0.1
Water	$H_2O \leftrightarrow OH^- + H^+$	14.00	1.0×10^{-14}	55.555

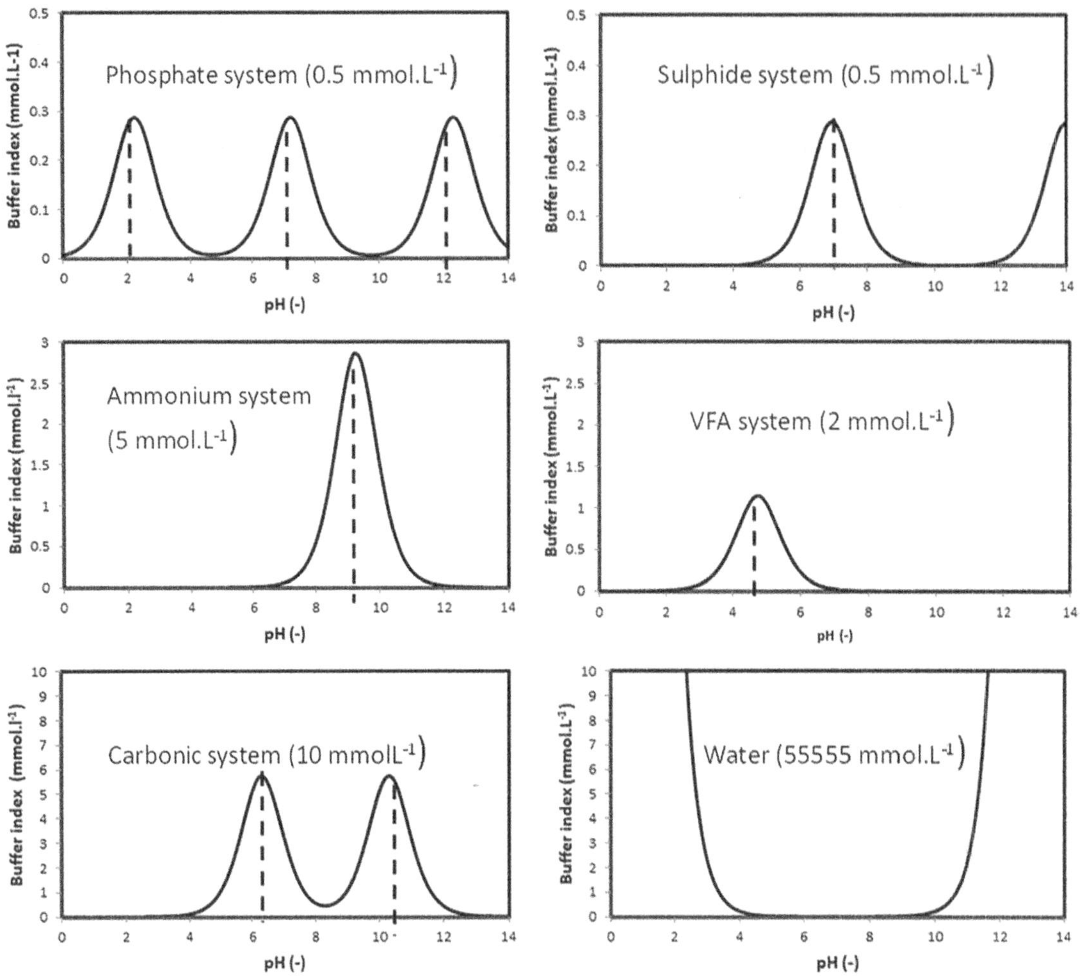

Figure 5.1 Buffer index values as function of pH for acid–base systems often encountered in municipal sewage.

- The buffer index and consequently the stability of pH is maximum when the pH value is equal to the pK_a value;
- The magnitude of the buffer index is directly proportional to the total concentration of the species of the acid–base system;
- The buffer index due to water hydrolysis is insignificant in the pH range between $5 < pH < 9$, which is the relevant range for biological treatment of wastewaters.

The total buffer index of the mixture of weak acid–base systems in municipal sewage is the sum of the individual buffer index values:

$$B_t = B_{H2O} + \sum_{j=1}^{M} B_j \tag{5.5}$$

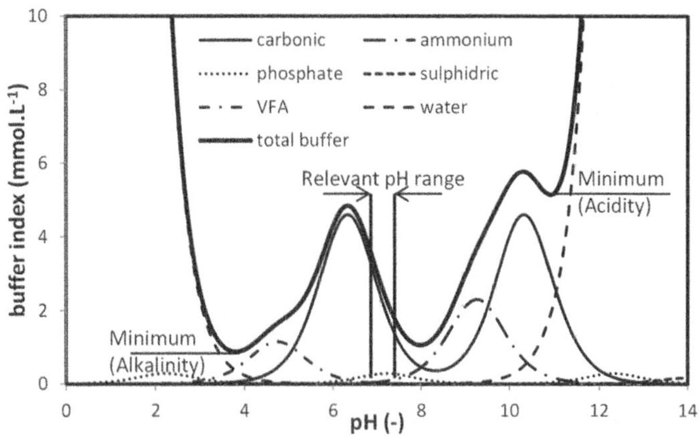

Figure 5.2 Buffer index as a function of pH for acid–base systems normally present in sewage treatment plants.

where:

B_t = total buffer index;
B_j = buffer index of system "j";
M = number of acid–base systems in the mixture.

Figure 5.2 shows the buffer index as a function of pH for a wastewater with the acid–base systems of Table 5.1 and Figure 5.1. With the aid of Figure 5.2 the following is concluded for anaerobic systems treating municipal sewage:

- In the narrow pH range for optimal anaerobic digestion, ammonia acts as if it were a strong base, i.e. it is almost completely in the form of NH_4^+.
- In the same range volatile fatty acids (mainly acetic acid and propionic acid) behave as strong acids and are almost completely dissociated.
- Among the acid–base systems with a pK_a value in the useful pH range of 6.5–7.5 (carbonic acid, phosphoric acid and hydrogen sulphide) the concentration of carbonic acid is much higher than the others and therefore contributes much more to total buffer index.

It follows that the carbonic system is predominant in the establishment of the pH value and stability of this value in biological sewage treatment systems. For this reason this system will be considered in more detail.

5.4 THE CARBONIC SYSTEM

In Section 5.3 it was established that the carbonic system normally determines the value and stability of the pH in anaerobic reactors treating municipal sewage. Ignoring the very small fraction of carbon dioxide that is associated with water to form carbonic acid, H_2CO_3 (less than 1%), the carbonic system in water system is determined by the concentrations (activities) of five species: CO_2, HCO_3^-, CO_3^{2-}, H^+ and OH^-. So, to fully define the system, five equations are required, three of which may be obtained directly from the dissociation equations:

$$K_1 = [HCO_3^-]\,[H^+]/[CO_2] \qquad (5.6)$$

$$K_2 = [CO_3^-]\,[H^+]/[HCO_3^-] \tag{5.7}$$

$$K_w = [H^+]\,[OH^-] \tag{5.8}$$

The numeric values of the constants have been determined as:

$K_1 \rightarrow pK_1 = 17052/T + 215.21 \log(T) - 0.12675T - 545.56$ (Harned & Davis, 1943)
$K_2 \rightarrow pK_2 = 2902.39/T + 0.0239T - 6.498$ (Harned & Scholes, 1943)
$K_w \rightarrow pK_w = 4787.3/T + 7.1321 \log(T) + 0.01037T - 22.801$ (Harned & Hamer, 1933)

A fourth equation is obtained by measuring the pH:

$$pH = -\log(H^+) \tag{5.9}$$

To define the carbonic system one more equation is needed. In principle, one could use the concentration of any of the carbonic species, or even the sum of the total species concentrations C_t, but none of these analytical tests is easy to carry out. For this reason in practice a different test is often used in which the alkalinity or the acidity is determined. The relationship between alkalinity, acidity and the carbonic system was established by Loewenthal and Marais (1976).

5.4.1 Alkalinity

If water were a pure solution of carbon dioxide in water, then depending on its concentration the pH of this solution would be around 4.2. This characteristic can be used to determine the alkalinity of a solution in water. For this, it is considered that this is a mixture of CO_2 and a (hypothetical) strong base. The concentration of the base is equal to the alkalinity of the solution and can be determined experimentally by titrating the solution with a strong acid to neutralize the strong base. At the end of the titration the pH will have a value equal to a pure solution of CO_2 in water, which is called the CO_2 equivalence point. This point is characterized by the fact that buffering is minimal so that the pH variation per unit addition of strong acid is maximum. This is clearly indicated in Figure 5.2. This point can be determined experimentally from a curve that relates the pH value to the concentration of acid added to the solution, the titration curve: the equivalence point is the pH at the maximum slope on the titration curve and the alkalinity is the acid concentration required to reach this pH value.

In practice the determination of the equivalence point is not routinely done, because it is known that for medium-alkalinity waters (3–10 $meq \cdot L^{-1}$ or 150–500 $mg \cdot L^{-1}$ of $CaCO_3$), the value of the equivalence point is always around 4.0 to 4.2, Thus, normally the titration is continued until a pH value of 4.2, without checking whether the slope of the curve is actually maximum.

An expression for alkalinity can be found by considering the dissociation reactions in pure water with a mixture of CO_2 and a strong base BsOH. The possible dissociations of the original compounds (H_2O, CO_2 and BsOH) are shown in Table 5.2. As the electroneutrality of the solution must be maintained, the sum of the negative charges of the generated anions is equal to the sum of the positive charges of the cations, which allows to equate the cation concentrations to the anion equations:

$$[Bs^+] + [H^+] = 2[CO_3^{2-}] + [HCO_3^-] + [OH^-] \text{ or}$$

$$Alk = 2[CO_3^{2-}] + [HCO_3^-] + [OH^-] - [H^+] \tag{5.10}$$

where:

Alk = alkalinity, i.e. concentration of strong base.

Note that the carbonate concentration is counted twice because its presence is the result of the dissociation of two protons from the original compound, CO_2 (i.e. H_2CO_3), first into bicarbonate (HCO_3^-)

Table 5.2 Anion and cation balance from the dissociation reactions of the original compounds present in the water.

Generated cations	H^+	H^+	Bs^+
	↑	↑	↑
Original compounds	H_2O	CO_2 (H_2CO_3)	BsOH
	↓	↓	↓
Generated anions	OH^-	HCO_3^-	OH^-
		↓	
		CO_3^{2-}	

and then carbonate (CO_3^{2-}). The name "bicarbonate" has an archaic origin, as the early chemists used it to distinguish between $NaHCO_3$ (sodium bicarbonate) and Na_2CO_3 (sodium carbonate). The prefix "bi" means twice and was given as sodium bicarbonate has twice the ratio of carbonate (CO_3) to sodium as does sodium carbonate. The correct chemical name is hydrogen carbonate, but this is not commonly used.

In an anaerobic reactor, where the pH must be near the neutral value of pH = 7, the concentrations of the ionic species of water dissociation ($[H^+] = [OH^-] = 10^{-7}\,mol \cdot L^{-1}$) and the carbonate concentration are all much smaller than the bicarbonate concentration. Hence, for all practical purposes the alkalinity of the system may be considered as equal to the bicarbonate concentration and for that reason is called the bicarbonate alkalinity:

$$Alk \approx [HCO_3^-] \; (6.5 < pH < 7.5) \tag{5.11}$$

In Eq. (5.10) the valence of the ions is considered so that the alkalinity is expressed not in molar concentrations but in milli-equivalents per litre ($meq \cdot L^{-1}$). In practice, the old-fashioned method of expressing alkalinity in terms of $mg \cdot L^{-1}$ $CaCO_3$ is still commonly used. The conversion can be made considering that 1 meq = 50 mg $CaCO_3$ (where 1 mmol is 100 mg $CaCO_3$). In this text preference is given to the more rational unit of $meq \cdot L^{-1}$.

Care must be taken when other acid–base systems are present with pK_a values in the pH range covered by the alkalinity titration curve. In that case part of the added acid will be used for conversion of these acid–base systems and the use of strong acid added to reach the equivalence point of CO_2 will be increased, and therefore is not indicative for the bicarbonate alkalinity. This acid consumption is commonly referred to as total alkalinity. One can calculate bicarbonate alkalinity from the total alkalinity by subtracting from the total consumption of strong acid the consumption associated with other acid–base systems.

The most common case is the titration of water containing both acetic acid and the carbonic system. In this case a correction must be made, because much of the acetate is converted to acetic acid when the pH drops from neutral to the end point of the titration; the equivalence point of CO_2. There are several methods to do this. These are discussed in more detail in Section 5.4.4. It is important to stress that only the bicarbonate alkalinity contributes to the buffering capacity in the neutral pH range, where the buffer index of the acetic acid–acetate system is practically zero.

5.4.2 Acidity

The definition of acidity is similar to that of alkalinity, but in this case, water is considered a solution of a mixture of carbonate and a (hypothetical) strong acid (HAs). The acidity is determined by titrating the solution with a strong base to reach the equivalence carbonate point of a pure carbonate solution. The expression that links acidity to the carbonic system is similar to Eq. (5.10):

$$Ac = 2[CO_2] + [HCO_3^-] + [H^+] - [OH^-] \tag{5.12}$$

In principle the determinations of Alkalinity and Acidity are equivalent: both provide the necessary information required to define the carbonic system. However, there are some practical considerations that favour the determination of alkalinity rather than acidity. These reasons are:

(1) The titration solution used to determine the alkalinity of a solution (a strong acid like sulphuric acid) is stable, whereas the solution for determining acidity (caustic soda) tends to be affected over time by absorption of CO_2 from the air.
(2) The titration end-point for alkalinity is the CO_2 equivalence point, where the index has a very clear minimum value, so that it is possible to determine the end point quite accurately. By contrast, the endpoint of the acidity titration is not so well defined, because the equivalence point of carbonate has no clear minimum, owing to the significant contribution of the water buffer index itself (see Figure 5.2).
(3) To reach the equivalence point of carbonate it is necessary to convert the ammonia buffer (normally the second largest after the carbonic buffer system, see Figure 5.1), so that it is also necessary to determine the concentration of ammonia to make the proper correction.
(4) Other reactions like the precipitation of $CaCO_3$ and $Mg(OH)_2$ may interfere near the acidity titration end point.
(5) As long as the CO_2 concentration is higher than the equilibrium concentration with air (at pH $\approx$ 8.5–9), CO_2 may desorb, but at the high pH values near the end of the titration CO_2 may instead be absorbed from the air. Thus, depending on the conditions of the titration, the experimental result may be lower or higher than the real value.

For these reasons, in practice only alkalinity is used to determine the carbonic system.

5.4.3 Ionic equilibrium

Alkalinity and acidity can be linked by the equations that define these parameters (Eqs. (5.10) and (5.12)). By eliminating the carbonic species using the dissociation equations (Eqs. (5.6) and (5.7)) and hydroxide using Eq. (5.8) and finally substituting the pH value from Eq. (5.9), this results in:

$$\text{Alk} = [\text{Ac} + 10^{(\text{pH}-\text{p}K1)} - 10^{-\text{pH}}] \cdot [1 + 2 \times 10^{(\text{pH}-\text{p}K2)}]/[1 + 2 \times 10^{(\text{p}K1-\text{pH})}] \quad (5.13)$$

If the pH is in the neutral range ($5 < \text{pH} < 9$) the influence of the carbonic system predominates over that of the dissociation of water, so that Eq. (5.13) can be simplified to:

$$\text{Alk} \approx \text{Ac}\,[1 + 2 \times 10^{(\text{pH}-\text{p}K2)}]/[1 + 2 \times 10^{(\text{p}K1-\text{pH})}] \quad (5 < \text{pH} < 9) \quad (5.13\text{a})$$

Equation (5.13a) shows that there is a linear relationship between alkalinity and acidity for any value of the pH. In Figure 5.3, the relationship of Eq. (5.13) is plotted for different pH values in a Deffeyes diagram (Deffeyes, 1956), who in 1953 first noted the relationship between the three parameters. The diagram is very useful to evaluate the changes of pH as function of changes in alkalinity or acidity, which often occur in anaerobic or other biological reactors. In the diagram, when two values of the three variables alkalinity, acidity or pH are known, this defines on the Deffeyes chart the ionic equilibrium that characterizes the liquid phase.

As an example, Figure 5.3 shows how the ionic equilibrium changes, when acids or bases are added or removed. For demonstration purposes a starting alkalinity and acidity of 10 and 15 meq $\cdot$ L^{-1} respectively were assumed, while the result of 5 meq $\cdot$ L^{-1} additions of different chemical is evaluated. Table 5.3 shows alkaline materials that can be added and the reaction equations that will affect the point of ionic equilibrium.

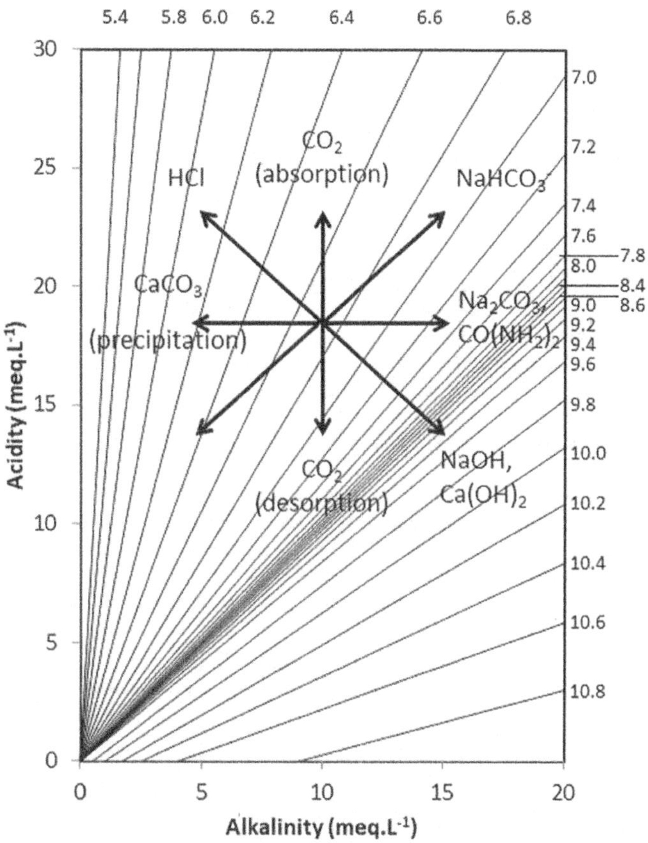

Figure 5.3 Deffeyes diagram for a temperature of 25°C ($pK_1 = 6.33$; $pK_2 = 10.33$; $pK_w = 14$).

Table 5.3 Effect of addition of chemical compounds, CO_2 desorption and $CaCO_3$ precipitation on alkalinity and acidity in anaerobic reactors.

Added Material	Reaction	ΔAlk (meq·mol^{-1})	ΔAc (meq·mol^{-1})
Caustic soda	$NaOH \rightarrow Na^+ + OH^-$	+1	−1
Lime (no precipitation)	$Ca(OH)_2 \rightarrow Ca^{2+} + 2\,OH^-$	+2	−2
Urea	$CO(NH_2)_2 + 3\,H_2O \rightarrow HCO_3^- + OH^- + 2\,NH_4^+$	+2	0
Soda (sodium carbonate)	$Na_2CO_3 \rightarrow 2Na^+ + CO_3^{2-}$	+2	0
Sodium bicarbonate	$NaHCO_3 \rightarrow Na^+ + HCO_3^-$	+1	+1
Hydrochloric acid	$HCl \rightarrow H^+ + Cl^-$	−1	+1
Desorption of CO_2	$CO_2\uparrow$	0	−2
Precipitation of $CaCO_3$	$Ca^{2+} + CO_3^{2-} \rightarrow CaCO_3\downarrow$	−2	0

EXAMPLE 5.1

To one litre of wastewater with pH = 7.2 and Alk = 6 meq·L^{-1} the following chemicals are added: 2 mL of 1N HCl (strong acid) and 336 mg·L^{-1} of sodium bicarbonate.

What are the final values of alkalinity, acidity and pH? And how much 0.5N NaOH must be added to return the pH to its original value of 7.2?

Solution

Point A of the Deffeyes diagram in Figure 5.4 characterizes the initial situation (pH = 7.2 and Alk = 6 meq·L^{-1}). From the diagram the corresponding acidity value can be determined as Ac = 7.5 meq·L^{-1}.

The addition of 2 ml of 1N HCl (this corresponds to 2 meq·L^{-1} of acid addition to the wastewater) results in an equivalent decrease of 2 meq·L^{-1} of alkalinity and an increase of 2 meq·L^{-1} of acidity, so that the ionic equilibrium moves from point A to B, causing the pH value to drop to about 6.45. The second addition consists of 336 mg·L^{-1} or 4 meq·L^{-1} of sodium bicarbonate. The addition of HCO_3^- results in an increase of the alkalinity and acidity of 4 meq·L^{-1} and moves the equilibrium from point B to C, corresponding to a pH value of 6.8.

To return to the original pH value of 7.2, a concentration of about 1.5 meq·L^{-1} of NaOH must be added, resulting in the displacement of the ionic equilibrium point from C to D. Note that while the pH is now back at its initial value, the alkalinity and acidity values are not. Table 5.4 summarizes the changes of alkalinity, acidity and pH resulting from each addition of chemical in Example 5.1.

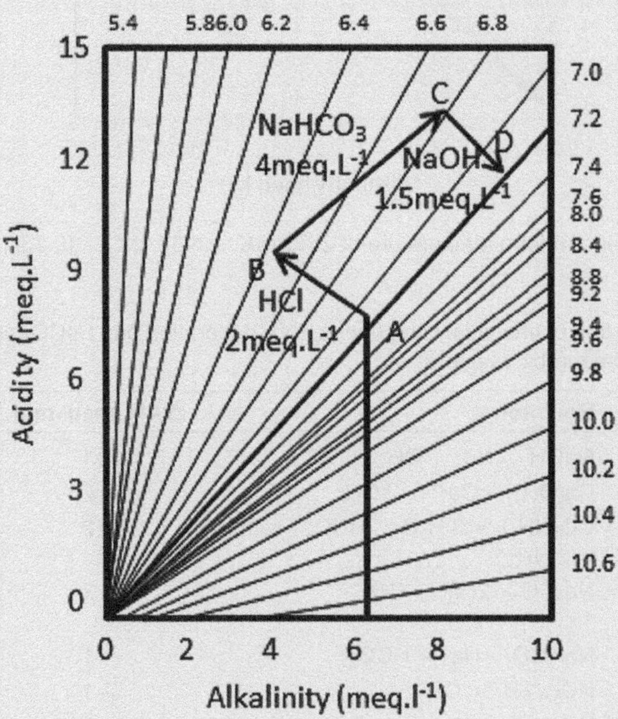

Figure 5.4 Displacement of the ionic equilibrium in the Deffeyes diagram according to Example 5.1.

Table 5.4 Change of the alkalinity, acidity and pH after additions of several chemicals that affect the ionic equilibrium as per Example 5.1.

Added Compound	Concentration ($meq \cdot L^{-1}$)	Alkalinity ($meq \cdot L^{-1}$)	Acidity ($meq \cdot L^{-1}$)	pH
(1) Initial situation		6	7.5	7.2
(2) Hydrochloric acid	2	4	9.5	6.45
(3) Sodium bicarbonate	4	8	13.5	6.8
(4) Caustic soda	1.5	9.5	12	7.2

EXAMPLE 5.2

Calculate the lowest value of the pH that can be reached in an anaerobic sewage treatment system with a starting alkalinity of 8 $meq \cdot L^{-1}$, pH = 7.3 and a biodegradable COD concentration of 320 $mg \cdot L^{-1}$.

Solution

In the Deffeyes diagram of Figure 5.5 it is established that for the given starting conditions of pH and alkalinity, the corresponding acidity is equal to 10 $meq \cdot L^{-1}$; point A is located at the intersection of pH = 7.3 and Alk = 8 $meq \cdot L^{-1}$, the initial ionic equilibrium point.

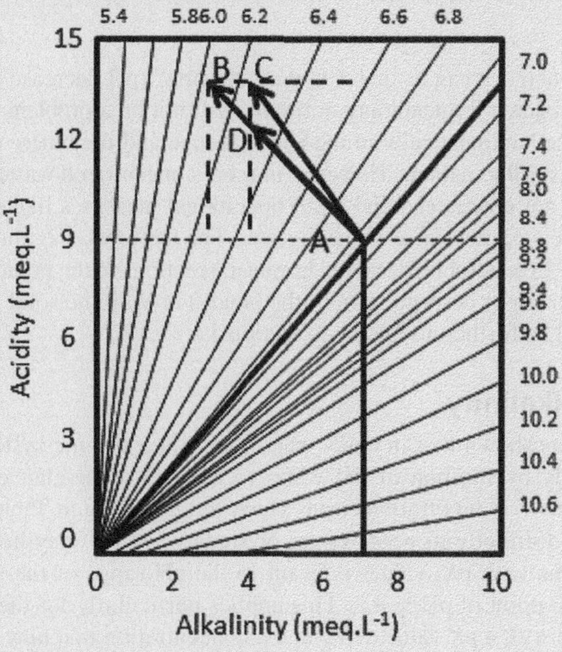

Figure 5.5 Deffeyes diagram for Example 5.2.

In the anaerobic reactor, the lowest pH value that can occur is when all biodegradable organic material is converted to acetic acid, that is, if there were only (complete) acid fermentation and no methanogenic fermentation. In this case, knowing that for a ratio of 1.067 mg COD·mg^{-1} HAc (see Chapter 1, Table 1.6), the acid fermentation of 320 mg·L^{-1} COD results in 320/1.067 = 300 mg HAc or 300/60 = 5 meq HAc. As the acetic acid acts as a strong acid in the neutral pH range, acid fermentation results in a decrease of alkalinity and an increase of acidity both of 5 meq·L^{-1}. Thus at least theoretically the alkalinity decreases to Point B where $Alk_e = Alk_i - 5 = 3$ meq·L^{-1}, while acidity increases to $Ac_e = Ac_i + 5 = 15$ meq·L^{-1}.

At the new ionic equilibrium point B, characterized by $Alk_e = 3$ meq·L^{-1} and $Ac_e = 15$ meq·L^{-1}, the pH that would develop can be read from the Deffeyes diagram (Figure 5.5) as $pH_e \approx 6.05$.

In Figure 5.5 it can be seen that the minimum pH value is 6.05, but in practice this would be higher due to several reasons:

(1) Only about 70% of the organic material in wastewaters is converted into acetic acid; the remainder converts to CO_2 and H_2 (Gujer & Zehnder, 1983); see Chapter 2, Figure 2.7). H_2 does not have an influence on alkalinity or acidity, whereas CO_2 produces acidity, but does not affect alkalinity (Eqs. (5.10) and (5.12). Hence the decrease in alkalinity would be only 70% of 5 meq·L^{-1} or 3.5 meq·L^{-1} and the equilibrium point would be displaced to point C with pH = 6.5.

(2) Part of the produced carbon dioxide would be released from the liquid phase, reducing the acidity. The loss of CO_2 is not known, but in Figure 5.5 a value of 20% was adopted as an estimate (point D), resulting in a decrease in acidity of 4 meq·L^{-1} and a pH value of 6.4.

(3) Alkalinity is generated in the anaerobic reactor by several other processes, of which an important one is ammonification due to mineralization of organic nitrogen, a process in which 1 equivalent of alkalinity is generated per mol of mineralized nitrogen. In addition methanogenesis will develop at least partially, which increases alkalinity and decreases acidity. Although the extent of the processes of processes of ammonification and methanogenesis is not known (as fraction of the maximum possible), it is expected that these processes will result in a final pH near the neutral point.

From the observations above it can be deduced that "souring" (pH decrease due to domination of acid fermentation over methanogenic fermentation) normally will not be a problem in sewage treatment. The pH value will be established automatically in the neutral range and the buffer index naturally present in the wastewater will avoid pH fluctuations. However, in more concentrated wastewaters pH stability in the neutral range is certainly not guaranteed and often operational measures like caustic soda addition will be required to achieve stable performance of the digester. The Deffeyes diagram is very useful but it only considers the liquid phase. In anaerobic digestion, however, reactions in the liquid phase often interact with the gas phase (CO_2 adsorption or desorption from the biogas) or with the solid phase ($CaCO_3$ dissolution or precipitation). This will be further discussed in Section 5.6.

5.4.4 Measuring alkalinity

In the section above it was shown that in cases where only the carbonic system is relevant, alkalinity can be determined simply by titration of the water to the CO_2 equivalence point, which in turn is determined by the derivative of the titration curve. However, as shown in Table 5.1, this simple titration method cannot be applied directly to wastewaters or sludges, which may have a significant presence of other acid–base systems with pK_a values relevant to the pH range of the titration, which normally is from about 7 to the end point of pH ≈ 4.2. This applies particularly for the acid–base system of the volatile fatty acids (VFA) with a pK value of 4.7 and a concentration that may be significant, especially for industrial wastewaters or in anaerobic reactors. In this case a different titration method may be used for the determination of not only alkalinity but also the concentration of VFA. This titration consists essentially of measuring the initial pH of the sample and the added volume of a known acid solution to two pH values: pH ≈ 5.5 and pH ≈ 4. With a known acid normality, the alkalinity and the concentration of the VFA can be calculated. The fraction of carbonic species present in the form of CO_2 depends on the pH value. Using Eq. (5.5):

$$[CO_2]/C_t = 1/[1 + 10^{(pH-pK1)}] \quad (5.14)$$

It is tacitly assumed that the initial pH is in the neutral range and that only the $CO_2 \leftrightarrow HCO_3^-$ equilibrium is relevant. Similarly the fraction of non-dissociated VFA is given by:

$$[HA]/C_a = 1/[1 + 10^{(pH-pKa)}] \quad (5.15)$$

where:

$pK_1 = 6.33$
$pK_a = 4.70$

In Figure 5.6 the curves representing Eqs. (5.14) and (5.15) have been drawn as functions of pH and the values for three pH values are indicated as an example: $pH_1 = 7.12$ (initial pH), $pH_2 = 5.53$ and $pH_3 = 4.02$. In Table 5.5 the calculated values of the CO_2 and HA fractions for the selected pH values are shown (f_{CO2} and f_{HA}), as well as the variations of the fractions (Δ_{CO2}) and (Δ_{HA}) when the pH decreases from pH_1 to pH_2 and finally to pH_3. Now the number of transformed moles of $[CO_2]$ and [HA] is equated to the number of moles of acid added during the titration (N). In the procedure it is tacitly assumed that all the acid is used for the transformation of HCO_3^- into CO_2 and A^- into HA, a supposition that is justified only if there are no other acid–base systems with pK_a values in the range covered by the titration.

$$V_{sample} \cdot (\Delta_{HA}\, C_a + \Delta_{CO2}\, C_t) = N_a\, V_a \quad (5.16)$$

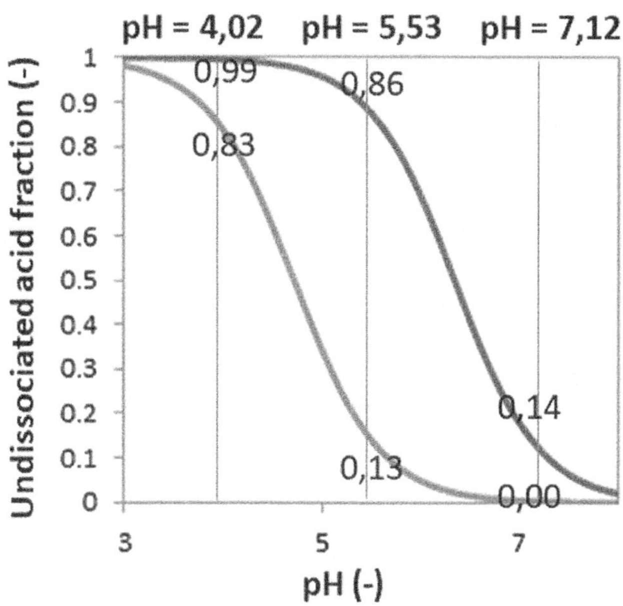

Figure 5.6 Distribution of the carbonic and VFA species as function of pH.

Table 5.5 Changes in the composition of carbonic and VFA species at different pH values.

$pH_1 = 7.12$	$pH_2 = 5.53$	$pH_3 = 4.02$
$f_{CO2} = 0.14$	$f_{CO2} = 0.86$	$f_{CO2} = 0.99$
$f_{HA} = 0.00$	$f_{HA} = 0.13$	$f_{HA} = 0.83$
	$\Delta_{CO2} = 0.86 - 0.14 = 0.72$	$\Delta_{CO2} = 0.99 - 0.14 = 0.85$
	$\Delta_{HA} = 0.13 - 0.00 = 0.13$	$\Delta_{HA} = 0.83 - 0.00 = 0.83$

where:

V_a = volume of added acid
V_{sample} = sample volume
C_a = VFA concentration
C_t = carbonic species concentration

For the two acid additions from pH_1 to pH_2 and from pH_2 to pH_3, the equations take the form of a set of linear equations:

$$c_1 \cdot C_a + c_2 \cdot C_t = c_3 \tag{5.17a}$$

$$d_1 \cdot C_a + d_2 \cdot C_t = d_3 \tag{5.17b}$$

where:

$c_1 = V_{sample}\ \Delta_{HA}$ for the titration from pH_1 to pH_2
$c_2 = V_{sample}\ \Delta_{CO2}$ for the titration from pH_1 to pH_2
$c_3 = N_a V_a$ for the titration from pH_1 to pH_2
$d_1 = V_{sample}\ \Delta_{HA}$ for the titration from pH_1 to pH_3
$d_2 = V_{sample}\ \Delta_{CO2}$ for the titration from pH_1 to pH_3
$d_3 = N_a V_a$ for the titration from pH_1 to pH_3.

The solution for the set of the two linear equations is:

$$C_t = c_3/(fc_1 + c_2) \tag{5.18a}$$

$$C_a = fC_t \tag{5.18b}$$

where: $f = (d_2/d_3 - c_2/c_3)/(c_1/c_3 - d_1/d_3)$.

In the initial sample the value of C_a is equal to the concentration of VFA, whereas the alkalinity is equal to the bicarbonate concentration:

$$\text{Alk} = (1 - f_{CO2}) \cdot C_t \tag{5.19a}$$

$$[\text{VFA}] = fC_t \tag{5.19b}$$

So both the alkalinity and the concentration of volatile fatty acids can be determined with a simple titration requiring the initial pH (sample pH) and the pH value after two additions of strong acid. The best result is obtained if the pH after the first addition (pH_2) is around 5.5 with most of the bicarbonate converted into CO_2 and still relatively little VFA transformed and the pH after the second addition (pH_3) equal to 4.0 with most of the bicarbonate and acetate converted into CO_2 and non-dissociated acid respectively.

EXAMPLE 5.3

A sample of 50 ml is titrated with an acid solution with a strength of 0.05 N. To reduce the pH from the original value of 7.12 to 5.53 (pH_2) an acid volume of 6.04 ml was required. To reduce the pH to 4.02 the required additional acid volume was 8.49 ml. Calculate the bicarbonate alkalinity and the VFA concentration in the sample.

Solution

Using Eqs. (5.17a) and (5.17b) and the values for Δ_{HA} and Δ_{CO2} supplied in Table 5.5 one has:

$c_1 = V_{sample} \cdot \Delta_{HA} = 0.05 \times 0.13 = 0.0065$ meq for the titration from pH_1 to pH_2
$c_2 = V_{sample} \cdot \Delta_{CO2} = 0.05 \times 0.72 = 0.036$ meq for the titration from pH_1 to pH_2
$c_3 = N_a V_a = 0.05 \times 6.04 = 0.302$ meq for the titration from pH_1 to pH_2
$d_1 = V_{sample} \cdot \Delta_{HA} = 0.05 \times 0.83 = 0.0415$ meq for the titration from pH_1 to pH_3
$d_2 = V_{sample} \cdot \Delta_{CO2} = 0.05 \times 0.85 = 0.0425$ meq for the titration from pH_1 to pH_3
$d_3 = N_a V_a = 0.05 \times 8.49 = 0.42$ meq for the titration from pH_1 to pH_3

Now with Eqs. (5.18a) and (5.18b):

$f = (d_2/d_3 - c_2/c_3)/(c_1/c_3 - d_1/d_3) = 0.25$
$C_t = c_3/(fc_1 + c_2) = 0.30/(0{\cdot}25 \times 0.13 + 0.72) = 8{\cdot}0\ meq \cdot L^{-1}$ and $C_a = fC_t = 0.25 \times 8 = 2.0\ meq \cdot L^{-1}$

And finally with Eqs. (5.19a) and (5.19b):

$Alk = (1 - f_{CO2})C_t = 0.86 \times 8 = 6.88\ meq \cdot L^{-1}$
$[VFA] = fC_t = 2.0\ meq \cdot L^{-1}$

5.5 REACTOR pH AND ALKALINITY DEMAND

The most important issue regarding the value and stability of the pH in an anaerobic digester is the question of whether the alkalinity present in the wastewater, along with eventually produced alkalinity in the digester, is enough to maintain the pH in the reactor above a certain minimum value, which ensures stable operation of the biological processes, especially that of methanogenesis. In this section, this issue will be evaluated assuming neither acidity nor alkalinity is affected by precipitation or dissolution of $CaCO_3$, which will be discussed in Section 5.6.

5.5.1 Mass balance of inorganic carbon and methane

It is convenient to formulate expressions of the mass flows of inorganic carbon and methane in the incoming and outgoing liquid and gas flows as well as the mass flows generated in the anaerobic reactor. With the help of Figure 5.7 one can describe the flow of carbonic species in the reactor as summarized in Table 5.6. If the biogas is considered as a mixture of only methane and carbon dioxide, its composition can be expressed by means of the partial pressure of the two components:

$$p_{CO2} = MC_{tg}/(MC_{tg} + MC_{mg}) = (MC_{ti} + MC_{td} - MC_{te})/(MC_{ti} + MC_{td} + MC_{md} - MC_{te} - MC_{me})$$
$$= (C_{ti} + C_{td} - C_{te})/(C_{ti} + C_{td} + C_{md} - C_{te} - C_{me}) \quad (5.20)$$

Equation (5.20) can be used in several ways. One important application is to evaluate if the sum of the influent alkalinity and the generated bicarbonate alkalinity is enough to maintain an adequate pH value for anaerobic digestion. If not, the same equation may be used to calculate the alkalinity demand to maintain a certain desired pH. To do this, first the terms of Eq. (5.20) must be expressed in such a way that they can be calculated or measured.

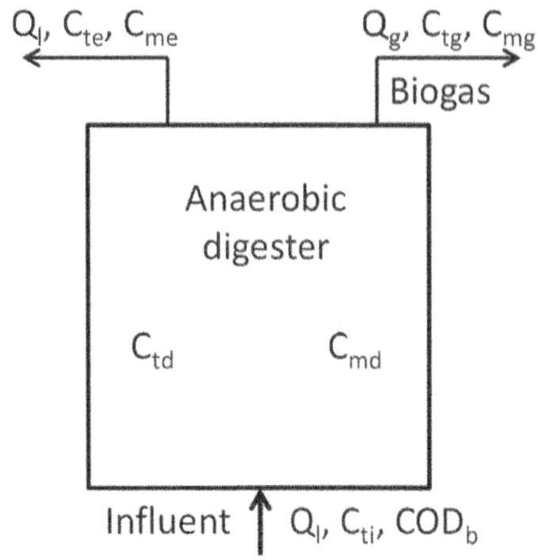

Figure 5.7 Representation of the mass fluxes of carbonic species and methane in an anaerobic digester.

(1) *Concentration of carbonic species in the influent (C_{ti})*
For specified (measured) values of the influent pH and alkalinity, the influent acidity can be easily read from the Deffeyes diagram (Figure 5.3). With known values of both alkalinity and acidity, the influent carbonic species concentration (C_{ti}) can be calculated through the addition of the definition equations Eqs. (5.10) and (5.12):

$$C_{ti} = (\text{Alk}_i + \text{Ac}_i)/2 \qquad (5.21)$$

(2) *Generated CH_4 and CO_2 concentrations (C_{md} and C_{td})*
In the EXCEPTIONAL case that the molecular formulas of the organic matter in the wastewater are known, the generated concentrations of methane (C_{md}) and carbonic species (C_{td}) can be directly calculated with Eq. (2.6). In all other cases these can be estimated from the concentration and composition of the organic material present in the wastewater. Methane production can be directly calculated from the digested COD concentration, knowing that 1 g or 1/16 mol CH_4 is formed upon digestion of 4 g of COD. The CO_2 production depends not only on the concentration, but also on the composition of the digested organic material. For digestion of carbohydrates the number of electrons transferred $N_{el} = 4$, as calculated with Eq. (1.14), and the molar CO_2 production is equal

Table 5.6 Fluxes of carbonic species and methane in anaerobic digesters.

(1) For carbonic species:	(2) For methane:
$MC_{ti} = Q_l C_{ti}$	$MC_{md} = Q_l C_{md}$
$MC_{te} = Q_l C_{tl}$	$MC_{me} = Q_l C_{me}$
$MC_{td} = Q_l C_{td}$	$MC_{mg} = Q_g p_{CH4}/RT$
$MC_{tg} = Q_g p_{CO2}/RT$	

where:
MC_t = mass flux of carbonic species
MC_m = mass flux of methane
Q = flow (liquid or gas)
Indices l and g refer to liquid and gas
Indices i, e, and d refer to influent, effluent and generated in the digester, respectively.

to the molar CH_4 production. This is also true in good approximation for the digestion of proteins. In the case of digestion of fats, the CO_2 production is lower than the CH_4 production ($N_{el} > 4$). Thus, a conservative estimate is that the CO_2 production is at least equal to the CH_4 production:

$$C_{td} \approx C_{md} = COD_d/64 \tag{5.22}$$

(3) *Alkalinity increase in the reactor ($Alk_e - Alk_i$)*
Changes in alkalinity in the digester are basically due to two processes: ammonification and removal of organic acids.
(a) **Ammonification**
If the ammonia concentration of the effluent is higher than that of the influent, the difference is due to ammonification of organic nitrogen, which occurs when organic material is degraded. This process produces alkalinity, as can be seen from the reaction equation:

$$RNH_2 + 2H_2O \rightarrow ROH + NH_4^+ + OH^- \tag{5.23}$$

Ammonification is essentially a hydrolysis reaction. The conversion of one "mol" TKN (14 g N) results in the formation of one equivalent of alkalinity, so that the alkalinity production from ammonification is $1/14 = 0.071$ meq·mg^{-1} N. In raw sewage the organic nitrogen concentration is often in the range between two-thirds and three-quarters of the total TKN. Significantly higher concentrations of organic nitrogen (2–4 g·L^{-1}) may be present in municipal sewage sludge and manure.
(b) **VFA removal**
A second source of alkalinity production in the reactor is the digestion of VFA. Since the VFA concentration in the influent usually is greater than in the effluent, there will be an increase in the bicarbonate alkalinity in the reactor, equal to the net reduction of VFA concentration. If there are no other reactions for the consumption or production of alkalinity, the effluent alkalinity can be estimated from the influent alkalinity of the wastewater and changes in the concentrations of VFA and NH_4^+ as:

$$Alk_e = Alk_i + VFA_i - VFA_e + (N_{ae} - N_{ai})\,0.071 \tag{5.24}$$

where:

N_{ai} = influent ammonium concentration (mg N·L^{-1})
N_{ae} = effluent ammonium concentration (mg N·L^{-1})

VFA are expressed in meq·L^{-1}
(4) *Concentration of carbonic species in the effluent (C_{te})*
The concentration of carbon species can be linked to the concentration of bicarbonate and hence to the alkalinity in the effluent. If the pH of the effluent is near the neutral value, then the carbonate concentration in the effluent is low and the concentration of carbonic species can be expressed as:

$$C_{te} = [CO_2] + [HCO_3^-] = [HCO_3^-](1 + [H+]/K_1) = Alk_e\cdot(1 + 10^{(pK1 - pH)}) = Alk_e(1 + f_e) \tag{5.25}$$

where:

$f_e = 10^{(pK1-pH)}$

(5) *Dissolved methane concentration (C_{me})*
The concentration of dissolved methane in the effluent will obey the Henry's law and therefore depends on the partial pressure of CH_4 and the value of Henry's law constant for methane (see Section 6.1). The solubility of methane at a partial pressure of 1 atm. is approximately 1 mmol·L^{-1} or 16 mg·L^{-1} in

the temperature range of 20–35°C. It then follows that the dissolved methane represents a maximum digested COD concentration of 64 mg·L^{-1} when the partial pressure of methane approaches 1 atm.

Thus, the dissolved methane concentration is only a significant fraction of the methane production, if the digested COD concentration is low. If the concentration of digested COD is high, the mass flow of dissolved methane is insignificant compared to the mass flow of methane in the gas phase. Considering the above, Eq. (5.20) can be rewritten as:

$$p_{CO2} = (C_{ti} + COD_d/64 - Alk_e(1 + f_e))/(C_{ti} + COD_d/32 - 0.001 - Alk_e(1 + f_e)) \quad (5.26a)$$

On the other hand the partial CO_2 pressure can also be expressed by Henry's law as:

$$p_{CO2} = [CO_2]/K_H = [HCO_3^-][H+]/(K_1\ K_H) = Alk_e f_e/K_H \quad (5.26b)$$

5.5.2 pH calculation in buffered wastewaters

If the buffer index of a wastewater is enough to maintain the pH in the neutral range, anaerobic digestion can be applied without addition of chemicals for pH correction. In this case Eqs. (5.26a) and (5.26b) form the basis for the calculations to determine the pH value that will develop in the anaerobic digester. Equating the two expressions yields:

$$(C_{ti} + COD_d/64 - Alk_e(1 + f_e))/(C_{ti} + COD_d/32 - 0.001 - Alk_e(1 + f_e) = [CO_2]/K_H = Alk_e f_e/K_H \quad (5.27)$$

The solution for this quadratic expression is:

$$f_e = 0.5[b - (b^2 - 4c)^{0.5}] \quad (5.28)$$

where:

$$b = (K_H + C_{ti} + C_{td} + C_{md} - C_{me} - Alk_e)/Alk_e$$
$$c = K_H \cdot (C_{ti} + C_{td} - Alk_e)/Alk_e^2$$

Having determined the value of factor f_e, the effluent pH can now easily be calculated with the aid of Eq. (5.25):

$$f_e = 10^{(pK1-pH)}$$

$$pH_e = pK_1 - \log f_e \quad (5.29)$$

EXAMPLE 5.4

Verify if sufficient alkalinity is present in a wastewater with an alkalinity of 8 meq·L^{-1} and a pH of 7.0 to apply anaerobic digestion under the following conditions: wastewater COD digestion = 320 mg·L^{-1} (i.e. a production of 0.005 mol CO_2 and 0.005 mol CH_4 per L of influent), whereas 2 meq·L^{-1} of alkalinity are generated (1 meq·L^{-1} due to ammonification and 1 meq·L^{-1} due to VFA removal) and the reactor temperature is 25°C (i.e. $K_H = 0.033$).

Solution

(1) From $Alk_i = 0.008$ and pH = 7.0, in Deffeyes diagram: $Ac_i = 0.012$ eq·L^{-1}
(2) $C_{ti} = (Alk_i + Ac_i)/2 = 0.010$ mol·L^{-1}
(3) $Alk_e = Alk_i + Alk_d = 0.008 + 0.002 = 0.010$ eq·L^{-1} (given)
(4) $C_{td} = C_{md} = 0.005$ mol·L^{-1} (given)

(5) $C_{me} = 0.001$ mol·L^{-1} (typical value)
(6) $pK_1 = 6.37$ and $pK_H = 1.47$ (25 °C)

Hence:

$$b = (0.034 + 0.01 + 0.005 + 0.005 - 0.001 - 0.010)/0.01 = 4.3$$
$$c = 0.034\,(0.01 + 0.005 - 0.010)/(0.01)2 = 1.70$$

so that:

$$f_e = 0.5[4.3 - (4.3^2 - 4 \times 1.70)^{0.5}] = 0.44 \quad (5.28)$$

Therefore the effluent pH will be:

$$pH_e = pK_1 - \log 0.44 = 6.37 + 0.36 = 6.73 \quad (5.29)$$

This pH value is acceptable because it is in the neutral range and allows the development of anaerobic digestion without the need for alkalinity addition. The conditions selected in the above example are fairly typical of domestic sewage in regions with hot climate. If the concentration of biodegradable organic material were much higher than the usual range for domestic sewage, there would be an alkalinity demand as will be shown in the next example.

EXAMPLE 5.5

Using the same conditions of Example 5.4 but for a biodegradable COD concentration of 3200 mg·L^{-1} instead of 320 mg·L^{-1}, predict the pH value of the effluent.

Solution

In this case one would have: $C_{td} = C_{md} = 0.05$ mol·L^{-1}, which results in a pH value of 5.7. This value is too low to allow for methanogenesis, which will be therefore be incomplete. The accumulation of acetic acid will lower the pH further. Assuming that acetogenesis is complete, the COD of 3200 mg·L^{-1} will generate 3000 mg HAc·l^{-1} or 3000/60 = 0.05 N HAc. Using the dissociation constant of acetic acid $pK_a = 4.7$, the resulting pH is about 3.3. This pH value is so low that even the acid fermentation bacteria do not develop completely. It is concluded that under these conditions there will be virtually no removal of organic material: acid fermentation would be partial and methanogenesis would not develop, unless alkalinity were added to increase the pH.

5.5.3 Alkalinity demand in wastewaters without enough buffer capacity

If the effluent pH calculated by the above procedure is too low (for example pH < 6.3), then the assumption that anaerobic digestion will remove the organic material is not correct. In this case, Eq. (5.27) can be used to calculate the effluent alkalinity required to maintain a certain target pH value in the anaerobic reactor. To do this Eq. (5.27) can be rewritten as:

$$(C_{ti} + C_{td} - \mathrm{Alk}_r(1 + f_r))/(C_{ti} + 2C_{td} - C_{me} - \mathrm{Alk}_r\,(1 + f_r)) = \mathrm{Alk}_r f_r/K_H \quad (5.30)$$

where:

$f_r = 10^{(pK1-pHr)}$
pH_r = pH value required for efficient methanogenesis
Alk_r = required reactor alkalinity for the target pH

When the pH value is selected, for example $pH_r = 6.8$ the required alkalinity can be calculated from Eq. (5.30), by solving the implicit quadratic expression:

$$Alk_r = 0.5[b - (b^2 - 4ac)^{0.5}]/a \tag{5.31}$$

where:

$a = (1 + f_r) f_r$
$b = (C_{ti} + 2C_{td} - C_{me}) f_r + k_H (1 + f_r)$
$c = k_H (C_{ti} + C_{td})$

Figure 5.8 shows Eq. (5.31) in graphical form where the required alkalinity is calculated as a function of the digested COD concentration, for different target pH values and for 25°C (left) and 35°C (right). The curves show the following aspects:

(1) The required alkalinity increases with the digested COD concentration. The demand for alkalinity is due to the dissociation of part of the dissolved CO_2 at the selected pH value, which is always in the neutral range, a process that consumes alkalinity in the form of hydroxyl ions:

$$CO_2 + OH^- \rightarrow HCO_3^- \tag{5.32}$$

(2) The required alkalinity is much smaller at 35°C than at 25°C, because the solubility of CO_2 decreases with increasing temperature.
(3) The required alkalinity is very strongly affected by the desired pH value, especially for an effluent pH > 7.

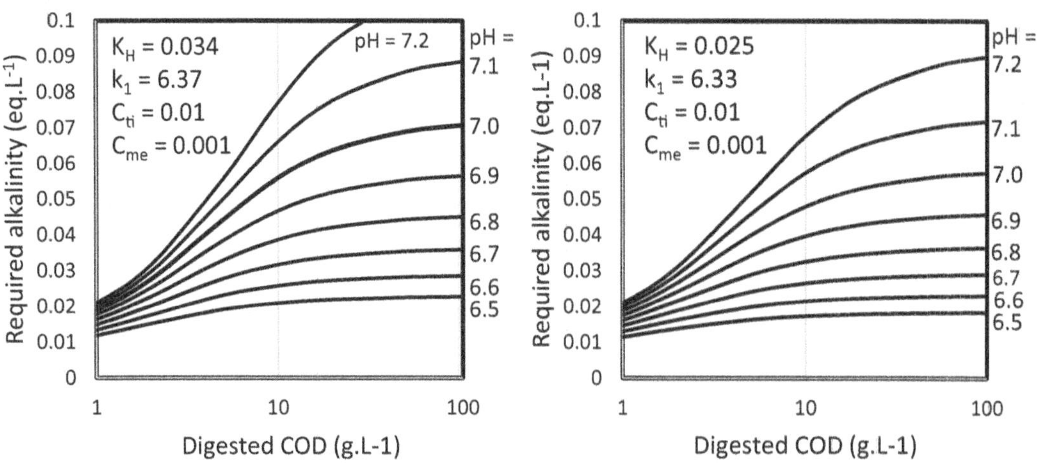

Figure 5.8 Required alkalinity in an anaerobic reactor as a function of the organic material concentration for different pH values and a temperature of 25°C (left) and 35°C (right).

EXAMPLE 5.6

Calculate the alkalinity demand to ensure stable anaerobic treatment at pH = 7.0 in the case of the wastewater of Example 5.5.

Solution

The required alkalinity is calculated with the aid of Eq. (5.31). For pH = 7 the following values are calculated: $f_r = 0.21$; $a = 0.26$; $b = 0.06$ and $c = 0.002$. By inserting the values in the quadratic expression the required alkalinity is calculated as: $Alk_r = 37\ meq \cdot L^{-1}$. If it is again assumed that the influent alkalinity is $10\ meq \cdot L^{-1}$ and that the alkalinity generated in the reactor is equal to $2\ meq \cdot L^{-1}$, then the alkalinity demand is $37 - 12 = 25\ meq \cdot L^{-1}$, which is equivalent to $25 \times 40 = 1000$ mg $NaOH \cdot L^{-1}$.

5.6 PHASE EQUILIBRIA IN ANAEROBIC DIGESTERS

5.6.1 Liquid–gas equilibrium

In an anaerobic reactor the produced biogas is released from the liquid phase once the saturation concentrations of carbon dioxide and methane are exceeded. According to Henry's Law, there is a proportional relationship between the partial pressure of a component in the gas phase and its saturation concentration in the liquid phase. Thus, for carbon dioxide one can write:

$$[CO_2]_s = K_H\, p_{CO2} \tag{5.33}$$

where:

$[CO_2]_s$ = saturation concentration of dissolved gaseous carbon dioxide
p_{CO2} = partial pressure of carbon dioxide in the biogas
K_H = Henry constant for CO_2, where $pK_H = 1.12 + 0.0138T$ for $0 < T < 35°C$ as determined by Capri and Marais (1974).

It is possible to link the partial pressure of CO_2 to the alkalinity by using the dissociation equation for bicarbonate, Eq. (5.6), and noting that for the pH range of interest the alkalinity is almost identical to the bicarbonate concentration as per Eq. (5.11). Hence:

$$p_{CO2} = \text{Alk}\ 10^{(pK1 + pKH - pH)} \tag{5.34}$$

Equation (5.34) shows that for any pH value there is a linear relationship between partial CO_2 pressure and bicarbonate alkalinity. This relationship is shown in Figure 5.9 where Eq. (5.34) was used to draw graphs correlating the partial CO_2 pressure in the biogas with the bicarbonate alkalinity as a function of the pH for different temperatures. The diagrams show that for the alkalinity expected to develop in the sewage (typically in the range of 5–10 $meq \cdot L^{-1}$), the partial CO_2 pressure will low at pH values around the neutral value.

As carbon dioxide and methane are produced in approximately equimolar concentrations in anaerobic digestion, it is concluded that a large part of the carbonic system remains dissolved in the liquid phase. Even for a high temperature of 35°C, i.e. at the lowest solubility of CO_2, the partial pressure is less than 10% of the total biogas pressure.

It was demonstrated previously that the alkalinity required to achieve a suitable pH value for anaerobic digestion can be calculated as a function of the concentration of digested organic material. Together with Eq. (5.34) this allows also the partial CO_2 pressure to be determined as a function of the digested COD

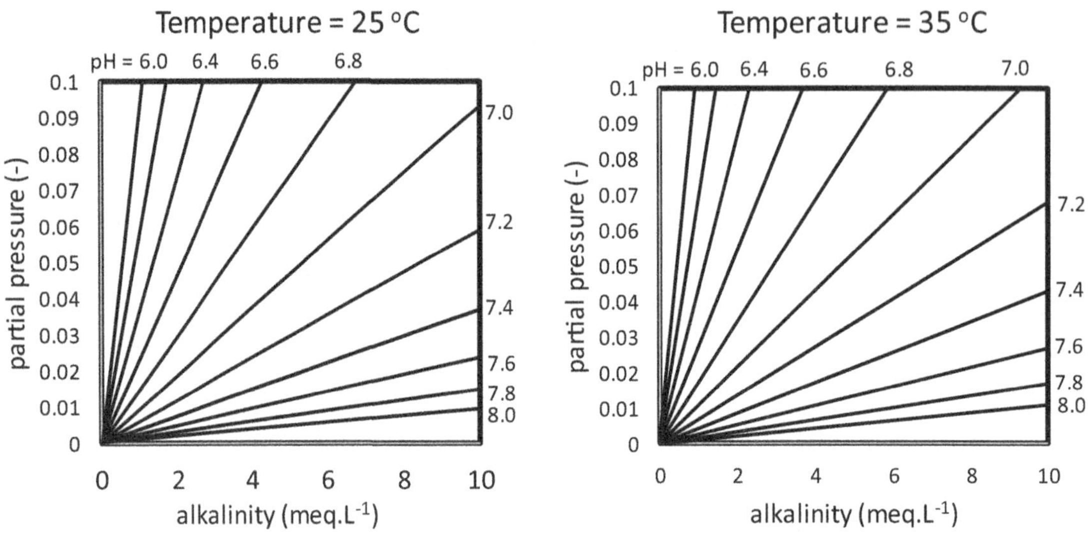

Figure 5.9 Partial CO_2 pressure in the biogas as function of the alkalinity for different pH values in the liquid phase and at different temperatures.

concentration, for any particular pH value, as is shown in Figure 5.10 for two temperatures: 25°C and 35°C. The graphs of Figures 5.9 and 5.10 highlight the following relations between digested COD concentration, bicarbonate alkalinity, pH and temperature:

(1) The biogas composition depends strongly on the digested COD concentration: for low COD concentrations (0.1 to 1 g $COD \cdot L^{-1}$), almost all of the CO_2 remains in the liquid phase as dissolved CO_2 or HCO_3^- and the biogas consists for the largest part of methane. When the COD concentration is high the CO_2 fraction in the biogas is relatively small, almost all of CO_2 that is produced will desorb and the partial pressure tends to its maximum value of 0.5, i.e. CH_4 and CO_2 have the same molar concentrations in the biogas.
(2) The pH value has a moderate influence on the biogas composition: at lower pH values, less bicarbonate is formed, and hence more CO_2 desorbs, resulting an increase of the partial CO_2 pressure.
(3) The influence of the temperature on the partial CO_2 pressure is small: as at lower temperatures the solubility of CO_2 in the liquid phase will increase, the partial CO_2 pressure in the biogas will be lower.

An important aspect of anaerobic digestion is the division of the CO_2 production between the liquid and the gas phase. Assuming a given concentration of carbonic species in the influent (C_{ti}), the production of CO_2 in the digester is calculated by Eq. (5.22) and the presence in the liquid phase from the required Alkalinity, which is equal to bicarbonate concentration:

$$C_{td} = COD_d/64$$

$$C_{te} = [CO_2] + [HCO_3^-] = (1 + 10^{(pK1-pH)}) \cdot [HCO_3^-] = (1 + 10^{(pK1-pH)}) \cdot Alk_e$$

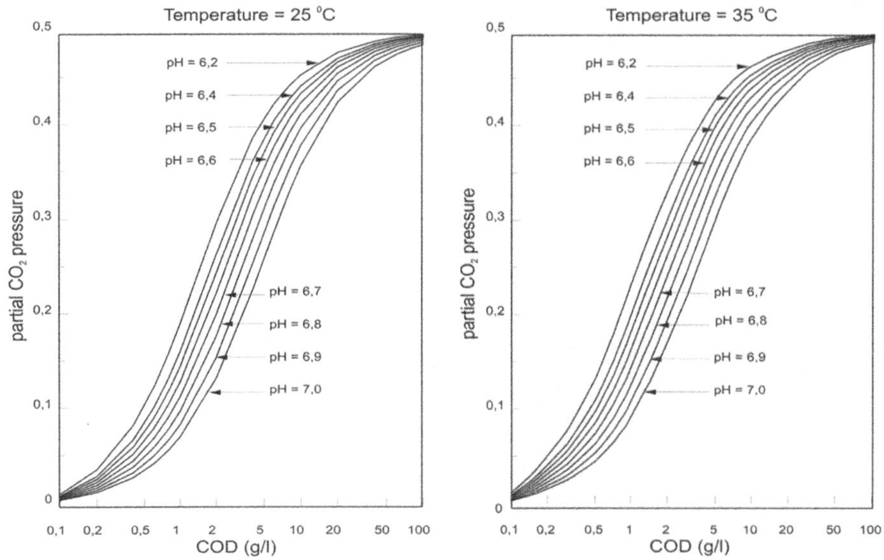

Figure 5.10 Partial CO_2 pressure in biogas as a function of the digested COD concentration at different pH values and for temperatures of 25°C (left) and 35°C (right).

Hence the concentration of released carbon dioxide is given by:

$$C_{tg} = C_{ti} + C_{td} - C_{te} = C_{ti} + COD_d/64 - (1 + 10^{(pK1-pH)})\ Alk_e \tag{5.35}$$

Figure 5.11 shows the fraction that remains of carbonic species remaining dissolved and leaving the digester with the effluent for two temperatures: 25 and 35°C.

5.6.2 Liquid–solid equilibrium

The alkalinity and pH influence the potential for calcium carbonate dissolution or precipitation in wastewater treatment systems. Calcium carbonate is a mineral with a low solubility:

$$K_{sp} = [Ca^{2+}]\ [CO_3^{2-}] \tag{5.36}$$

where:

K_{sp} = solubility constant
$pK_{sp} = 8.03 + 0.01183T$ (from Truesdale and Jones (1973), T in °C)

If the ionic product $[Ca^{2+}][CO_3^{2-}]$ exceeds the solubility constant, the solution is oversaturated and there will be tendency for $CaCO_3$ precipitation. This precipitation will continue until the ionic product is equal to the solubility constant, at which point the solution will be thermodynamically stable. If the ionic product is less than the solubility constant, the solution is undersaturated and calcium ions and/or carbonate can dissolve from solids in contact with the liquid phase: the solution has a corrosive effect on the concrete used in the wastewater treatment system and can be described as "aggressive".

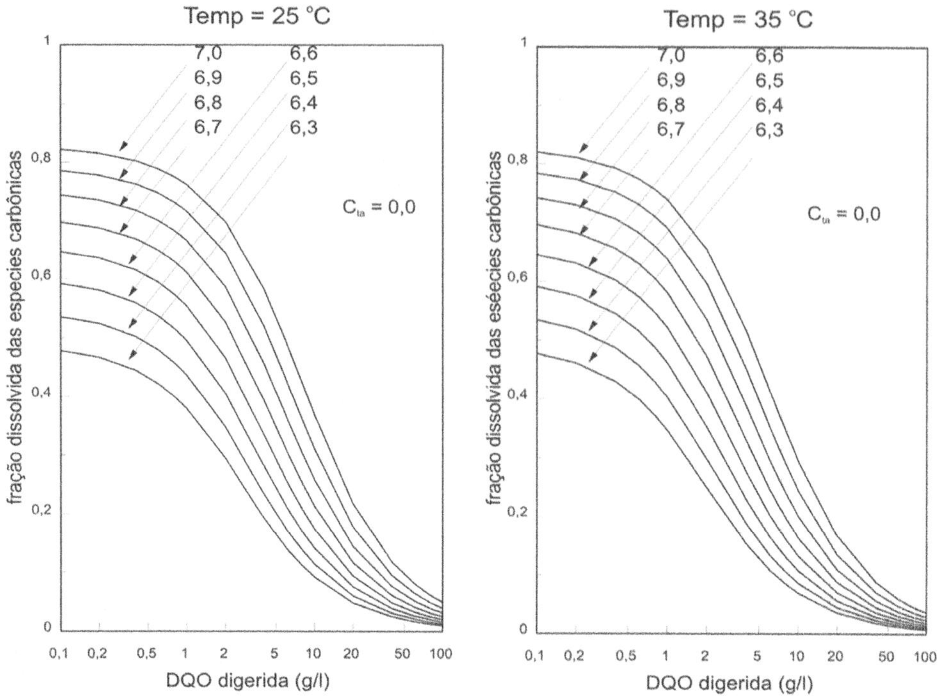

Figure 5.11 Fraction of the carbonic species that remain dissolved in an anaerobic digester as a function of the digested COD concentration for different pH values.

The potential for precipitation or dissolution of $CaCO_3$ in an anaerobic reactor can be calculated from the values of pH, alkalinity and the calcium concentration. For example, the minimum calcium concentration required to have a saturated solution of calcium carbonate in an anaerobic digester with pH = 6.9 and Alk = 10 meq $\cdot$ L^{-1} can be calculated as follows: the alkalinity at a pH 6.9 is virtually identical to the concentration of bicarbonate, which allows the carbonate concentration can be calculated as:

$$[CO_3^{2-}] = [HCO_3^-]K_2/[H^+] = Alk_e \times 10^{(pH-pK2)} = 0.01 \times 10^{(6.9-10.4)} = 3.2 \times 10^{-6}\ mol \cdot L^{-1}$$

The calcium concentration of the saturated solution can now be calculated with the aid of the solubility constant:

$$pK_{sp} = 8.03 - 0.1183 \times 25 = 8.33 \text{ and } K_{sp} = 4.7 \times 10^{-9}$$

Hence:

$$[Ca^{2+}]_{sat} = K_{sp}/[CO_3^{2-}] = 4.7 \times 10^{-9}/3.2 \times 10^{-6} = 1.5 \times 10^{-3}\ mol \cdot L^{-1}$$

In practice the concentration of calcium in wastewater can be much higher or lower than the calculated equilibrium concentration of 1.5 mmol $\cdot$ L^{-1} value, depending on the hardness of the water supply used. A calcium concentration of 1.5 mmol $\cdot$ L^{-1} corresponds to a hardness of 150 ppm $CaCO_3$, i.e. a moderately hard water. So depending on the characteristics of the wastewater, the environment in the digester may be subsaturated (aggressive), saturated (stable) or supersaturated (crusting).

It has been established (Merill & Jordan, 1975) that the dissolution or precipitation rates are so low that they can hardly be detected if the degree of sub- or supersaturation is not high. Thus apparent or pseudo constants may be introduced, such that the pseudo constant for dissolution indicates the minimum ionic product that does not result in a perceptible rate of calcium dissolution, while the pseudo constant for precipitation indicates the maximum value of the ionic product that can be kept in solution with no perceptible rate of precipitation:

$$K_{pp} = \{[Ca^{2+}]_{max}\,[CO_3^{2-}]_{max}\}_{rp=0} \tag{5.37a}$$

and

$$K_{pd} = \{[Ca^{2+}]_{min}\,[CO_3^{2-}]_{min}\}_{rd=0} \tag{5.37b}$$

where:

K_{pp} = pseudo precipitation constant of calcium carbonate
K_{pd} = pseudo dissolution constant of calcium carbonate
r_p = precipitation rate
r_d = dissolution rate.

This means that if the ionic product of calcium and carbonate has a value between the two pseudo constant, there would be no measurable precipitation or dissolution of calcium carbonate, although strictly speaking it is not a thermodynamically stable solution.

Table 5.7 shows published values of the K_{pp} constant for different qualities of water. The experimental data clearly indicate that the value of the K_{pp} constant tends to deviate more from its thermodynamic equilibrium value as the concentration of (organic) impurities increase. Capri and Marais (1974) showed that in anaerobic digesters solutions can be maintained with a high degree of supersaturation if phosphate is present at sufficiently high concentrations (>3 $mmol \cdot L^{-1}$).

In contrast, there seems to be no experimental data published on the value of K_{pd} or about the factors that influence it, although the practical importance of this pseudo constant is very great. For example, it would allow evaluation of the possibility of corrosion of concrete structures in anaerobic treatment systems. The danger of damage to concrete in anaerobic systems is real and has been observed in practice: in many cases, corrosion caused a very significant decrease to the wall thickness of concrete reactors and phase separator elements after only few years of operation, for example in Cali, Columbia, where the corrosion reduced wall thickness by 10 mm in no more than four years.

Table 5.7 Experimental values of pseudo constants for calcium carbonate precipitation for different types of waters (thermodynamic equilibrium value $pK_{pp} = pK_{sp} = 8.32$; Truesdale & Jones, 1973).

Type of Water	Reference	pK_{pp} Value
Fountain water	Sheen and Woodruff (1944)	7.5
Secondary effluent	Wiechers (1977)	7.4
Primary effluent	Wiechers (1977)	6.8
Primary and secondary effluent	Menar and Jenkins (1972)	6.7
Raw sewage	Merrill and Jordan (1975)	6.4
Supernatant of a sludge digester	Ferguson and McCarty (1969)	6.3

As the value of the pseudo constant is not known, one also cannot calculate what would be the minimum demand of calcium and/or alkalinity (e.g. lime addition) that should be added to prevent future damage from an undersaturated wastewater. On the other hand it is possible to calculate what concentration of lime is required to produce a saturated environment in the digester. The calculation to determine the necessary addition of lime to remove the aggressiveness of the water and obtain a thermodynamically stable solution requires an iterative procedure, as the lime addition does not only change the concentration of calcium, but also the alkalinity and therefore the pH and hence increases the carbonate concentration. This calculation procedure will be demonstrated in Example 5.7. A direct graphical solution is possible by using the modified Caldwell–Lawrence diagram, developed by Loewenthal and Marais (1976), but this will not be discussed in this text.

EXAMPLE 5.7

Determine the lime demand to produce a thermodynamically stable solution of water starting with a calcium concentration of 1 mmol·L^{-1}, alkalinity of 10 meq·L^{-1} and a pH of 6.7.

Solution

First it must be verified whether the water is indeed aggressive by calculating the ionic product and comparing it with the solubility constant. For the adopted values the carbonate concentration can be calculated as:

$$[CO_3^{2-}] = [HCO_3^-]K_2/[H^+] = Alk_e \cdot 10^{(pH-pK2)} = 2.5 \times 10^{-6}\,mol \cdot L^{-1}$$

Hence the ionic product is given by:

$$[Ca^{2+}][CO_3^{2-}] = 1.0 \times 10^{-3} \times 2.5 \times 10^{-6} = 2.5 \times 10^{-9} < K_{sp} = 4.7 \times 10^{-9}$$

Since the ionic product is indeed smaller than the solubility constant the water is aggressive although it will probably not exhibit actual aggressiveness because the degree of undersaturation is small.

The lime addition required to produce a thermodynamically stable solution is calculated by an iterative procedure which contains the following steps:

(1) Select a value for the lime addition.
(2) Determine the new values of alkalinity and calcium concentration under the assumption that there is no $CaCO_3$ precipitation, which is not unreasonable as, if the dosing of lime is not excessive, we are moving from mildly aggressive conditions towards stable equilibrium).
(3) Determine the new pH.
(4) Determine the new carbonate concentration.
(5) Verify if the new ionic product $[Ca^{2+}][CO_3^{2-}]$ is equal to the solubility constant K_{sp}.
(6) If not, increase the lime dosing and repeat this calculation procedure until the ionic product is equal to the value of the solubility constant.

If an addition of 0.5 mmol·L^{-1} of lime is assumed (equivalent to 28 mg·L^{-1} of CaO), the Ca concentration increases by 0.5 mmol·L^{-1} as well, while the alkalinity increases by 2 × 0.5 = 1 meq·L^{-1}. Hence the new effluent alkalinity will be 11 meq·L^{-1} or 0.011 eq·L^{-1} and the corresponding pH after digestion (assuming no $CaCO_3$ precipitation occurs) can be determined with the aid of Eq. (5.28): $f_e = 0.32$ and $pH_e = 6.37 - \log(0.32) = 6.86$. Hence the carbonate concentration may be estimated as:

$$[CO_3^{2-}] = [HCO_3^-]\,K_2/[H^+] = Alk_e 10^{(pH-pK2)} = 3.6 \times 10^{-6}\,mol \cdot L^{-1} \text{ and}$$

$$[Ca^{2+}] = 0.001 + 0.0005 = 1.5 \times 10^{-3}\,mol \cdot L^{-1}$$

Therefore the ionic product is equal to:

$[Ca^{2+}][CO_3^{2-}] = 1.5 \times 10^{-3} \times 3.6{\times}10^{-6} = 5.4 \times 10^{-9} > K_{sp} = 4.7 \times 10^{-9}$

As the ionic product now exceeds the $CaCO_3$ solubility constant, a solution becomes supersaturated and precipitation may occur with time. Knowing that an addition of 0.5 mmol·L^{-1} of lime leads to a marginally supersaturated solution, the calculation procedure is repeated for a smaller lime addition until the ionic product becomes equal to solubility constant. Under the given conditions this occurs for a lime addition of 0.4 mmol·L^{-1}, as can be verified:

(1) New alkalinity: $Alk_e = 10 + 2 \times 0.4 = 10.8$ meq·L^{-1} = $[HCO_3^-]$
(it is assumed that the pH remains in the neutral range, so that $Alk_e = [HCO_3^-]$)
(2) New pH (assume no precipitation): $f_e = 0.33$ and $pH_e = 6.83$. It is confirmed that indeed the pH is in the neutral range
(3) New carbonate concentration: $[CO_3^{2-}] = Alk_e 10^{(pHe-pK2)} = 0.0108 \times 10^{(6.83-10.33)} = 3.4 \times 10^{-6}$
(4) New calcium concentration: $[Ca] = 1.0 + 0.4 = 1.4$ mmol·L^{-1}
(5) Compare the ionic product with the solubility constant:

$[Ca]\,[CO_3^{2-}] = 1.4 \times 10^{-3} \times 3.4 \times 10^{-6} = 4.7 \times 10^{-9} = K_{sp}$

Because the ionic product is now equal to the solubility constant, it is concluded that the dose of 0.4 mmol·L^{-1} of lime would lead to a saturated effluent and avoid the corrosion of the concrete reactor. In economic terms this addition is not very significant; much larger additions are often necessary in (nitrifying) activated sludge systems and water treatment. The biggest problem however is that the addition of lime is a relatively difficult operation due to the aggressiveness of this base and low solubility in water (1200 mg·L^{-1}), so that the inherent simplicity of the UASB reactor is compromised.

In Brazil there are no reports of systems where lime addition is used to prevent corrosion of CONCRETE in the UASB reactor. The most frequently applied solution is to use a special grade of cement that is better resistant to corrosion or to apply a protective layer (for example epoxy paint or lining) to prevent direct contact between the aggressive water and concrete. In practice these solutions have not always given the desired result and in many cases serious corrosion problems can still be observed even after only a few years of operation.

Another cause of corrosion is found at the UASB reactor overflow level when hydrogen sulphide generated in the digester is in contact with air, forming sulphuric acid, like the problems frequently arising in sewer systems. Actually, the hydrogen sulphide formed in the anaerobic reactor may first transfer into the gas phase and then redissolve in the water layer that forms on the surfaces of the concrete in contact with air, owing to condensation of water vapour present in the gas phase. As a result, the pH of the condensate is very low (pH < 1 has been measured!). So when it is decided to cover the UASB, it is important to adequately protect the concrete surface that is in direct contact with the gas phase above the liquid phase.

Chapter 6

Biogas generation in anaerobic digesters

6.1 INTRODUCTION

The production of biogas is the most characteristic feature of anaerobic treatment. Biogas is composed mainly of methane and carbon dioxide, which are decomposition products of organic material and have a limited solubility in water. There are other components, whose magnitude depends on the operating conditions in the digester and the characteristics of the wastewater. In this chapter it will be shown that sewage treatment gives rise to contributions of nitrogen and water vapour in the biogas that may exceed that of carbon dioxide. If sulphate is present in the wastewater hydrogen sulphide will also be formed; it generally has a low concentration, but is nonetheless important because it causes several problems in anaerobic digesters: (1) hydrogen sulphide has a strong and unpleasant odour of rotten eggs, (2) at high concentrations it is a toxic compound for methanogens, and (3) upon biogas combustion the hydrogen sulphide may be converted into sulphuric acid, which corrodes metals. If it is not converted into sulphuric acid after combustion, it enters the atmosphere as sulphur dioxide.

The destination of biogas depends on the quantity and quality that is produced. When enough biogas is produced, it is possible to use biogas in a productive way, generating electricity and heat in combined heat and power (CHP) units, or refining and pressurizing it for subsequent use, for example for automotive purposes or injection in the gas grid. The technical feasibility of using biogas for power generation is not in question: there are many manufacturers who produce generators specifically designed for the use of biogas as fuel, with power generation ranging from 30 kW to over 5 MW. The transformation efficiency of these units is in the range of 35–40%. For very large biogas flows (>5 MW_{el}) there is the possibility of using gas turbines, which have a high initial cost, but an efficiency of more than 50%.

A more complicated application is the upgrading of biogas for producing pure methane. In this case the purified biogas must be compressed to a pressure high enough for grid injection (typically > 16 bar) or to an even higher pressure of 200–250 bars to be used in cars which are adapted for use of CNG (compressed natural gas). The conversion of biogas into methane has also been demonstrated, but this

Editor(s): Adrianus van Haandel and Jeroen van der Lubbe
doi: 10.2166/9781780409627_0221

alternative depends on the price of conventional fuel, which actually is too low for widespread application of CNG, at least without incentives.

Biogas treatment processes are well established, and their economic feasibility depends mainly on the benefit of the productive use, i.e. the cost of fuel that otherwise would have to be purchased. In small treatment systems it is usually not economical to use the produced biogas for energy production and it is preferable to burn it with a flare as it is being produced. It must be noted that the combustion in itself is important and that its omission can result in a situation in which anaerobic treatment makes a higher contribution to the generation of greenhouse gases (GHG) than aerobic treatment, bearing in mind that methane is a much more potent greenhouse gas: it has a carbon dioxide equivalence of 21:1, i.e. expressed per mole of gas, methane is 21 times more harmful than carbon dioxide.

EXAMPLE 6.1

Estimate the per capita production of gases contributing to the greenhouse effect in aerobic and anaerobic systems for a per capita contribution of 100 g COD · d^{-1} and under the following assumptions:

- *Aerobic system:* oxygen consumption = 0.72 g O_2 · g^{-1} COD; aeration efficiency = 1 kWh · kg^{-1} O_2; fuel consumption for power generation = 0.2 kg CH_4 · d^{-1} per kW of electric power.
- *Anaerobic system:* anaerobic digestion efficiency = 70%

Solution

First it must be established how much electric energy can be obtained from methane combustion. In Table 1.8 (bottom line) the energy per mole of methane combustion is 796 kJ, so per g CH_4 the produced chemical energy is about 50 kJ or 13.8 Wh (1 Ws = 1 J). Since the best generators have an efficiency of 35–40% the electric energy produced is 4.8–5.5 kWh ≈ 5 kWh per kg used CH_4. Hence for a power production of 1 kW the required methane consumption is about 0.2 kg CH_4 · h^{-1} or 5 kg CH_4 · d^{-1}. Conversely 1 kg CH_4 · d^{-1} can produce about 5 kWh of electric energy.

Assuming that carbon dioxide is produced in equimolar ratio with the consumption of oxygen in the aerobic system, it follows that the oxygen consumption of 72 g O_2 · inh^{-1} · d^{-1} or 72/32 = 2.25 mol · inh^{-1} · d^{-1} generates 2.25 mol · inh^{-1} · d^{-1} of CO_2. On the other hand the energy consumed in oxygen transfer (1 kWh · kg^{-1} O_2) will be around 72 Wh · d^{-1} or 3 W. For a power generation of 1 W, a consumption of 5 gCH_4 · d^{-1} is required and this generates 5/16 = 0.31 mol · d^{-1} CO_2 (other fuels have roughly the same contribution). For a power generation of 3 W, the production CO_2 is therefore 3 × 0.93 mol · inh^{-1} · d^{-1}, leading to a total of 2.25 + 0.93 = 3.2 mole CO_2 · inh^{-1} · d^{-1} in the aerobic system.

In anaerobic systems the per capita production is 70% × 100/4 = 17.5 g CH_4 · d^{-1} or 17.5/16 = 1.1 mole CH_4 · d^{-1}. If this amount is not burned, it will eventually escape into the atmosphere where it contributes to the greenhouse gas effect. Not considering the production of CO_2 in anaerobic digestion (largely it does not pass into the atmosphere, but remains dissolved), the methane production is equivalent to 21 × 1.1 = 23 mole CO_2 · inh^{-1} · d^{-1}, a factor 23/3.2 = 7 times greater than the calculated value for the aerobic generation system! By burning the methane production of biogas, the CO_2 production will be 1.1 mol · inh^{-1} · d^{-1}, about 33% of the production in aerobic systems. Even if the methane in the biogas is combusted, there is still the dissolved CH_4 in the effluent, which amounts to about 20 mg CH_4 · L^{-1} that originates from the digestion of 80 mg COD · L^{-1}. If the dissolved CH_4 is not metabolized, for example by methanotrophic bacteria, it will eventually be released into the atmosphere. If a per capita sewage contribution of 120 L · d^{-1} is assumed, the dissolved methane production would be 20/16 × 120 = 150 mmol CH_4 · inh^{-1}, equivalent to 3.1 mole CO_2 · Inh^{-1}, so that the dissolved methane has a greenhouse effect about equal to the production in aerobic systems.

It is concluded that even if all the biogas is combusted, the CO_2 production of the combusted biogas and the release of dissolved methane leads to at least the same contribution of GHG by anaerobic treatment systems as is calculated for aerobic treatment systems. So actually, the sustainability of anaerobic systems for sewage treatment is questionable.

In the above analysis the anaerobic digestion of the produced aerobic excess sludge is not considered. In practice some 30% of the organic load present in the feed is converted in aerobic systems into methane, which will increase the output of GHG gases, so that in the end the two systems produce a more or less equivalent amount. On the other hand, if the methane can be used productively, anaerobic systems become more attractive from the point of view of sustainability, because it offsets the use of fossil fuels. In Section 6.7 this subject is discussed in more detail.

6.2 SOLUBILITY OF GASES

Biogas is a mixture of various gases and vapours that desorb from the liquid phase in the anaerobic digester, because of the limited solubility of these products in water. To estimate the production of biogas it is necessary to know the production of these gases and their solubility. To carry out these calculations, first it must be assumed that there is equilibrium between the gaseous phase (biogas) and liquid (the liquid phase in the reactor is assumed to have the same composition as the effluent) and that there is a uniform composition in both phases. To good approximation this will normally be the case, even though the solid phase in the reactor is highly stratified.

The solubility of gases in the liquid phase can be expressed with Henry's law, which states that if a liquid and a gaseous phase are in equilibrium, there is a proportional relationship between the saturation (maximum) concentration of a gas component in the liquid phase and its concentration in the gas phase:

$$C_s = k_d \cdot C_g \tag{6.1}$$

where:

C_s = saturation concentration of the dissolved gas ($mg \cdot L^{-1}$ or $mmol \cdot L^{-1}$)
C_g = gas concentration in the gas phase ($mg \cdot L^{-1}$ or $mmol \cdot L^{-1}$)
k_d = distribution coefficient (non-dimensional)

The distribution coefficient indicates the degree of gas solubility: the higher the value of the coefficient, the higher its solubility. Table 6.1 shows the values of the distribution coefficients at different temperatures for the gases that are most important in biogas. Concentrations may be expressed in any unit ($g \cdot L^{-1}$, $mol \cdot L^{-1}$, etc.). Eq. (6.1) can also be written in a different form:

$$C_s = k_h \cdot P_p \tag{6.2}$$

k_h = Henry's constant
P_p = partial pressure of the component in the biogas (in $g \cdot m^{-3} \cdot Pa$ or $g \cdot J^{-1}$)

Table 6.1 Distribution coefficients of several biogas components.

Gas	Formula	Molar Weight ($g \cdot mol^{-1}$)	Temperature (°C)		
			k_{d20}	k_{d25}	k_{d30}
Methane	CH_4	16	0.034	0.032	0.031
Carbon dioxide	CO_2	44	0.94	0.83	0.74
Nitrogen	N_2	28	0.017	0.016	0.015
Oxygen	O_2	32	0.034	0.032	0.030
Ammonium	NH_3	17	0.76	0.71	0.67
Hydrogen sulphide	H_2S	34	2.87	2.55	2.27

The relationship between the distribution coefficient and the Henry constant is:

$$k_h = k_d \cdot W_m / R \cdot T \ (\text{g} \cdot \text{J}^{-1}) \quad (6.3)$$

where:

W_m = molar weight of the gas
R = universal gas constant = 8.3 J · kmol^{-1}
T = temperature in K.

6.3 BIOGAS PRODUCTION

To estimate the flow of biogas generated in anaerobic treatment systems, expressions must first be derived for the contributions of the gases that have a significant contribution in terms of total volume. In Figure 6.1 the gases normally present in biogas are presented. Along with methane, which is the most voluminous component, significant contributions of carbon dioxide, water and nitrogen are to be expected. In addition there are other compounds which are less voluminous, but whose presence is important because of their nature: O_2, NH_3 and H_2S.

Regarding hydrogen sulphide, the contribution of this component depends on the concentration of sulphate in the water supply that generates the sewage, on industrial contributions and on infiltration of seawater in the sewage system. Usually this concentration will be small but there are regions where water resources with a high sulphate concentration are the only source for public water supply. In these cases the sulphide production may still not be high, but due to sulphate reduction, there is consumption of organic material and the methane production may be strongly reduced (see Sections 6.5 and 3.7). Nitrogen, oxygen and ammonium are dissolved in the influent whereas methane, carbon dioxide and hydrogen sulphide gas are generated in the anaerobic treatment unit. Being a vapour and not a gas at ambient temperature, the partial pressure of H_2O is constant for a given temperature and is independent of the pressure of the gas phase.

6.3.1 Methane production potential

Methane production depends directly on the applied organic load and the digestion efficiency. Stoichiometric methane production is calculated as 0.25 kg of CH_4 per kg of digested COD. When it is assumed that about 10–15% of influent COD is transformed into organic sludge, the production of methane can be calculated from the applied organic load, if the removal efficiency of the organic material is known. As discussed

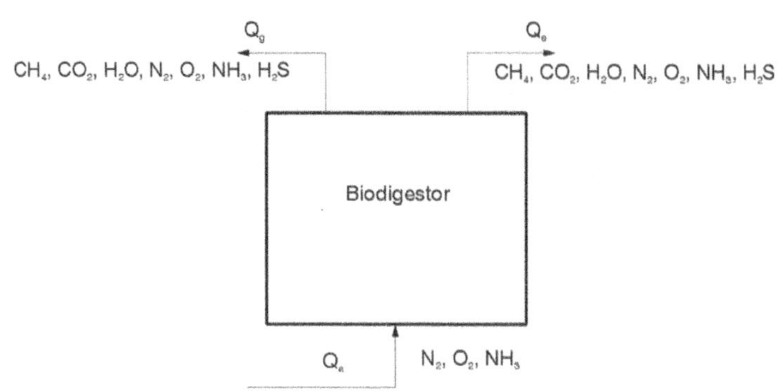

Figure 6.1 Schematic representation of the different components of biogas and the equilibrium between the gas and liquid phases.

in Chapter 3, this removal efficiency depends on a variety of factors: the nature of the organic material, environmental conditions, operating conditions and reactor quality in terms of design and construction. The daily mass of methane produced may be expressed as:

$$MCH_4 = (MS_{ti} - MS_{te} - MS_{xv})/4 \tag{6.4}$$

To make a first estimate of the methane production, it can be considered that all the methane that is produced will desorb from the liquid phase. Knowing that the molar volume of a gas at atmospheric pressure is given by 22.4(273 + T)/273 (T in °C) and that one mole of methane has a mass of 16 g, the volume of methane produced is calculated as:

$$\begin{aligned} V_{CH4} &= 22.4(273 + T)/273 \cdot MCH_4/16 \\ &= 22.4(273 + T)/273 \cdot (MS_{ti} - MS_{te} - MS_{xv})/64 \end{aligned} \tag{6.5}$$

where:

V_{CH4} = first estimate of methane volume released to the biogas ($L \cdot d^{-1}$)
MCH_4 = mass of produced methane ($g \cdot d^{-1}$)
MS_{ti} = mass of applied COD ($g \cdot d^{-1}$)
MS_{te} = mass of COD in the effluent ($g \cdot d^{-1}$)
MS_{xv} = COD mass converted into organic sludge ($g \cdot d^{-1}$)

EXAMPLE 6.2

Determine the potential for per capita methane production from sanitary sewage by anaerobic digestion, making the following assumptions: (1) per capita organic matter production = 100 g COD $\cdot$ inh$^{-1} \cdot$ d^{-1}, (2) the organic material removal efficiency = 80%, and (3) the conversion into volatile sludge = 10%.

Solution

From Eq. (6.4):

$$MCH_4 = (MS_{ti} - MS_{te} - MS_{xv})/4 = (100 - 20 - 10)/4 = 17.5 \text{ g } CH_4 \cdot inh^{-1} \cdot d^{-1}.$$

From Eq. (6.5) at a temperature of 25°C:

$$\begin{aligned} V_{CH4} &= 22.4(273 + T)/273 \times MCH_4/16 \\ &= 22.4(273 + 25)/273 \times 17.5/16 = 26.7 \text{L CH} \cdot inh^{-1} \cdot d^{-1} \end{aligned}$$

However, the estimated volume of methane collected in the biogas is now overestimated, as part of the methane produced will remain dissolved in the liquid phase. The solubility of methane is low as shown in Table 6.1: the value of the distribution coefficient k_d is only 0.03. To make a first estimate of the dissolved methane concentration in the effluent it can be assumed that biogas is composed only of methane. As one mole of methane (or any other gas) has a volume of 24.4 L at a temperature of 25°C and atmospheric pressure, its concentration is 1/24.4 mol $\cdot$ L^{-1} = 41 mmol $\cdot$ L^{-1}. Now applying Eq. (6.1) for the assumption of having only methane in the biogas the value of C_s is calculated as 41 × 0.03 = 1.2 mmol $\cdot$ L^{-1} = 20 mg $CH_4 \cdot L^{-1}$.

To produce 1 mg CH_4 a mass of 4 mg of COD must be digested, hence it is estimated that a maximum of 80 mg COD $\cdot$ L^{-1} will be dissolved as methane after anaerobic digestion. This concentration is not negligible in the case of municipal sewage, which normally has a concentration of digestible COD in the range 300–500 mg $\cdot$ L^{-1}. In Example 6.2, for a per capita contribution of 120 L $\cdot$ d^{-1}, the methane loss in the effluent would be 0.020 g$CH_4 \cdot L^{-1}$ × 120 L $\cdot$ inh$^{-1} \cdot$ d^{-1} = 2.4 g CH4 $\cdot$ inh$^{-1} \cdot$ d^{-1} or 2.4/17.5 = 14%.

So the best estimate of the CH_4 production in the biogas would MCH_4 = 17.5 − 2.4 = 15.1 g $\cdot$ inh$^{-1} \cdot$ d^{-1}, while the per capita volume is V_{CH4} = 22.4 (273 + 27) /273 × 15.1/16 = 23.1 L $CH_4 \cdot d^{-1}$. It is important to notice that as the temperature decreases, the digested COD fraction reduces while the dissolved methane concentration increases. As a result the dissolved methane fraction increases and may be as high as 50% in case with a low influent COD and a low temperature.

EXAMPLE 6.3

Estimate the fraction of produced methane that remains dissolved in a UASB reactor operating at a temperature of 15°C and a sludge age of 75 days, treating sewage with a COD of 500 mg · L^{-1} in the influent.

Solution

The fractions of the influent COD which are discharged in the effluent and transformed into sludge can be estimated from Eqs. (3.23 and 3.24): $f_s = 0.2$ and $f_x = 0.11$. Hence the digested fraction is $f_d = 1 - f_s - f_x = 0.67$. Therefore the digested COD is $0.67 \times 500 = 330$ mg · L^{-1}, equivalent to $330/4 = 82$ mg $CH_4 \cdot L^{-1}$. On the other hand at 15°C the solubility of methane in water is about 23 mg · L^{-1} ($k_d = 0.036$), so that a fraction of $23/82 = 28\%$ is retained in the liquid phase and 72% is released in the biogas. Figure 6.2 shows the methane fraction in the gas phase for different temperatures and influent COD concentration as a function of the sludge age.

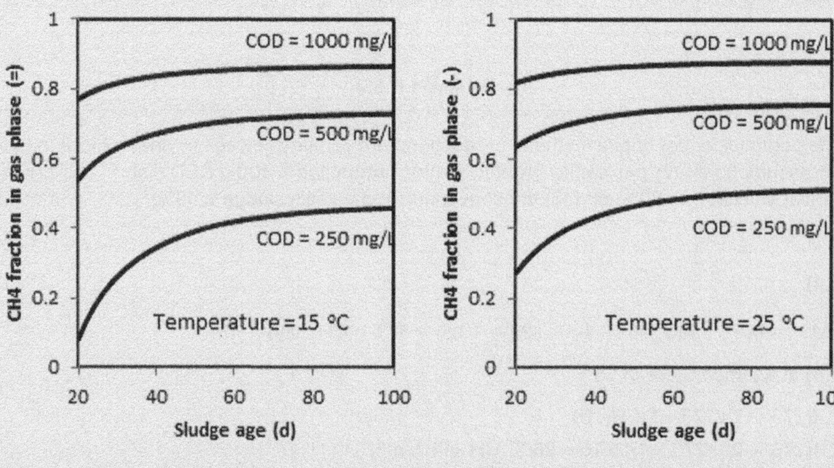

Figure 6.2 Fraction of the methane production that desorbs from the liquid phase as a function of the sludge age at different temperatures and for influent COD concentrations.

To be able to make a more accurate estimate of the biogas volume and the partial pressures of methane and other components in the biogas, it is necessary to apply an iterative calculation, considering also the contributions of other gases. First the contribution of each component individually will be considered and then the iterative calculation procedure will be completed.

6.3.2 Carbon dioxide

The presence of CO_2 in the biogas is due to its production in the anaerobic digestion process. It can be assumed that the CO_2 production in molar terms is equivalent to the methane production, but its presence in the gas phase is much smaller because of its greater solubility. The partial CO_2 pressure can be estimated from the equilibrium of the carbonic system:

$$CO_2 + H_2O \leftrightarrow H^+ + HCO_3^- \text{ or}$$
$$[HCO_3^-][H^+]/CO_2] = k_1 \text{ or}$$
$$[HCO_3^-][H^+]/(k_d \cdot [CO_2]_g) = k_1$$

With the aid of Eq (6.3) this can also be written as:

$$p_{CO2} = [HCO_3^-] \times 10^{(pk1+pkH-pH)} \quad (6.6)$$

where:

$pX = -\log[X]$
k_1 = first dissociation constant of the carbonic system = 4.6×10^{-7}
k_H = Henry constant for CO_2
= $0.034\ mol \cdot L^{-1} \cdot atm^{-1}$ at 25°C; i.e. $pK_H = 1.47$ (Loewenthal & Marais, 1976)

Since at pH ≈ 7 the alkalinity is practically equal to the bicarbonate concentration, Eq. (6.6) can also be written as:

$$p_{CO2} = Alk \times 10^{(pk1+pkH-pH)} \quad (6.6a)$$

It is concluded that in the case of sewage treatment the CO_2 partial pressure will be low: for an alkalinity of 350 ppm or 7 $meq \cdot L^{-1}$, a typical value for domestic sewage, and a pH of 7.0, the partial pressure is only 0.04 at 25°. The CO_2 pressure can be calculated provided there are data for alkalinity and pH, as well as the temperature. The pH value can be calculated by applying the procedure outlined in Section 5.5. The presence of CO_2 in the biogas normally is not problematic. In contrast, dissolved CO_2 in the liquid phase results in "aggressive" water, which can lead to serious concrete corrosion problems, as was discussed in Section 5.6.

6.3.3 Water vapour

The pressure of the water vapour is directly related to temperature. At 25°C the water saturation pressure is equal to 23.7 mm of mercury (31 Pa) or, stated differently, at a pressure of 760 mm or 1.013 kPa the partial pressure is 0.031 bar. Table 6.2 gives some values of water vapour pressure for different temperatures. The presence of water vapour in the biogas can cause problems, particularly when the ambient temperature varies greatly. During periods of low ambient temperature water may condense in biogas piping, because the biogas is generated in the relatively warm liquid phase. The biogas will contain water vapour at the pressure of the liquid phase in the reactor, but when it leaves the reactor the biogas temperature decreases

Table 6.2 Values of the water vapour saturation pressure at different temperatures.

	Saturation Pressure			Saturation Pressure	
Temperature (°C)	**mm Hg**	**Partial Pressure**	**Temperature (°C)**	**mm Hg**	**Partial Pressure**
0	4.6	0.006	20	17.5	0.023
5	6.5	0.009	25	23.7	0.031
10	9.2	0.012	30	31.7	0.042
15	12.8	0.017	35	41	0.054

and water vapour will tend to become supersaturated and condense. Apart from causing blockages, the condensed water tends to cause corrosion problems, especially if hydrogen sulphide is also the present in the biogas. For those reasons there will usually be a condensate retention device in the biogas transport system, which may be the hydraulic seal of the UASB reactor, if located at the lowest point of biogas piping.

6.3.4 Nitrogen

In many cases it may be assumed that wastewaters are saturated with atmospheric nitrogen (which has a partial pressure of 0.8 atm). In that case the saturation concentration is 14.8 $mg \cdot L^{-1}$ at 25°C (using the value of the distribution coefficient listed in Table 6.1). If it is assumed that nitrogen is not produced or lost during anaerobic treatment, the influent nitrogen will leave the system in the biogas and in the effluent:

$$Q_l \cdot C_i = Q_l \cdot C_s + Q_g \cdot C_g \text{ or } C_i = C_s + (Q_g/Q_l) \cdot C_g \quad (6.7)$$

Hence:

$$C_g = C_i/[k_d + (Q_g/Q_l)] \text{ and} \quad (6.7a)$$

$$C_s = k_d \cdot C_i/[1 + (Q_g/Q_l)] \quad (6.7b)$$

where:

C_i = concentration in the influent
Q_g/Q_l = ratio between the biogas and wastewater flow rate.

In order to calculate nitrogen concentrations in the effluent and in the biogas, it is necessary to know the concentration in the influent and the ratio of the flows of biogas and wastewater.

EXAMPLE 6.4

For the conditions of Example 6.2 and for a per capita sewage contribution of 120 $L \cdot d^{-1}$, estimate the partial pressure of nitrogen in the biogas.

Solution

The concentration of nitrogen in the influent is estimated from the distribution constant ($k_d = 0.016$, Table 6.1). At a partial pressure of 0.8 bar, the concentration of nitrogen in the air is $41 \times 0.8 = 34$ $mmol \cdot L^{-1}$. Now the saturation concentration of nitrogen is calculated as:

$C_s = 0.016 \times 34 = 0.54$ $mmol \cdot L^{-1}$ and $C_i = 14.8$ $mg\,N \cdot L^{-1}$

As $Q_g/Q_l = 26.7/120 = 0.22$ (Example 6.2), the nitrogen concentration in the biogas is calculated with Eq. (6.7a):

$C_g = C_i/[k_d + (Q_g/Q_l)] = 14.8/(0.016 + 0.22) = 63$ $mg \cdot L^{-1}$ or $63/28 = 2.2$ $mmol \cdot L^{-1}$.

Hence $C_s = k_d \cdot C_g = 0.016 \times 63 = 1.0$ $mg \cdot L^{-1}$.

The calculations indicate that nitrogen is efficiently stripped in the anaerobic digester and almost all transferred to the gas phase, due to the very low solubility of the gas. At a temperature of 25°C and a pressure of 1 bar the concentration of a gas is 41 $mmol \cdot L^{-1}$. Hence, when the nitrogen concentration is 2.2 $mmol \cdot L^{-1}$ in the gas phase, its partial pressure is $2.2/41 = 0.05$ atm.

Comparing the calculation results of the partial pressures of carbon dioxide, water and nitrogen (Sections 6.3.2 and 6.3.3), one comes to the maybe somewhat surprising conclusion that for anaerobic digestion of normal municipal sewage, the partial pressures for carbon dioxide, nitrogen and water vapour in the produced biogas are all of the same order of magnitude.

6.3.5 Oxygen

In contrast to nitrogen, the oxygen concentration in influents is very probably low, because in case of dissolution of the atmospheric gas, it is consumed by micro-organisms present in the liquid phase. Even when this consumption does not occur during transportation of the wastewater in the sewerage network, its elimination by facultative hydrolytic and acidogenic bacteria in the anaerobic reactor will cause rapid depletion. In the case of a reactor influent distribution system that causes a lot of air to be drawn into the digester, it might be possible that the methane production is reduced, because the organic material will selectively be used to be oxidized by oxygen at the cost of methanogenesis. Therefore it is important that air is not dragged into the feeding system, although the main reason is temporary obstruction to flow when accumulated air bubbles travel back up the feed line.

Nitrate $\left(NO_3^-\right)$ and nitrate $\left(NO_2^-\right)$ also consume COD at the expense of reduced methane production: these components however are not normally present in high concentrations in municipal sewage, but can be present in industrial effluent due to their use as cleaning agents in the food industry (nitrous and nitric acid)

6.3.6 Ammonium

Ammonium is present in the influent, but it is also generated in the anaerobic digester, as a result of mineralization of organic nitrogen, when proteins are hydrolysed and acidified. In sewage the concentration normally is in the range of 40–60 mg $N \cdot L^{-1}$ or 3–4 $mmol \cdot L^{-1}$. As in the case of nitrogen, ammonium is divided between the gas and liquid phases. The ratio between the number of moles per day leaving the system in the gas and in the liquid phase can be expressed as:

$$N_g/N_e = (Q_g \cdot C_g)/(Q_l \cdot C_s) \qquad (6.8)$$

With the aid of Eq. (6.1) one has $N_g/N_e = (Q_g/Q_l)/k_d$ where N_g and N_e = the number of moles of NH_3 in the biogas and in the effluent (the sum of $NH_3 + NH_4^+$) respectively.

EXAMPLE 6.5

For the conditions of Example 6.2 and for a per capita sewage contribution of 120 $L \cdot d^{-1}$, estimate the partial pressure of ammonia in the biogas.

Solution

Maintaining a proportion of gas and liquid flows of $(Q_g/Q_e) = 0.22$ from the previous example, and knowing the value of the distribution coefficient ($k_d \approx 0.7$ in Table 6.1) it is estimated that:

$$N_g/N_e = 0.19/0.7 = 0.3$$

On the other hand in the neutral pH range the fraction of ammonium that is present in in the liquid phase in gaseous state as ammonia (NH_3) is small (as $pK_{NH3} = 9.1$ at 25°C, Table 5.1, so almost all ammonium – about 99% at a neutral pH – is in the ionic form (NH_4^+). Therefore the ammonium fraction in the biogas will be very small, in the range of 0.3% of the total ammoniacal mass in the system. Thus for practical purposes it can be considered that ammonium remains in the liquid phase in anaerobic reactors.

Even in concentrated wastes where both the ammonium concentration in the liquid phase and the ratio (Q_g/Q_l) are high, the ammonia fraction in the biogas will not be dominant. For example in the case of anaerobic digestion of manure or sludge, ammonium concentrations of $C_l = 3\ g\ N \cdot L^{-1}$ and $(Q_g/Q_l) = 3$ may be found, so that $N_g/N_l = 3/0.7 = 4.3$ and $N_g = 4.3N_l$. The NH_3 concentration in the effluent will be 1% of the total ammoniacal concentration: $N_l = 0.03\ g \cdot L^{-1}$ and the NH_3 concentration in the biogas = $4.3 \times 0.03/3 = 0.043\ g \cdot L^{-1} = 3\ mmol \cdot L^{-1}$. This means that still almost all ammonium is in the liquid phase, but the partial pressure of ammonia is no longer insignificant: $p_{NH3} = 3/41 = 0.07 = 7\%$.

6.3.7 Hydrogen sulphide

As discussed in Chapters 2 and 3, sulphate reduction by sulphate-reducing bacteria (SRB) is a microbiological process which develops in anaerobic reactors and usually predominates over methanogenesis. The sulphate reduction in anaerobic systems may have advantages and disadvantages (Lens & Hulshof Pol, 2000). Among the many drawbacks the following can be highlighted:

- Increased BOD and COD of the effluent because of the oxygen demand of the sulphide
- Corrosion problems: the hydrogen sulphide can be oxidized to sulphuric acid and result in corrosion of metals and concrete
- In the event of considerable presence of metals: inorganic sludge accumulation in the form of precipitated sulphides
- Methane production decreases due to oxidative catabolism of organic material
- Problems of odour in the treatment unit
- Toxicity for methanogens (and for operating personnel for that matter)
- Deterioration of aerobic post-treatment (filamentous sludge, nitrification inhibition).

On the other hand, there are also potential advantages that the sulphate reduction can bring about that could be mentioned:

- Removal of sulphate with the potential for biological transformation into elemental sulphur and subsequent water recovery
- Degradation of xenobiotic compounds that otherwise could not be removed biologically
- Possibility of removal and recovery of heavy metals from the liquid phase
- The precipitated metal sulphide may serve as granulation nucleus.

The sulphide generated in the anaerobic reactor leaves through three pathways: (1) in the biogas, (2) in the anaerobic effluent, and (3) through emissions from the surface of anaerobic reactors or other points of escape. To prevent odours from spreading due to gas release from the surface, the anaerobic reactor can be covered while a partial vacuum (1–2 cm water column (1–2 mbar) is sufficient) is applied by forced ventilation, which removes the gas phase including any desorbing biogas. This mixture must be treated, for example in a biofilter, to remove hydrogen sulphide. In quantitative terms the mass of hydrogen sulphide present in the contaminated air is small.

The sulphide generated in the digester will be distributed among the three forms of reduced sulphur: dissolved hydrogen sulphide (H_2S), bisulphide (HS^-) and sulphide ions (S^{2-}) and can be estimated from the chemical equilibrium:

$$H_2S \leftrightarrow HS^- + H^+ \leftrightarrow S^{2-} + 2H^+ \tag{6.9}$$

Hydrogen sulphide is a weak acid, as its dissociation constants indicate: $pK_1 = 7.0$ and $pK_2 = 14.0$. With the aid of the dissociation constants the division of the three species can be calculated. In Figure 6.3 the division between sulphide, bisulphide and sulphide in water is plotted as a function of pH. Figure 6.4 is the corresponding pH–pX diagram. It can be seen that in water the form of non-dissociated hydrogen sulphide prevails when the pH is lower than the neutral value. The sulphide ion (S^{2-}) only appears in a significant quantity when the pH is extremely high (pH $>$ 12). The ratio of the concentration of dissociated bisulphide to dissolved hydrogen sulphide can be calculated by the dissociation equation:

$$C_{hs}/C_{h2s} = 10^{(pH-pK1)} \tag{6.10}$$

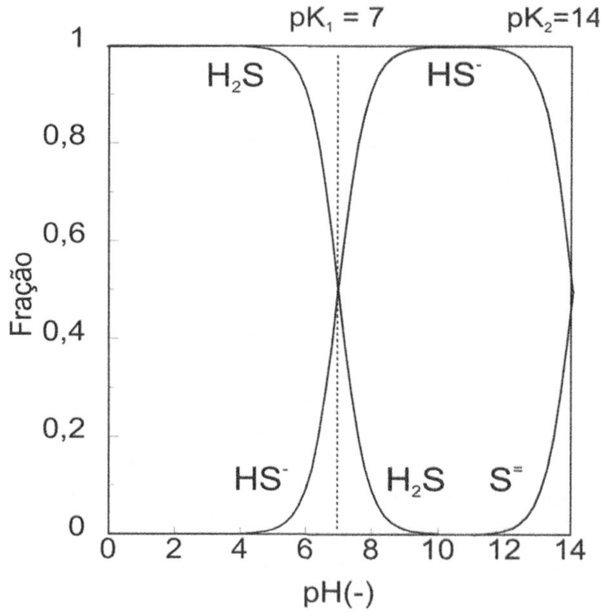

Figure 6.3 Fractions of hydrogen sulphide, bisulphide and sulphide as function of the pH.

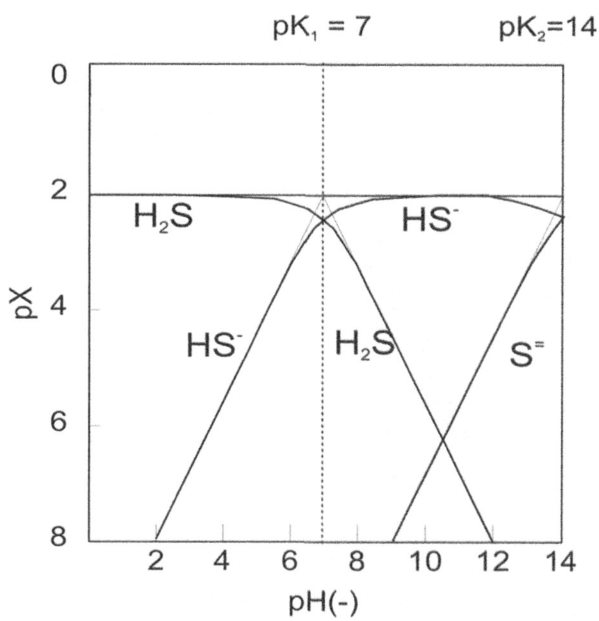

Figure 6.4 pX-pH plot for the sulphidic system (10 mmol · L^{-1} of reduced sulphur).

where:

C_{hs}, C_{h2S} = concentration of bisulphide (HS^-) and dissolved hydrogen sulphide (H_2S) in the effluent (and in the liquid phase of the digester).

Hydrogen sulphide gas, being a volatile compound, will be in equilibrium between the liquid phase (the content of the reactor) and the gas phase (biogas). When it is assumed that the hydrogen sulphide in the produced biogas and the digester contents is in equilibrium according to Henry's law (Eq. (6.1)), the concentration of hydrogen sulphide (H_2S without the dissociated forms HS^- and S^{2-}) can be calculated as:

$$C_s = (N_{he}/Q_e) = k_d \cdot C_g = k_d \cdot (N_g/Q_g) \text{ or } N_g/N_{he} = (Q_g/Q_e)/k_d \tag{6.11}$$

where:

N_{he}, N_g = flux (daily number of moles) of H_2S in the effluent and the biogas.

Hence the proportion between the molar flows of H_2S in the biogas and in the effluent depends on two factors: (1) the Q_g/Q_l ratio and (2) the value of the distribution coefficient. In Figure 6.5 the division between the species of sulphide has been plotted as a function of the Q_g/Q_l ratio for different pH values. It shows that the fraction of the sulphide that ends up in the biogas depends heavily on the Q_g/Q_l ratio. In the case of wastewaters with a low concentration of organic material, such as sewage ($Q_g/Q_l \approx 0.2$), almost all sulphide remains dissolved, but in concentrated wastes ($Q_g/Q_l \approx 10$), a large fraction of the sulphide migrates to the gas phase. The curves were calculated with the aid of the following equations:

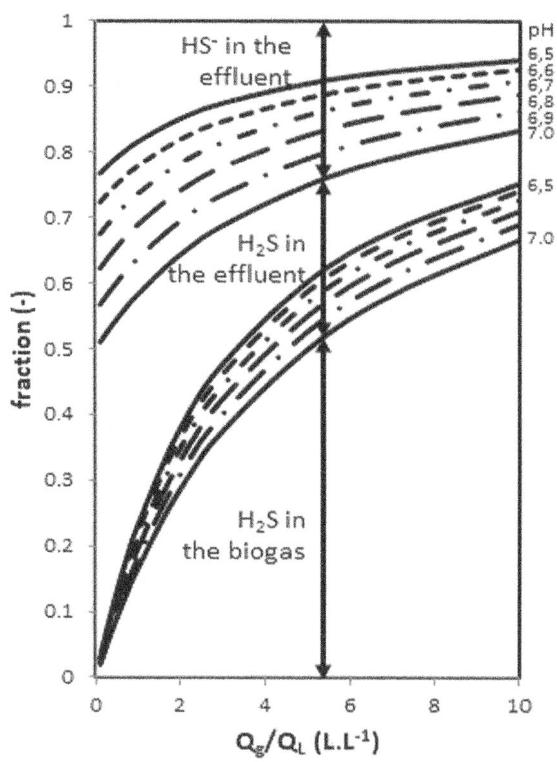

Figure 6.5 Fraction of the sulphide species as a function of the Q_g/Q_l ratio for different pH values.

$$N_{be}/N_{he} = 10^{(pH-pK1)} \tag{6.10}$$

$$\begin{aligned} N_t &= N_g + N_{he} + N_{be} = N_g + N_{he}\cdot(1+10^{(pH-pK1)}) \\ &= N_{he}\cdot\left[(Q_g/Q_e)/k_d + 1 + 10^{(pH-pK1)}\right] \end{aligned} \tag{6.12}$$

$$N_{he}/N_t = 1/[(Q_g/Q_e)/k_d + 1 + 10^{(pH-pK1)}] \tag{6.13}$$

$$N_g/N_t = (Q_g/Q_e)/k_d/[(Q_g/Q_e)/k_d + 1 + 10^{(pH-pK1)}] \tag{6.14}$$

$$N_{be}/N_t = 1 - (N_g/N_t) - (N_{he}/N_t) \tag{6.15}$$

$$\begin{aligned} N_e/N_t &= (N_{be} + N_{he})/N_t \\ &= (1 + 10^{(pH-pK1)})/\left[(Q_g/Q_e)/k_d + 1 + 10^{(pH-pK1)}\right] \end{aligned} \tag{6.16}$$

where:

N_t = total number of reduced sulphur moles
N_g = moles of H_2S in biogas
N_{he} = moles of H_2S in the effluent
N_{be} = moles of HS^- in the effluent.

EXAMPLE 6.6

For the conditions of Example 6.2 estimate the fraction of the generated hydrogen sulphide gas remaining in the effluent, when the sulphate concentration in the influent is 20 mg S · L^{-1} and all sulphate in the digester is reduced.

Solution

It was estimated that the ratio $(Q_g/Q_i) = 23/120 = 0.19$ and for a distribution coefficient $k_d = 2.5$ (Table 6.1) it is calculated that:

$$N_g/N_e = (Q_g/Q_e)/k_d = 0.19/2.5 = 0.08. \tag{6.11}$$

Therefore, the number of hydrogen sulphide moles in the biogas is equal 8% of the number of H_2S moles in the effluent. Furthermore, in the liquid phase the pH is approximately neutral, so that there is a partial dissociation of the hydrogen sulphide gas into bisulphide:

$$H_2S \leftrightarrow HS^- + H^+ \ (pK_1 = 7 \text{ at } 25°C)$$

At pH = 7, the number of dissolved hydrogen sulphide moles will be equal to the number of disulphide moles. Thus the fraction of hydrogen sulphide in the biogas is estimated as:

$$\begin{aligned} f_g &= N_g/N_t = N_g/(N_g + 2N_e) \\ &= N_g/N_e/(N_g/N_e + 2) = 0.08/2.08 = 0.04 \end{aligned} \tag{6.14}$$

and $f_l = 1 - f_g = 0.96$.

Therefore, for a generated sulphide concentration in the digester of 20 mg S · L^{-1}, the concentration of sulphide in the effluent will be 20 × 0.96 = 19.2 mg · L^{-1}. The concentration of hydrogen sulphide in the biogas $C_g = C_s/k_d = 0.5 \times 19.2/2.5 = 3.8$ mg · L^{-1} or 3.8/32 = 0.12 mmol · L^{-1}. As in the biogas total concentration of gases is equal to 41 mmol · L^{-1}, the partial pressure of hydrogen sulphide $p_{H2S} = 0.12/41 = 0.003$.

EXAMPLE 6.7

For the conditions of Example 6.4, but for a ratio $Q_g/Q_l = 10$, estimate the fraction of the hydrogen sulphide gas which remains in the effluent, when the sulphate concentration in the influent is 20 mg S · L^{-1} and all the sulphate is reduced in the digester (pH = 6.8)

Solution

For $(Q_g/Q_l) = 10$ and a distribution coefficient $k_d = 2.5$ (Table 6.1) it is calculated that:

$$N_g/N_l = (Q_g/Q_l)/k_d = 10/2.5 = 4.$$

Hence the number of moles per day that is present in the biogas is four times larger than that in the effluent. On the other hand for pH = 6.8 with Eq. (6.10): $N_{be} = 0.63\ N_{he}$ and one can estimate the fraction of sulphide that will end up in the biogas as:

$$f_g = N_g/N_t = N_g/(N_g + 1.63 \cdot N_l) = (N_g/N_l)/(N_g/N_l + 1.63) = 4/(4 + 1.63) = 0.70 \tag{6.14}$$

and $f_l = 1 - f_g = 0.30$

Hence, for a generated sulphide concentration in the digester 20 mg S · L^{-1}, the concentration of sulphide in the effluent will be 20 × 0.3 = 6 mg · L^{-1}. The concentration of hydrogen sulphide in the biogas $C_g = C_s/k_d = 0.5 \times 6/2.5 = 1.2$ mg · L^{-1} or 1.2/32 = 0.04 mmol · L^{-1}. As in the biogas the total concentration of gases is 41 mmol · L^{-1}, the partial pressure of hydrogen sulphide is $p_{H2S} = 0.04/41 = 0.001$.

Note that much more H_2S is stripped from the liquid phase than in Example 6.5 (70% against 4%), but the partial pressure is actually smaller (0.001 rather than 0.003), i.e. the biogas contains a lower H_2S concentration, because there is so much more methane and carbon dioxide produced.

Although some production of sulphide may be generated from mineralization of proteins, generally the major part comes from sulphate reduction by organic material:

$$8e + 8H^+ + SO_4^{2-} \rightarrow S^{2-} + 4H_2O$$

In this half-reaction sulphate acts as an oxidant in the oxidative catabolic reaction in a similar way to oxygen in an aerobic environment:

$$4e + 4H^+ + O_2 \rightarrow 2H_2O$$

It is concluded that 1 mole of sulphate (32 g S) has the same oxidation capacity as 2 moles of oxygen (64 g), or in other words 1 g of sulphur in sulphate form is equivalent to 2 g of oxygen and can therefore oxidize 2 g of COD, when it is reduced to hydrogen sulphide. As 2 g of COD have the potential of producing 0.5 g of CH_4 it follows that per mg of sulphide formed there is a reduction of methane production of 0.5 mg CH_4.

The sulphate concentration in natural waters and therefore in municipal wastewater varies widely: from less than 10 mg S·L^{-1} to more than 100 mg S·L^{-1}, sometimes even higher. This has a very important effect on anaerobic treatment: not only is the methane production potential reduced, but the concentration of hydrogen sulphide in the biogas will increase, complicating making productive use of it. The presence of hydrogen sulphide in the liquid phase is not only potentially toxic to the methanogens, but the sulphide in the effluent is also toxic to many aquatic organisms and exerts an oxygen demand to effect oxidation to sulphur or sulphate, which may reduce the dissolved oxygen in the receiving water body. Thus post-treatment of the effluent to remove or reduce the sulphide concentration is almost always necessary when the sulphate concentration in the sewage is high. The recovery of sulphur from the effluent may be interesting if water must be reused in industrial processes like cellulose pulp production.

6.4 CALCULATION PROCEDURE FOR BIOGAS PRODUCTION AND COMPOSITION

In the previous sections the partial pressures of different components of biogas were estimated. These preliminary calculations contain two flaws: (1) it was tacitly admitted that the methane fraction in the biogas was very high, and (2) it was supposed that sulphide formation did not have a significant repercussion on the methane production potential. In this section an iterative calculation procedure is developed that corrects these deficiencies and takes into account that the methane fraction in the biogas is variable and that there may be considerable oxidation of the organic material by sulphate reduction instead of methane production. One can distinguish essentially two cases: (1) the sulphate concentration is low and its reduction does not significantly affect the methane production, or (2) the sulphate concentration is high and has a significant effect on the entire process of anaerobic digestion.

6.4.1 Low sulphate concentration

In this case it is not necessary to consider the reduction in methane production resulting from the oxidation of organic material by sulphate, which simplifies the calculations. The iterative calculation procedure for estimating the biogas composition is as follows (data of Examples 6.2, 6.4 and 6.5):

6.4.1.1 New estimate of the biogas composition

Having estimated the partial pressures of CO_2 (0.04), H_2O (0.03) and N_2 (0.06), a new estimate can be made of the methane pressure. In the previous examples:

$$p_{CH4} = 1 - p_{CO2} - p_{H2O} - p_{N2} = 1 - 0.04 - 0.03 - 0.06 = 0.87$$

6.4.1.2 New estimate of the methane production in the biogas

The new estimate of the methane partial pressure leads to a new estimate of the methane solubility in the liquid phase. In the example above the methane partial pressure was 0.87 and the methane gas concentration in the biogas is now $0.87 \times 41 = 36\ \text{mmol}\cdot\text{L}^{-1}$, so that the saturation concentration in the liquid phase will be $C_s = 0.03 \times 36 = 1.08\ \text{mmol}\cdot\text{L}^{-1}$ or $17\ \text{mg}\cdot\text{L}^{-1}$. Therefore, the COD equivalent of the dissolved methane remaining in the liquid phase is $4 \times 17 = 68\ \text{mg COD}\cdot\text{L}^{-1}$.

For a per capita contribution of $120\ \text{L}\cdot\text{d}^{-1}$, a mass of $68 \times 120 = 8160\ \text{mg COD}\cdot\text{d}^{-1}$, equivalent to $MCH_{4e} = 8.16/4 = 2.04\ \text{g}\ CH_4\cdot\text{d}^{-1}$, remains in the effluent. Moreover, for the assumed sulphate concentration of $20\ \text{mg S}\cdot\text{L}^{-1}$, the oxidation of $2 \times 20 = 40\ \text{mg}\cdot\text{L}^{-1}$ COD or $0.04 \times 120 = 4.8$ g $\text{COD}\cdot\text{d}^{-1}$ is used for sulphate reduction (MS_{oxid}), which reduces the production of methane by $4.8/4 = 1.2\ \text{g}\ CH_4\cdot\text{d}^{-1}$. Therefore the mass of methane produced and collected in the biogas can now be expressed as:

$$\begin{aligned} MCH_4 &= (MS_{ti} - MS_{te} - MS_{xv} - MS_{CH4e} - MS_{oxid})/4 \\ &= (100 - 20 - 10 - 8.2 - 4.8)/4 = 57/4 = 14.25\,\text{g}\,CH_4 \cdot \text{inh}^{-1}\text{d}^{-1} \end{aligned} \quad (6.16)$$

The new collected methane volume can now be calculated, duly discounting the dissolved methane and the prevented methane production from oxidation of organic material by sulphate:

$$V'_{CH4} = 22.4(273 + T)/273 \times MCH_4/16 = 22.4(273 + T)/273 \times 14.25/16 = 23.2\ \text{L}\cdot\text{inh}^{-1}\cdot\text{d}^{-1}$$

6.4.1.3 New estimate of the biogas/effluent flow ratio

Using the new estimates of the methane flow and partial pressure, the biogas to effluent flow ratio, Q_g/Q_i can now be calculated:

$$Q_g = V'_{CH4}/p_{CH4} = 23.2/0.87 = 26.7\ \text{L} \cdot \text{inh}^{-1} \cdot \text{d}^{-1} \text{ and}$$

$$Q_g/Q_l = 26.7/120 = 0.22\ \text{L} \cdot \text{L}^{-1}$$

6.4.1.4 New estimate of the partial pressure of nitrogen

Having a new estimate for the Q_g/Q_l ratio, the new value for the partial nitrogen pressure is obtained, using the same procedure as in Example 6.3:

$$p_{N2} = C_i/(k_d + (Q_g/Q_l))/28/41 = 0.05$$

6.4.1.5 Compare the recalculated value of the composition and flow of the biogas with the values obtained in the previous iteration

Knowing that the values of the partial pressure of CO_2 (=0.04) and H_2O (=0.03) do not change, because they depend only on the pH and alkalinity (for CO_2) and the temperature (for H_2O), the new value of the partial pressure of CH_4 (=1 − 0.04 − 0.03 − 0.05 = 0.87) and the biogas volume (=26.7 L·inh^{-1}·d^{-1}) are now compared with the values of the previous iteration (1.0 and 26.7 L·inh^{-1}·d^{-1} respectively).

6.4.1.6 Continue iterative calculations until a constant value is obtained for the variables

Normally two or three iterations are sufficient. Table 6.3 shows a series of iterative calculations relative to the data of Examples 6.2 and 6.4. It can be noted that the values become constant after the third iteration (bold), when four decimal numbers are used. The values of the partial pressures of CH_4 and N_2 converge slowest.

6.4.2 High sulphate concentration

When the sulphate concentration is high, the methane production is affected, because the organic material will be preferentially oxidized. In that case the interactive calculation procedure above is extended in two aspects:

Table 6.3 Iterative calculation results to establish the final values of volume and composition of biogas for case (1) low sulphate concentration (A_q = 120 L · inh^{-1} · d^{-1}; T = 25°C; MS_{inh} = 100 g COD · d^{-1} (70% digestion); methane production = 17.5 g CH_4 · inh^{-1} · d^{-1}; p_{CO2} = 0.04; p_{H2O} = 0.03).

Iteration	pCH_4	MCH_4	Q_g	Q_g/Q_l	pN_2
0	1.0000	17.500	26.6875	0.2224	0.0526
1	0.8732	14.2500	21.7313	0.2074	0.0562
2	0.8697	14.2500	21.7313	0.2082	0.0559
3	**0.8699**	**14.2500**	**21.7313**	**0.2082**	**0.0560**
4	0.8699	14.2500	21.7313	0.2082	0.0560

- Methane production is now estimated, decreasing the availability of organic material to anaerobic digestion with 2 mg COD·mg^{-1} S of sulphate. For example, at a concentration of 100 mg·L^{-1} of SO_4^-S or 120/1000 × 100 = 12 g S·inh^{-1}·d^{-1}

$$V'_{CH4} = 22.4(273 + T)/273(MS_{ti} - MS_{te} - MS_{xv} - 2MS_{SO4})/64$$
$$= 24.6(100 - 20 - 10 - 2 \times 12) = 17.7\ L \cdot inh^{-1} \cdot d^{-1}$$

- Part of the generated hydrogen sulphide will pass to the gas phase, but the magnitude of this depends on the ratio of the flow of biogas (which one seeks to establish) and the effluent flow. Figure 6.4 shows clearly that the fraction of hydrogen sulphide in the biogas gas depends on the Q_g/Q_l ratio and the pH. Once the sulphide fraction which desorbs to the gaseous phase has been established, the volume and the partial pressure of hydrogen sulphide can be calculated, as well as the concentration of this component in the effluent.

EXAMPLE 6.8

Estimate for the data in previous Examples the production of methane and biogas and the residual sulphide concentration in the liquid phase as a function of the sulphate concentration over the range from 0 to 200 mg S · L^{-1}.

Solution

For each sulphate concentration, the problem requires an iterative calculation procedure that consists of the following steps:

(1) Initially it is assumed that the sulphate is converted to sulphide, the corresponding organic material demand to perform this redox process is estimated and the corresponding reduction in methane production is calculated. (Column 2 of Table 6.4)

$$MS_{oxid} = 2 \times MSO_4$$
$$MS_{CH4tot} = (MS_{ti} - MS_{te} - MS_{xv} - MS_{oxid})/4 \qquad (6.16)$$

In practice this estimate is imperfect, because the sulphate-reducing bacteria will grow and in this anabolic process the SRBs, being heterotrophic, use organic material. At high SO_4 concentrations, this growth will not be insignificant, because oxidative catabolism occurs, which generates much energy and thus anabolism (growth) is considerable.

(2) Estimate the partial pressure of methane in the biogas using the values of the partial pressures of CO_2, H_2O, N_2 and H_2S. In the case of N_2 and H_2S this will require an iterative calculation procedure

(3) Estimate the methane mass remaining in the effluent as the difference between the total methane production and the mass of methane that is transferred to the gaseous phase (column 4):

$$MS_{CH4e} = MS_{CH4tot} - p_{CH4} \times 0.020 \times 120 \text{ and } MCH_{4e} = MS_{CH4e}/4$$

(4) Estimate the methane volume corresponding to the daily mass of produced methane in the gas phase (column 5)

$$V_{CH4} = 22.4(273 + T)/273(MS_{CH4tot} - MS_{CH4e})/64$$

(5) With the aid of the estimates of the pressure and volume of methane, calculate the ratio between the gas and liquid flows, Q_g/Q_l, (column 6):

$$Q_g/Q_l = (V_{CH4}/p_{CH4})/Q_l$$

(6) Knowing the Q_g/Q_l ratio, calculate the partial pressure of nitrogen and hydrogen sulphide (columns 7 and 8)

(7) Knowing that the partial pressures of CO_2 and water vapour do not depend on the flow of biogas, calculate the partial pressure of these two components.

Table 6.4 Estimates of the production of biogas and the partial pressures of methane, nitrogen and hydrogen sulphide as functions of the reduced sulphate concentration for the same suppositions as in Table 6.3.

1	2	3	4	5	6	7	8
SO_4^{2}	MS_{CH4tot}	p_{CH4}	MS_{CH4g}	V_{CH4}	Q_g/Q_l	p_{N2}	p_{H2S}
$mg\ S \cdot L^{-1}$	$g \cdot inh^{-1}/d^{-1}$	(–)	$g \cdot inh^{-1} \cdot d^{-1}$	$L \cdot inh^{-1} \cdot d^{-1}$	(–)	(–)	(–)
0	17.5	0.873	15.40	23.54	0.1962	0.0569	0.0000
20	16.3	0.866	14.22	21.73	0.1811	0.0613	0.0026
40	15.1	0.858	13.04	19.93	0.1661	0.0663	0.0053
60	13.9	0.849	11.86	18.12	0.1510	0.0723	0.0079
80	12.7	0.840	10.68	16.33	0.1361	0.0794	0.0106
100	11.5	0.828	9.51	14.53	0.1211	0.0881	0.0134
120	10.3	0.815	8.34	12.75	0.1063	0.0988	0.0161
140	9.1	0.798	7.18	10.98	0.0915	0.1123	0.0189
160	7.9	0.778	6.03	9.22	0.0768	0.1301	0.0217
180	6.7	0.751	4.90	7.48	0.0624	0.1541	0.0246
200	5.5	0.714	3.79	5.79	0.0482	0.1881	0.0275

(8) Now, having the estimates for the partial pressures of CO_2, H_2O, N_2 and H_2S, calculate a new estimate for the partial pressure of CH_4.

(9) With the aid of the new p_{CH4}, for each sulphate concentration repeat steps 2 to 8 until the same result is obtained in two consecutive iterations. At the end of each iterative calculation the flow and composition of the biogas is obtained.

Table 6.4 shows the results of the iterative calculations for different sulphate concentrations, assuming full conversion to sulphide. Figures 6.6 and 6.7 are graphical representations of Table 6.4. When analysing the numbers of Table 6.4 and Figures 6.6 and 6.7, the dramatic influence of sulphate reduction on the production of methane and the biogas composition can be observed: the mass of produced methane is reduced by a factor 4 when the concentration of sulphate is increased from 0 to 200 mg $S \cdot L^{-1}$.

For the typical per capita sewage production of 100 g COD $\cdot inh^{-1} \cdot d^{-1}$ and 120 L $\cdot d^{-1}$, the calculated COD concentration of the influent is $100/120 = 0.833$ g $\cdot L^{-1}$ and a biodegradable concentration of $0.7 \times 833 = 583$ (70% efficiency). The data indicate that for a ratio of S-SO_4/COD in the influent greater than $200/583 \approx 1/3$ mg S $\cdot mg^{-1}$ COD, methane production is not feasible, which has also been the experience in practice. Under these conditions all the organic material is used for sulphate reduction.

As for the estimation of the concentration of hydrogen sulphide in the liquid phase, one can first calculate the mass of sulphur in the biogas: for estimated influent concentration of 200 mg S $\cdot L^{-1}$ the partial pressure of $H_2S = 0.0275$ and the concentration in the gas is $41 \times 0.0275 = 1.15$ mmol $\cdot L^{-1}$ or 37 mg $\cdot L^{-1}$ biogas.

As the biogas is only a factor $Q_g/Q_l = 0.048$ of the sewage flow (column 6 of Table 6.4), the daily production per capita is $0.048 \times 120 = 5.8$ L $\cdot inh^{-1} \cdot d^{-1}$ with a sulphur mass of $37 \times 5.8 = 213$ mg S $\cdot d^{-1}$. This is 213/(120 × 200) or less than 1% of the influent sulphur mass. For this reason the mass of sulphur in the effluent will, to good approximation, be equal to the mass of sulphide generated in the anaerobic reactor or present in the influent: 200 mg S $\cdot L^{-1}$. This concentration in the effluent must still be divided between bisulphide and hydrogen sulphide. At pH = 7.0, this ratio is about equal, so $C_{H2Se} = C_{HS-} = 100$ mg $\cdot L^{-1}$.

The concentration of hydrogen sulphide in the effluent and thus also in the reactor is very high: many researchers (Buisman *et al.*, 1990; Rinzema, 1988) have shown that a concentration of 100 mg $\cdot L^{-1}$ H_2S severely inhibits methane production in anaerobic digesters. Thus the production of methane would tend to be even lower than the calculations indicate, as they were carried out without taking into account the toxic effect of hydrogen sulphide gas.

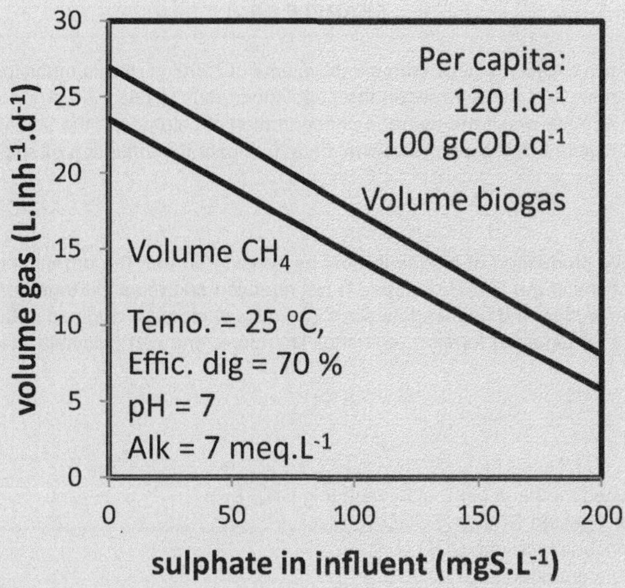

Figure 6.6 Influence of the sulphate concentration on the biogas production (see also Table 6.4).

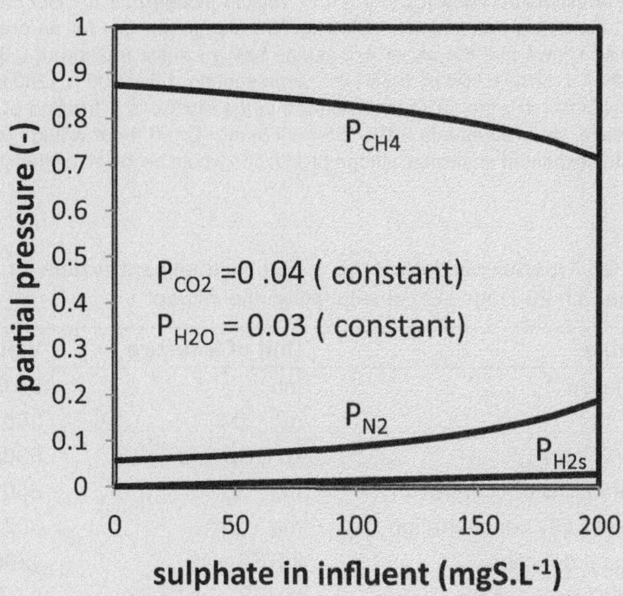

Figure 6.7 Influence of the sulphate concentration (reduced to sulphide) on the production and composition of biogas (see also Table 6.4).

EXAMPLE 6.9

A system composed of two UASB reactors, with a total volume of 7500 m^3, treats municipal sewage. The water supply system is precarious and for that reason the population tends to use also ground water, which is very rich in sulphate (1 g · L^{-1}). As a result the average concentration of sulphate in the sewage is 200 mg · L^{-1} (as SO_4^{2-}) or 67 mg S · L^{-1}. Table 6.5 shows operational data. Evaluate the influence of sulphate on the methane production.

Solution

To estimate the effective production of biogas it must be considered that the organic material of the influent is divided into four fractions in the UASB reactor: (1) not removed and released together with the effluent; (2) transformed into volatile sludge and released as such in excess sludge; (3) oxidized in the redox process with sulphate reduction; and (4) digested forming methane. Therefore, one can calculate the methane production potential as:

$$CH_4 = S_d/4 = S_{ti} - S_{te} - S_{xv} - S_{oxid})/4 \tag{6.16}$$

where:

CH_4 = mass of produced methane per L of influent (mg CH_4 · L^{-1})
S_d = concentration of digested COD (mg COD · L^{-1})
S_{ti} = influent COD concentration (mg COD · L^{-1})
S_{te} = effluent COD concentration (mg COD · L^{-1})
S_{xv} = COD transformed into volatile sludge in the UASB reactor (mg COD · L^{-1})
S_{oxid} = oxidized COD by sulphate (mg COD · L^{-1}).

If in the influent there is 200 mgSO_4^{2-} · L^{-1} or 200/3 = 67 mgSO_4^{2-} − S · L^{-1}, with an oxidation capacity of 2 × 67 = 133 mg COD · L^{-1}, it can prevent the production of 133/4 = 33 mg CH_4 · L^{-1}.
The S_{xv} concentration is estimated knowing that 1 g of volatile sludge has a COD of approximately 1.5 g. According to Table 6.5 there is a production of 1200 kg TSS sludge per day for an organic load of 6300 kg COD · d^{-1}. Table 6.5 also shows that the anaerobic sludge has a volatile fraction of 0.67 so that the volatile sludge production is 0.67 × 1200 = 800 kg VSS · d^{-1}, representing 1.5 × 800 = 1200 kg COD · d^{-1}. As the applied load is 6300 kg COD · d^{-1}, the COD incorporated in the sludge is a fraction of 1200/6300 = 0.19 of the COD fed to the system, so it represents 0.19 × 650 = 124 mg · L^{-1}. If there would not have been sulphate present in the influent, the expected anaerobic sludge production would be lower, as the sludge yield generated

Table 6.5 Operational data of the anaerobic treatment system for sewage with 200 mg · L^{-1} of sulphate in the influent.

Variable	Unit of Measure	Value
Population	Inh	90.000
Flow	m^3 · d^{-1}	9000
Organic load	kg COD · d^{-1}	6300
Influent COD concentration	mg · L^{-1}	650
Effluent COD concentration	mg · L^{-1}	307
Sludge production	kgTSS · d^{-1}	1200
Volatile fraction in the sludge	g VSS · g^{-1} TSS	0.67
Sulphate in the effluent	mg · L^{-1} SO_4	≈0

from sulphate reduction is much greater (oxidative catabolism). Hence, the methane production can now be calculated as:

$$CH_4 = S_d/4 = (S_{ti} - S_{te} - S_{xv} - S_{oxid})/4 = (650 - 307 - 124 - 133)/4 = 86/4 = 22 \text{ mg } CH_4 \cdot L^{-1}$$

However, not all of the methane produced will desorb: methane solubility in water is about 20 mg · L^{-1}, if the partial pressure of methane in the gas phase is 1 atm. Therefore, in that case the concentration of methane that can desorb is only:

$$CH_{4g} = CH_4 - 20 = 22 - 20 = 2 \text{ mg} \cdot L^{-1}$$

Knowing that the density of methane is approximately 0.67 kg $CH_4 \cdot m^{-3}$, it is calculated that the produced methane volume is:

$$V_{CH4} = 1.5 \times CH4_g = 3 \text{ mL } CH_4 \cdot L^{-1} \text{ influent}$$

It is concluded that there is virtually no biogas production, not only due to the obvious direct cause – the oxidation of organic material by sulphate, but also for several indirect reasons. The first is that organic material is also lost to the larger sludge production. Furthermore, because of the reduction in biogas production the mixing of the reactor contents by rising biogas bubbles decreases. The generation of biogas is the main contributing factor in reactor mixing and facilitates uniform distribution of influent organic material and its contact with the sludge mass. If reactor mixing is reduced, it is conceivable that preferential channels will be formed, resulting in hydraulic short circuiting. Under these conditions the part of the sludge that is present in stagnant zones receives little substrate and operates below its full capacity, while in other parts overloading occurs. Both features tend to result in a reduction of organic material removal efficiency and particularly of methanogenesis. On the other hand, the concentration of 67 mg · L^{-1} of sulphide itself that may be formed is toxic to methanogens, especially if the pH is somewhat reduced, which would lead to a further decrease of methanogenic activity.

It is concluded that the presence of 200 mg · L^{-1} of sulphate has a disastrous effect on the efficiency of the anaerobic sewage digesters. The problems extend beyond the performance of the UASB reactor itself: post-treatment in either polishing ponds or aerobic reactors is very complicated. The high sulphide concentration makes algae growth in polishing ponds almost impossible and exerts a large oxygen demand in aerobic reactors, increasing energy consumption. In fact if aerobic post-treatment is applied, it would be preferable not to use the UASB reactors but apply the aerobic treatment directly to the influent.

EXAMPLE 6.10

In ethanol distilleries a wastewater called stillage or vinasse is also produced in a proportion of 15 L per L of ethanol. The organic material in vinasse is mostly soluble, almost completely biodegradable and typically has a COD of 30 g · L^{-1}, so anaerobic digestion with an efficiency of greater than 90% is feasible. The vinasse has a sulphate concentration of 320 mg SO_4-S · L^{-1}. In addition there is cane wash water generated at equal rate, but with low concentrations of organic material and sulphate. A distillery produces approximately 90 L of ethanol per ton of cane (t_c) and requires 30 kWh/t_c electricity. Vinasse production is 90 × 15 L or 1.35 m^3 per ton of cane.

(1) Determine the loss of methane due to sulphate reduction
(2) Evaluate if the potential for energy production from the methane generated by the digestion of stillage would be sufficient to meet the industry's demand for energy, knowing that a power generator can produce about 5 kWh · kg^{-1} CH_4.
(3) Evaluate the effect of mixing cane washing with vinasse.

Solution

The organic load of stillage is 15 × 30 = 450 kg COD · m^{-3} of ethanol or 90 × 450/1000 = 40 kg COD/t_c. For the assumed digestion efficiency of 90% and without considering sulphate reduction the methane production potential is 0.9 × 40/4 = 9 kg CH_4 or 9 × 5 = 45 kWh per ton of cane, amply sufficient to meet the demand of 30 kWh/t_c, required for operation of the distillery.

If there is complete reduction of sulphate of 320 g S · m^{-3}, methane production is reduced by 160 g CH_4 · m^{-3} or 0.16 × 1.35 = 0.21 kg CH_4/t_c. Thus, the potential production is affected only marginally: 0.21/9 = 2%. However, there is potentially another problem: 320 mg · L^{-1} of sulphide may lead to toxicity in the methanogenic digester and thus reduce the production rate and the organic material removal efficiency. Therefore the concentration of hydrogen sulphide in the liquid phase must be estimated. It is considered that the production of methane is 0.98 × 0.9 × 30/4 = 6.7 kg CH_4 · m^{-3} of vinasse or 10 m^3 · m^{-3} of vinasse (methane solubility does not affect the calculations for the high concentration of organic material). The partial pressure of CO_2 is calculated by Eq. (6.6):

$$p_{CO2} = \text{Alk} \times 10^{(pk1+pkH-pH)}$$

In Figure 6.8 one can determine that to achieve a stable pH of 7.0 in a digester operated at 35°C and for a digestible COD concentration of 30 g · L^{-1}, the alkalinity must be maintained at approximately 50 meq · L^{-1}, so that $p_{CO2} = 0.050 \times 10^{(6.37+1.47-7)} = 0.35$ bar. Furthermore, from Table 6.2 it is known that the vapour pressure of water at 35°C is $p_{H2O} = 0.05$ bar. If the insignificant contributions of N_2 and H_2S are neglected, the partial methane pressure is estimated at $p_{CH4} = 1 - 0.35 - 0.05 = 0.6$ bar. Hence the biogas production is calculated at 10/0.6 = 16.7 m^3 biogas per m^3 of vinasse. The ratio between the number of moles of hydrogen sulphide in the biogas and in the effluent is given as:

$$N_g/N_e = (Q_g/Q_l)/k_d = 16.7/2 = 8.3$$

As the number of moles of HS^- and H_2S in the effluent are equal (at pH = 7.0), the production of reduced sulphur is $N_g + 2N_e = 10.3N_e$. The number of moles of sulphide produced per L of vinasse is 320/32 = 10 mmol · L^{-1}. Therefore the number of moles of H_2S and of HS^- present in the effluent is $N_e = 10/10.3 = 1$ mmol or 32 mg S per L of vinasse for both H_2S and HS^-. This concentration may lead to significant toxicity: Rinzema and Lettinga (1988) determined a reduction of 50% of the maximum methanogenic activity for a concentration of 50 mg · L^{-1} of H_2S.

The number of moles of hydrogen sulphide in the biogas will be 8 mmol per L of vinasse. It is concluded that 80% of the reduced sulphur will end up in the biogas and only 20% will remain in the liquid phase. However, since the biogas production is 16.7 times higher than the flow rate of vinasse, the concentration in the gas phase is only 8/16.7 = 0.5 mmol · L^{-1} biogas or 16 mg · L^{-1}. Although this concentration is not very high, it can cause corrosion problems if the biogas is used for electricity generation and it will be necessary to treat the biogas to reduce the sulphur content.

Mixing the vinasse with the cane wash water dilutes the concentration of both contaminants (sulphate and organic material) by half, which has several consequences for the digester:

- The fraction of oxidized organic material in the sulphate remains 2%.
- The production of methane is now 6.7/2 = 3.3 kg CH_4 · m^{-3} or 5 m^3 · m^{-3}
- The alkalinity required to keep the pH at 7.0 is 45 meq · L^{-1} (Figure 6.8), but after mixing the wash water and stillage the number of litres will be doubled.
- The partial pressure of CO_2 will be: $p_{CO2} = 0.045 \times 10^{(6.37+1.47-7)} = 0.31$
- $p_{CH4} = 1 - 0.31 - 0.05 = 0.64$
- $N_g/N_e = (5/0.64)/2 = 3.9$ and $N_g + 2N_e = 5.9N_e$, whereas the total production of reduced sulphate is $N_t = 10/2 = 5$ mmol · L^{-1}
- Therefore: $N_e = 5/5.9 = 0.85$ mmol · L^{-1} or 27 mg · L^{-1} of H_2S and HS^- so that toxicity will probably be a lesser problem than with vinasse alone. The number of moles in the biogas is calculated at $N_g = 3.9 \times 0.85 = 3.3$ mmol · L^{-1} vinasse or 3.3/(5/0.64) = 0.4 mmol · L^{-1} or 12 mg · L^{-1} biogas.

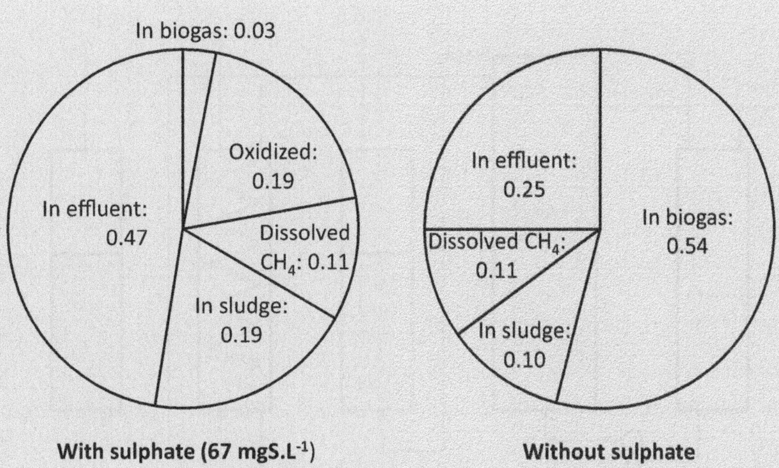

Figure 6.8 Division of the influent COD over the different COD fractions with (left) and without (right) sulphate.

Tables 6.6 and Figure 6.7 show the dramatic influence of a high sulphate concentration in the sewage on the performance of the anaerobic treatment process. Whereas without sulphate at least 65% of the influent COD is converted into methane, of which 11% remains as dissolved methane in the liquid phase, in the case with a high sulphate concentration there is virtually no biogas production.

6.5 TREATMENT OF SULPHIDE PRODUCED IN ANAEROBIC DIGESTERS

It is apparent from Figure 6.4 that the division of reduced sulphur mass depends primarily on the ratio Q_g/Q_l and also on the pH in the digester. As the ratio Q_g/Q_l depends on the concentration of organic material, this

Table 6.6 Comparison of the performance of UASB reactors treating sewage with and without presence of 67 mg SO_4^--S · L^{-1} in the influent, assuming 100% SO_4 conversion.

Mass of COD	With Sulphate (67 mg S · L^{-1})		No Sulphate		Assumptions
	kg COD · d^{-1}	Fraction	kg COD · d^{-1}	Fraction	
In the influent	6300	100%	6300	100%	
In the effluent	2980	47%	1575	25%	Efficiency = 75%
Sludge production	1200	19%	630	10%	Anaerobic yield = 10%
Dissolved CH_4	720	11%	720	11%	20 mg CH_4 · L^{-1}
Oxidized	1200	19%	0	0%	2 kg COD · kg^{-1} S
In the biogas	200	3%	3375	54%	
Methane mass	50 kg · d^{-1}		844 kg · d^{-1}		0.25 kg CH_4 · kg^{-1} COD
Methane volume	75 m^3 · d^{-1}		1265 m^3 · d^{-1}		1.5 m^3 · kg^{-1} CH_4
Electrical power	10 kW		169 kW		0.2 kW · kg^{-1} CH_4 · d^{-1})

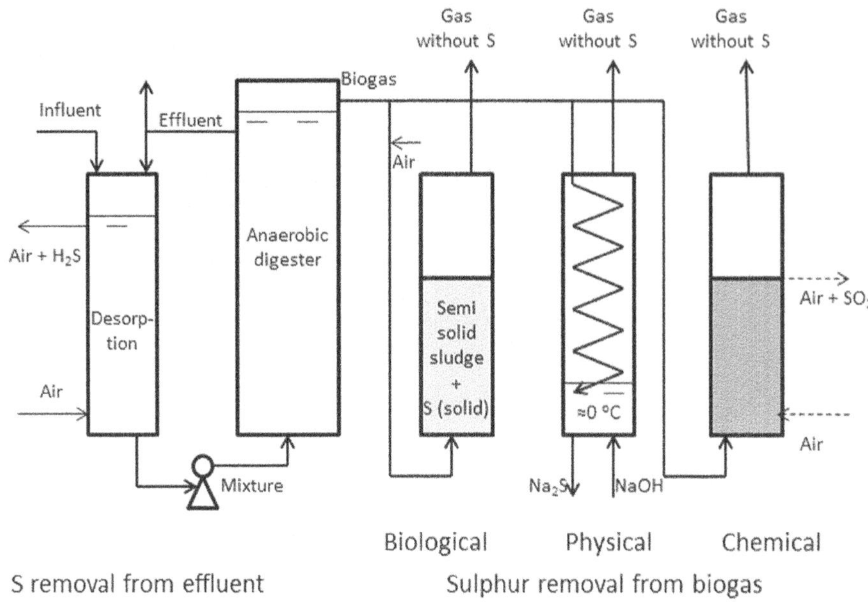

Figure 6.9 Schematic representation of an anaerobic digestion system with possible auxiliary units for sulphide removal from biogas.

will be the main factor that determines whether the generated sulphide is predominantly in the liquid phase or in the gas phase. Depending on the concentration of organic material it may be necessary to remove the sulphide from the biogas or the effluent. These two alternatives will be discussed below. The removal of sulphide from contaminated air resulting from removal of the air above the liquid surface in the digester is also discussed. Figure 6.9 shows schematically the different alternatives to treat reduced sulphur generated in the anaerobic reactor; for the treatment of biogas three methods are distinguished: biological, physical and chemical. For removal of sulphide in the liquid phase the same methods can be applied.

6.5.1 Sulphide in biogas

The most widely applied units for biological hydrogen sulphide removal from biogas are probably the Thiopaques (Paques) and Sulfothane (Veolia) processes, which are both based on adsorption of H_2S in a mildly alkaline liquor, with subsequent partial oxidation in an aerated bioreactor, in which the hydrogen sulphide is transformed into elemental sulphur.

The process is composed of three steps (see Figure 6.10):

(1) In a counter-current scrubber the biogas is contacted with an alkaline solution and the H_2S in the biogas is absorbed and dissociates

$$H_2S + OH^- \rightarrow HS^- + H_2O \tag{6.17a}$$

(2) In a biological unit and a controlled aerobic environment HS^- is oxidized to sulphur:

$$HS^- + ½\,O_2 \rightarrow S^0\downarrow + OH^- \tag{6.17b}$$

(3) The solid sulphur is separated from the liquid phase and becomes available as a nutrient or raw material.

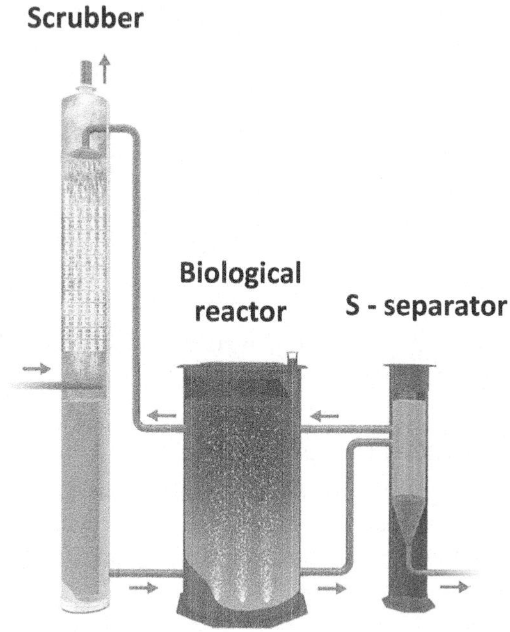

Figure 6.10 Removal of H_2S from biogas by absorption and partial oxidation.

An alternative to remove smaller quantities of hydrogen sulphide in biogas is to add a small amount of oxygen to the mixture of biogas and then pass it through or over a solid body to which a biomass of sulphide-oxidizing bacteria is attached, thus converting the H_2S in the biogas into sulphur, while oxygen is consumed. Physically, this biological treatment unit can be located in the biogas headspace zone above the liquid phase in the digester. This option is regularly used as a low-cost option for cleaning biogas in manure digesters, where biogas may be used directly to power generators for electricity production. Boards or other objects with a rough surface and a large superficial area for adhesion of biomass are placed in the area for biogas treatment. The sulphide absorbed onto the solid surface that accommodates the biomass will be oxidized by oxygen as per Eqs. (6.17a) and (6.17b). The formed sulphur accumulates on the solid surface and occasionally chunks of sulphur fall off and can be withdrawn for use as recovered sulphur or mixed with the effluent for use as a nutrient in irrigation. Care must be taken not to add too much oxygen since it can make a potentially explosive mixture with the biogas when the volumetric ratio is above $O_2/CH_4 = 1:15$ to 1:20. The oxygen demand for the removal of the sulphide from the biogas is much smaller than this critical proportion.

To inoculate biomass for the process, the solid surface can be covered with liquid manure, which always contains bacteria that oxidize hydrogen sulphide. The size of the hydrogen sulphide oxidation unit is small relative to the size of the digester. For a 1000 m^3 digester just a few square metres are enough to remove the hydrogen sulphide in the biogas.

Physical removal of sulphides is possible because of the fact that the hydrogen sulphide solubility in water increases with decreasing temperature. When the water vapour in the biogas is to be removed, the biogas may be cooled to near 0°C: condensation of water will occur and hydrogen sulphide from the biogas will tend to dissolve in the formed liquid phase.

To facilitate this dissolution, alkali is added to the condensed water. This alternative is attractive in regions with a cold climate, where ambient temperature is low for much of the year.

The chemical removal process (iron sponges process) makes use of the property that iron oxide reacts with hydrogen sulphide and produces ferrous sulphide. In a reactor with partially oxidized iron filings one has:

$$FeO + H_2S \rightarrow FeS + H_2O \tag{6.18a}$$

When the removal capacity is exhausted, due to the conversion of ferrous oxide to ferrous sulphide, it can be regenerated by oxidation with air as indicated in Figure 6.9.

$$FeS + {}^{3}\!/_{2}O_2 \rightarrow FeO + SO_2{}^{\uparrow} \tag{6.18b}$$

Other oxidants may also be used for chemical treatment of the biogas, for example hypochloride solutions, but in these cases, regeneration is usually not possible.

6.5.2 Sulphide removal from the effluent

Several methods can be employed to remove hydrogen sulphide from the effluent: biological, physical and chemical. The biological method is oxidation of sulphide to sulphur with oxygen. The physical method is the application of a desorption unit to transfer the hydrogen sulphide dissolved in recycled effluent to air in an open tank or in a closed unit with air in counter current.

For sulphide oxidation in the anaerobic reactor it is possible to add oxygen directly to the reactor contents through micro-aeration. In this case after some time in the reactor a population of sulphide-oxidizing bacteria (*Thiobacillus* spp.) is established, which oxidizes the sulphide in the liquid phase to elemental sulphur. The sulphur suspension that is formed is mixed with the sludge mass in the reactor and then discharged together with the excess sludge. This way accumulation of a considerable concentration of sulphide in the liquid phase is avoided and thus indirectly the desorption of hydrogen sulphide to biogas will be limited. This method is particularly useful if much of the sulphur remains in the liquid phase, i.e. when the ratio Q_g/Q_l is small and when the sulphate concentration is high.

An alternative is to operate a separate reactor where a low concentration of dissolved oxygen is maintained to develop the required *Thiobacillus* spp. Accurate control of the dissolved oxygen concentration is required to avoid oxidation to sulphate, which requires four times more oxygen (and therefore energy) and does not remove the sulphur from the liquid phase. Janssen *et al.* (1999) demonstrated the advantage of using the redox potential of the reactor to control the concentration of dissolved oxygen. If the oxygen is introduced in such a way that there is not much turbulence, sulphur will grow in large flocs that are removable by simple settling. High turbulence tends to generate a colloidal suspension and separation of the sulphur from the liquid phase may then become problematic.

The removal of sulphide from the liquid phase can also be affected by desorption, a physical process that develops in a separate unit of the digester, where the digester effluent is transferred and placed in counter current with air. This method is particularly effective when the influent is acidic. In that case mixing effluent (neutral pH) and influent (low pH) produces a mixture with an intermediate pH value, in which the HS^- present in the anaerobic effluent is converted into H_2S. Like carbon dioxide, the hydrogen sulphide gas will be supersaturated in the mixture and part is transferred to the air phase. This unit may have the physical form of an industrial unit (tray aerator or scrubber) or simply be an accumulation tank, in which influent and effluent are mixed – but in this case the H_2S stripping is normally incomplete. Mechanical stirring or bubble aeration may be applied to accelerate the desorption of volatile supersaturated compounds (CO_2 and H_2S). In cases with a high sulphide concentration in the effluent, the desorption unit may be closed to avoid spreading of bad odours. This implies that a solution must be provided for the hydrogen sulphide transferred to the air. In the desorption unit there is also oxygen transfer to the liquid phase of the mixture. This micro-aeration can facilitate the oxidation of sulphide in the digester itself when the influent plus recirculation mixture is introduced. Any sulphur (in colloidal form) that may be formed will eventually be expelled together with the effluent or in excess sludge.

Chemical removal of sulphide from the liquid phase has been applied mostly for wastewaters with a high sulphate concentration (Song *et al.*, 2004). Ferric chloride may be added to the influent to precipitate as ferrous sulphide; in addition it will act as a coagulant improving the settleability of the anaerobic sludge.

6.5.3 Hydrogen sulphide in the air used for odour collection

An important source of odour due to hydrogen sulphide may be the liquid surface of the UASB reactor, where gas may desorb from the liquid phase. For this reason sometimes the UASB reactor is covered. In this case it is possible to apply a small vacuum (1–2 mbar) and treat the air containing desorbed biogas. The amount of air that is required to reduce odours depends on the vacuum that is applied, but should not be greater than 1% of the wastewater flow. A small fan can pump the air to the odour removal unit. Another important source of odour problems is the discharge of effluent, especially if this discharge leads to turbulence due to pressure drops.

The treatment of the contaminated air essentially consists of H_2S removal. There are different mechanisms to achieve odour elimination: (1) use of contaminated air as part of the oxidation air for combustion of the biogas; (2) biological oxidation of sulphide in a fixed bed; (3) use of iron oxide for sulphide oxidation; and (4) use of limestone (sea shells).

In the first option, the air is mixed with the biogas for its combustion. This option is particularly attractive when the biogas is being burnt off in a flare rather being used as a fuel. In this case the hydrogen sulphide in the biogas and the contaminated air will be oxidized to sulphur dioxide, SO_2:

$$2H_2S + 3O_2 \rightarrow 2SO_2 + 2H_2O$$

This option is less suitable in cases of productive use of biogas, because it introduces an amount of sulphide (albeit not large), which tends to accelerate corrosion of the generator engine.

The biological oxidation can also be carried out in a fixed bed with peat or bagasse, which develops a specific bacterial mass under adequate conditions of humidity. The unit is simple and compact and can handle more than $10\ m^3 \cdot h^{-1}$ of air per m^3 reactor. To inoculate the bed material with a population of oxidizing bacteria, it can be mixed with manure.

Chemical oxidation of hydrogen sulphide with iron oxide is a spontaneous process that occurs at ambient temperature when biogas passes through a solid bed which contains iron oxide (FeO), for example the oxidized form of iron filings or soil containing iron oxide as per Eq. (6.18a). The oxidation capacity of the unit can be recovered by passing air (oxygen) through the solid, oxidizing sulphide to SO_2. The experiences from a full-scale system with a soil rich in iron at Pirassununga, Brazil is that it requires a very small area (about 100 m^2 for 100,000 inhabitant equivalents), obtaining a removal of more than 90% of H_2S and practically eliminating the problem of odours that was serious at the plant site. The soil depth is 0.5 m and air is introduced by perforated PVC pipes placed with spacing of 0.5 m. Another elegant method was applied successfully in Bucaramanga (Colombia), where sea shells in a suspension were used to fix H_2S. The oxygen in the air oxidizes the hydrogen sulphide to sulphuric acid and the acid reacts with the calcium carbonate that is the main component of shells, forming gypsum:

$$H_2S + 2O_2 + CaCO_3 \rightarrow CaSO_4 + H_2O + CO_2$$

The only operation in this case is the occasional withdrawal of $CaSO_4$ to replace it with shells.

Regardless of the manner of removing odours from air in the case of sewage treatment, a large part of the sulphide will not desorb from the liquid phase and will eventually be discharged from the UASB reactor with the reactor effluent for the post-treatment, which can be aerobic. In this case the formed sulphide will be oxidized back to sulphate. This oxidation is rapid and complete. If ponds are used for post-treatment, the presence of sulphide may reduce the growth of algae and hence the photosynthetic production of oxygen, which is an essential aspect of the pond performance, as will be shown in Chapter 9. On the other hand under normal conditions the produced oxygen in the pond will oxidize the sulphide, forming sulphur or sulphate.

6.6 BIOGAS COLLECTION

The main driver for the implementation of anaerobic treatment is the reduction of the costs of wastewater treatment. This is mainly due to the significant decrease in energy demand compared to conventional activated sludge systems, as the oxygen consumption for carbon oxidation in the aerobic post-treatment system is typically reduced by 65–80% for a well designed and operated UASB. Furthermore, energy can be generated from the methane that is produced in the UASB. Both the reduction in energy demand and the potential for energy production reduce the carbon footprint of wastewater treatment plants. For municipal sewage, at 15–30°C and at an anaerobic sludge age of 30–60 days, typically about 40–60% of the influent COD is converted into methane. The methane can be used in a variety of ways:

- In gas boilers (typically at 80–85% thermal efficiency) for the generation of heat: e.g. for steam production, for heating of (sludge) digesters or for sludge drying;
- In electricity generation with combustion motors (typical 35–38% net electrical efficiency); the overall energy efficiency can be significantly increased when the hot exhaust gases are used to heat steam or water in combined heat and power (CHP) generation;
- As cooking fuel, by connecting to a gas distribution grid in a nearby residential area; suitable gas burners should be used, i.e. capable of handling a variable methane content in the cooking gas.

While electricity and/or heat generation are proven technologies, the safety issues related to the construction of a biogas distribution grid, at least when western safety standards are applied, are considerable. This means this option is mainly attractive for small-scale community-based UASBs near to residential areas, or even on the level of individual households, when the manure from livestock can be co-digested.

Perhaps surprisingly given the large potential benefits, at present it is still by no means common practice to utilize the biogas produced from anaerobic sewage treatment. For example, of the ten large scale UASBs visited by Van der Lubbe *et al.* (2010a), with design capacities ranging from 30,000 to 165,000 $m^3 \cdot d^{-1}$, not even one was equipped with working equipment for biogas utilization, although in one plant a gas motor was being installed at the time of visit. The most commonly observed method of biogas disposal was by means of flaring, although several of the installations were in such a poor state that methane was being emitted directly to the atmosphere. This should be prevented at all costs, as the impact of methane emissions to the greenhouse effect is much higher than that of carbon dioxide.

These disappointing findings are very different from the practices observed in industrial anaerobic wastewater treatment plants or in municipal sludge digesters in developed countries, where energy production from biogas is much more common. It may be partly attributed to a lack of knowledge or the unavailability of appropriate equipment in some countries, while in other countries the costs of fossil energy are (still) so low that it is simply not economically viable to use the methane for energy generation. It can be concluded that significant improvement can and should be made in the area of energy recovery.

The biogas generated in the UASB reactor has a pressure of approximately 0.3–0.5 m water column (30–50 mbar), which is normally sufficient for direct use in power generators. The gas collection system in each phase separator element is composed of a network of rigid PVC pipes with diameters much larger than necessary for the biogas transport, but necessary to avoid blockages due to introduction of solid particles from the reactor into the collection system. For this reason it is important that the biogas outlet from the gas chamber situated as far as possible from the gas–liquid interface in the phase separator. It is convenient that washing of the pipes is foreseen to remove solids periodically from the interior of the pipes.

The biogas pressure is usually controlled by a hydraulic seal or pressure control valve and used to set the level of the liquid–gas interface in the separator elements as shown in Figures 4.38 and 4.39. In the case of domestic sewage treatment, biogas is normally burned in a biogas flare, which can be a single unit designed to serve several UASB reactors as shown in Figure 6.11. The flare shown in Figure 6.11 is designed to dispose of the biogas generated in a system for 70,000 inhabitants. It is concluded that it can be a low-cost unit.

If the goal is to use the biogas for the production of electric energy, it is advisable to install a biogas holder to accommodate the daily variations in production rate. The volume of this unit should be at least such that it can accommodate one hour of gas production. Traditionally biogas has been stored in a gasholder which was constructed of polyester reinforced with fiberglass or steel covered with material to resists corrosion. These units are rigid and the gas volume varied with changing production and usage rates. On top of the gasholder a weight was placed so that the desired pressure is automatically set. Nowadays semi-spheres of inflatable plastic material ("biogas balloons") are used more frequently as they have lower construction costs. These units have an outer layer that is kept inflated by a small fan to the desired pressure (typically 30 mbar) and an inner gas membrane impermeable to methane. This inner membrane expands or contracts inside the outer sphere depending on the biogas production variations. Figure 6.12 shows an example of a biogas holder.

Figure 6.11 Biogas flare of the UASB reactor in Matão (SP – Brazil).

Figure 6.12 Plastic biogas holder or "balloon" in Tegucigalpa (Honduras).

6.7 GENERATION OF ELECTRICAL ENERGY FROM BIOGAS

Ideally all produced biogas should be used in generators, but due to variations in biogas production over the day it is difficult to operate the generators continuously and use exactly the amount of biogas that is produced. Therefore it is realistic to expect that the average biogas production rate will be higher than the average consumption rate by the generators. In this case it is necessary to have a flare to burn the excess production during peak load. The biogas flare must be equipped with a device to automatically initiate combustion every time biogas starts to flow through it: this is typically regulated based on the biogas pressure. The flare may be made in masonry or more typically in stainless steel. Its burning capacity should be enough to burn all biogas, when no electric energy is generated.

Power generation from biogas produced in the treatment plant is particularly attractive because the cost of generation will be low. Normally the major cost element is the operational cost, i.e. the cost of operation of the generator. In the case of biogas generation in a wastewater treatment system, in principle, the fuel is free. Energy production becomes particularly interesting when there is a direct use in the treatment system, for example to provide electrical power to pumps or aerators in post-treatment units. Also there is a strong incentive for electricity production if there is legislation that stimulates sales of electricity to distributors or if carbon credits can be obtained.

The electric power production potential is directly proportional to the methane production, which in turn depends on the organic load and the efficiency of anaerobic digestion. To calculate the potential for power generation, the so-called lower combustion heat is used, which is the combustion energy when one considers that the water that is formed in the combustion is not condensed and therefore the condensation energy will be lost. The lower methane combustion heat is 50.4 $MJ \cdot kg^{-1}$ CH_4 or 50,000/3600 = 13.9 $kWh \cdot kg^{-1}$ CH_4. However, only part of the methane combustion energy is effectively converted into electrical energy in a power generator. The most commonly used equipment for power generation is a gas motor in which biogas and air are directly injected. Modern units have a power conversion efficiency of 35 (small units) to 40% (for units of more than 1 MW). For this range the efficiency of electricity production is 4.9–5.6 kWh per kg CH_4.

EXAMPLE 6.11

Estimate the per capita production potential of electric energy based on the following data:

- Per capita contribution: 100 g COD · d^{-1}, 120 L · d^{-1};
- Anaerobic digestion efficiency: 70% of the COD present in the feed;
- Losses due to dissolved methane in the effluent: 20 mg · L^{-1};
- Biogas losses due to leaks, collection, transport and storage: 30%;
- Electrical conversion efficiency of the generator: 40%

Solution

Methane production is one-quarter of the digested COD mass, so the per capita production of methane is equal to $0.7 \times 100/4 = 17.5$ g $CH_4 \cdot d^{-1}$. For a per capita contribution of 120 L · d^{-1} the dissolved methane lost in the effluent is $20 \times 0.120 = 2.4$ g $CH_4 \cdot d^{-1}$. Hence the biogas production will be $17.5 - 2.4 = 15.1$ g $CH_4 \cdot inh^{-1} \cdot d^{-1}$. Now taking into account the losses of 30%, the remaining production available for generating electrical energy is $0.7 \times 15.1 = 10.6$ g $CH_4 \cdot inh^{-1} \cdot d^{-1}$.

For an electrical efficiency of 40%, the power production is estimated as 5 kWh · kg^{-1} CH_4 or 5 Wh · g^{-1} CH_4. As there is a per capita collected methane production of 10.6 g $CH_4 \cdot d^{-1}$, per capita energy production will be $10.6 \times 5 = 53$ Wh · d^{-1}, i.e., there is a potential to produce a power of $53/24 = 2.2$ W · inh^{-1}. This power is usually sufficient to meet the entire demand of a combined anaerobic–aerobic sewage treatment system, including aerobic post-treatment.

The potential for energy production may not be enough for a water company to warrant exploration, unless it is a very large system. One of the largest anaerobic sewage treatment systems in the world is the Onça plant at Belo Horizonte, Brazil, with a volume of 48,000 m^3. This system operates with a residence time of 8 hours so the treated volume is close to 150,000 $m^3 \cdot d^{-1}$ with an organic load estimated at 120 ton COD·d^{-1} (assumed COD = 800 mg·l^{-1}). Considering that 60% of this load will be effectively used for power generation one would have a mass of $0.6 \times 80/4 = 12$ ton $CH_4 \cdot d^{-1}$ or about 60,000 kWh·d^{-1}, i.e. an electrical power of 60,000/24 = 2500 kW. In this case the generation of electrical power should be economically feasible. In many cases smaller systems probably do not have economic feasibility in the Brazilian context, where the price for energy is low, thanks to ample possibilities for hydropower generation.

6.8 GREENHOUSE GAS EMISSIONS AND CARBON FOOTPRINT

It is interesting to evaluate and compare the carbon footprint of different wastewater treatment processes. First, consider the biological processes. The organic matter in wastewater may be considered as a renewable source of energy. As will be demonstrated below, there is little difference between anaerobic and aerobic treatment in this aspect, provided the effluent quality is the same, which will be the case when aerobic systems are compared to anaerobic–aerobic systems and when the formed methane is oxidized (with or without energy generation). The catabolic reaction (energy production) for aerobic treatment is:

$$C_xH_yO_z + {}^1/_4\,(4x + y - 2z)O_2 \rightarrow xCO_2 + {}^1\!/_2 yH_2O \tag{6.19}$$

It does not matter if nitrate or another oxidant is used instead of oxygen, as in this section it is not oxygen consumption but carbon dioxide production that is of interest in these metabolic reactions.

For anaerobic treatment the catabolic reaction is:

$$C_xH_yO_z + {}^1/_4(4x + y - 2z)H_2O \rightarrow {}^1/_8(4x - y + 2z)\,CO_2 + {}^1/_8(4x + y - 2z)\,CH_4 \quad (2.6)$$

From the anaerobic reaction equation, it can be observed that 50% of the carbon atoms are used for methane production instead of carbon dioxide production. However, in the subsequent oxidation of methane (combustion), this fraction will also be fully converted into carbon dioxide. Apart from the catabolic reactions discussed above, the anabolic reaction (biomass generation) should be considered as well. Often it is assumed that the excess sludge production of anaerobic systems is much smaller than that of aerobic systems, but for municipal sewage this is actually only true when secondary treatment (COD removal) is considered.

At the high sludge age required for nutrient removal, the increase in endogenous respiration will decrease the sludge production of an aerobic system significantly. Furthermore, anaerobic sludge digestion is nowadays often applied, which further reduces the sludge production from aerobic systems. In the end, all the sludge eventually ends up as CO_2, either directly (combustion) or indirectly by biological and/or chemical degradation process (when disposed on land or in a landfill). Thus, in the end there is no real difference in carbon dioxide production between modern aerobic and anaerobic–aerobic systems when the overall biological degradation of the organic load is concerned. Therefore, it may be concluded that the main difference in carbon footprint between aerobic and anaerobic processes is due to differences in the energy consumption (aeration, dewatering) and the production of energy. There are two important exceptions: (1) if the methane produced from digested sludge in a landfill is not collected and flared, then the contribution to greenhouse gas emissions will be *very* large, which benefits the anaerobic process due to its lower production of excess sludge; and (2) the loss of dissolved methane with the anaerobic effluent.

Methane is roughly twenty-one times more potent a greenhouse gas than carbon dioxide, or in other words, the CO_2 equivalence of methane is equal to 21. In developed countries the disposal of excess sludge at landfills is no longer permitted, while generally the biogas of existing landfills is collected and either utilized or flared. For example, the most common route for sludge disposal in the Netherlands consists of anaerobic digestion, dewatering and sludge drying followed by incineration. In many developing countries the situation is different and the application of poorly stabilized excess sludge on agricultural land or in (uncontrolled) landfills is often the largest contributor to the overall emission of greenhouse gases in the wastewater treatment process.

To evaluate the impact on the carbon footprint of uncontrolled digestion of stabilized excess sludge following disposal in a landfill, then in principle it is not unrealistic to assume that, given enough time, the stabilized sludge is almost completely mineralized under anaerobic conditions. When a composition of the stabilized organic sludge of $C_7H_{10}O_3N$ is assumed ($f_{cv} = 1.48$ mg COD·mg^{-1} COD, the carbon mass fraction in the sludge is equal to 54%. In the anaerobic digestion of this sludge, equal molar quantities of methane and carbon dioxide are produced. Hence, the maximum methane production per g VSS of stabilized sludge disposed is equal to $0.54 \times 0.5 \times 16/12 = 0.36$ g $CH_4 \cdot g^{-1}$ VSS or $21 \times 0.36 = 7.56$ g $CO_2 \cdot g^{-1}$ VSS when the carbon equivalence of methane is considered.

However, the guidelines of the Integrated Pollution Prevention and Control committee (IPCC, 2007) suggest a default value of 50% anaerobic degradation in a landfill. Furthermore, it is assumed that part of the produced methane (40%) is oxidized to carbon dioxide by aerobic bacteria present in the upper layers of the landfill. When the IPPC values are applied, then the expected methane emission is $0.5(1 - 0.4) \times 0.36 = 0.11$ g $CH_4 \cdot g^{-1}$ VSS or 2.27 g $CO_2 \cdot g^{-1}$ VSS.

As to the wastewater treatment process itself, every kWh of consumed electricity results in an emission of CO_2, which depends on the means of electricity production: coal-fired plants, nuclear plants, gas-powered plants etc. Often country-specific emission factors are used that correspond to the mix of energy production methods in use. For example, for the US the CO_2 emission factor used in 2009 is equal to 0.718 kg CO_2 per kWh consumed (EPA, 2003). In comparison, if methane is used as a fuel source and a

35% combustion efficiency is assumed, then per kWh generated (=3600 kJ) about 3600/35% = 10,300 kJ is required which is equivalent to 10,300/50,400 = 0.204 kg CH_4. Upon combustion, 1 kg of CH_4 produces 44/16 = 2.75 kg of CO_2. Thus, the CO_2 emission factor per kWh produced is 0.204 × 2.75 = 0.56 kg CO_2. The average US number is higher because it is based on a mix of different fuel sources and it includes the energy losses from the transport of electricity from power plant to final consumer.

EXAMPLE 6.12

Compare the energy balance and carbon footprint of a conventional activated sludge plant designed for secondary treatment (including a heated digester) with that of a combined anaerobic–aerobic sewage treatment plant. The feed flow is 50,000 $m^3 \cdot d^{-1}$ of raw sewage containing 548 mg COD $\cdot L^{-1}$ (i.e. 10,000 ton COD $\cdot yr^{-1}$) at a temperature of 25°C.

Use the system performance specified in Table 6.7, which is fairly typical for the system configurations involved and is based on a COD removal efficiency of 70% in the anaerobic reactor and recovery of all of the methane produced. Assume an oxygen transfer efficiency of 1.2 kg $O_2 \cdot kWh^{-1}$ and an electrical efficiency of the gas motor of 35%.

To estimate the greenhouse gas emissions resulting from the energy demand for aeration, use the typical US figure of 0.718 kg $CO_2 \cdot kWh^{-1}$.

Solution

Consider the conventional activated sludge system first. On a yearly basis, the total oxygen demand will be 0.5 × 10,000 = 5000 ton O_2, which requires 5000/1.2 = 4167 MWh. Using the US emission factor of 0.718 kg $CO_2 \cdot kWh^{-1}$, the yearly CO_2 emission is equal to 2990 ton. On the other hand, methane is produced from the digested sludge: 0.15 × 10,000/4 = 375 ton $CH_4 \cdot yr^{-1}$, which can be used to produce energy: 35% × 375 × 50,400/3600 = 1840 MWh $\cdot yr^{-1}$, equivalent to an averted CO_2 emission of 1840 × 0.718 = 1320 ton $\cdot yr^{-1}$. Thus, the net CO_2 emission of the aerobic system is 2990 − 1320 = 1670 ton $CO_2 \cdot yr^{-1}$.

As to the performance of the combined anaerobic–aerobic treatment system, the oxygen consumption is much lower (1500 ton $O_2 \cdot yr^{-1}$), corresponding to an electrical power consumption of 1250 MWh $\cdot yr^{-1}$ and resulting in a CO_2 emission of 900 ton $\cdot yr^{-1}$. Furthermore, due to the presence of the anaerobic reactor the methane production is increased to 1500 ton $CH_4 \cdot yr^{-1}$, from which 7350 MWh $\cdot yr^{-1}$ of electricity can be produced. This reduces the CO_2 emission by 5280 ton $CO_2 \cdot yr^{-1}$. The net CO_2 emission is therefore equal to 900 − 5280 = −4380 ton $CO_2 \cdot yr^{-1}$, i.e. the anaerobic–aerobic system actually *reduces* (or averts) the emission of CO_2!

Table 6.7 Performance of the conventional aerobic and the combined anaerobic–aerobic system for secondary treatment of Example 6.10.

COD Fraction	Symbol	Aerobic System	Anaerobic–aerobic
Oxidized	mS_o	50%	15%
In effluent	mS_{te}	10%	10%
In stabilized sludge	mS_{xve}	25%	15%
Digested	mS_d	15%	60%

Clearly the anaerobic–aerobic system is much more sustainable than a conventional activated sludge system. However, there are several additional factors to consider such as: (1) the potential emission of methane following disposal of the stabilized sludge, which has been discussed earlier in this section and will be evaluated below; (2) the impact of nitrogen removal on overall power consumption and carbon emissions; and (3) the loss of dissolved methane with the anaerobic effluent, discussed in the next section. Based on the conditions of a specific design case, the performance with regards to the carbon footprint might even be inverted.

6.9 LOSS OF METHANE IN THE EFFLUENT

Unfortunately, even if leakages are minimized, not all of the methane produced in the UASB can be collected, as part of the methane will remain dissolved in the anaerobic effluent. In anaerobic reactors treating high-strength wastewater or slurries/waste sludge, this is not a problem as the wastewater flow rate will be low and therefore the loss of methane with the effluent will only be a small fraction of the methane produced per litre of wastewater. On the other hand, for diluted wastewaters such as municipal sewage the methane fraction lost with the effluent can be as high as 20–40%. Apart from the obvious reduction in energy production, methane is also released into the atmosphere, resulting in a considerable contribution to the greenhouse gas emissions. To address this issue, it is important to known what the methane concentration in the UASB effluent at the level of the overflow weir will be, i.e. corresponding to the non-collected methane production.

A theoretical approach can be made using Henry's law to calculate the theoretical equilibrium dissolved methane concentration as a function of the temperature and the partial methane pressure, using the temperature coefficients $A = 675.74$ and $B = 6.88$ (Metcalf & Eddy, 2013) and knowing that one litre of water contains 55.6 moles:

$$K_H = 10^{(-A/(273+T)+B)} \tag{6.20}$$

$$x_{CH4} = p_{ch4}/K_H \tag{6.21}$$

$$[CH_4]_{eq} = 1000 \times x_{CH4} \times 55.6 \times 16 \tag{6.22}$$

where

K_H = Henry coefficient (atm.)
x_{CH4} = mole fraction of dissolved methane gas in water ($mol \cdot mol^{-1}$)
T = temperature (°C)
p_{CH4} = partial methane pressure (at.): the product of actual pressure (including the effect of liquid depth) and the molar (volume) fraction of the methane in the gas phase
$[CH_4]_{eq}$ = equilibrium dissolved methane concentration ($mg\ CH_4 \cdot L^{-1}$)

For example, using Eqs. (6.20)–(6.22) the equilibrium methane concentration at atmospheric conditions, at 70% methane content and 25°C can be calculated as:

$K_H = 10^{(-675.74/(273+25)+6.88)} = 40{,}965$ atm.
$x_{CH4} = 0.7/40695 = 1.71 \times 10^{-5}\ mol \cdot mol^{-1}$
$[CH_4]_{eq} = 1000 \times 1.71 \times 10^{-5} \times 55.6 \times 16 = 15.2\ mg\ CH_4 \cdot L^{-1}$

However, this theoretical approach yields insufficient information to predict the expected methane concentration in the UASB effluent, first of all because the value of the Henry coefficient is based on solubility of methane in pure water, which might very well be different from that in anaerobic effluent. Furthermore, it is likely that a certain degree of oversaturation exists. Methane is produced in the sludge blanket, which for most municipal UASB reactors will be located at 2.5–5.0 m liquid depth.

The formation of biogas bubbles is normally not observed in the settler section of municipal sewage UASB reactors, even though the effluent slowly travels upward to the overflow weirs, resulting in a gradual release of the liquid pressure. This fact alone already shows that oversaturation of methane in the anaerobic effluent will exist.

In this book the assumption is made that the methane concentration in the top liquid layer (i.e. the effluent) of the UASB is *at least* equal to the equilibrium methane concentration at 2.5 m submergence, which is a typical depth of the deflector. Above this level, i.e. when the liquid enters the settler section, no

Table 6.8 Calculated effluent methane concentration of an UASB reactor as function of the temperature and for different methane fractions in the biogas, based on the *equilibrium* methane concentration at 2.5 m liquid depth.

Temperature (°C)	60% CH_4 (mg $CH_4 \cdot L^{-1}$)	70% CH_4 (mg $CH_4 \cdot L^{-1}$)	80% CH_4 (mg $CH_4 \cdot L^{-1}$)
15	19.2	22.4	25.6
20	17.5	20.4	23.3
25	16.0	18.7	21.3
30	14.7	17.1	19.6
35	13.5	15.7	18.0

Table 6.9 Comparison between measured and theoretical methane concentration, based on the data from Souza *et al.* (2010a, b), for $T = 25°C$ and $p = 0.91$ atm.

Parameter	Units	Case 1	Case 2	Case 3
HRT	h	5.0	7.0	12.0
Reactor	(–)	pilot	pilot	demo
Methane content	%	63%	70%	74%
Measured methane concentration	$mg \cdot L^{-1}$	19.6	22.0	19.2
Theoretical methane concentration[1]	$mg \cdot L^{-1}$	15.6	17.4	18.3
Degree of oversaturation	%	125%	127%	105%

[1]*Based on the theory presented in this section for a liquid depth of the deflector of 2.5 m.*

more biogas will be produced nor collected. Table 6.8 shows the calculated effluent methane concentration at this depth as function of the liquid temperature.

Data on measured effluent methane concentrations was published by Souza *et al.* (2011a, b). A pilot-scale UASB reactor (0.35 m^3 with 4.0 m liquid height) and a demo-scale UASB (14 m^3 with 4.5 m liquid height), i.e. both with a liquid height comparable with full-scale reactors, were operated at the test site of the Federal University of Minas Gerais (UFMG) and the state water authority (COPASA), located at the Arrudas Sewage Treatment Plant. In the experiment the methane concentration was measured just below the effluent overflow level. The temperature during the experiments was around 25°C, while the methane content in the biogas varied between 63 and 74%, depending on the applied hydraulic retention time. The results are shown in Table 6.9, where it can be observed that the measured effluent methane concentration ranged between 19.2 and 22 mg $CH_4 \cdot L^{-1}$. This coincides quite well with the values predicted by our theoretical approach (15.6–18.3 mg $CH_4 \cdot L^{-1}$), although our values tend to be slightly but consistently lower. This is indeed a strong indication that some degree of oversaturation occurs. An interesting finding from Table 6.9, albeit based on a limited amount of data, is that the degree of oversaturation at a lower hydraulic retention time is considerably higher (around 125% at 5–7 hours) than at a higher hydraulic retention time (105% at 12 hours). A possible explanation might be that the increase in hydraulic retention time allows for more time to reach the equilibrium methane concentration in the liquid phase, but this remains to be validated. Furthermore, it can be observed that there is a clear trend toward higher methane fractions in the biogas at increasing hydraulic residence time. An increase in the retention time (and thus in sludge age) results in increased conversion of organic material into methane. As pointed out by Cakir and Stenstrom (2005), this will reduce the nitrogen concentration in the biogas, as the mass of dissolved nitrogen gas entering the reactor does not change. It is interesting to observe that the nitrogen fraction in the biogas can be quite substantial in the case of anaerobic treatment of low-strength wastewater such as municipal sewage.

Chapter 7

Dewatering, Dehydration and drying of sludge

7.1 INTRODUCTION

The sludge produced in anaerobic sewage treatment systems has two undesirable features: a low solids content and a high concentration of pathogenic organisms. The sludge generated in UASB systems treating sewage is a more viscous fluid than water itself and – if discharged from the bottom of the reactor – can reach a solids content of up to 50–100 TSS $g \cdot L^{-1}$, i.e. a humidity of 90–95%. If sludge is discharged from the top of the sludge bed this concentration will be much lower (20–30 g $TSS \cdot L^{-1}$). Before transporting the sludge to its final destination, it should be treated to increase the fraction of solids, making it more easily manageable and reducing both mass and volume. As for the hygienic quality of sludge, the worst problem is that many helminth eggs in sewage accumulate in the sludge, while its survival rate in the anaerobic reactor environment is high. In addition, the extent of decay of pathogenic micro-organisms in the water retained in the sludge is only partial. The objective of sludge treatment is to remove both water and pathogens from the waste sludge, thus generating a final product that may have productive use (for example, as an organic fertilizer or as a component for brick production) or that can be disposed of (for example in a landfill), without creating environmental problems and a low operating cost.

Four types of processes can be distinguished for reduction of the water content of sludge:

(1) Dewatering: processes that use gravitational processes such as settling and flotation to produce a fluid with 2–8% solids;
(2) Dehydration: mechanical processes consuming energy to produce a sludge cake of 15–20% solids;
(3) Drying: thermal processes that produce sludge with high solids content (50–95%);
(4) Incineration: which destroys the organic material present in the sludge, leaving only a dry inorganic residue.

Editor(s): Adrianus van Haandel and Jeroen van der Lubbe
doi: 10.2166/9781780409627_0257

The main difference between the first three types of processes is the energy consumption, which amounts to respectively 1 kWh per ton of dry solids for dewatering, 1 $MWh \cdot t^{-1}$ for dehydration and 1 $GWh \cdot t^{-1}$ for thermal drying. Incineration of *dry* sludge releases energy and may turn the whole sewage treatment process into an energy-producing operation. With respect to the destination of sludge there are basically three possibilities: (1) application in agriculture for the productive use of nutrients in the sludge, (2) incineration, and (3) landfill. Countries that incinerate a large part of their waste sludge generally have both a high population density and the technical and financial resources for this process. Landfill is rarely used anymore in most countries. In most developing countries the application of waste sludge in agriculture is the most common destination.

In practice, settling or flotation are commonly used to thicken waste sludge with a relatively low solids content, like that from activated sludge systems. In the case of anaerobic sludge systems these processes have less applicability, because the waste sludge concentration is often already high and the possibility of further increasing this concentration is very limited.

Important mechanical dehydration processes are filtration and centrifugation. The solids concentration from these systems is much higher than in the dewatering systems and depends on several factors: (1) the nature of the sludge, (2) operational conditions, notably the applied load and mechanical force, and (3) the application of auxiliary materials, especially polyelectrolytes. The final product is a sludge cake, whose utilization is still problematic, not only because of its semi-fluid constitution, but also owing to the presence of pathogens. A further disadvantage is that both the investment and operating costs are significant. Filters and centrifuges are expensive equipment, while operational costs comprise the consumption of electrical energy (in the order of 1 $MWh \cdot t^{-1}$ dry solids) to carry out the mechanical dehydration operation and the addition of polyelectrolytes, usually necessary to increase the solids fraction of the cake.

Natural drying normally involves the use of drying beds, which in essence function as granular filters of sludge batches, generally discharged without addition of polyelectrolyte. The removal of water takes place by two distinct mechanisms: percolation and evaporation. Usually most of the water is removed by percolation, a process that ceases when the sludge reaches a humidity of about 80%. To achieve a lower humidity, it is required to evaporate part of the remaining water. In practice it is important to reduce the maximum water content in the sludge: the mass of water increases the cost of transport and final disposal, and the survival and even the development of pathogenic organisms is avoided or at least supressed at lower humidity. The final humidity in sludge drying beds may be very low: less than 10%. Under these conditions most pathogenic organisms do not survive. Often dehydration is applied before discharging sludge on beds, so that a smaller area is required, and the bed surface may be solid instead of granular, since there is no percolation, so that the collection of the dry sludge is easier.

Sludge drying beds have important advantages compared with the mechanized methods for sludge dewatering and dehydration: in addition to being a very simple method, the cost of installation and operation of the bed is generally low and it is possible to attain a much lower final humidity than when mechanized methods are used. An important aspect concerns the hygienic quality of the sludge in case of agricultural use: in tropical regions 'solar pasteurization' can be applied to virtually eliminate pathogens, including helminth eggs. A major disadvantage of drying beds is that the required surface area is large. This area depends on the percolation and evaporation rates which, in turn, depend on several factors. The percolation is influenced by sludge filterability and the applied solid load. The evaporation rate depends on the weather conditions (particularly temperature, wind and humidity), which in turn determine the water evaporation rate, and also on the applied solid load and the desired final moisture content of the sludge.

In view of the complexity of these processes and the large number of variables that influence them, it is not possible to develop a simple process model for sludge drying. Therefore an empirical model is suggested in this chapter, which can be used for rational design of sludge drying beds.

The main reason for the relatively low evaporation rate is that the applied solar energy is mainly used for superficial heating of the sludge layer, which for this reason dries out quickly, forming hard chunks of sludge at the surface, acting as a barrier against further solar energy transfer to lower layers and reducing the rate of water evaporation from the lower layers. Moreover, due to the relatively high surface temperature of the bed, much of the absorbed solar energy is radiated back into the atmosphere, preventing its use as a heat source for water evaporation. As a result, the water in the lower layers evaporates at a low rate, allowing survival of pathogenic organisms, since neither sludge humidity nor temperature are unfavourable. Depending on the desired final humidity of the sludge and the applied solids load, the evaporation rate of drying beds is only 15–30% of that of pure water under the same conditions.

Recently, equipment has been introduced on the market by IST Anlagenbau in 2013 that repeatedly rakes and turns the sludge layer on the bed, effectively turning it upside down, thereby reducing thermal stratification and increasing the evaporation rate. In this chapter experimental data are presented to assess the performance of sludge tilling on drying beds to accelerate sludge drying on beds. Beds with sludge tilling may be open or closed.

Closed drying beds can be used to capture solar energy and thus increase the temperature of the sludge, which increases the water saturation vapour pressure. However, to have continued evaporation in closed drying beds, it is necessary to renew the air which is confined above the bed to, so that air with large amounts of water vapour can be replaced by ambient air with less water vapour, but also with a lower temperature. Thus, the withdrawal of hot air with water vapour reduces the temperature in the bed. Thus there is an optimal temperature (or air humidity) that must be maintained to maximize the evaporation rate.

It is possible to reduce the required area for drying beds by application of an external energy source (not only the solar energy applied to the bed) to accelerate the evaporation of water, which in practice is the process that takes most time and thus determines the required area. Interesting energy sources are solar energy and combustion of biogas. In the Northeast of Brazil solar energy produces an average hourly output of at least 200 $W \cdot m^{-2}$, or 5 $kWh \cdot m^{2} \cdot d^{-1}$, so that the required area for solar energy capture in addition to the area of the sludge drying bed can readily be calculated. In systems with anaerobic digestion it is possible to take advantage of the combustion heat of the collected biogas. The potential for energy generation with biogas in anaerobic sewage treatment systems is more than enough to supply the heat required to evaporate the residual water from percolated sludge (with 80% humidity) as will be shown in Section 7.6.

An important aspect is that the use of an external energy source eliminates the need to use shallow drying beds. It is perfectly feasible to operate drying towers with a considerable bed depth, sending hot air through the sludge in upward direction. Therefore these units can be implemented on small areas, but this advantage must be weighed against the extra investment and operating cost.

When an external energy source is available, mixed dehydration may be applied, first using mechanical dehydration, followed by application of the resulting sludge cake in a drying bed or tower, for additional water removal. In this case there is no need to use a filter medium, because after mechanical removal of water there will be no further possibility of percolation. So the sludge cake can be released on a concrete or asphalt bed or other solid materials, which greatly facilitates its collection after drying is finished. Sequentially thermal pathogen removal can be carried out with an external energy source, if closed beds are used.

Sludge incineration is usually applied only when the sludge has a high solids content after previous mechanical and thermal treatment, so that high fuel consumption can be avoided. Incineration may be an energy-producing operation if the combustion heat of the volatile sludge fraction is greater than the evaporation heat of the residual water. Sludge incineration requires very expensive equipment and normally is only used in regions with a very high population density (where land is very expensive) or in cases where other solids (like industrial solid wastes) must be incinerated. In fact in countries with a hot climate, incinerators are used primarily for dangerous or toxic solid wastes and if sludge is used, it is mainly as a source of cheap fuel.

7.2 SLUDGE PRODUCTION AND COMPOSITION

The sludge in anaerobic treatment systems is a suspension that is slightly more viscous than water. Depending on the composition of the wastewater, the efficiency of the pre-treatment (grit removal) and the sludge age in the digester, the anaerobic sludge generally has an inorganic fraction of 30–50% of TSS. The organic material is a mixture of biological sludge (new biomass generated by the metabolic processes), inert sludge from the influent organic material (non-biodegradable), and biodegradable organic material from the influent which has not been hydrolysed in the anaerobic treatment process. In Chapter 3 it was shown that the generated excess sludge in UASB reactors treating sewage amounts to 0.1–0.2 g $TSS \cdot g^{-1}$ COD and is produced at a concentration of 50–100 $g \cdot L^{-1}$. Thus the sludge flow is small compared with the sewage flow, as shown in Example 7.1.

EXAMPLE 7.1

For a sludge production of 0.15 g $TSS \cdot g^{-1}$ COD, an influent COD concentration 0.75 $g \cdot L^{-1}$ and an excess sludge concentration of 50 g $TSS \cdot L^{-1}$, estimate the ratio between the flows of excess sludge and waste water (q/Q_i).

Solution

For a production of 0.15 kg $TSS \cdot kg^{-1}$ COD and a sludge concentration of 50 $kg \cdot m^{-3}$, the specific sludge production volume is 0.15/50 = 0.003 $m^3 \cdot kg^{-1}$ COD = 3 $L \cdot kg^{-1}$ COD (a sludge density of 1 $kg \cdot L^{-1}$ is assumed). As the concentration of the influent COD is 0.750 $g \cdot L^{-1}$, the volume of 1 kg COD is 1/0.75 = 1.33 m^3. Therefore one has a production of 3 L of excess sludge for 1.33 m^3 of sewage, i.e. q/Q_i = 3/1.33 = 2.25 $L \cdot m^{-3}$, so that the sludge flow is only 0.22% of the sewage flow.

To avoid the discharge of excess sludge together with the effluent, the sludge produced should be withdrawn separately from the reactor and then treated. In the case of using mechanical dehydration equipment this discharge of sludge will usually be (semi-) continuous, so that the dehydration unit can be working for extended periods at a relatively low flow rate and thus the unit capacity need not be very large. In the case of drying beds, the produced excess sludge is removed with periodic discharges, filling at least one drying bed with each discharge. The sludge discharge strategy was discussed in more detail in Section 3.8.2.

A batch of excess sludge should never be discharged to a bed where sludge has previously been discharged without prior cleaning: any partially dried sludge will form a barrier to the percolation of water from the newly applied load. Sludge dehydration results not only in a reduction of the volume but there is also transformation of the fluid suspension into a solid. Figure 7.1 shows the relation between the sludge volume and the solids fraction (or humidity) for initial solids contents ranging from 50 to 100 $g \cdot L^{-1}$, which are the maximum and minimum values, respectively, that can be expected in UASB reactors.

It is observed that initially the sludge volume decreases rapidly, when the percentage of water is reduced. This removal proceeds until the humidity reaches about 20%. Additional water reduction can be obtained by evaporation.

Figure 7.1 also shows how the solid content influences the mechanical properties (rheology) of the sludge. As the solids concentration of the initial suspension increases, the sludge becomes more viscous. Below a humidity of 75%, it loses the characteristics of a fluid and becomes a semi-solid cake.

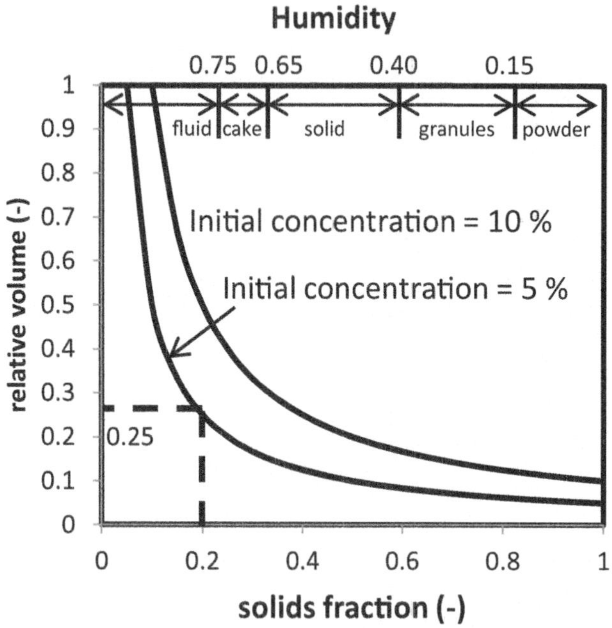

Figure 7.1 Relationship between the volume of a sludge batch and its solids fraction or humidity.

As more water is lost, the cake solidifies and at a humidity of 65% turns into a hard solid, although almost two-thirds of its mass is still water. Further drying causes the sludge to break up into granules when the humidity reaches a value around 40% and eventually it disintegrates, forming a fine powder at a humidity lower than 15%.

EXAMPLE 7.2

Sludge with solids concentration of 50 g TSS · L^{-1} is to be dried on a bed. After percolation the solids concentration increases to 200 g TSS · L^{-1}, forming a sludge cake. Subsequent evaporation results in an increase of the solids content to 50%, resulting in the formation of granules. Estimate the volumetric fractions of percolated and evaporated water.

Solution

As the diagram presented in Figure 7.2 shows, basically three situations can be distinguished: (1) the initial situation with 50 g TSS · L^{-1} and consequently 950 mL of water; (2) the situation after percolation with 50 g of solids forming a cake with 20% solids (80% humidity), so that the cake mass decreases to 250 g and the retained water is 200 mL, with percolation of 750 mL · L^{-1}; and (3) the situation after evaporation, when 50 g of TSS forms a 50% solid sludge, which therefore also has 50 mL of water, so that there is evaporation of 150 mL · L^{-1} of water.

It is concluded that the drying process requires the removal of 900 mL · L^{-1} of water, of which five-sixths (750 mL · L^{-1}) is by percolation and one-sixth (150 mL · L^{-1}) is by evaporation. Although the removal by percolation is five times higher than that by evaporation, the latter is critical to obtain a high final fraction of solids, since percolation is not feasible beyond approximately 20% solids content.

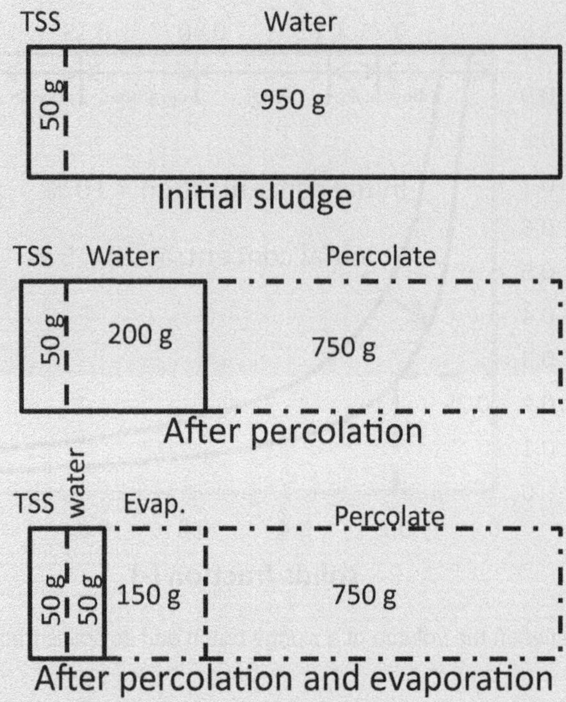

Figure 7.2 Water losses after percolation and evaporation.

7.3 MECHANICAL SLUDGE DEHYDRATION

There are basically three types of systems for mechanical solid–liquid separation commercially available: belt filters, decanter centrifuges, and pressure or vacuum filters. The three systems produce a cake with a high residual humidity and a solid fraction of only 15–25%, depending on the characteristics of the sludge, the use of auxiliary materials (coagulants and flocculants) and the selected operational conditions. In the following sections the characteristics of each of the three systems are briefly described.

7.3.1 Belt filter press

Figure 7.3 schematically shows a belt filter press, while Figure 7.4 shows a unit in operation. The sludge is (continuously) introduced as a fluid onto a conveyor belt which moves over a set of rollers. The belt is porous, and the water passes through it while the solids are retained. Three sections of the rotary conveyor path can be distinguished: (1) gravity drainage zone, (2) low-pressure zone, and (3) high-pressure zone. In the gravity drainage zone, most of the free water released from sludge flocculation (due to the addition of polyelectrolytes) drains through the screen pores. Then the sludge flow passes to the low-pressure zone, removing more water and giving a certain structure to the sludge. In the high-pressure zone the sludge is compressed between the upper and lower straps through rollers. This roller system is specifically constructed to progressively increase the pressure, removing more and more sludge water until eventually

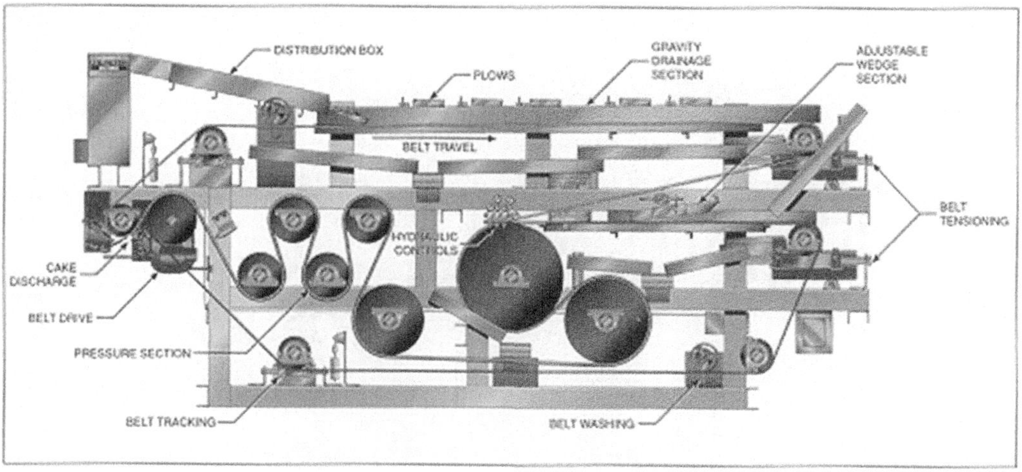

Figure 7.3 Schematic representation of a belt filter press (from US Filter).

a cake is produced that is scraped off the belt before it returns to the head of the unit to receive a new load of sludge. The belt filter press is capable of producing a semi-solid cake. Little vibration is generated, and this results in a process with low noise. On the other hand, the belt filter press is sensitive to variations in the sludge quality and chemical conditioning. Furthermore this process requires a significant amount of water for washing the fabric of the belt. There may also be odour generation, as the equipment is not enclosed.

7.3.2 Decanter centrifuges

The most commonly applied dewatering system is the decanter centrifuge. This continuously fed process involves the separation of solid and liquid phases by applying a centrifugal force on the pre-conditioned sludge (which involves prior coagulation and/or flocculation), which accelerates the separation of the relatively light liquid and slightly denser solid fractions. This process involves both clarification (solids separation from the liquid phase) and compaction (thickening). Figure 7.5 shows a schematic representation.

Figure 7.4 Belt filter press at WWTP-Franca, Brazil (Courtesy of Prof. Pedro Alem Sobrinho).

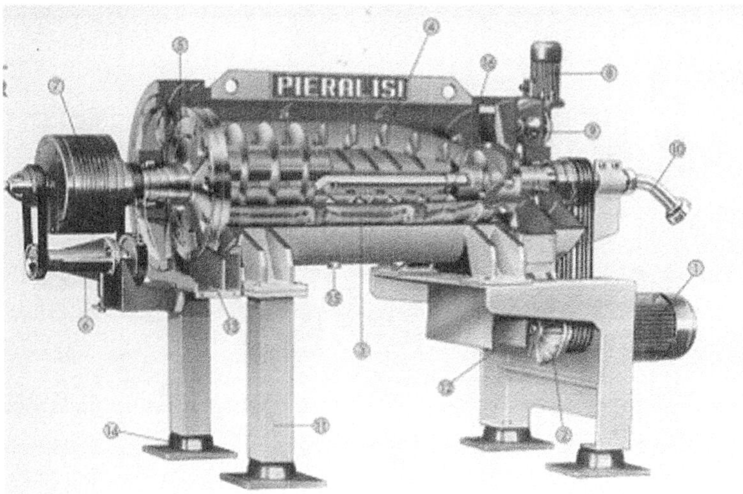

1 - main motor
2 - hydraulic coupling
3 - drum
4 - screw
5 - liquid discharge
6 - transmission
7 - reducer
8 - solids scraper
9 - scraper chamber
10 - feeding tube
11 - supports
12 - solids discharge
13 - liquid discharge
14 - vibration reducers
15 - cleaning drains
16 - washing device

Figure 7.5 Schematic representation of a decanter centrifuge (Pieralisi catalogue).

The centrifuge separates the dewatered sludge cake from the clarified water. This separation occurs due to the difference in density between the sludge solids and the water, like that occurring in a gravity settler. However in this case it utilizes a centrifugal force between 500–3000 times greater than the gravitational force.

The centrifuge has a cylindrical rotating drum which develops the centrifugal force responsible for the separation between the solid and liquid phase. Upon entering the interior of the centrifuge, the sludge is deposited on the cylindrical wall due to this force. Inside the centrifuge, a screw conveyor is mounted, with a small clearance to the wall of the drum. The rotation speed differential between the drum and the screw results in the transport of the solids to the outlet. The more sophisticated centrifuge models allow for automated control of the differential torque, thus providing opportunity for automatic optimizing the operation of the centrifuge, without the need for manual adjustments.

The unit is typically enclosed and can be connected to a vent gas extraction system. Decanter centrifuges can reach a dry solids content in the cake between 15–25%, depending on the quality of the feed sludge and the extent of chemical conditioning. The units are quite expensive but require less maintenance than the belt filter press. A real disadvantage is the amount of noise they produce.

7.3.3 Filter press

Figures 7.6 and 7.7 show respectively a schematic representation and a photograph of an installed filter press. The operation of this equipment is mainly based on pressure generated by the sludge feed pump, providing enough force to pressure the liquid through a filter cloth or lining, while the solids are retained.

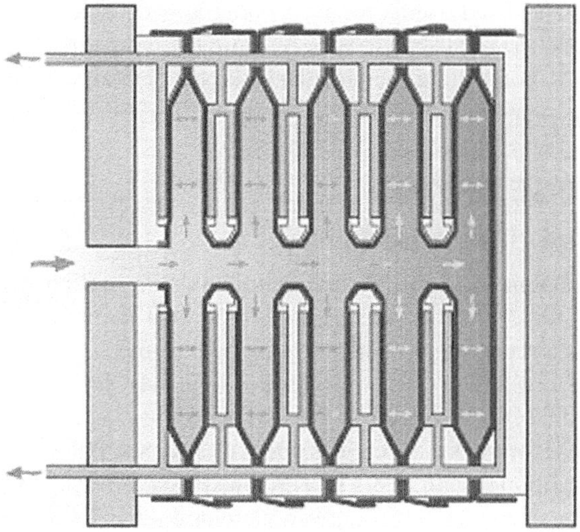

Figure 7.6 Schematic representation of a filter press.

At the start of the filtration cycle, the filtration chambers are filled with sludge and initially the filtrate runs easily through the filter medium, since the medium itself offers little resistance for liquid flow. As filtration continues, a sludge cake layer builds up and thus the pressure drop increases and the filtration rate decreases. As filtration proceeds, the solids retention improves, but as consequence of the decrease in filtration rate, at some point filtration needs to be interrupted. After removal of the accumulated filter cake a new batch cycle can be initiated.

Traditionally ferric chloride and lime have been used for sludge conditioning, which decreases the cake resistance to filtration. While the use of these conditioners has yielded good results in terms of humidity

Figure 7.7 Large filter press at the WWTP of Buarueri, São Paulo, Brazil. Courtesy of Prof. Pedro Alem Sobrinho.

reduction, they have the great disadvantage of incorporating a considerable amount of additional dry solids ('chemical sludge') into the final sludge cake. Because the inorganic chemical conditioning process is also carried out in batches, controlling the chemicals dosage is relatively simple. Alternatively, sludge conditioning can be done using polyelectrolytes.

7.4 NATURAL SLUDGE DRYING

7.4.1 Types of sludge drying beds

Sludge drying beds are different from mechanical solid–liquid separation methods by the fact that the former use no mechanical force to effect the separation, so that there is no consumption of electrical energy. Furthermore the area required for sludge drying beds is much greater than that for the compact mechanical systems. Therefore it is more applicable for relatively small treatment systems in smaller communities, where the availability of land is ample and hence the cost presumably is lower.

Initially sludge drying beds were constructed as open-air systems. Physically these drying beds do not differ substantially from traditional sand filters in a confined area of brickwork. Figure 7.8a shows a plan view at different levels of the drying beds. From the bottom up Figure 7.8 shows left to right the following parts (see also Figure 7.8b for a sectional overview):

(1) A drainage system with perforated PVC pipes in a gravel bed is supported on a concrete or brickwork bottom, constructed with a slight inclination to facilitate the collection and removal of percolated water.
(2) Layers of gravel with decreasing grain size from bottom to top form the base of the drying bed, which serves as a support for the middle layer of sand, which is the actual filter medium of the sludge.
(3) Bricks are placed on top of the upper layer of sand, leaving spaces between the bricks to facilitate sludge removal after drying. The openings between are approximately 1 cm wide and are filled with sand, which allows the percolation of water. Figure 7.9a shows an example.
(4) Excess sludge is discharged on to the bed of sand + bricks. To prevent sludge discharge causing damage to the sand layer, there is usually a concrete slab which serves to absorb the impact of the flow and to ensure uniform sludge distribution (Figure 7.9a). The sludge layer that is applied determines the solids load (kgTSS $\cdot$ m^{-2}), which will be identified as the most important operational

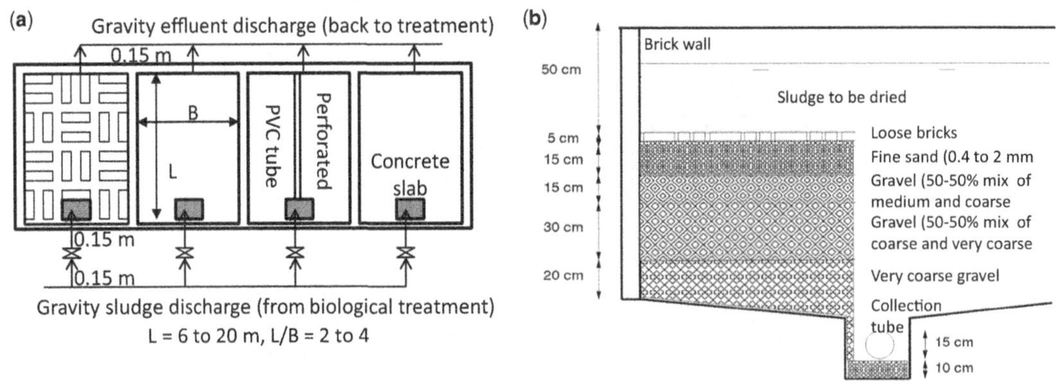

Figure 7.8 (a) Typical layout of a sludge drying bed, (b) Section of the drainage layer of a sludge drying bed.

Figure 7.9 (a) Conventional (open) sludge drying beds (empty) at Ponta Grossa (SANEPAR) (b) Sludge drying bed in operation at Cambuí (SANEPAR).

variable in design and optimization of drying beds. Figure 7.9b shows an example of multiple beds operated in parallel.

(5) The bed is surrounded by an outer wall of masonry or concrete with a height greater than the maximum height of sludge discharged in the unit.

After the application of beds in the open air, covered beds were introduced to keep rain water from mixing with the exposed sludge. Figure 7.10 shows some examples of covered drying beds. This can be an interesting step in regions with a high precipitation rate. To evaluate the feasibility of covering beds, the cost of building the cover must be compared with the advantage of preventing rainfall on the beds. These aspects will be discussed in the following sections.

A third type of bed is the closed variant where the sludge layer is isolated from the outside air by means of a transparent roof and walls as shown in Figure 7.11. Traditionally glass was used for both roof and side walls (as in greenhouses), but the application of plastic materials is becoming more usual, while borosilicate glass is presumed to be the best material for solar energy accumulation. In closed beds solar radiation accumulates as heat in the sludge layer and results in an increase in temperature. If the thermal insulation of the cover is good and radiation intense, the temperature can rise significantly which can even eliminate pathogenic organisms in the sludge, mainly helminth eggs.

Figure 7.10 Covered drying beds (a) before sludge application (left) and in operation (right).

Figure 7.11 Experimental closed drying bed using plastic film. (Courtesy of Prof. Pedro Alem Sobrinho).

Without proper control the rate of evaporation in closed beds will be lower than in open beds, because the cover and walls hinder or even avoid the removal of moist air saturated with water vapour at high temperature. Therefore these beds must have a strategy to increase the temperature to kill pathogens, but also to refresh the air, so that the evaporation is not impaired. The latter may be done by using either the air temperature or the air humidity as a control variable for air ventilation. Usually this is done sequentially: first the water is evaporated and removed as fast as possible and then the temperature can rise. A high temperature is much easier to attain when there is little water present with its very high thermal capacity that works as an impairment against temperature rise. Thus evaporation and solar sterilisation are often carried out sequentially.

A fourth type of drying bed takes into consideration that stratification always develops: the top layer of sludge in the bed is subjected to solar energy and heats up before the lower layers do, thus favouring evaporation in this layer. Also the evaporation rate is relatively high, because the water is closer to the surface and thus easier to desorb. Thus both thermal and material stratification develops with a higher average temperature and lower humidity at the top. The result of this stratification is that the top layer quickly turns into a hard solid layer covering the sludge and reducing the rate of water diffusion from the underlying layers. Thus the evaporation rate in a sludge bed is much smaller than the rate in a comparable layer containing only water for two main reasons: (1) the relatively high temperature of the upper layer of sludge results in the transfer of part of the absorbed solar energy back to the air above the sludge, which therefore is not used for evaporation of water and (2) the difficulty of water diffusion from the lower layers to the surface itself is another limiting factor of the evaporation rate.

Recently equipment has become available to minimize the adverse effects of stratification: the sludge is mechanically tilled, so that the sludge with higher humidity is transported to the top of the layer and the drier sludge is transferred to the bottom. These beds with revolving sludge layers can significantly increase the evaporation rate of the sludge. Turning the sludge layer is only possible when the humidity is lower than 80%, so that this can only be applied after percolation or mechanical dehydration. Normally these beds are closed and the ventilation of air and the consequential release of water vapour are automatically controlled. Figures 7.12 and 7.13 show respectively a schematic representation and a full-scale example of a closed sludge drying bed with equipment for sludge tilling and ventilation.

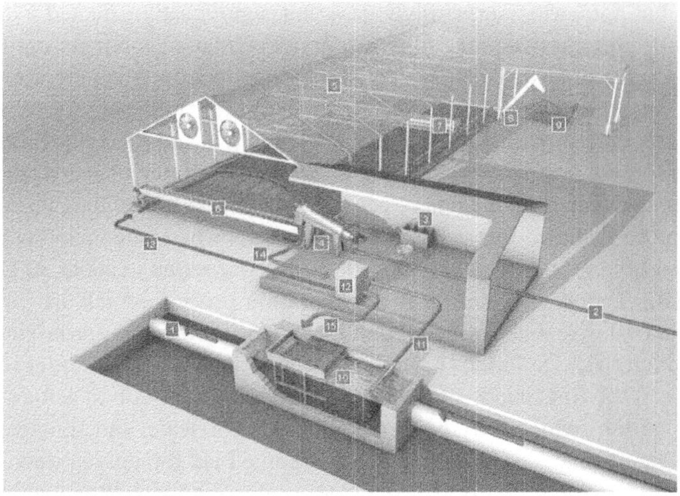

(1) Wastewater treatment plant effluent
(2) Liquid sludge
(3) Polymer station
(4) Sludge dewatering screw pump
(5) Closed bed with sludge tilling and turning
(6) Distribution screw for dewatered sludge
(7) Travelling sludge turner

Figure 7.12 Schematic representation of closed drying bed with mechanical equipment to turn the sludge during the drying process (courtesy Huber Technology).

7.4.2 Experimental determination of required percolation and evaporation time of open sludge drying beds

When sludge is discharged on a bed, the percolation (filtration) process and the evaporation of the water present in the sludge starts immediately. The percolated water is collected by the drain and recirculated to the wastewater treatment system. The percolation rate depends on the nature of the sludge (well stabilized

Figure 7.13 Full-scale example of a closed sludge drying bed with mechanical equipment to turn the sludge during the drying process (courtesy IST Anlagenbau).

or not) and the applied load, but somewhat surprisingly, it is not significantly affected by the initial concentration of the sludge or the particle size of the filter medium. Percolation continues until the solids fraction in the sludge reaches a value of approximately 20%. From this point on, the only way to remove additional water is by evaporation. Evaporation proceeds at a much lower rate than percolation and for this reason it is the process that largely determines the duration of a drying cycle and hence the minimum area required for a particular sludge production. This, even though the amount of percolated water usually is much higher than that of evaporated water. As the rate of percolation is much higher than the evaporation rate, to good approximation one can assume that evaporation only starts when percolation is completed, which is when the residual humidity of the percolated sludge layer has decreased to about 80%.

Knowing that there are many factors involved in the evaporation process which are variable and unpredictable, it is not possible to derive a general equation for dimensioning and optimizing a drying bed. One can however find an empirical solution through an experimental study in order to establish the influence of solids loading on the time needed for percolation and evaporation to achieve a desired final humidity. In this investigation, which may be performed at pilot scale, the time required for evaporation and percolation is determined as a function of the organic load for a given sludge and for a set of climatic conditions. In the following sections such a study is described, and the obtained results are discussed. Later the important aspects of the local climate are discussed: the water evaporation rate and rainfall.

7.4.2.1 Determination of the time required for percolation

The time required for percolation can be established experimentally as a function of the solids load on the sludge bed with the aid of simple equipment consisting of a PVC pipe, which is a model of a drying bed, in which layers of gravel and sand are placed. After the sludge is applied, the volume of percolated water is determined by weighing it as a function of time until percolation is complete. Figure 7.14 shows a photo of the necessary test equipment.

An important aspect to be considered is that the resistance of the filtration medium (sand) is small in comparison to the resistance of the layer of percolated sludge that builds up on the drying bed as percolation proceeds. As the resistance of the filtration is concentrated in the layer of sludge that is already percolated, the available area of the filter medium has no influence on the percolation rate. For this reason it is not counterproductive to place bricks on the bed, as long as sufficient space remains available to drain the percolated water, which is shown in Figure 7.15.

In Figure 7.16 experimental results are presented, where the percolation time (T_2) in days is plotted as a function of the solids load for sludge from a UASB reactor treating raw sewage. To correlate the two parameters, the following empirical equation is suggested to calculate the percolation time:

$$T_2 = C_s^2/320 + 0.5 \tag{7.1}$$

where C_s is the applied solids load (valid for $15 < C_s < 50$ kg TSS·m^{-2}).

It can be observed that the initial sludge concentration does not seem to influence the percolation time. Therefore there is no need to apply methods to thicken the sludge before applying it onto a drying bed: a higher layer of a more dilute suspension can be discharged and percolation will be complete at the same time as a thinner layer of more concentrated sludge, if the applied solids load is equal. Figure 7.16 shows that sludge samples with different initial sludge concentrations all comply with the same empirical curve of percolation time versus applied solids load.

Figure 7.14 Equipment to determine the percolation rate of sludge as a function of the applied load concentration.

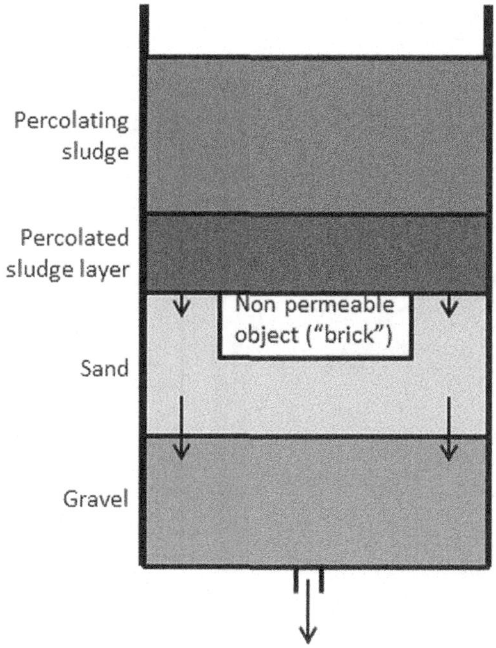

Figure 7.15 Covering part of the medium surface (with bricks) does not affect the percolation rate.

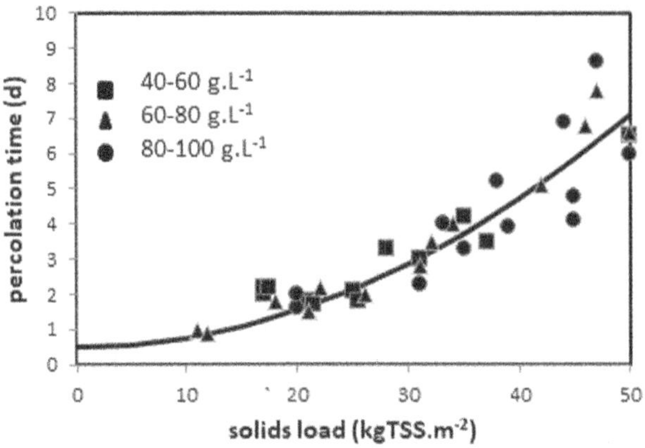

Figure 7.16 Experimental results for the percolation time of batches of anaerobic sludge as function of the applied solids load for different initial sludge concentrations.

7.4.2.2 Determination of the evaporation rate

For rational design of the drying bed area it is necessary to know the water evaporation rate from the sludge, since this process is slow and significantly influences the duration of a complete drying cycle. After completion of the percolation, the evaporation rate can be determined from the weight loss of a model drying bed containing a batch of percolated sludge as shown in Figure 7.17. If after percolation a batch of sludge is placed in a model drying bed, the evaporation rate can be determined by repeated

Figure 7.17 Pilot models of sludge drying beds (plastic rings of 0.25 m diameter) and schematic representation of their operation. Front: pilot models in the open air; back: covered pilot models.

measurements of the weight loss with time. The evaporation rate will depend on the humidity at the time of the determination of the sludge mass loss rate for two reasons which were identified in the previous section: (1) thermal stratification by which part of the absorbed solar energy is reflected back to the atmosphere and therefore is not used for evaporation, and (2) the crust of dry sludge at the top of the sludge batch limiting water evaporation by making it difficult for the water to be transported to the surface, although this is partly countered by the appearance of cracks in the sludge beds as the drying process proceeds.

To evaluate the limitations of the evaporation of water from a sludge batch, the evaporation rate in a batch with sludge can be compared with the water evaporation rate in a bed with only water, equipped with a black bottom for maximum absorption of solar energy. The relative evaporation rate is defined as the ratio between the water evaporation rate from the sludge batch on the bed model and the evaporation rate of the water on an identical model.

By introducing the concept of relative evaporation rate, the influence of the climate (determining the water evaporation rate) is separated from the influence of the sludge batch on the evaporation rate. As the water evaporation rate for a specific region is generally known, then with the aid of the relative evaporation rate it becomes possible to estimate the evaporation rate from a sludge batch, and the influence of the average humidity of the sludge bed at any time during drying on the evaporation rate can be determined.

The drying bed model for evaporation must represent as reliably as possible the evaporation rates, but the size of the model must be limited so that the daily mass losses due to evaporation of water can easily be measured by weighing the model: the weight loss directly translates into a value for the evaporation rate. A suitable solution is to use a small model where both bottom and walls are thermally insulated, for example with layers of polystyrene. When the sludge is being dried, its volume decreases, so that in an initially full bed after a period of evaporation the sludge surface will be well below the height of the side wall, which will have a detrimental effect on the evaporation rate. To eliminate this effect, the side walls are constructed with removable segments. Each segment is removed from the model as soon as it does no longer contain any sludge. Figure 7.17 indicates the operation of the drying bed models.

Table 7.1 shows experimental data of one experimental study to determine the evaporation rate by repeated measurement of the weight of models after different evaporation times (column 1). From the decrease of the mass of the sludge batch in time (column 2) and knowing the dry mass of the applied solids (0.356 kg TSS) and the model area (0.049 m^2), the variation of humidity (column 3) as well as the evaporation rate of water in the bed (column 4) were calculated. Simultaneously also the weight loss of a model with only water was measured, and for this batch the evaporation rate was determined (column 5). Finally the relative evaporation rate (column 6) was calculated as the ratio of the evaporation rates in columns 4 and 5 (with and without sludge). This relative evaporation rate (column 6) can then be related to the residual humidity (column 3) for the applied solids load (7.3 kg TSS $\cdot m^{-2}$).

7.4.2.3 Determination of the relative evaporation rate

Following the above procedure, data were obtained for a series of applied solids loads between 10 and 35 kg $\cdot m^{-2}$. In Figure 7.18 the relative rate of evaporation (R_{rel}) is plotted against the residual sludge humidity (U_e) for several different applied solids loads in open drying bed models. The results clearly show that the relative evaporation rate in open sludge drying beds depends on the residual humidity at any time, but surprisingly it is not dependent on the applied sludge load. The experimental data indicate that to good approximation the relative evaporation rate may be expressed by the following empirical equation:

$$R_{rel} = R_{sb}/R_{H2O} = (1/0.8)U_e = 1.25U_e \tag{7.2}$$

Table 7.1 Weight loss of the sludge batch on a model of an open drying bed with an area of 0.049 m^2 and a solids load of 7.3 kg TSS · m^{-2} (dry solids content of the batch = 0.356 kg TSS).

1	2	3	4	5	6
Evaporation Time (d)	**Batch Mass (kg)**	**Humidity (%)**	**Evaporation Rate (mm · d^{-1})**		**Relative Evaporation Rate (–)**
			With Sludge	**Only Water**	
0	2.542	86.0			
1	2.416	85.3	2.57	1.20	2.14
2	2.262	84.3	3.14	3.77	0.83
3	2.066	82.8	4.00	3.53	1.13
4	1.868	80.9	4.04	3.45	1.17
5	1.656	78.5	4.33	4.24	1.02
6	1.454	75.5	4.12	3.57	1.15
7	1.342	73.5	2.29	2.29	1.00
8	1.066	66.6	5.63	6.12	0.92
9	0.866	58.9	4.08	4.98	0.82
10	0.68	47.7	3.80	5.27	0.72
11	0.49	27.4	3.88	11.08	0.35
12	0.408	12.8	1.67	5.98	0.28
13	0.38	6.3	0.57	3.81	0.15
14	0.378	5.9	0.04	3.37	0.01

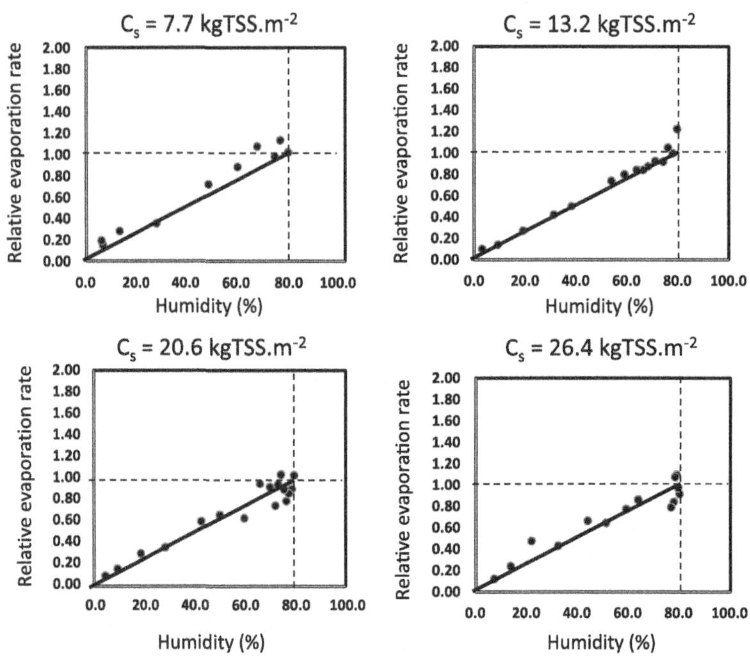

Figure 7.18 Relative evaporation rate as a function of the residual humidity in models of open drying beds for different applied solids loads.

where:

R_{rel} = relative evaporation rate
R_{sb} = evaporation rate in the model with the sludge batch
R_{H2O} = evaporation rate in the model with only water
U_e = average residual humidity of the sludge.

There is a good correlation between the experimental data and the empirical equation over the entire range of humidities from 80% directly after percolation to very low values (<10%) at the end of the drying period, for all the applied solids loads. From the research study the following conclusions can be drawn:

(1) The initial evaporation rate in sludge drying beds is larger than the evaporation rate of pure water under comparable conditions. Possibly this is due to the presence of a thin layer of water over the sludge particles at the surface, so that the available contact area with the air is in fact larger than the flat surface of pure water. Thus the desorption of water from the granular surface is larger than that from a flat surface.
(2) As the humidity in a sludge batch on the bed decreases, so does the evaporation rate, so that the relative water evaporation rate decreases in models with sludge.
(3) The reduction of the evaporation rate as humidity decreases is largely independent of the applied solids load.
(4) For a low residual humidity (10–30%) the evaporation rate on the beds is only 20–35% of the rate observed in beds with only water.

The low evaporation rate in the sludge batch in models of open or covered drying beds is due to their inherent inefficiency. For models that contain only water, the absorbed solar energy is used primarily to promote the evaporation of water while the water temperature rises only marginally. In contrast, for sludge beds solar energy absorption causes stratification of the temperature: during the day time (when evaporation is more intense), the temperature in the top layer is much higher than in the bottom sludge layer. The increased temperature at the sludge surface in turn promotes increased heat irradiation from this level into the atmosphere, reducing the amount of heat available for evaporation of the water. Thus it is apparent that a deficiency in heat conduction promotes the reduction of solar energy effectively absorbed, thereby reducing the evaporation rate of water from the sludge and increasing the time required to achieve a given desired residual humidity. Another aspect is the limitation of the diffusion of water on its way out of the sludge bed. It mostly must diffuse through a layer of sludge that gets denser and more difficult to pass as evaporation proceeds. The sum of these two effects causes the low rate of efficiency that is observed in tests such as described above and reflected quantitatively in Figure 7.18.

7.4.2.4 Determination of the time required for evaporation

The time required for evaporation in an open sludge drying bed can be calculated by considering that the rate of weight loss is equal to the rate of evaporation:

$$(\mathrm{d}P_a/\mathrm{d}t) = R_{sb} = (1/U_i)\, R_{H2O}\, U_e = (1/U_i)\, R_{H2O}\, P_a/(P_a + P_s) \qquad (7.3a)$$

or

$$\int[(p_a + p_s)/p_a]\, \mathrm{d}p_a = \int(1/U_i)\, R_{H2O}\, \mathrm{d}t \qquad (7.3b)$$

For the evaporation of water from a sludge batch with an initial humidity $U_i = 0.8$ to a residual humidity U_e, the solution of the differential equation is:

$$P_{ai} - P_{ae} + P_s \ln(P_{ai}/P_{ae}) = R_{H2O}\, T_3/(1 - U_i) \tag{7.4}$$

This allows expression of the evaporation time T_3 as:

$$T_3 = C_s\, [(U_i - U_e)/(1 - U_i)(1 - U_e) + \ln((1 - U_e)U_i/(1 - U_i)U_e)]/(R_{H2O}/U_i) \tag{7.5}$$

where:

T_3 = required evaporation time to reduce the average sludge humidity from the initial value U_i to the desired final value U_e
P_{ai} = initial weight of water in the batch per unit area
P_{ae} = final weight of water in the batch per unit area
U_i = humidity after percolation (=0.8)
U_e = desired final humidity.

In Figure 7.19 the required evaporation time can be observed as a function of desired residual humidity and for water evaporation rates of 2.5 and 5 mm·d^{-1}, which may be considered respectively as low and high for tropical regions. As can be expected, the time for evaporation increases with an increasing applied solids load.

7.4.3 Determination of time required for evaporation in covered beds

The material used to cover drying beds may either be transparent or not. Transparent materials like plastic sheets, glass or borosilicate have the advantage that solar energy passes at least partially and thus can be used for evaporation of sludge water. Non-transparent material may have the advantage of being more robust and cheaper but on the other hand the solar energy cannot be absorbed by the drying sludge. The suitability of the different materials depends on the climate. If storms are frequent the use of plastic is not feasible and in the case of occasional hailstorms glass can only be used if the thickness is adequate, as in

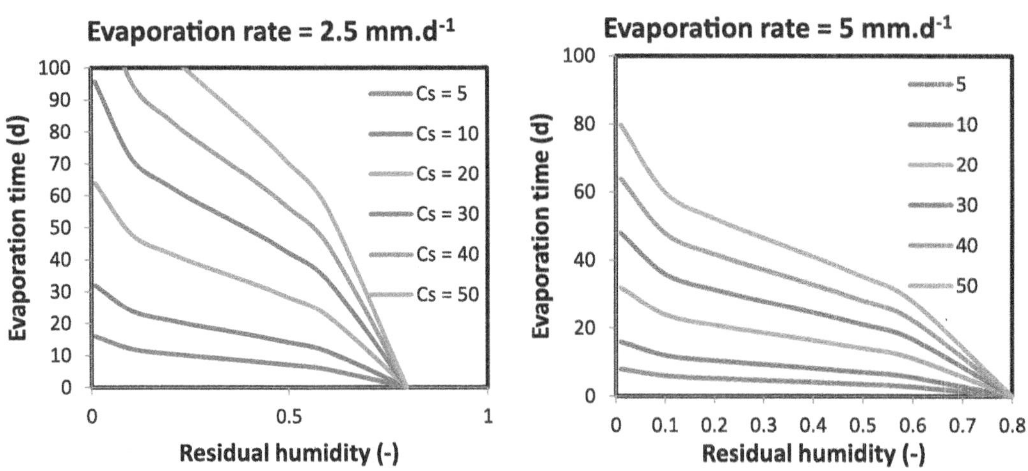

Figure 7.19 Required evaporation time as a function of the desired residual humidity for different applied solids loads and water evaporation rates.

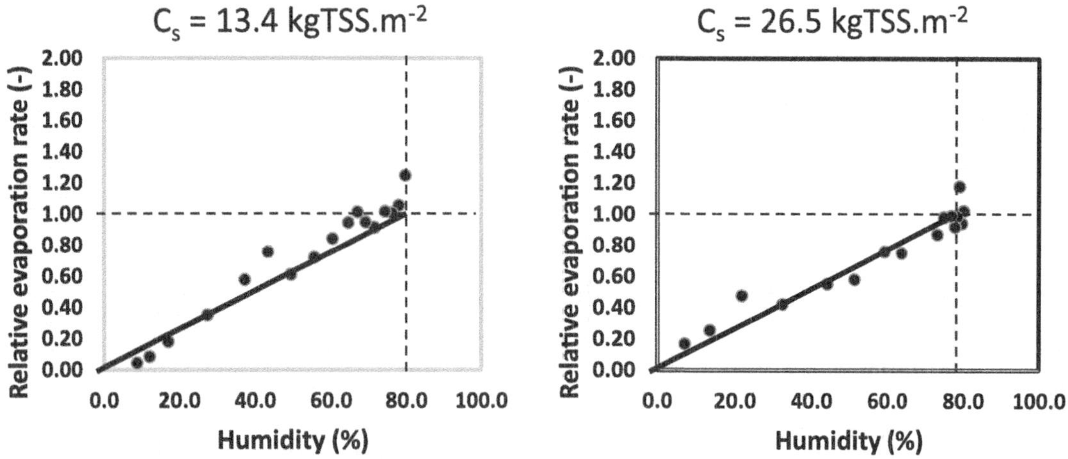

Figure 7.20 Relative evaporation rate as a function of the residual humidity in models of covered drying beds with transparent material, for different applied solids loads.

greenhouses. Borosilicate glass has superior characteristics, but its price is prohibitively high. In view of the problems and costs of transparent materials in practice non-transparent materials are mostly applied as in Figure 7.10, using materials commonly applied in civil construction.

The suitability of applying covered drying beds can be assessed by comparing the positive aspect of protection against rainwater against their negative aspects: the fact that direct solar energy is not – or only partially – used in covered beds, which reduces the evaporation rate, and the higher costs of construction. In Figure 7.20 experimental values of the relative evaporation rate are plotted as a function of the residual humidity for covered drying beds with a transparent cover, for solids loads of 13.4 and 26.5 kg TSS $\cdot$ m^{-2}. The evaporation rate can be described with the same equation as developed for open drying beds – Eq. (7.2). Consequently the formula used to calculate the evaporation time for these beds is also expressed by Eq. (7.5).

In contrast, the evaporation rate for covered beds with opaque (non-transparent) material is much lower. Under the experimental conditions the evaporation rate of beds covered with opaque material and filled only with water was only about 50% of the value in open air. Figure 7.21 shows the relative evaporation rates for these beds. In this case the relative evaporation rate can be described by

$$R_{rel} = R_{sb}/R'_{H2O} = (1/0.8)U_e = 1.25U_e \tag{7.6}$$

where R'_{H2O} is the water evaporation rate from a bed with an opaque cover.

As the water evaporation rate (R'_{H2O}) was only half the value of the rate in an open bed (R_{H2O}), the evaporation rates from sludge batches in beds with an opaque cover will also be only half of the evaporation rates of open beds. It must be remembered that this very significant reduction of 50% was observed under the conditions of the Northeast of Brazil, where sunshine is plentiful and intense. In regions with less favourable conditions the reduction may be less. In any case, covering drying beds with opaque material is much less attractive. To decide if covering the bed with opaque material is advantageous, the fraction of rainwater falling on an open bed that is actually absorbed by the sludge should also be considered. In Section 7.4 it will be shown that a large part of the rainwater percolates quickly through the bed and does not need to be evaporated.

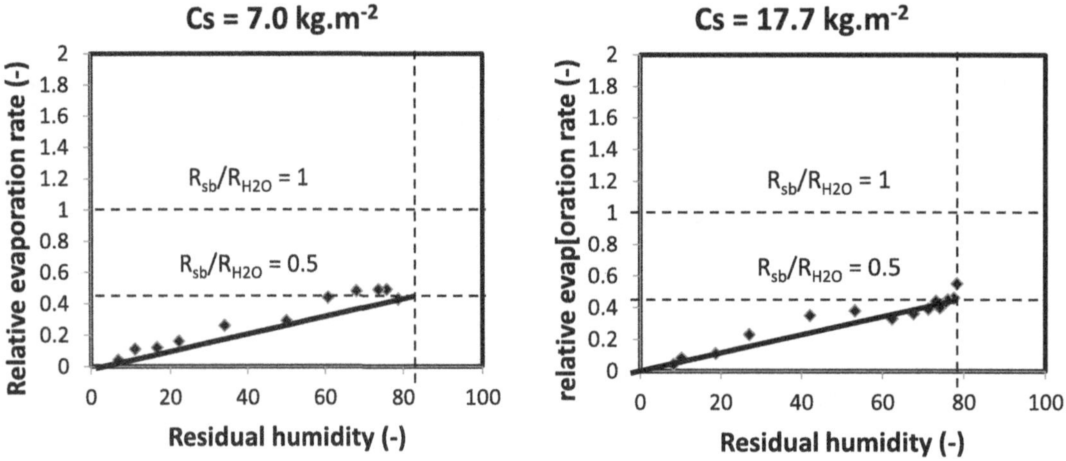

Figure 7.21 Relative evaporation rate as a function of the residual humidity in models of covered drying beds with opaque material, for different applied solids loads.

The time for evaporation for a covered sludge drying bed with opaque material is expressed as:

$$T_3 = C_s[(U_i - U_e)/((1 - U_i)(1 - U_e)) + \ln\big((1 - U_e)U_i/(1 - U_i)U_e)]/(R'_{H2O}/U_i)\big) \tag{7.7}$$

In Figure 7.22 the evaporation times in beds with opaque material are shown as functions of the residual humidity for different solids loads and water evaporation rates. When these diagrams are compared with Figure 7.19, there is indeed a much longer evaporation time required if a cover with opaque material is placed on a drying bed. It can be noted that the time for evaporation in Figure 7.22 is twice the value of Figure 7.19, because the opaque cover reduces the water evaporation rate by 50%.

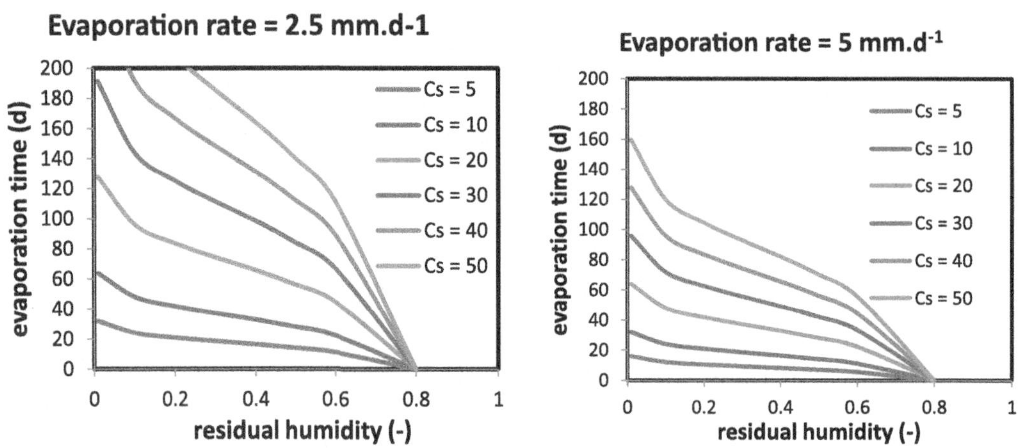

Figure 7.22 Evaporation time as a function of the desired residual humidity for different applied solids loads and water evaporation rates in drying beds with opaque covers, that reduce the water evaporation rate by 50%.

7.4.4 Determination of time required for evaporation in open beds with sludge tilling

If daily sludge tilling is applied, the rate of evaporation is affected as shown in Figure 7.23, where the relative evaporation rates are shown as a function of the humidity of sludge batches for two applied loads: 12.9 and 32.8 kg TSS · m^{-2}. As shown in Figure 7.23, the relative evaporation rate remains essentially constant, at a value equal to or even slightly greater than the evaporation rate of pure water, until the humidity has decreased to 40%. From this point onwards, the evaporation rate gradually decreases to very low values. Thus the relative evaporation rate can be expressed as:

For humidities between 40–80%: $R_{rel} = 1 = R_{sb}/R_{H2O}$

For humidities < 40%: $R_{rel} = (1/0.4)R_{H2O}U_e = 2.5R_{H2O}U_e$ (7.8)

The data in Figure 7.23 reflect operation of the beds where tilling was manually applied as soon as the sludge did not stick, which occurred at a humidity of 0.7.

Figure 7.24 shows the evaporation time as a function of the residual humidity for different loads. When the data is compared to that in Figure 7.19, it is clear that the time for evaporation is reduced. It is quite possible that more frequent and more efficient sludge tilling, as is applied in full scale units, would have led to an even larger reduction of the evaporation time.

Equation (7.9) can be used to calculate the evaporation time as a function of the residual humidity for different values of the applied solids load:

$$T_3 = T_{3a} + T_{3b} \tag{7.9}$$

where:

$T_{3a} = C_s(U_{0.8}/(1 - U_{0.8}) - U_{0.4}/(1 - U_{0.4}))/R_{H2O}$
$T_{3b} = C_s[(U_{0.4} - U_e)/(1 - U_{0.4})(1 - U_e) + \ln((1 - U_e)U_{0.4}/(1 - U_{0.4})U_e)]/(R_{H2O}/U_{0.4})$
T_{3a} = duration of evaporation period when the humidity is greater than 40% and $R_{rel} = 1$
T_{3b} = duration of evaporation period when humidity is between 40% and the desired residual humidity U_e.

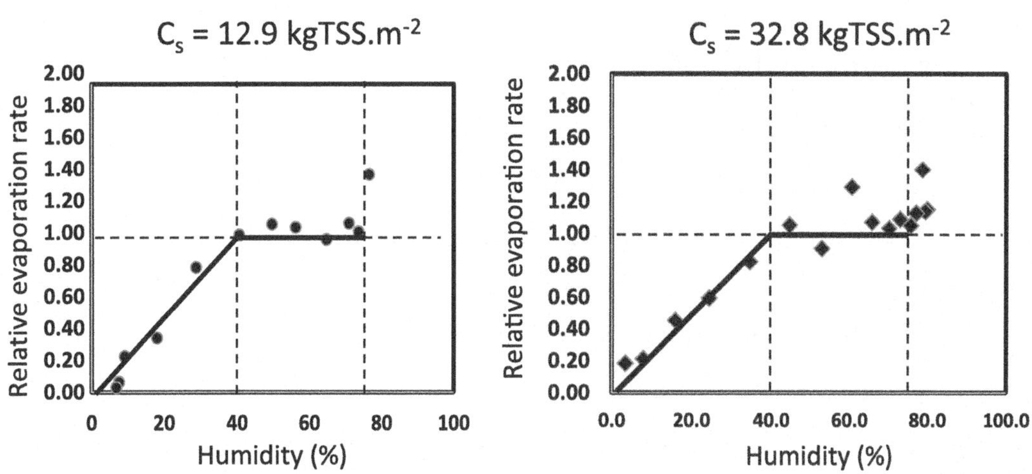

Figure 7.23 Experimental values of the relative evaporation rate as function of the humidity and for different applied solids loads for sludge batches applied on drying bed models when daily sludge tilling is applied.

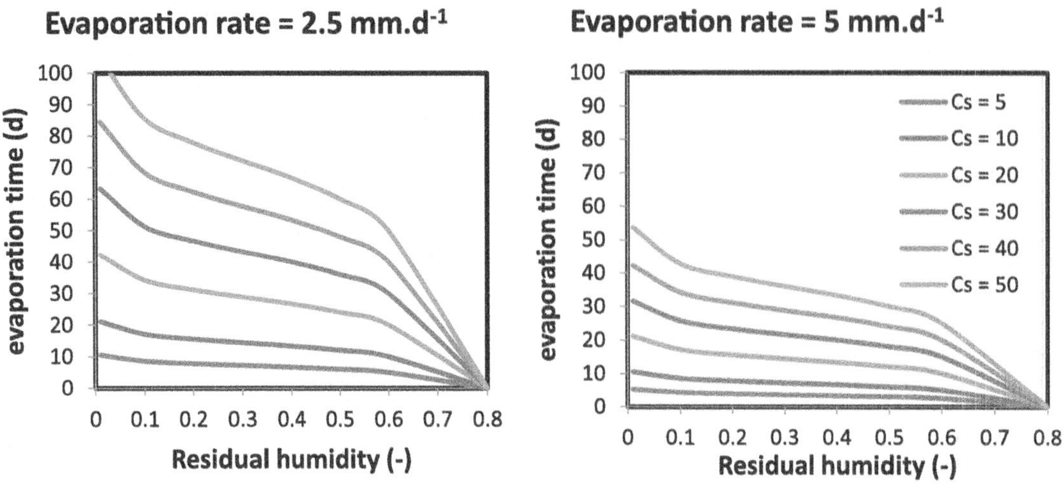

Figure 7.24 Evaporation time for open beds with sludge tilling as function of the residual humidity for different solids loads and water evaporation rates.

7.4.5 Design and optimization of drying beds

7.4.5.1 Determination of the bed productivity

In order to be able to make a rational estimate of the required area for drying beds, it is necessary to know the expected productivity of the bed, defined as the mass of solids that can be processed per unit area and per unit time to produce a sludge with a desired residual humidity. This productivity is defined as the ratio between the applied solids loading and the time required to complete a full drying cycle:

$$P_r = C_s/T_c \tag{7.10}$$

where:

P_r = bed productivity (kg TSS · m^{-2}·d^{-1})
C_s = applied solids load (kg TSS · m^{-2})
T_c = total time of a drying cycle (days).

The total time for a drying cycle consists of four different parts:

T_1 = period required to prepare the bed and discharge the sludge to be dried.
T_2 = percolation period
T_3 = evaporation period
T_4 = period to remove the dried sludge and clean the bed.

$$T_c = T_1 + T_2 + T_3 + T_4 \tag{7.11}$$

The duration of periods T_1 and T_4 will depend primarily on factors related to the availability of equipment and skilled labour to operate the beds and are relatively short. In contrast the duration of periods T_2 and T_3 mainly depend on external factors: the nature of the sludge to be dried and the local weather (humidity, temperature, wind speed).

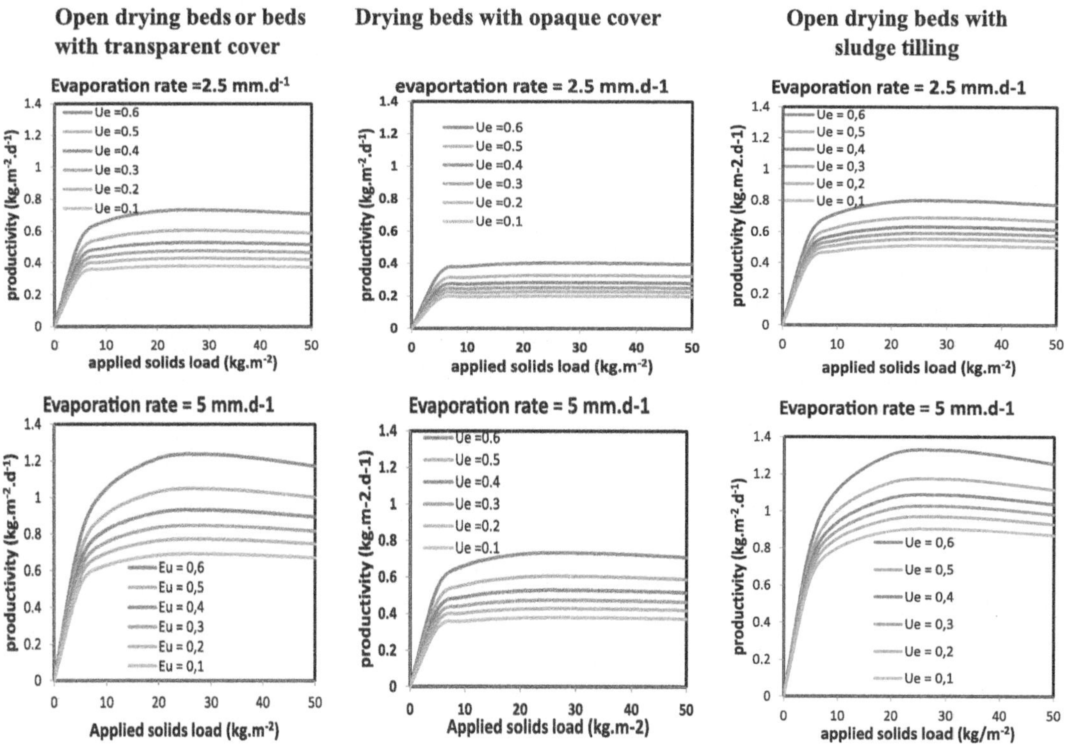

Figure 7.25 Productivity of different types of drying beds as a function of the applied solids load for different residual humidities and for low and high water evaporation rates.

A key operational variable is the solids loading which is the applied mass of dry solids per unit bed area. The percolation time can be estimated from Eq. (7.1) and is independent of the type of drying bed that is used. In the previous section, expressions have been developed for the evaporation time for the different types of drying beds (Eqs. (7.5), (7.7) and (7.9)).

Once the time required to complete a full cycle has been determined, the productivity can be calculated as a function of the solids load for different values of the desired residual humidity. In Figure 7.25 productivity curves have been plotted for the different types of drying beds. It can be observed that:

(1) The productivity curve of a drying bed has a maximum, indicating that there is an optimal solids loading rate. It is clear from the curves that near optimal loads lead to virtually the same productivity.
(2) The rate of water evaporation has a huge influence on the productivity of a drying bed. It is concluded that the area needed to dry sludge depends very strongly on the climate of the region where the drying bed is operating. The influence of rainfall on productivity will be considered later (Section 7.5).
(3) The productivity of the drying bed is strongly influenced by the desired solids fraction after drying. A low residual humidity (<30%) is important, not only to increase the value of the final product as a fertilizer for agricultural use, but also to reduce the costs of transportation to the destination. There might also be a certain reduction in pathogen count when the humidity is very low.

(4) Covering the drying beds with opaque material reduces the productivity of a bed very considerably and is only advantageous if the water evaporation rate is low and the rainfall is plentiful (these two factors usually go together).
(5) Sludge tilling can increase the bed productivity considerably, even when the bed is not covered.

7.4.5.2 Design of open drying beds

After having established the productivity of a drying bed for a desired residual humidity, the steps for optimal design can be summarized as follows:

(1) Determine the daily sludge production in the WWTP. The daily excess sludge production depends on the applied organic loading of the WWTP and the operating conditions and sewage characteristics of the treatment system, notably the sludge age and temperature. The spreadsheet that will be introduced in Chapter 10 calculates the excess sludge production as a function of these variables.
(2) Estimate the minimum required drying bed area, which is defined as the ratio between the daily sludge production and the expected productivity for the desired residual humidity.
(3) Establish the maximum fraction of the sludge mass present in the reactor that can be released in one discharge. For anaerobic reactors this is typically in the range of 30–50% of the sludge mass in the reactor or digester (see also Section 3.8.2).
(4) With the aid of the excess sludge production rate and the maximum discharge fraction, determine the minimum frequency of sludge discharges.
(5) Determine the required number of drying bed beds.

EXAMPLE 7.3

Estimate the number of drying beds and their required area for a sludge drying facility treating the excess sludge produced in a UASB reactor of 1500 m^3, treating 6000 $m^3 \cdot d^{-1}$ wastewater with an organic load of 4500 kg $COD \cdot d^{-1}$. The maximum allowed residual humidity is 30%. Adopt a specific sludge production rate of 0.15 kg $TSS \cdot kg^{-1}$ COD applied and an excess sludge concentration of 60 $g \cdot L^{-1}$. The average concentration in the UASB is 20 $g \cdot L^{-1}$, so that the reactor holds a mass of $1500 \times 20 = 30$ tons of sludge. To ensure stable performance of the UASB reactor discharges, do not admit sludge discharge batches larger than 50% of the available sludge mass. The water evaporation rate is 5 mm per day. The time for filling and removal of dry sludge ($T_1 + T_4$) is given as 3 days. Taking into consideration rainfall and other adverse conditions (including repair and maintenance of the beds), provide for the construction of a reserve capacity of around 50%.

Solution

For the provided waste sludge concentration of 60 kg $TSS \cdot m^{-3}$, the initial sludge layer applied to the sludge bed is equal to $20/60 = 0.33$ m. The sludge mass after drying is $1/(1 - 0.3) = 1.4$ times the dry solids mass (for the desired final humidity of 30%) or $1.4 \times 20 = 28$ $kg \cdot m^{-2}$. Consequently, the dried sludge layer will have a thickness of 0.028 $m^3 \cdot m^{-2}$ or 28 mm, when a dry sludge density of 1 $kg \cdot L^{-1}$. Actually, dry sludge density tends to be considerably lower, in the range of 0.75–0.85 $kg \cdot L^{-1}$.

The sludge age can be calculated based on the sludge production of $0.15 \times 4500 = 675$ kg $TSS \cdot d^{-1}$ and the anaerobic sludge mass present in the system of 30 ton TSS as $30{,}000/675 = 44$ days, which is a suitable value for an anaerobic treatment system at a temperature of 25°C. When Figure 7.25 is evaluated for different evaporation rates and drying beds types, it is concluded that the maximum bed productivity is obtained for an applied solids load of about 30 $kg \cdot m^{-2} \cdot d^{-1}$, although the productivity in the range of 20–40 $kg \cdot m^{-2} \cdot d^{-1}$ is only marginally lower than the maximum value. The calculation procedure requires several steps:

(1) Calculate T_2, the required time for percolation, from Eq. (7.1)
(2) The time for evaporation is calculated with the aid of Eqs. (7.5), (7.7) and (7.9)
  (a) Open bed:

$$T_3 = C_s[(U_i - U_e)/((1 - U_i)(1 - U_e)) + \ln((1 - U_e)U_i/(1 - U_i)U_e)]/(R_{H2O}/U_i) \quad (7.5)$$

  (b) Opaque cover:

$$T_3 = C_s[(U_i - U_e)/((1 - U_i)(1 - U_e)) + \ln\big((1 - U_e)U_i/(1 - U_i)U_e\big)]/(R'_{H2O}/U_i)\big) \quad (7.7)$$

  (c) Sludge tilling:

$$T_3 = T_{3a} + T_{3b}$$

$$T_{3a} = C_s(U_{0.8}/(1 - U_{0.8}) - U_{0.4}/(1 - U_{0.4}))/R_{H2O} \quad (7.9)$$

$$T_{3b} = C_s[(U_{0.4} - U_e)/(1 - U_{0.4})(1 - U_e) + \ln((1 - U_e)U_{0.4}/(1 - U_{0.4}) \cdot U_e)]/(R_{H2O}/U_{0.4})$$

(3) Calculate productivity with the aid of Eq. (7.10) for the different drying beds
(4) Determine the minimum required area of the drying beds as the ratio of the sludge production rate (675 kg·d^{-1}) and the productivity. Table 7.2 shows the values for a high and a low water evaporation rate (2.5 and 5 mm·d^{-1}) and different types of drying beds.

The maximum allowed interval between sludge discharges can be determined from the condition that a single discharge cannot exceed more than 50% of the sludge mass in the reactor. As the daily sludge production is 675 kg TSS·d^{-1} or 675/30,000 = 2.25% of the total sludge mass per day, the maximum period between discharges is 50/2.25 = 22 days.

The only variables that remain to be determined are the frequency of excess sludge discharge from the UASB reactor and the required number of drying beds. Insofar as the discharge frequency is concerned, it is convenient to have a constant interval between discharges, for example weekly or multiples of a week. In this case, since the maximum interval between discharges is 22 d, a discharge frequency of once every three weeks (21 days), or every fortnight (14 days) or every week (7 days) may be adopted.

Taking as example the open drying bed for a water evaporation rate of 2.5 mm·d^{-1}, the total cycle time is presented in Table 7.2 as 63 days. Hence the excess sludge discharged at a solids load of 30 kg·m^{-2} will occupy the drying bed for a period of 63 days or 9 weeks. If an interval of one discharge per 3 weeks is adopted, the minimum number of beds can be calculated as 9/3 = 3; in other words the sludge drying cycle time comprises 9 weeks while every 9 weeks a new batch of excess sludge must be discharged . For a total discharge mass of 21 × 675 = 14,175 kg sludge at a solids load of 30 kg·m^{-2}, the required area is equal to 14,175/30 = 472 m^2.

Table 7.2 Evaporation time (T_3), total cycle time(T_c), productivity (P_r) and minimum area (A_{min}) for a sludge production of 675 kg TSS·d^{-1} for different water evaporation rates and bed types (open, opaque cover and with sludge tilling).

Water Evaporation Rate (low) = 2.5 mm·d^{-1}				
Bed Type	**T_3 (d)**	**T_c (d)**	**P_r (kg·m^{-2}·d^{-1})**	**A_{min} (m^{-2})**
Open air	55.7	63	0.47	1436
Opaque cover	111	119	0.25	2700
Sludge tilling	43.2	51	0.59	1144
Water Evaporation Rate (high) = 5 mm·d–1				
Bed Type	**T_3 (d)**	**T_c (d)**	**P_r (kg·m^{-2}·d^{-1})**	**A_{min} (m^{-2})**
Open air	27.8	23	0.84	804
Opaque cover	55.6	42	0.43	1569
Sludge tilling	21.5	29	1.01	655

Table 7.3 Number of sludge drying beds and their dimensions for different bed types for the conditions of Example 7.3, for different water evaporation rates and for a sludge discharge intervals of 3 weeks.

Type of Bed	Evaporation Rate (low) = 2.5 mm·d^{-1}				Evaporation Rate (high) = 5 mm·d^{-1}			
	# in Operation	# stand-by	Bed Area	Total Area	# in Operation	# stand-by	Bed Area	Total Area
Open	3	2	472	2362	2	1	392	1176
Opaque	6	3	445	4050	3	2	472	2360
Tilling	2	1	382	1910	1	1	472	944

Considering the need for spare capacity, a total of 5 beds of each 472 m^2 are to be built, 3 in operation and 2 as stand-by to cope with adverse conditions, resulting in a total area of 2362 m^2.

If the bed is built with an opaque cover, then according to Table 7.2 the cycle time T_c increases to 119 days. However 119 days is not a multiple of 21 days, which is the time interval for discharges. So the next largest multiple is chosen: 126 d (as 126/7 = 18). For a 3 week discharge cycle the number of beds in operation is now 18/3 = 6, but as the cycle time is now 126 days instead of 119 days, the bed area can be reduced by a factor 126/119 = 1.06 from 472 to 445 m^2. For a reserve capacity of 50% there must be 3 additional beds, so the total number is equal to 9 and the total area is 9 × 445 = 4050 m^2.

If additionally sludge tilling (once per day as per Section 7.3.4) is applied to the open bed, the cycle time T_c = 51 days. Again this is not a multiple of the operation cycle of 21 days so an operation cycle of 63 days (=3 × 21) is selected, requiring 3 beds in operation with an area of (51/63) × 472 = 382 m^2 each. With 2 beds in reserve the total area is now 5 × 382 = 1910 m^2.

In Table 7.3 the number of beds in operation and the stand-by capacity is calculated for the three types of drying beds discussed in this example and for a high (5 mm·d^{-1}) and a low (2.5 mm·d^{-1}) water evaporation rate, for a sludge discharge interval of 21 days. The values for other sludge discharge intervals can be calculated using the procedure as explained above. The results presented in Table 7.3 allow the following to be concluded:

(1) The required area for the sludge drying beds is not very large. In Table 7.3 the total required area for open drying bed is 2362 m^2 when the water evaporation rate is low (2.5 mm·d^{-1}). In this example, for an organic load in the sewage of 4,500 kg COD·d^{-1}, representing 45,000 inhabitants with a per capita contribution of 0.1 kg COD·d^{-1}, the required drying bed area per capita is 2362/45,000 = 0.052 m^2 or 529 cm^2. This can be compared to the required UASB area for an assumed depth of 4 m, which is 1500/4 = 375 m^2, which results in per capita UASB area of 375 m^2/45,000 = 0.0083 m^2. Thus the area of the bed would be about six times larger than the area of the UASB reactor under these unfavourable conditions. If the evaporation rate of water is higher, the bed area can be decreased correspondingly.

(2) Covering the bed with an opaque material increases the required bed area very significantly and is only recommendable if rainfall is intense and frequent (refer to Section 7.5).

(3) Under the conditions of the example, sludge tilling did not result in a significant reduction of the required bed area, but this may be due to a low frequency of tilling (once per day).

7.5 INFLUENCE OF RAINFALL ON DRYING BED PRODUCTIVITY

To analyse the influence of rainfall on drying bed productivity, it is required to calculate the average rate of water removal on a drying bed and compare this value with the additional water load to the bed resulting from rainfall. For a productivity P_r, the rate at which water is applied to the bed (as part of the wet sludge) is given by:

$$W_{ai} = P_r U_i/(1 - U_i). \tag{7.12}$$

Similarly the rate of water removal together with the dried sludge can be expressed as:

$$W_{af} = P_r U_e/(1 - U_e) \tag{7.13}$$

Hence the rate of water removal on the bed is:

$$W_a = W_{ai} - W_{af} = P_r[U_i/(1 - U_i) - U_e/(1 - U_e)] \tag{7.14}$$

where:

W_{ai} = bed loading rate with water
W_{af} = water removal rate from the bed with the dried sludge
W_a = water evaporation rate

EXAMPLE 7.4

Sludge is applied with an initial humidity of 95%, while the required final humidity is 50%. Calculate the rate of water removal in the bed for a productivity of 0.6 kg TSS·m^{-2}·d^{-1}.

Solution

For the productivity of the bed of 0.6 kg TSS·m^{-1}·d^{-1}, the average rate of water removal is given by $P_a = 0.6[0.95/0.05 - 0.5/0.5] = 12$ L·m^{-2}·d^{-1} or 0.012 m·d^{-1} = 12 mm·d^{-1} or around 4,000 mm·$year^{-1}$.

The rate of water removal calculated in above example is very much higher than the annual rainfall precipitation in most regions. One way to compensate for the addition of rainwater on the bed would be to increase the area in the same proportion as the ratio of the rainwater and the annual rate of water removal. So in the above example, for an annual rate of 1500 mm of rain, the area of the bed would need to increase by 1500/4000 = 37%. However, in practice the required correction may be considerably lower, as it can be observed visually that rainwater readily percolates through a sludge drying bed when the exposed sludge is already quite dry (i.e. when most of the water has already been removed by percolation). This is because the capacity of the dry sludge to absorb rainwater is very limited. To analyse the extent of rainwater retention in drying beds an experimental study was carried out with bed models like those shown in Figure 7.17, but with a discharge point at the bottom, so that any water could flow out of the bed (Moreira *et al.*, 2006). It was observed that when the sludge is still in fluid form (above 80% humidity), rainwater simply will mix with the sludge so that the water retention is 100%. Therefore the study was limited to evaluation of already percolated sludge with different initial humidities and at two solids loading rates: 20 and 40 kg TSS · m^{-2}. To measure the percentage of rainfall water retained, bed models constructed in plywood were used, which absorbs no water, with an area of 0.075 m^2, in which percolated sludge was applied with an initial humidity of 80%. On the perforated bottom of the beds a layer of geotextile (Bidim XT-4) was placed to replace the drainage system (gravel and sand) of a normal drying bed. To obtain sludge of different humidities the following experimental procedure was adopted:

(1) Initially for each solids loading rate to be studied (20 and 40 kg TSS · m^{-2}), a series of seven beds was prepared, using percolated sludge with a high humidity (80%), the value of which was determined experimentally. The weight of solids was then calculated as:

$$M_s = M_1(1 - U_i) \tag{7.15}$$

where:

M_s = weight of the applied solids
M_l = initial sludge mass (with water)
U_i = initial humidity of the sludge samples in the beds.

(2) Based on the initial weight and humidity of the sludge samples on the beds, it was calculated what the weight would become after drying the sludge until the desired final humidity was reached (in the range of 45–75%), using the following equation:

$$M_d = M_s/(1 - U_d) \tag{7.16}$$

where:

U_d = desired humidity after drying and weight (range of 45–75%)
M_d = weight of the sample when dried to desired test humidity.

(3) Evaporation of water from the beds was continued until the sample weight M_d was attained that corresponded to the final humidity desired for the beds.
(4) When this weight was attained, the drying bed was wrapped in a plastic bag that did not allow additional evaporation, thus preserving the residual humidity at the desired value;
(5) By doing so the crust at the top of the bed was preserved. This was shown to be critically important: without the hard and cracked top layer, much less water percolated.
(6) After this preparatory procedure seven drying beds had been prepared, each of which had been subjected to the same solids loading rate (series with 20 or 40 kg TSS $\cdot$ m^{-2}), but with varying humidities: 75, 70, 65, 60, 55, 50 and 45%.

To evaluate the retention of rainfall water on beds with different humidities, rainfall was simulated by sprinkling on each bed of 0.075 m^2 a water volume of 750 ml, representing 10 mm of 'rainfall'. Then the mass of percolating water was determined from the measurement of the mass of the drying bed over time. During the experiment the weight of each bed was measured six times per day. The first measurement was done before the addition of the 750 mL of water, the second one hour after, and the third, fourth and fifth after 3, 7 and 11 hours and finally the sixth measurement was performed just before 24 hours, when a new artificial 'rainfall event' was scheduled. During the experiment, four consecutive days of rainfall were simulated on all beds.

The effect of the successive rainfall events, each representing 10 mm of artificial rain, was evaluated by measurement of the water volume collected below the beds, which allowed calculation of the percolated water layer (i.e. the percolated fraction of the added water volume), as a function of time for different initial sludge humidities. Figure 7.26 shows the water column height collected after the rainfall events for different initial humidities and for applied loads of 20 kg TSS $\cdot$ m^{-2} (top) and 40 kg TSS $\cdot$ m^{-2} (bottom).

It can be observed that most of the 10 mm of 'rain' that was applied to the bed quickly percolated, at least when the humidity was lower than 65%. In contrast, much less percolated through the bed for humidities higher than 70%, especially at the higher value of the solids loading rate (40 kg TSS $\cdot$ m^{-2}).

Figure 7.27 summarizes the results from the experimental study for the full range of humidities evaluated (40–75%). For both loads the rate of rainwater accumulation was low for a humidity below 65%, whereas for higher humidities the rainwater will almost completely accumulate. The humidity of 65% represents the transition value of a fluid to a semi-solid cake phase as shown in Figure 7.1. Whereas for values lower than 65% most rainwater quickly percolates through the semi-solid cake, above 65% humidity the rainwater mixes with the fluid sludge and accumulates. It is also observed in Figure 7.27 that 80–90% of the rainwater percolates at low sludge humidity, with the higher value corresponding to the lower solids load.

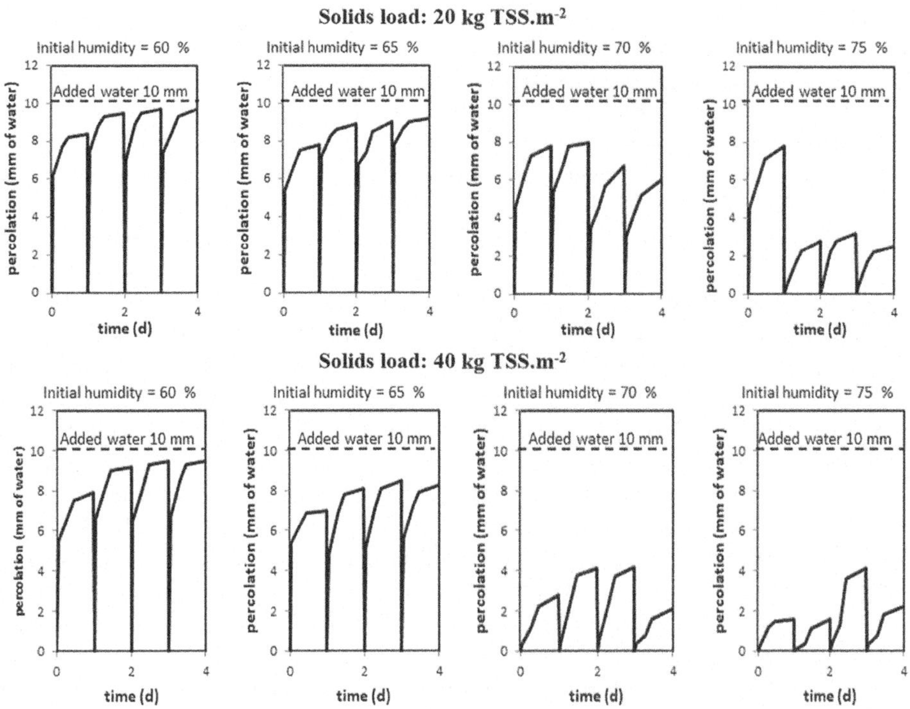

Figure 7.26 Fraction of rainfall percolating through the bed (expressed in mm water percolated) in less than 24 hours after application of consecutive artificial daily rain of 10 mm, for solids loads of 20 (top) and 40 kg TSS·m^{-2} (bottom) and for sludge humidities ranging between 60–75%.

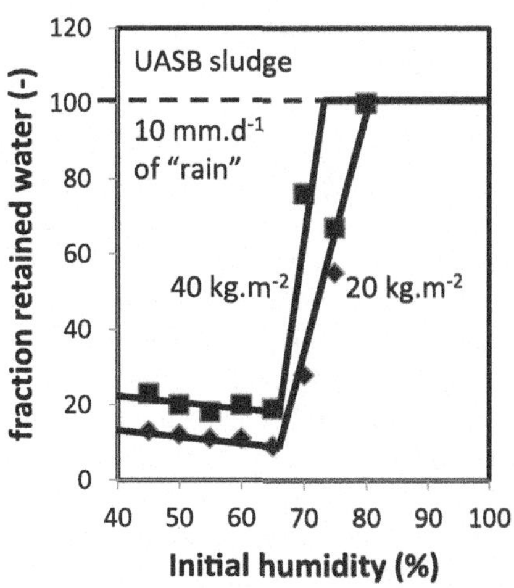

Figure 7.27 Fraction of retained water after artificial rain (10 mm) on sludges in drying beds as a function of the humidity at the start of the rainfall event.

The percolation time is also important. Figure 7.26 shows that about 80% of the percolated water is collected during the first hour following the rainfall event and 95% during the first 7 hours. The percolation of rainwater ceases completely after 24 hours for all beds with sludge humidity less than 65%.

As seen in Figure 7.27, the rainwater fraction that percolates in drying beds tends to be higher as the applied solids load is smaller, but the influence of the loading rate is not strong. The experimentally determined fraction of retained rainwater varied between 12% for 20 kg $TSS \cdot m^{-2}$ and 18% for 40 kg $TSS \cdot m^{-2}$ for humidities below 65%.

The experimental results clearly indicate that the influence of rain on the sludge drying time is very limited.

If the humidity is less than 65%, the fraction of the precipitation that will percolate is at least 80%. In practice the sludge discharged on a drying bed will have an initial humidity of 90–98% and percolation occurs relatively quickly, removing most of the water. When the humidity reaches approximately 80% percolation ceases and evaporation becomes the only drying process. The percolation time depends on the type of bed and the applied load, as exemplified in Eqs. (7.5), (7.7) and (7.9). Previously it was shown that the productivity of a bed (without rain) does not vary much for a wide range of applied loads (Figure 7.25). Knowing that in case of frequent rainfall it is important that the drying sludge has a humidity lower than 65%, it is important to minimize the time necessary for the sludge humidity to decrease from its initial value to less than 65%. Naturally this period decreases when the applied solids load is smaller. Thus it is advantageous to apply a relatively low solids load, especially when rainfall is abundant. Based on the above analysis it is suggested that in regions with frequent rainfall the sludge drying beds are equipped with moving covers, similar to the way these covers are used for cacao drying: the cover is placed over the bed directly after receipt of a sludge batch, and this cover would be moved to a different bed when this in turn receives the next batch of excess sludge. This way, the beds in early stage of drying, where the humidity is still above 65% would be covered, preventing interference of rainfall, while in the remaining beds (typically with humidities < 65%), the rainwater would largely percolate.

7.6 THERMAL SLUDGE DRYING AND INCINERATION

When drying beds are completely enclosed with transparent material (Figures 7.12 and 7.13), more of the solar heat is retained resulting in an increase of temperature of both the sludge mass and the enclosed air in the bed, resulting in a higher evaporation rate and a higher saturation pressure of water vapour in the air. The temperature will tend towards an equilibrium in which the absorbed solar energy equals the energy loss by heat irradiation from the hot sludge and the enclosed air. In the air the vapour pressure will tend towards the saturation value corresponding to the air temperature and when equilibrium is reached, further evaporation will stop. From this point onwards, additional water can only evaporate if the air in the confined space is refreshed, which will reduce the temperature, but is necessary to allow for continued evaporation. The ventilation rate of air is a key control variable.

Depending on the potential use and the cost of final disposal of the dried sludge, it may be interesting to increase the solids fraction beyond the value that can be used by mechanical methods (15–25%). If the required area for drying beds is not available, thermal drying may be applied, which is a process that produces a final sludge with no or very little humidity. It requires an external energy source, which can supply the heat required to evaporate the water from the sludge cake. Thermal drying can be completed with combustion of the dried sludge with air in an incinerator. In that case the energy source of the process is the combustion heat of the volatile sludge itself (Vaxelaire *et al.*, 2010).

The potential for producing electrical energy from the dried sludge is limited compared to that from biogas generation as the following analysis indicates: in anaerobic treatment systems roughly two-thirds of the influent organic material is transformed into biogas and the remainder is more or less equally divided between effluent and sludge, both being one-sixth of the influent organic load. Therefore the fraction of organic material transformed into biogas (methane) is about four times greater than the fraction transformed into volatile sludge. When it is desired to use energy to generate a usable product like electric energy the difference becomes even greater for several reasons.

The first reason is that the efficiency of transforming biogas into electrical power is greater than that of sludge, because biogas is a gaseous fuel that can be used in an Otto-type explosion engine with a typical efficiency of 35–40%. In contrast, in the case of sludge combustion it is necessary to combine a boiler (to generate steam for drying) with a turbine (to generate electricity). The efficiency of the boiler–turbine combination is at most about 15%.

Knowing that the chemical energy per unit mass of COD is around 13.7 MJ (see Figure 1.10), it is estimated that per unit mass of COD applied the potential of electricity generation is $2/3 \times (0.35\text{–}0.40) \times 13.7 \approx 3.4\ \mathrm{MJ \cdot kg^{-1}}$ COD, while the energy generated from sludge is no more than $1/6 \times 0.15 \times 13.7 = 0.34\ \mathrm{MJ \cdot kg^{-1}}$ COD.

Therefore, it is concluded that the potential for electrical energy generation from the biogas is about ten times higher than that generated from the sludge combustion. In fact the available sludge combustion energy is even lower because the sludge after drying still has a residual humidity, which must be removed and consumes around $2.5\ \mathrm{MJ \cdot kg^{-1}\ H_2O}$. Thus if the sludge after drying has a humidity U_e and a volatile fraction f_v, the available combustion heat is calculated by:

$$Q_c = (1 - U_e) f_v f_{cv} \times 13.7/6\ \mathrm{MJ \cdot kg^{-1}\ COD} \tag{7.17}$$

where f_{cv} is the ratio of COD to VSS or biomass ($=1.5\ \mathrm{g\ VSS \cdot g^{-1}\ COD}$).

On the other hand, the heat required to evaporate the water from the dried sludge with a residual humidity U_e is given by:

$$Q_e = 2.5 U_e/6(1 - f_v)\ \mathrm{MJ/kg^{-1}\ COD}. \tag{7.18}$$

In Figure 7.28, the heat production by combustion (Q_c) from dried sludge and the heat demand from water evaporation (Q_e) are plotted as functions of the residual humidity U_e for organic sludge fractions of 60 and 70%. Also indicated are the Q_c curve at 80% heat utilization efficiency and the net heat production,

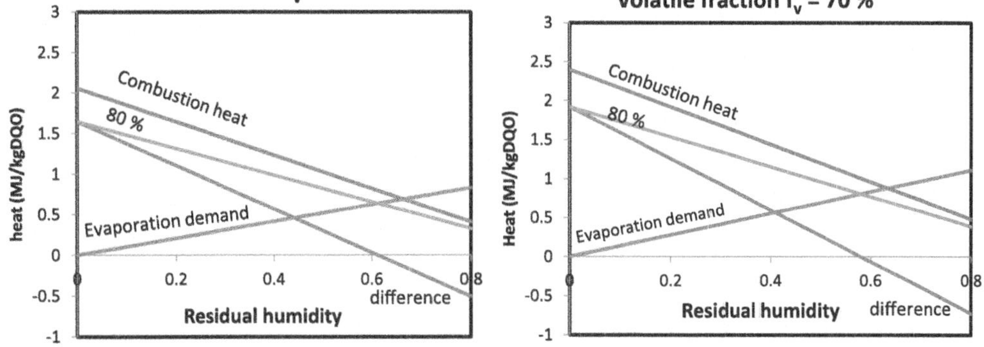

Figure 7.28 Evaporation and combustion energy upon combustion of sludge cake as a function of humidity. The energy balance for 80% efficiency is also indicated, showing the need for external energy for a residual humidity of more than about 60%.

i.e. the difference between the effective heat generated considering the efficiency and the heat demand for evaporation. It can be observed that the net heat production is positive for a residual humidity of about 60%, i.e. the heat generated is enough to sustain sludge combustion when the residual moisture is less than 60%. This value can be easily obtained in drying beds, but not with the aid of mechanical methods of sludge dehydration alone.

EXAMPLE 7.5

Evaluate if the biogas production in a UASB reactor is enough to further dry (to 100% dry solids) the produced sludge after initial mechanical dehydration to 20% dry solids (80% humidity). Adopt a sludge production of 0.1 g $VSS \cdot g^{-1}$ COD, an organic fraction of 0.67 g $VSS \cdot g^{-1}$ TSS and a digestion efficiency of 70%. Evaporation heat of water is 2.5 $MJ \cdot L^{-1}$ and combustion heat is 13.7 $MJ \cdot kg^{-1}$ COD.

Solution

For the residual sludge humidity of 80% after mechanical dewatering and a solids production of 0.1 g VSS or 0.15 g TSS per g COD, this results in a production of $0.15/(1 - 0.8) = 0.75$ kg of wet sludge (20% DS) per kg COD applied. The required heat for evaporation of $(0.75 - 0.15) = 0.6$ kg H_2O per kg COD is $0.6 \times 2.5 = 1.5$ MJ per kg COD.

On the other hand the digested COD mass is 0.7 g per g COD, and assuming all methane is recovered this will generate a combustion heat for the formed methane of $0.7 \times 13.7 \approx 10$ $MJ \cdot kg^{-1}$ COD. However typically only about 70% of this biogas is captured and used, so that a more realistic energy production of 7 $MJ \cdot kg^{-1}$ COD is estimated. It is concluded that the combustion heat of the biogas is about five times greater than the energy demand for evaporation of the water in the sludge cake, so that even with a low heat transmission efficiency it should be possible to dry the sludge using only the combustion heat of the biogas. However, this does not seem to be applied in practice.

EXAMPLE 7.6

Evaluate if the heat production from sludge combustion from a UASB reactor is enough to have auto-sufficiency in an incinerator. Adopt the same assumptions as in the previous example.

Solution

The required heat for evaporation is the same as in the previous example. The combustion heat depends on the fraction of the sludge that is volatile. In Figure 7.28 the evaporation heat and the combustion heat are shown as functions of the sludge humidity. Also shown in Figure 7.28 is the energy balance for an assumed energy transmission efficiency of 80%. The figures show that under these conditions the maximum humidity of the sludge is about 60%, depending on the VSS/TSS ratio, which is much less than the humidity obtained from mechanical dehydration. So some external energy source must be applied if it is desired that completely dry sludge is produced for maximum heat production by combustion. This external energy can be biogas, in which case the energy balance is positive even for a very high initial humidity of the sludge.

EXAMPLE 7.7

After drying, the dried sludge is combusted to generate steam to feed a turbine for the production of electric energy. Estimate the production potential per capita using a top quality boiler–turbine combination, which has an efficiency of 16%. The combustion heat of volatile sludge is 13.7 $MJ \cdot kg^{-1}$ COD = 20 $MJ \cdot kg^{-1}$ VSS. Volatile sludge production is 0.1 g $VSS \cdot g^{-1}$ COD. Per capita contribution: 100 g $COD \cdot d^{-1}$.

Solution

The volatile sludge production per capita is 100 × 0.1/1000 = 0.01 kg VSS·inh^{-1}·d^{-1} with a per capita combustion heat of 20 × 0.01 = 0.2 MJ·d^{-1} = 2.3 W. If the conversion efficiency is 16%, the per capita production potential is 0.16 × 2.3 = 0.37 W. This production potential is much smaller than the potential from gas, not only because the organic material converted into sludge is much smaller (roughly 16% of the influent COD against 68% into methane), but also because the conversion efficiency of the solid fuel (16% for dry sludge) is much smaller than that for the gaseous fuel (40% for methane).

An important question that now arises is what would actually be a good value for the desired residual humidity of the sludge after drying? Naturally, the ideal situation would be that all water would be removed from the sludge, but this may take a very long time and therefore a large area is required, which may not be available or have a very high cost. The desired residual humidity also depends on the destination of the dried sludge: if the sludge can be discharged on land as an organic fertilizer the hygienic quality is important, which is favoured by a low humidity. Also the agricultural area may be far away from large urban centres where much sludge is produced, so that transport of the sludge is an important cost issue. In practice in order to stimulate the use of sludge as an organic fertilizer it is often mixed with lime, so that the final product is not only a fertilizer but is also instrumental for correction of the pH of the soil.

If sludge can be disposed of in landfills, the overriding consideration for the ideal residual humidity is economic: a low humidity means a relatively low cost for transport and discharge of the dry sludge in the landfill, but the cost of producing the sludge is higher as the humidity is smaller, so that there is an economic optimum.

If sludge combustion is applied, the ideal final humidity of the dried sludge again depends on the use: if combustion is only applied to reduce the sludge mass that is produced, the residual humidity may be high, as long as it is smaller than 60% (Figure 7.28); but if the thermal energy is to be used for economic purposes like electricity generation, it is important that the humidity of the dried sludge is as low as possible. It is concluded that the ideal value for the residual sludge humidity depends on the intended use and on several local issues related to economic considerations, such as the cost of land for a sludge drying bed and the cost of transport of the dry sludge to its destination.

It is also important to consider that the legal limitation for the use of dry sludge may change with time, particularly in developing countries. While in these countries agricultural use or discharge in landfills is the rule and sludge combustion is highly exceptional, in developing countries legal limitations on agricultural use and discharge in landfills are stricter, which make combustion the rule rather than the exception. It is likely that developing countries will follow suit and place restrictions on the use of sludge on land or disposal in landfills.

Chapter 8

Aerobic post-treatment of anaerobic effluents

8.1 INTRODUCTION

The UASB reactor has been successfully used for the treatment of sewage, especially in hot climate regions (van Haandel & Lettinga, 1994; van Lier *et al.*, 2010). Although it is efficient, often the concentrations of biodegradable organic material in terms of BOD and total suspended solids (TSS) in the UASB reactor effluent need to be reduced to comply with environmental standards. In such cases there will be a need for post-treatment. In this chapter the applicability of aerobic treatment of effluent from UASB reactors and other anaerobic systems is examined. In aerobic wastewater treatment systems it is possible to obtain a good effluent quality with low concentrations of organic material and suspended solids, but both the investment and operating costs of such systems are higher than those of anaerobic systems. One can then consider the possibility of introducing an anaerobic pre-treatment to reduce costs and yet maintain the high effluent quality that allows reuse or discharge of the effluent without adverse environmental impact.

Unfortunately, anaerobic pre-treatment has a major impact on the feasibility of biological nutrient removal in aerobic post-treatment. There is insufficient organic material in the UASB effluent for classic nitrogen removal, and phosphorus removal is even less feasible. In principle, there is the possibility of applying a relatively new process: autotrophic denitrification. In this process partial nitrification is applied on the UASB effluent, converting part of the ammonia to nitrite. This nitrite is used to oxidize the remaining ammonia to molecular nitrogen, but requires the mediation of specific bacteria: Anammox (Mulder *et al.*, 1995). The difficulty of this process is to carry out nitritation (to NO_2^-) without subsequent nitration (to NO_3^-). It has not yet been possible to reliably produce nitrite without nitrate at lower temperatures (<15–20°C), but nitrification kinetics show that at higher temperatures production of nitrite without nitrate is feasible. Up to now the Anammox process has been mainly applied to sidestreams of wastewater treatment systems, like the liquid phase (reject water) of sludge digesters, although there are also several references on anaerobically pre-treated industrial wastewater. In both cases however, the reactor is operated at a temperature of 30–35°C.

 Anaerobic Sewage Treatment: Optimization of Process and Physical Design of Anaerobic and Complementary Processes
Editor(s): Adrianus van Haandel and Jeroen van der Lubbe
doi: 10.2166/9781780409627_0293

In aerobic treatment systems the oxygen demand is satisfied by mechanical aeration, transferring atmospheric oxygen to the liquid phase. On this point aerobic systems differ from polishing ponds, where oxygen is generated in the liquid phase by algae (or by diffusion from the atmosphere), as will be shown in Chapter 9. Aerobic treatment is convenient when, owing to lack of space, it is not possible to use photosynthetic ponds, or when the presence of algae in the effluent is undesirable. Lack of an area of proper size and topography is the problem commonly encountered in urban regions with large population. Algae in the effluent are undesirable when reuse of the effluent is foreseen as industrial water or when the effluent is released into a brackish water estuary where the algae from the pond cannot survive and, upon dying, will release organic material (BOD) that can damage the ecosystem of the receiving body.

8.2 TYPES OF AEROBIC TREATMENT

Aerobic systems can be divided into different types according to the principle of aeration that is applied. There are three basic system configurations, which currently are all applied for the post-treatment of anaerobic effluent: (1) activated sludge systems, (2) bio- or trickling filters, and (3) submerged aerobic filters. In Figure 8.1 the basic flow schemes of the three basic alternatives are presented. In activated sludge systems, the most widely applied aerobic treatment configuration, the sludge is contained in one or more reactors where the oxygen is transferred by aerators and the sludge is present as the suspension of macroscopic flocs, with a density greater than water. The suspension of sludge flocs in the liquid phase is called mixed liquor. The density difference is used to separate the flocs from the liquid phase prior to discharge of the effluent by the application of simple settling.

In the biofilter and the submerged aerobic filter the sludge is fixed to support material, forming a film on the surface of the macro bodies present in the reactor. When in the latter two systems parts of the biofilm are sheared from the support material, they are carried by the liquid to the final settler where separation of the solids takes place. The retained flocs are returned to the reactor and the liquid effluent is discharged essentially without suspended solids. As the sludge concentration in the liquid phase of the fixed-film systems is much lower than in systems with suspended biomass (activated sludge), settling is relatively simple.

The main problem with fixed-film systems is that it is not possible to directly influence the mass of sludge that remains in the system, which is the most important variable in system performance. The sludge age, which is defined as the ratio of retained aerobic sludge mass in the reactor to the daily discharge, is the fundamental parameter of all biological treatment systems. In this regard the activated sludge system has a decisive advantage because the sludge mass can be controlled by planned sludge discharges. This fact explains why the activated sludge system is much more widely applied than aerobic systems with fixed sludge film.

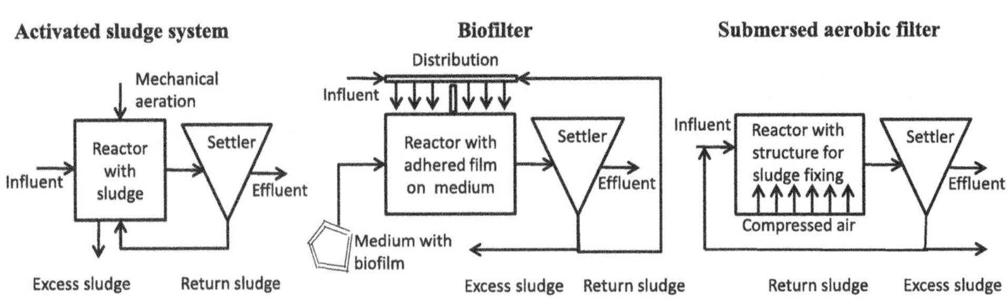

Figure 8.1 Schematic representation of basic aerobic treatment systems.

8.2.1 Activated sludge systems (AS)

In activated sludge systems (Figure 8.1, left), settling can occur in the reactor, by temporarily interrupting the aeration, or in a separate unit, the final settler. The first option is used in systems with sequential batch feeding (SBR systems), while the second method is applied in systems with continuous feed. The settled sludge (the return sludge) is returned to the reactor and the effluent, essentially free from solids, is discharged. Owing to the mechanism of sludge retention by settling, the residence time of the solids – known as the sludge age – is higher than the liquid retention time. To compensate for the sludge growth in the aerobic system, there is a requirement for waste or excess sludge discharge to avoid the biomass concentration of the mixed liquor increasing to a point where separation by settling becomes difficult. The purpose of excess sludge discharge is to maintain an adequate values of sludge age and sludge concentration in the treatment system. As both sludge growth and yield in aerobic systems are much higher than in anaerobic systems, excess sludge discharges are more frequent; in many cases daily or even continuous sludge discharges are applied. The sludge age is defined as the ratio between sludge mass present in the system and the daily excess sludge discharge:

$$R_s = MX_t/ME_t \tag{8.1}$$

where:

R_s = sludge age
MX_t = sludge mass in the activated sludge system
ME_t = excess sludge mass discharged from the system.

Aeration can be applied with compressed air introduced as bubbles at the bottom of the reactor or with surface aerators which introduce air bubbles into the liquid phase as they drag drops from the liquid phase into the air thus effecting the oxygen transfer. In current practice the application of compressed air prevails, because these systems tend to have a higher oxygen transfer efficiency. Aeration not only provides oxygen for bacterial metabolism, but also supplies enough turbulence to keep the sludge flocs in suspension. In SBR systems the aeration is stopped at the end of the aeration cycle, when biological treatment is complete, allowing sludge settling in the reactor itself, so that no settler is necessary which simplifies the construction and reduces the costs.

As in the case of anaerobic treatment (Section 3.9), the parameter of fundamental importance to describe the behaviour of aerobic systems is the sludge age (Van Haandel & Van der Lubbe, 2012). The optimum value of this variable depends on the purpose of the post-treatment, which may be only the removal of the residual concentrations of TSS, BOD and COD from the UASB reactor effluent or it may include oxidation of ammonia through the nitrification process. For the removal of organic material and suspended solids, the required sludge age may be short, in the range of 3–5 days depending on the temperature. The minimum sludge age for nitrification depends heavily on temperature, but in general will not be less than 8–10 days in regions of tropical climate (sewage temperature >18°C).

8.2.2 Biofilter (BF)

In the biofilter (Figure 8.1, centre), aeration occurs when the influent is distributed (sprayed) on a bed filled with filter medium. The filter medium is generally composed of large stones of size 5–10 cm (Pacheco & Pessoa, 2017), but more efficient industrial media are also available (Metcalf & Eddy Inc, 2013). The bacterial mass attaches in the form of a thin biofilm a couple of millimetres thick on the support material. When the wastewater passes over the film, the bacterial mass metabolizes the organic material, consuming dissolved oxygen in the process. Oxygen is transferred by dissolution from the air present in the void spaces of the filter. A flow of air tends to pass through the filter, which can be in an upward or downward direction, depending on the difference in temperature between the air and the liquid phase. In cases of high organic loading rates induced ventilation is sometimes applied. In its percolation path through the medium, the liquid phase entrains part of the sludge film as suspended flocs. For this reason the effluent passes through a final settler, separating the solids from the liquid. The settled sludge can be returned to the distribution system to be used again in the bed. This recirculation may also become necessary to introduce more oxygen into the system. The energy required for the pumping of the recirculation is of the same order of magnitude as the energy for aeration in the activated sludge system. Part of the sludge should be discarded to avoid excessive build-up of sludge in the filter that eventually would cause clogging as well as increasing the oxygen demand.

The biofilter was little used for decades because it was inferior to other alternatives, notably the activated sludge system, but curiously it has re-emerged, especially to serve as a post-treatment system for the effluent from several anaerobic systems in Brazil. Apparently, this was initially due to a fear that the anaerobic effluent, which would no longer have the inorganic solids component of the raw sewage, would result in a "light" sludge with poor settling characteristics, impairing the successful application of the activated sludge system. In this chapter experimental results will be presented showing that in reality the sludge that develops in the SBR variant of the activated sludge system when applied to UASB reactor effluent has excellent settleability, much better than those of conventional activated sludge systems treating raw sewage in continuous-flow configurations. After several years of experimental studies in Campina Grande there has never been a single case of filamentous sludge bulking in post-treatment systems operated at the research facility. Also in full-scale units there are no problems with the sludge settleability, neither in SBR nor in continuous-flow systems. However, even though it has been established that in fact sludge from post-treatment activated sludge systems has a superior settleability, application of the biofilter persists. The main problem of the biofilter is the impossibility of controlling the sludge age, which is precisely the most important operational parameter of aerobic systems. The experimental results obtained so far (Chernicharo, 2006; Gonçalves *et al.*, 1999) show that the aerobic filter is reasonably efficient with regards to the removal of organic material, but cannot reliably perform nitrification unless a very long retention time (8–12 h) is applied.

The aerobic filter has other disadvantages such as excessive volume, inefficient aeration, high construction cost and operational instability due to failures in influent distribution and bed clogging. For these reasons it is not considered in this text as a realistic alternative for aerobic post-treatment of anaerobic effluents. The fact that the biofilter for post-treatment of the UASB reactor is applied only in Brazil, brings to mind a saying that Brazilians have: "If something exists only in Brazil and it is not *jabuticaba* fruit [a black berry], it is advisable to distrust it".

An interesting variant of the biofilter, which generally uses stones or plastic medium, is the system of hanging sponges, where polyurethane sponges stiffened with plastic rings are used. In this case the sludge mass – and therefore the treatment capacity – is higher, which implies a greater oxygen demand that is introduced by compressed air (Uemura & Harada, 2010). Thus, this system is similar to AS (mechanical aeration), but it has no adequate sludge age control.

8.2.3 Submerged aerobic filter (SAF)

The submerged aerobic filter (Figure 8.1, right) is similar to the biological filter but has a normal engineered aeration system and is, unlike the biofilter, completely filled with water. Instead of depending on the aeration of the liquid while being distributed (sprayed) over the filter bed as in the aerobic filter, aeration is carried out with compressed air, so that the design oxygen demand in the reactor can always be met.

The bed of stones has been replaced by an industrial medium of rigid and rough plastic bodies with large specific surface for fixation of the bacterial mass.

Due to these improvements the SAF is a system of greater treatment capacity and performance reliability than the biofilter. However, the main problem remains that there is no easy control of sludge age. Sludge age can to a certain extent be controlled by forced discharges, increasing the aeration intensity and releasing film particles from the bacterial mass, but this operation is too inaccurate to be used to establish the selected sludge age. For this reason also this system will not receive more attention in this text.

8.3 MODELLING OF AEROBIC POST-TREATMENT

The most important aspect of modelling aerobic treatment for UASB effluent is the description of the fate of the organic material. In the anaerobic model, the organic material present in the influent is divided into three fractions:

(1) Discharged into the UASB effluent (mS_{eu});
(2) Transformed into sludge in the UASB reactor (mS_{xu});
(3) Digested to methane (mS_d).

In Chapters 3 and 5 it was shown that sulphate can oxidize part of the organic material, which for that reason is not considered in the calculations of the anaerobic digestion process.

In Chapter 3 it was shown that the values of these fractions depend fundamentally on the anaerobic sludge age and the temperature, and empirical expressions were developed to calculate them. In aerobic post-treatment of UASB reactor effluent, the remaining biodegradable organic material is almost completely metabolized, leaving only the non-biodegradable and soluble fraction, mS_e, to be discharged as final effluent. The biodegradable material is converted into inorganic compounds by oxidation, mS_o, or into aerobic sludge, mS_{xae}. Figure 8.2 schematically shows the combined anaerobic–aerobic treatment system. When comparing this system with the conventional activated sludge system of Figure 8.3, it can be observed that the anaerobic pre-treatment could in principle have the following important effects on the overall process:

(1) The energy cost is affected not only by the decrease of the organic load in the post-treatment, but also by the possibility of generating electric energy from the biogas, and may leave the anaerobic–aerobic system independent of external energy or even generate surplus energy to be sold.

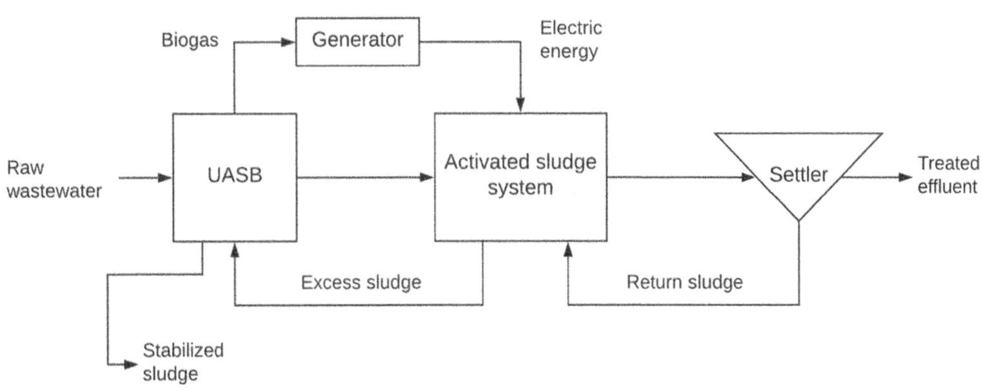

Figure 8.2 Anaerobic–aerobic sewage treatment system.

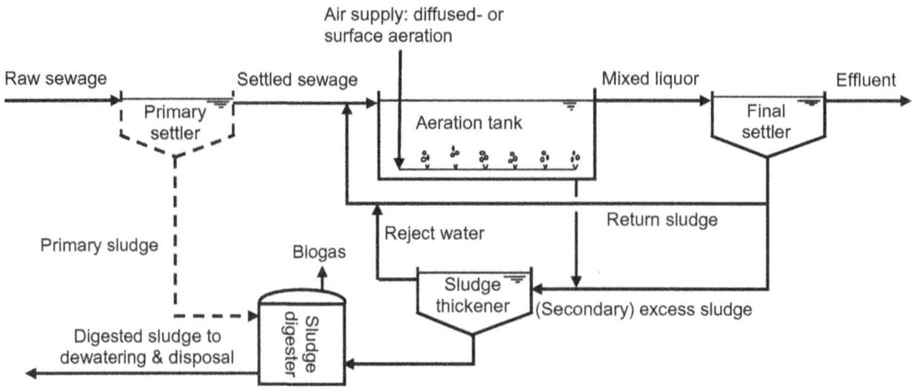

Figure 8.3 Conventional activated sludge system with sludge digestion.

(2) Due to the large-volume organic load that can be applied in pre-anaerobic treatment, there is the possibility of operating anaerobic–aerobic systems with a much lower volume than conventional aerobic systems, leading to a reduction in investment costs.
(3) The stabilization of aerobic sludge from the post-treatment could be carried out in the anaerobic reactor of the pre-treatment, dispensing in this case with the need to build and operate units for sludge thickening and digestion, which are part of conventional aerobic systems.
(4) As anaerobic systems produce less sludge than aerobic systems, the excess sludge production of the anaerobic–aerobic system will tend to be smaller than that of the purely aerobic system, even if these are submitted to stabilization before dewatering.

All these four aspects show considerable potential for cost reduction and therefore it is important to evaluate their feasibility. Therefore the following questions need to be answered:

(1) Can the aerobic treatment of UASB reactor effluent be described with the same basic model developed for conventional activated sludge systems treating raw sewage?
(2) What are the values of the reduction of energy demand in post-treatment and how does it compare to the potential of energy generation in anaerobic pre-treatment?
(3) How great is the reduction of sludge production when applying efficient anaerobic pre-treatment compared with only aerobic treatment?
(4) Is it feasible to operate anaerobic-aerobic systems with a smaller volume than a comparable conventional aerobic system?
(5) Is it feasible to stabilize waste sludge from the aerobic post-treatment system in the anaerobic pre-treatment reactor?

On the other hand anaerobic pre-treatment also has a negative aspect: the removal of biodegradable organic material reduces the feasibility of nitrogen and phosphorus removal by respectively nitrification–denitrification and biological excess phosphate removal processes. Hence there are two additional questions to answer:

(6) Are there possibilities for biological nitrogen removal when the concentration of biodegradable organic matter is low or non-existent?
(7) What are the alternatives for phosphorus removal if biological removal is not possible due to lack of biodegradable organic material?

8.3.1 Steady-state AS model applied to aerobic post-treatment

Figure 8.4 shows a diagram with the processes related to organic material, which are developed in an aerobic treatment system according to the theory of activated sludge as described by Marais and Ekama (1976). Subsequently the work of Marais and Ekama became the basis of the Activated Sludge Model produced by the International Water Association (IWA) (Henze *et al.*, 1987). It can be observed that the organic material of the influent is divided into three parts. One part (f_{ns}), equal to the non-biodegradable and soluble fraction, is discharged along with the effluent without being modified. A second part (f_{np}), equal to the non-biodegradable and particulate fraction, is flocculated and forms the inert sludge fraction.

The third part, $(1 - f_{ns} - f_{np})$ is biodegradable and is metabolized, partially transforming into active sludge (living mass), while the remainder is oxidized, generating oxygen demand (exogenous respiration). At the same time, active sludge decays, and a part (f) of the active sludge forms the endogenous residue, while the remainder $(1 - f)$ is oxidized, generating additional oxygen demand: the endogenous respiration. Van Haandel and Van der Lubbe (2012) presented the following equations for the three basic fractions:

$$mS_e = f_{ns} \tag{8.2}$$

$$mS_{xae} = f_{cv}\,(1 - f_{ns} - f_{np})(1 + f\,b_h R_{sa})Y_{ae}/(1 + b_h\,R_{sa}) + f_{np} \tag{8.3}$$

$$\begin{aligned} mS_o &= mS_{ex} + mS_{end} \\ &= (1 - f_{ns} - f_{np})(1 - f_{cv}Y_{ae} + f_{cv}b_h(1 - f)Y_{ae}R_{sa}/(1 + b_h R_{sa})) \end{aligned} \tag{8.4}$$

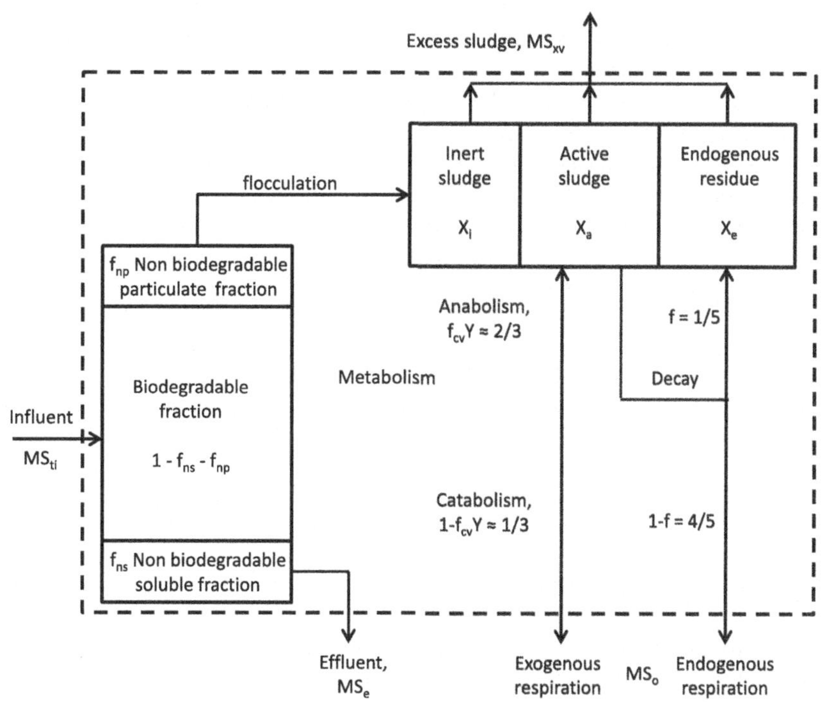

Figure 8.4 Schematic representation of the interaction between the influent organic material and the sludge in aerobic treatment systems.

where:

mS_e = influent COD fraction leaving the system as non-biodegradable soluble COD
mS_{xae} = influent COD fraction transformed into sludge
mS_o = influent COD fraction oxidized in the system
mS_{ex} = influent COD fraction oxidized due to utilization of biodegradable material
mS_{end} = influent COD fraction oxidized due to endogenous respiration
f_{ns} = non-biodegradable soluble influent COD fraction
f_{np} = non-biodegradable particulate influent COD fraction
Y_{ae} = yield coefficient in an aerobic environment = 0.45 mg VSS · mg^{-1} COD (Marais & Ekama, 1976)
f_{cv} = ratio between COD and the volatile sludge mass = 1.5 mg COD · mg^{-1} VSS (Marais & Ekama, 1976)
f = sludge fraction remaining as endogenous residue after decay of active sludge = 0.2 (Marais & Ekama, 1976)
b_h = decay constant = $0.24(1.04)^{(T-20)}$ d^{-1} (Marais & Ekama, 1976)
R_{sa} = sludge age in the aerobic system (days).

In this model the fundamental operational variable is the sludge age which determines the degree of decay of active sludge. Van Haandel and Van der Lubbe (2012) have used their model to estimate the minimum sludge age for the essentially complete removal of biodegradable organic material. This sludge age is in fact very short: only 3, 1.2 and 0.6 days for temperatures of 14, 20 and 26°C respectively. Usually the sludge age in an activated sludge system will be (much) higher than the minimum value and for this reason the utilization of biodegradable material will be virtually complete. If there is biodegradable material present in the effluent, it will almost always be the result of imperfect functioning of the final settler, resulting in the discharge of small sludge particles or individual microorganisms in the effluent. A problem is that soaps and surfactants are not removed by anaerobic pre-treatment and in aerobic post-treatment their removal requires a relatively long sludge age (5 to 6 days at 25°C). If this material is not removed, extensive foam formation may occur.

Apart from the visual and odorous nuisance that is generated, this foam may eventually escape the reactor together with part of the biomass, so that the sludge age becomes undetermined. Thus the removal of surfactants is necessary in aerobic post-treatment systems and often sets the lower value of the sludge age that can be applied in practice. Equations (8.2) to (8.4) are of fundamental importance in quantitatively describing the system of activated sludge, because they define exactly the most important points:

- The COD concentration in the effluent (which determines the effluent quality);
- The COD converted into sludge which determines sludge production and therefore both the required volume of the aerobic reactor and the required capacity of the units for waste sludge treatment and final disposal;
- The oxidized COD, which determines the demand for oxygen and therefore the required aeration capacity.

The validity of the above expressions has been tested and proven in aerobic systems with suspended sludge of all types and under most diverse operating conditions (temperature, sludge age, number of reactors etc.), but always using raw or settled sewage as influent.

In the next section it is demonstrated that the equations are also valid to describe the behaviour of activated sludge systems treating effluent from anaerobic systems. The validity of the model can be checked by operating an aerobic post-treatment system and measuring three variables once steady state has been established:

(1) The difference between the organic material concentration in the influent of the post-treatment system and in the final effluent;
(2) The concentration of volatile sludge which is the sum of the inert organic material present in the feed, the active sludge and the endogenous residue;
(3) The rate of oxygen consumption, the sum of exogenous and endogenous respiration.

With these parameters the experimental fractions of the organic material (COD) ending up in the effluent, in the excess sludge and as oxidized material can be expressed as:

$$\mathrm{m}S'_{\mathrm{e}} = S_{\mathrm{e}}/S_{\mathrm{eu}} \tag{8.5}$$

$$\mathrm{m}S'_{\mathrm{xae}} = f_{\mathrm{cv}}\mathrm{M}X_{\mathrm{v}}/(R_{\mathrm{s}}Q_{\mathrm{a}}S_{\mathrm{eu}}) = R_{\mathrm{h}}f_{\mathrm{cv}}X_{\mathrm{v}}/(R_{\mathrm{s}}S_{\mathrm{eu}}) \tag{8.6}$$

$$\mathrm{m}S'_{\mathrm{o}} = \mathrm{M}O_{\mathrm{c}}/(Q_{\mathrm{i}}S_{\mathrm{eu}}) = R_{\mathrm{h}}\mathrm{OUR}_{\mathrm{c}}/S_{\mathrm{eu}} \tag{8.7}$$

where:

$\mathrm{m}S'_{\mathrm{e}}$ = influent COD fraction leaving the system in the effluent (experimental value)
$\mathrm{m}S'_{\mathrm{xae}}$ = influent COD fraction transformed into volatile sludge (experimental value)
$\mathrm{m}S'_{\mathrm{o}}$ = oxidized influent COD fraction (experimental value)
S_{e} = final effluent COD
S_{eu} = COD in the UASB effluent
$\mathrm{M}X_{\mathrm{v}}$ = volatile sludge mass
$\mathrm{M}O_{\mathrm{c}}$ = mass of daily consumed oxygen for the oxidation of organic material
$\mathrm{OUR}_{\mathrm{c}}$ = oxygen uptake rate
Q_{i} = sewage flow rate
R_{h} = hydraulic retention time in the aerobic post-treatment unit.

The value of the parameters defined by Eqs. (8.5) to (8.7) can be experimentally determined and therefore the values of the three fractions can be calculated for any set of operating conditions. These can then be compared with the theoretical values of Eqs. (8.2) to (8.4) respectively. By trial and error one can modify the constants f_{ns} and f_{np} until the theoretical values $\mathrm{m}S_{\mathrm{e}}$, $\mathrm{m}S_{\mathrm{xae}}$ and $\mathrm{m}S_{\mathrm{o}}$ are equal to the experimental values $\mathrm{m}S'_{\mathrm{e}}$, $\mathrm{m}S'_{\mathrm{xae}}$ and $\mathrm{m}S'_{\mathrm{o}}$ respectively. When this equality is established for several sludge ages and other operational conditions, then this not only constitutes proof that the model adequately describes the behaviour of the system, but with the model thus validated the important aspects of the system (COD effluent, sludge production and oxygen demand) can be calculated for any set of operating conditions.

Clearly the composition of the organic material in the anaerobic effluent will be very different from that of raw sewage. In raw sewage the fraction of the non-biodegradable and soluble material (f_{ns}) and the non-biodegradable and particulate material (f_{np}) are both in the range of 0.08–0.12 (Van Haandel & Van der Lubbe, 2012). In the anaerobic effluent the concentration of biodegradable material will be much lower than in the raw sewage and for this reason the fraction of non-biodegradable soluble material will be relatively large, as presumably its concentration did not change in the anaerobic pre-treatment. Therefore an increase in the numerical value of f_{ns} is expected. On the other hand, it can be expected that much of the non-biodegradable particulate material of the raw sewage will be retained in the anaerobic pre-treatment and therefore the concentration of inert particulate organic material in its effluent will be smaller than the value for raw sewage.

Not indicated in Figure 8.5 is the presence of chemical oxygen demand in the form of sulphide, generated in the UASB upon reduction of sulphate. Normally however for municipal sewage the sulphate concentration should be low and when it is not, then the application of UASB (or any anaerobic

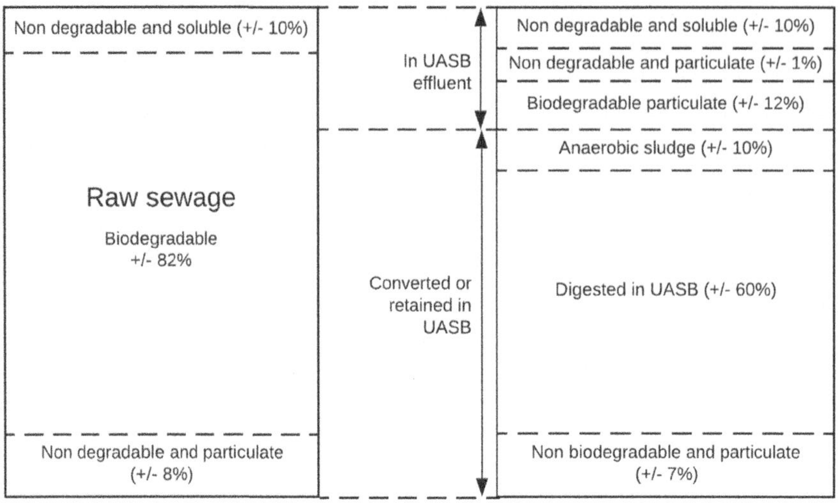

Figure 8.5 Representation of the effect of anaerobic pre-treatment on the composition of organic material in a typical sewage (low in sulphate).

technology) should be re-evaluated. It is important to note that the reduction of sulphate not only reduces the methane production, as pointed out in Example 6.7, but in the aerobic post-treatment system there will be an additional oxygen demand to satisfy the requirement for sulphide to sulphate oxidation, which is stoichiometrically the same as the initial concentration of the oxidized COD by sulphate reduction. Depending on the initial sulphate concentration, this demand may actually be greater than the demand for metabolism of the residual biodegradable COD in the UASB effluent.

Figure 8.5 shows the typical division of the organic material in sewage before and after anaerobic treatment. It can be observed that the effluent COD concentration of the anaerobic system is about 26–32% of the original COD concentration in the raw sewage (depending on the fraction of the produced anaerobic sludge being discharged in the effluent). The inert soluble fraction f'_{ns} in the anaerobic effluent can be calculated as in the range of 0.10/(0.26 to 0.32) = 0.33 to 0.40 and the inert particulate fraction f'_{np} in the range of 0.01/(0.26 to 0.32) = 0.03 to 0.04.

8.3.2 Validation of the steady-state model for digested sewage

Several experimental studies (Derks, 2007; Guimarães *et al.*, 2003; Leite, 2016) were carried out using anaerobic effluent as feed to an aerobic system. The aerobic system was operated at different sludge ages, using both continuous flow through (CFT) and sequential batch reactor (SBR) configurations at a temperature of 25°C.

It was concluded that sludge ages above 8 days were enough for both complete nitrification and operation without foaming issues. A longer sludge age was shown to be necessary when the sulphide concentration in the UASB effluent was high (Guimarães *et al.*, 2003), which is due to the high degree of toxicity of sulphide to nitrifiers. With a sludge age of less than 5 days, nitrification was not possible and problems with foaming appeared: apparently the applied sludge age of 5 days was too short to remove surfactants reliably. Table 8.1 summarizes the results of the experimental investigations and from this data it can be verified whether the IWA model for activated sludge applies when the influent of the aerobic system is digested sewage after UASB treatment.

Table 8.1 Experimental values from an aerobic system for treatment of UASB effluent, at a temperature of 25°C.

R_s (d)	S_{tu} (mg · L^{-1})	S_e (mg · L^{-1})	X_v (g · L^{-1})	R_h h	f_{ns} (-)	f_{np} (-)
Sequential batch (SBR) system						
5	222	65	2.05	2.4	0.293	0.037
8	250	70	2.98	2.4	0.280	0.025
12	236	72	3.75	2.4	0.298	0.033
				Average	0.29	0.031
Continuous flow through (CFT) system						
15	244	75	3.1	3.8	0.308	0.032

EXAMPLE 8.1

An aerobic SBR is operated with a hydraulic retention time of 2.4 h, a sludge age of 8 d and a temperature of 25°C. The system is fed with effluent from an UASB reactor, with a COD of 250 mg · L^{-1}. The final effluent has a COD of 70 mg · L^{-1}. The volatile sludge concentration in the reactor is 3 g · L^{-1}.

Determine the fractions of the influent COD of the SBR that are discharged in the effluent, transformed into sludge or oxidized. Calculate the non-biodegradable soluble and particulate fractions in the influent to the aerobic reactor (i.e. the anaerobic effluent).

Solution

Using Eqs. (8.5) and (8.6) and the data in Table 8.1, the experimentally determined COD fractions in the effluent and transformed into sludge can be calculated:

$$mS_e' = S_e/S_{eu} = 70/250 = 0.28$$

$$mS_{xae}' = R_h f_{cv} X_v/(R_s S_{eu}) = 0.1 \times 1.5 \times 3/(8 \times 0.25) = 0.225$$

Hence the oxidized fraction $mS_o' = 1 - 0.28 - 0.22 = 0.5$. The value of f_{ns} is equal to the value of mS_e' calculated above. Finally the non-biodegradable particulate COD fraction can be calculated with Eq. (8.3):

$$mS_e = 0.28 = f_{ns}$$

$$mS_{xae} = 0.225 = f_{cv}(1 - f_{ns} - f_{np})(1 + f\, b_h\, R_{sa})Y_{ae}/(1 + b_h\, R_{sa}) + f_{np}.$$

Hence $f_{np} = 0.025$.

For the data obtained in the experimental studies listed in Table 8.1, the non-biodegradable COD fractions of the UASB effluent can be calculated similarly to the method presented in Example 8.1 above. The fractions mS_e and mS_{xae} of the activated sludge system can be calculated with Eqs. (8.2) and (8.3) and compared with the experimental fractions mS_e' and mS_{xae}' calculated with the aid of Eqs. (8.5) and (8.6). In Figure 8.6 the theoretical curves and the experimental data are shown. The theoretical curves were calculated based on the average f_{ns} and f_{np} data in Table 8.1. It can be observed that these averages are virtually the same for the SBR and CFT systems. Figure 8.6 thus clearly shows

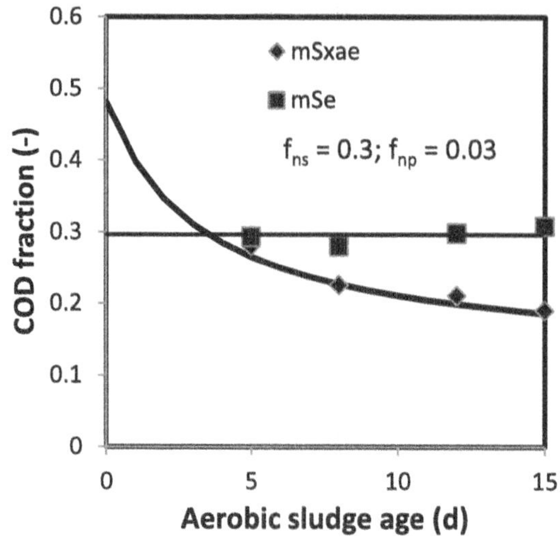

Figure 8.6 Experimental and theoretical values of mS_e and mS_{xae} as function of the sludge age.

that it is possible to simulate the main parameters of the activated sludge system (concentration of volatile sludge in the reactor, COD of the final effluent) as function of the sludge age: the experimental and theoretical values are very similar for all sludge ages. This is accepted as convincing evidence that the IWA steady-state model for COD removal is applicable to describe the behaviour of the activated sludge system treating UASB effluent, provided that the correct values of the non-biodegradable fraction f_{ns} and f_{np} are used.

8.3.3 Modelling the sequential batch reactor (SBR)

During the experimental studies discussed in the previous section, both continuous flow-through and sequencing batch systems were operated. It was shown that under comparable conditions the settleability of the sludge in the SBR systems was much better than that of the continuous flow-through system. For that reason there is a considerable advantage to use the SBR rather than the CFT system for aerobic post-treatment. In order to develop a model of the activated sludge system in a sequential batch regime, it is considered that the treatment is carried out by following a series of steps that all occur sequentially in a single reactor:

(1) *Reaction phase*: Aeration of a mixture of aerobic sludge and a batch V_b of effluent from the UASB reactor applied in a reactor with volume V_r. The aeration time depends on the composition and concentration of the biodegradable organic material in the UASB effluent and the operational conditions in the reactor, especially the active sludge concentration and the temperature.

(2) *Settling phase:* Gravity settling is applied to separate the sludge from the treated final effluent. The settling time depends on the settling height in the reactor, the sludge settleability and the sludge concentration.

(3) *Effluent discharge*: Using gravity discharge or pumping, the clarified supernatant is discharged after settling. The discharge time depends on the equipment used and the size of the batch, but typically is much smaller than the times for aeration or settling.

(4) *Feeding of a new batch:* Using gravity feed or pumping, a new batch of UASB effluent is introduced after the final treated effluent has been discharged. Feeding time basically depends on the equipment used. It is possible to feed the influent at the bottom of the reactor and simultaneously discharge the effluent at the top of the reactor. The advantage is that both occur at the same time, reducing the cycle time, but care must be taken that short circuiting of influent directly to the effluent is prevented.

Sequential batch systems typically are constructed as two or more parallel reactors that are fed alternatingly, so that there may be no need for an intermediate buffer tank. Only when there is only one SBR reactor is an intermediate buffer tank required in which the UASB effluent is stored during the period the SBR is not fed.

In the previous section it was shown that the IWA steady-state model is valid for aerobic post-treatment, so the following expressions can be used to calculate the sludge components:

$$\begin{aligned} X_a &= Y_{ae}R_{sa}/(1 + b_h R_{sa})S_{bu}/R_h \\ &= (1 - f_{ns} - f_{np})Y_{ae}R_{sa}/(1 + b_h R_{sa})S_{eu}/R_h \end{aligned} \tag{8.8}$$

$$X_e = f\,b_h R_{sa} X_a \tag{8.9}$$

$$X_i = (f_{np}/f_{cv})R_{sa}/R_h \tag{8.10}$$

$$\begin{aligned} X_v &= X_a + X_e + X_i \\ &= [(1 - f_{ns} - f_{np})(1 + f b_h R_{sa})Y_{ae}R_{sa}/(1 + b_h R_{sa}) + f_{np}R_{sa}/f_{cv}]S_{ti}/R_h \end{aligned} \tag{8.11}$$

$$X_t = X_v/f_v \tag{8.12}$$

where:

X_a = active sludge concentration (sludge composed of live bacterial mass)
X_e = endogenous residue concentration
X_i = inert organic sludge concentration (= flocculated non-biodegradable particulate material from the influent)
X_v = volatile sludge concentration = $X_i + X_a + X_e$
X_t = total sludge concentration (volatile + mineral) = X_v/f_v
R_h = hydraulic retention time = V_r/Q_i
S_{eu} = COD concentration in the influent of the post-treatment
S_{bu} = biodegradable COD concentration in the influent of the aerobic reactor.

8.3.3.1 Calculation of the minimum aeration time

When the objective is the production of water for reuse in agriculture, then the main objective of biological treatment in an aerobic post-treatment system will be the removal of the residual biodegradable organic material. The use of organic material by active sludge can be described by the Monod equation (Van Haandel & Van der Lubbe, 2012):

$$r_u = k_m S_b/(S_b + K_s)X_a = k_m S_{bu}/(S_{bu} + K_s)(Y_{ae}R_s/(1 + b_h R_S)S_{bu}/R_h \tag{8.13}$$

where:

r_u = utilization rate of biodegradable material during the treatment of a batch
k_m = utilization constant ($g\ COD \cdot g^{-1}\ X_a \cdot d^{-1}$)

S_b = biodegradable COD concentration in the SBR reactor
K_s = half saturation constant
X_a = active sludge concentration (live bacterial mass)
S_{bu} = biodegradable COD in the UASB effluent
R_h = hydraulic retention time in the SBR reactor = $V_r/Q_i = V_r/(NV_b)$
V_r = volume of the SBR reactor
Q_i = daily treated flow in the SBR reactor
N = number of daily treated batches
V_b = batch volume

When the biodegradable COD concentration is high ($S_b \gg K_s$), the ratio $S_b/(S_b + K_s) \approx 1$ and the utilization rate is maximum, independent of the biodegradable organic material concentration:

$$r_{um} = k_m \cdot X_a \tag{8.13a}$$

where

r_{um} = maximum utilization rate of biodegradable organic material.

During aeration part of the organic material metabolized by the active sludge is oxidized or catabolized, while the remainder is incorporated as cell mass. When it is considered that Y_{ae} g of active sludge is generated per g of metabolized COD and this sludge has a COD content of $f_{cv}Y_{ae}$, then a fraction of $f_{cv}Y_{ae}$ of the biodegradable COD is anabolized and the remainder $(1 - f_{cv}Y_{ae})$ is catabolized, requiring $(1 - f_{cv}Y_{ae})$ oxygen. Therefore the exogenous OUR is linked directly to the utilisation rate of the biodegradable organic material:

$$\mathrm{OUR}_{ex} = (1 - f_{cv}Y_{ae})r_u \tag{8.14}$$

The oxygen uptake rate (OUR) in an activated sludge reactor can be determined experimentally with a respirometer, an instrument that determines the rate of decrease of the dissolved oxygen concentration, when the oxidation is interrupted (Van Haandel & Van der Lubbe, 2012). The automated respirometer generates a respirogram that shows the value of the OUR as a function of time. Figure 8.7 shows a typical respirogram that was generated after the introduction of a batch of UASB reactor effluent into an SBR reactor with activated sludge. From the respirogram the following can be deduced: before the introduction of UASB effluent into the SBR reactor the OUR tends to a minimum value, corresponding to the endogenous aerobic sludge respiration rate (OUR_{min}). After the introduction of the UASB effluent the OUR increases rapidly to a maximum value (OUR_{max}), which is sustained for some time and then gradually decreases and again establishes the minimum value of the endogenous respiration.

The active sludge concentration value (live bacterial mass) can be calculated from the respirogram knowing that the endogenous OUR is given by (Van Haandel & Van der Lubbe, 2012):

$$\mathrm{OUR}_{min} = (1 - f) f_{cv}\, b_h\, X_a \tag{8.15}$$

The maximum rate of utilization of oxygen for catabolism is the difference between the maximum and minimum OUR values in the respirogram:

$$\mathrm{OUR}_{exm} = \mathrm{OUR}_{max} - \mathrm{OUR}_{min} \tag{8.15a}$$

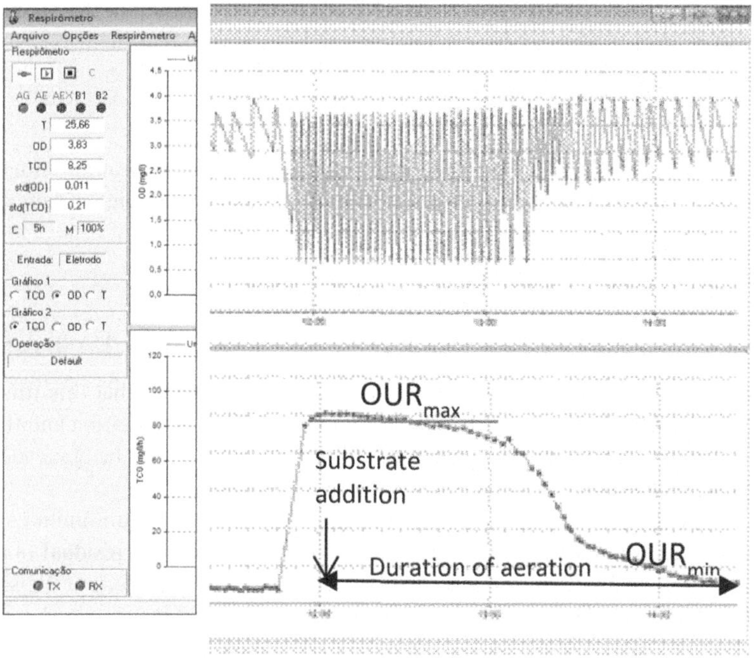

Figure 8.7 Typical respirogram obtained from an activated sludge system operated in sequential batch regime, upon receiving a batch of UASB effluent.

The value of the utilization constant can be calculated from the maximum rate of the exogenous OUR, knowing the proportionality between OUR_{ex} and r_u (Eq. (8.13a)). Directly after feeding the new feed batch the OUR remains at the maximum value:

$$OUR_{exm} = (1 - f_{cv}Y_{ae})r_{um} = (1 - f_{cv}Y_{ae})k_m X_a$$

where:

OUR_{min} = endogenous respiration rate
OUR_{exm} = maximum exogenous respiration rate
$X_a = OUR_{min}/[(1 - f_{cv}Y_{ae})f_{cv}b_h]$
$K_m = OUR_{exm}/[(1 - f_{cv}Y_{ae})X_a]$

The initial concentration of biodegradable organic material in the reactor can be estimated knowing that upon metabolism there is oxidation of a fraction $(1 - f_{cv}Y_{ae})$ of COD while a fraction $f_{cv}Y_{ae}$ is incorporated into new biomass. As the area under the OUR_{ex} curve and above the OUR_{min} baseline is the exogenous OUR, the concentration of the biodegradable organic material metabolized in the treatment can be calculated as:

$$S_{bi} = A_s/(1 - f_{cv}Y_{ae}) \qquad (8.16)$$

where:

S_{bi} = concentration of biodegradable organic material in the reactor directly after feed batch addition
A_s = area between the OUR_{ex} and the OUR_{end} curves.

It is important to recognize that the biodegradable COD at the beginning of the batch treatment is not equal to the concentration of biodegradable organic material in the UASB effluent, S_{bu}, since S_{bu} is diluted by a factor V_b/V_r when the batch is added into the reactor. Therefore:

$$S_{bi} = (V_b/V_r)S_{bu} \tag{8.17}$$

Knowing the values of S_{bi}, k_m and X_a, it now becomes possible to calculate the minimum required aeration time, when it is considered that the concentration S_{bi} cannot be used at a rate higher than r_{um} so that:

$$t_{aem} = S_{bi}/r_{um} = (V_b/V_r)S_{bu}/(k_m X_a) \tag{8.18}$$

where

t_{aem} = minimum time required for the treatment of the effluent batch of the UASB reactor.

Another way of estimating the time required for aeration is to consider that this time is equal to the time for the OUR to decrease until it is again at the level of endogenous respiration and there is no further demand for catabolism of biodegradable organic material. In Figure 8.7 it can be observed that this occurs around 2.3 hours after feeding the batch.

It is important to note that when the actual aeration time is less than the minimum calculated value in Eq. (8.18), the effluent quality does not necessarily decrease: as most of the residual organic material in the UASB effluent is particulate, it will be incorporated (adsorbed or enmeshed) in the sludge. Therefore, stopping the aeration before completing the utilization of organic material may not lead to an increase of the final effluent COD, but it will result in an accumulation of particulate organic material in the sludge, hence resulting in an increase in sludge production and a decrease in energy demand for aeration. However there is a limitation: unless enough reactor volume is available to allow for an increase in sludge concentration, the fraction of biodegradable organic material in the sludge will increase at the expense of the active sludge concentration. As a result, the utilization rate will decrease as well, so that the required aeration time will increase even further: a vicious cycle will result and ultimately the aerobic treatment may fail. As a secondary effect, a high fraction of non-metabolized material may affect the sludge settleability adversely.

EXAMPLE 8.2

Estimate the value of the utilization rate constant k_m from the respirogram of Figure 8.7.

Solution

The oxygen uptake rate in the respirogram of Figure 8.7 varies approximately between a minimum of $10\ mg \cdot L^{-1} \cdot h^{-1}$ (=OUR_{min}) to a maximum of $100\ mg \cdot L^{-1} \cdot h^{-1}$ (=$OUR_{exm} + OUR_{min}$). Using Eq. (8.15): $OUR_{min} = (1 - f)f_{cv}b_h X_a = 10 = 0.8 \times 1.5 \times 0.3 \times X_a$, whence the value of X_a can be calculated as $X_a = 667\ mg \cdot L^{-1}$. Now the utilization rate constant can be calculated from $OUR_{exm} = (100 - 10) \times 24 = (1 - f_{cv}Y_{ae})k_m X_a = 0.67 k_m X_a$, i.e. $k_m = 4.8\ mg\ COD \cdot mg^{-1}\ X_a \cdot d^{-1}$.

This means that the active sludge can metabolize up to 4.8 mg COD per mg of bacterial mass per day, which is a typical value for active sludge under the operational temperature of 25°C at the time of the experiment (Van Haandel & Van der Lubbe, 2012). Values of k_m larger than $10\ mg\ COD \cdot mg^{-1}\ X_a \cdot d^{-1}$ indicate a sludge with a large utilization rate. Values of less than $3\ mg\ COD \cdot mg^{-1}\ X_a \cdot d^{-1}$ are an indication of a slow utilization rate, which could be due to toxicity (for example sulphide) or some other problem that limits the sludge activity.

The reader should not confuse the value of k_m with that of the maximum F/M value that can be applied (which has the same units of measure), as the F/M ratio is based on the volatile sludge concentration X_v, which will include a significant fraction of inert sludge (X_i) and endogenous residue (X_e). Naturally the maximum F/M value will thus be lower.

8.3.3.2 Calculation of the settling time

When the utilization of the organic material is complete, the aeration of the batch is interrupted and settling of the sludge begins. The settling rate can be expressed by the Vesilind equation (Vesilind, 1968);

$$v = v_0 \exp(-kX_t) \tag{8.19}$$

where

v = settling velocity of the sludge
X_t = initial concentration of settling sludge
k, v_0 = settling constants.

The values of the settling constants v_0 and k can be determined experimentally by specific tests (Appendix A2). The time required for settling L meters (where L is the height of the supernatant to be discharged after settling) with a velocity v is given by:

$$t_{se} = L/v = H_r(V_b/V_r)/(v_0 \exp(-kX_t)) \tag{8.20}$$

8.3.4 Optimization procedure for aerobic post-treatment

In the optimization procedure for aerobic post-treatment, a sequence of steps will lead to the desired result: minimization of the costs to produce a final effluent essentially free of suspended solids and biodegradable organic material. The main cost factors are:

- The hydraulic retention time that determines the volume of the aerobic reactor and therefore a large part of the investment cost;
- The oxygen demand that determines large part of the equipment cost and the energy required for aeration;
- The production of excess sludge that determines the cost of final disposal.

For the optimization procedure it is recommended to carry out an experimental study, which can be at laboratory or pilot scale, to determine the values of the relevant variables which are:

- The sludge age required for aerobic post-treatment;
- The concentration and the composition of the COD in the UASB reactor effluent;
- The utilization constant of the biodegradable organic material k_m;
- The settleability constants of the aerobic sludge, k and v_o.

However, in many cases the effluent is not available, not all data are or can be provided or the client is not willing to pay for experimental work as is often the case. Then assumptions will have to be made. For example data from nearby sewage treatment plants could be used, or typical values could be assumed: luckily the variations in the composition of municipal sewage are not as large as those in industrial wastewaters so that as a last resort the values used in this book could be used, especially for design of aerobic post-treatment, as the preceding anaerobic treatment has reduced at least part of the variations of the influent.

The procedure starts with the selection of the design sludge age for the aerobic post-treatment. To reduce operating costs, the minimum sludge age must be chosen, ensuring that the basic objective is met, which is the efficient removal of biodegradable organic material. In principle this sludge age can be very short, only a few days depending on the temperature. However, in practice a second issue arises: the anaerobic pre-treatment is inefficient in the removal of soaps and other surfactants.

For this reason the sludge age should be sufficiently high to remove these materials and thus avoid the appearance of foam, which impairs both the operational stability of the process and the quality of the effluent. The sludge tends to attach to the produced foam forming a thick layer on the surface of the reactor and eventually this mixture can exit the reactor and spread over surrounding areas. The result is not only that the sludge age will be undefined, but also that the release of foam around the treatment system creates an unpleasant, dangerous and unhealthy environment for operators. The minimum sludge age of degradation of the surfactants depends on the temperature. Our experience, although limited to the sewage from Campina Grande, Brazil, is that at 25°C the foam layer starts to form when the sludge age is shorter than 5 days.

After selection of the sludge age, the next step is the characterization of the organic material in the UASB effluent to be treated in the post-treatment unit. In order to do so, it is necessary to operate the SBR system at a constant sludge age (not necessarily the minimum sludge age, but in general a short sludge age will reduce the duration of the experiment as equilibrium is reached more quickly) in order to establish the experimental values of fractions of the COD of the UASB effluent that end up in the effluent or in the excess sludge, using Eqs (8.5) and (8.6). In the next step Eqs (8.2) and (8.3) are used to determine the non-biodegradable fractions in the UASB reactor effluent, f_{ns} (soluble) and f_{np} (particulate), as shown in Example 8.1. The biodegradable COD to be removed (S_{bu}) is also calculated. Table 8.2 summarizes all parameters whose values must be known or determined before the optimization procedure can start. The parameters can be classified as follows:

- Concentration and composition of the COD to be treated (S_{bu}, f_{ns}, f_{np});
- Operating conditions (V_b, R_s, temperature, feed and discharge times);
- Sludge mass parameters (Y_{ae}, f, f_{cv}, b_h, f_v);
- Properties of the generated sludge (k_m, v_0, k);
- Reactor dimensions (V_r, H_r).

While in pilot-scale reactors the feed and discharge times can be very short (<0.1 h), in large reactors they may take up a considerable part of the total cycle time (>0.25 h), unless a very large pump is used.

Table 8.2 Required parameters and constants with their attributed values for the optimization procedure of aerobic post-treatment in an SBR reactor.

Sludge mass parameters			**Operational conditions**		
Y_{ae}	0.45	g VSS · g⁻¹ COD	V_b	0.5	m^3
f	0.2	g VSS · g^{-1} VSS	R_s	5	d
f_{cv}	1.5	g COD · g^{-1} VSS	Temperature	25	°C
b_h	0.29	d^{-1}	t_{feed}	0.1	h
f_v	0.8	g VSS · g^{-1} TSS	t_{disc}	0.1	h
COD composition			**Properties of the generated sludge**		
S_{bu}	200	mg · L^{-1}	k_m	5	g COD · g^{-1} X_a · d^{-1}
f_{ns}	0.4	–	v_0	300	m · d^{-1}
f_{np}	0.05	–	k	0.3	L · g^{-1} TSS
Reactor dimensions			**Batch volume**		
V_r	1	m^3	V_b	0.5	m^3
H_r	6	m			

Table 8.3 Optimization procedure for an SBR system for aerobic post-treatment of UASB effluent, using the data of Table 8.2 ($t_{feed} = t_{disch} = 0.3$).

1	2	3	4	5	6	7	8	9	10
R_h	Q_i	X_a	X_t	t_{ae}	t_{se}	t_c	N	Q_{max}	$Q_{max} - Q_i$
d	$m^3 \cdot d^{-1}$	$mg \cdot L^{-1}$	$mg \cdot L^{-1}$	h	h	H	d^{-1}	$m^3 \cdot d^{-1}$	$m^3 \cdot d^{-1}$
0.14	7.14	707	1365	0.75	0.36	1.71	14.05	7.02	−0.12
0.141	7.09	702	1356	0.75	0.36	1.71	14.01	7.01	−0.08
0.142	7.04	697	1346	0.76	0.36	1.72	13.98	6.99	−0.05
0.143	6.99	692	1337	0.76	0.36	1.72	13.94	6.97	−0.02
0.144	6.94	687	1328	0.77	0.36	1.73	13.91	6.95	0.01

The numerical values listed in Table 8.2 are the results of an experimental investigation carried out in Campina Grande treating UASB reactor effluent in an SBR system at a sludge age of 8 days and at 25°C (Leite, 2016). Once the values of all required variables and constants are known, the optimization procedure is relatively simple and consists essentially of comparing the required treatment capacity in the SBR reactor with the treatment capacity effectively available in the reactor, for different values of the hydraulic retention time. The following sequence of steps must be followed (see Table 8.3):

(1) First, assign a starting value for the volume of the reactor, for example 1 m^3.
(2) Calculate for different values of the hydraulic retention time in the aerobic post-treatment system (column 1) the anaerobic effluent flow to be treated (Q_i, column 2).
(3) The minimum aeration and settling times required for the SBR system are calculated using the data in Table 8.2:
    (a) Use Eq. (8.8) to calculate first the active sludge concentration X_a and then use Eq. (8.12) to calculate the total sludge concentration X_t (columns 3 and 4);
    (b) With the calculated values of X_a and X_t, the required aeration time t_{ae} is calculated with Eq. (8.18) and the settling time t_{se} with Eq. (8.20) (columns 5 and 6).
(4) Calculate the minimum required cycle time: $t_t = t_{feed} + t_{ae} + t_{se} + t_{disc}$, using the specified values in Table 8.2 for feed and discharge times (column 7).
(5) Calculate N_b, the number of batches that can be treated per day: $N_b = 1/t_t$ (column 8).
(6) Calculate the maximum flow rate that can be treated: $Q_{max} = V_b N_b$ (column 9).
(7) Calculate the difference between the maximum flow rate that can be treated (Q_{max}, column 9) and the adopted flow rate (Q_i, column 2). The difference is listed in Column 10.

If ($Q_{max} - Q_i$) has a positive value, then the flow that can be treated (Q_{max}) is higher than the selected flow (Q_i), i.e. the reactor is underloaded and can be operated at a shorter retention time. If the difference is negative, the treatment capacity is insufficient to treat the selected flow rate. In that case the flow must be reduced, i.e. the retention time must be increased.

As an example, Table 8.3 shows a series of retention times for which the corresponding values of the feed flow Q_i were calculated (for $V_r = 1$ m^3), under the conditions specified in Table 8.2. The minimum retention time can now be calculated by trial and error. For each value of the feed (column 2), the maximum flow rate that can be treated (column 9) is calculated following the above procedure. It can be observed that for small values of the retention time (for example 0.14 days), the flow that can be treated (Q_{max}) is smaller than the selected flow rate Q_i and the difference between them is negative. But when R_h reaches a value of 0.144 days, the treatment capacity becomes sufficient as ($Q_{max} - Q_i$) becomes positive. Therefore, for the conditions specified in Table 8.2, it can be concluded that the maximum flow rate that can be treated in a post-treatment SBR is equal to 7.0 $m^3 \cdot d^{-1}$ per m^3 reactor volume, i.e. the minimum hydraulic retention time is 0.144 days or 3.5 hours. Once the required retention time in the SBR reactor has been determined, the required reactor volume of the aerobic post-treatment system is readily calculated. Taking as an example the full-scale data from Table 1.5, which was discussed in Chapter 1, the daily volume to be treated is 91,220 $m^3 \cdot d^{-1}$. The volume of the post-treatment can now be estimated as follows: for the optimal retention time of 3.5 hours) the number of batches per day is 14 (Table 8.3).

Thus if only one aerobic SBR is used, the batch volume would be 91220/14 = 6515 m^3 and the reactor volume = 2 × 6515 = 13,030 m^3 (in the calculations of Table 8.3 it was assumed for convenience that the batch volume is equal to half the reactor volume). This results in a very large tank and probably it is more convenient to have several reactors in parallel of each about 2000 m^3. For six reactors the resulting volume of each is 2170 m^3 in which 14 batches of 6515/6 = 1100 m^3 would be discharged per day. At any rate the daily variation of the sewage flow will persist, so that an equalization tank would be required. The equalization tank is also very useful as a settling tank to separate the solids in the UASB effluent, thus minimizing the load on the post-treatment unit. For the data in Tables 1–5, the volume of the equalization tank was calculated at 16,000 m^3 (see Example 1.1).

In the calculations of Table 8.3 it is assumed that the four times for feeding, treating, settling and discharging batches are constant. In practice it is possible to optimize the aerobic post-treatment further by installing sensors that indicate the end of the post-treatment steps: a liquid level sensor for the feed and discharge stops, a DO sensor for the reaction phase and an optical sensor (to detect turbidity) for the settling step. Thus the aerobic post-treatment is then operated with the minimum time for each step of each batch. In practice the times for feeding, settling and discharging are almost constant, so the greatest advantage of automation is to have a sensor that indicates the end of the aeration period, as the load to be treated may vary considerably over the day. Due to this optimization, there may be periods when there is not enough digested sewage in the equalization tank. In that case, to save energy, the aeration of the reactor which is to receive the next batch can simply be switched off until enough digested sewage has accumulated, and the reactor can be fed.

8.4 INTEGRATION OF THE COMBINED ANAEROBIC–AEROBIC TREATMENT SYSTEM

Having developed an optimization procedure for the aerobic post-treatment in the previous sections, this section focuses on the optimization of the integrated anaerobic–aerobic treatment process. Figure 8.8 shows the basic flow diagram of such a treatment system, which has the following advantages compared to a conventional aerobic treatment system:

(1) Drastic reduction of treatment system volume;
(2) Reduction of energy consumption for aeration: auto sufficiency is feasible;
(3) Significant reduction of total sludge production;
(4) Potentially, stabilization of aerobic excess sludge in the anaerobic pre-treatment system.

The above advantages are discussed in more detail in Sections 8.3.1 to 8.3.4.

8.4.1 Potential for reduction of total treatment volume

In order to compare the combined volume of the anaerobic–aerobic system with that of a conventional aerobic system, it is necessary first to consider the design sludge concentration of both the anaerobic and

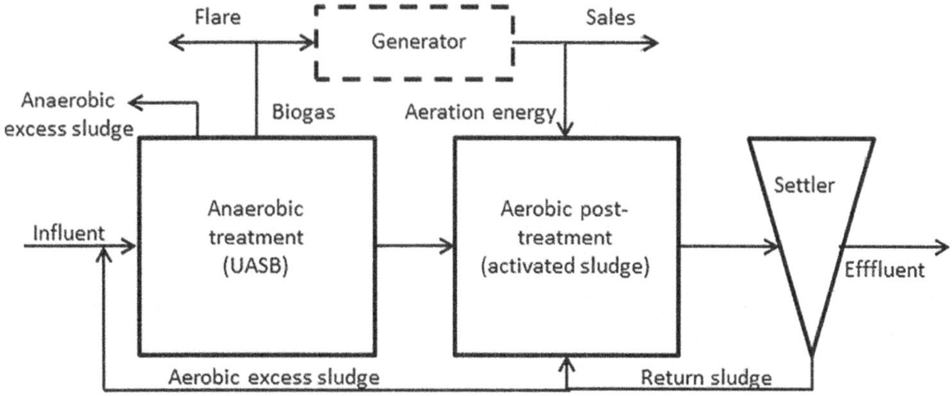

Figure 8.8 Flow sheet of a combined anaerobic–aerobic treatment system.

aerobic systems. In Chapters 3 and 4 it was established that in UASB reactors the average concentration of the sludge is in the range of 15–30 g · L^{-1}, with typically an organic fraction of about 60%. Therefore the average concentration of the organic sludge can be estimated at approximately 15 g · L^{-1}. In aerobic systems the sludge concentration of sludge is typically in the range of 3.5 to 4.5 g · L^{-1} with an organic fraction of approximately 75%, thus resulting in about 3 g · L^{-1} of organic sludge. The second operational factor that has a major influence on the required reactor volume is the sludge age.

EXAMPLE 8.3

Compare the required total volumes of a combined anaerobic–aerobic system and of a conventional aerobic system for a per capita contribution of $S_{ti} = 100$ g COD · d^{-1} and $Q_i = 120$ L · d^{-1}. Assume an organic sludge concentration $X_{van} = 15$ g VSS · L^{-1} in the anaerobic pre-treatment and $X_{vae} = 3$ g VSS · L^{-1} in the aerobic post-treatment. The aerobic post-treatment is operated at $R_{sae} = 5$ d. Adopt a value of 0.03 for the non-biodegradable particulate COD fraction in the UASB effluent (i.e. the part of f_x that ends up in the effluent as solids).

Solution

In Section 3.9.2 it was established that the COD fractions that end up in the effluent of a UASB reactor or will be transformed into sludge can be expressed as:

$$f_s = f_{ns} + f_{bs}$$
$$= f_{ns35} \times 1.037^{(35-T)} + 0.20 \times 1.011^{(35-T)} \exp[-0.04(R_{san} - 3.8 \times 1.06^{(35-T)})] \quad (3.24)$$

$$f_x = f_{np} + f_{bp}$$
$$= f_{np35} \times 1.015^{(35-T)} + 0.16 \times 1.014^{(35-T)} \exp[-0.04(R_{san} - 3.8 \times 1.06^{(35-T)})] \quad (3.25)$$

Furthermore, an expression was derived for the retention time in the UASB reactor as a function of the anaerobic sludge age:

$$R_h = R_s f_x S_{ti}/(f_{cv} X_v) \quad (3.32)$$

Figure 8.9 shows the retention time in the UASB reactor as a function of the applied sludge age in the anaerobic reactor for temperatures of 15°C (left) and 25°C (right). Figure 8.9 also shows the retention time in the

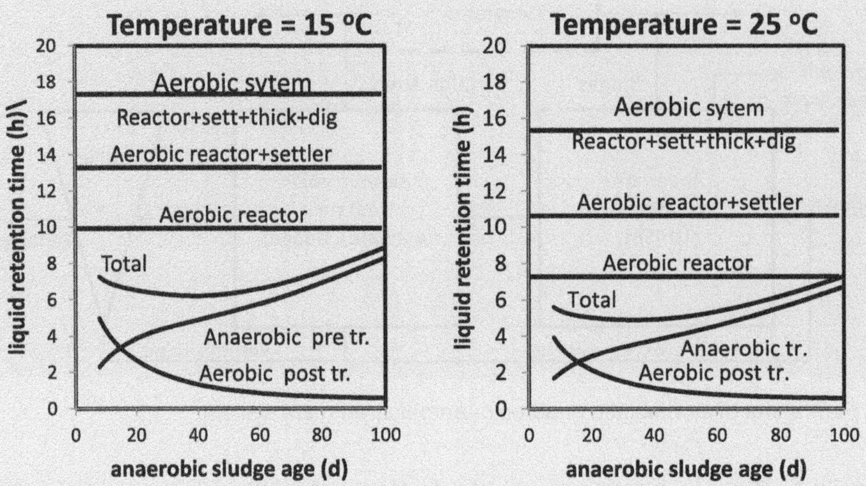

Figure 8.9 Hydraulic retention time in a combined anaerobic–aerobic treatment system as function of the anaerobic sludge age. The retention time in the purely aerobic system is also indicated for a sludge age of 5 days.

aerobic reactor for a sludge age of 5 days, again for both temperatures. In this example the production of sludge was calculated assuming that two-thirds of the biodegradable material in the influent is oxidized and one-third is converted into organic sludge.

To calculate the concentration of inert organic sludge in the aerobic reactor, the presence of a fraction of non-biodegradable particulate organic material in the UASB effluent equal to 0.03 of the COD of the raw sewage must be considered. In other words, the accumulation of organic sludge in the reactor is based on $f_x - 0.03$.

The required aerobic retention time decreases with increasing anaerobic sludge age, because the UASB reactor becomes more efficient as its sludge age increases and hence removes more organic material. The combined retention time, which is also indicated in Figure 8.9, passes through a minimum at medium sludge age (40–60 days).

At very short or very long anaerobic sludge ages the total retention time increases, because the UASB reactor is inefficient and consequently the aerobic reactor will be large. At very long sludge ages the UASB itself is efficient, but very large. In the same figure the estimate of the retention time of a purely aerobic system is also indicated, calculated with the basic theory of activated sludge presented by Van Haandel and Van der Lubbe (2012). An aerobic sludge age of 5 d is considered and the optimized aerobic system is composed of a reactor, a settler, a thickener and a sludge digester.

Figure 8.9 clearly shows that it is possible to greatly reduce the volume of a conventional aerobic activated sludge system by replacing it with a combined anaerobic–aerobic system. Only if the sludge age in the anaerobic pre-treatment is very long (>100 days), then the retention time in the combined anaerobic–aerobic system will be higher than that of the aerobic reactor. Such high sludge retention times in practice would only be used at temperatures of less than 18–20°C (van Lier, 2008) to avoid the reduction of unstable sludge. However, even though in terms of total volume the combined anaerobic–aerobic system loses its advantage at these lower temperatures, the benefits in terms of reduced operating costs (less sludge production, less power consumption) still apply, as will be demonstrated in the next sections.

The figures also show that the total retention time tends to increase slowly with the anaerobic sludge age so that, when only the point of view of volume reduction is considered, it would be interesting to use the minimum possible anaerobic sludge age (in the range of 30–40 d). However, to determine the optimum anaerobic sludge age, energy and (especially) quantity and quality of anaerobic excess sludge production must also be considered, as they are important operational cost factors. These factors are analysed in the following sections.

8.4.2 Feasibility of energy self-sufficiency

In this section the impact of anaerobic pre-treatment with complementary aerobic treatment on energy consumption is discussed. In the anaerobic pre-treatment system methane is generated which, after hydrogen sulphide is removed, can be used directly for the generation of electric energy. An additional source of material for digestion may be the aerobic sludge from the post-treatment, which can be recirculated to the anaerobic reactor. In Section 8.4.4 the feasibility of the stabilization of excess sludge from aerobic post-treatment of UASB effluent in the UASB reactor will be evaluated and the potential for energy production will be estimated.

In the aerobic post-treatment the energy demand is reduced compared to the conventional aerobic system because there is relatively little biodegradable organic material remaining in the UASB effluent. The steady-state model of the activated sludge system presented in Section 8.3.1 shows that energy consumption increases with the aerobic sludge age, because the oxygen demand for endogenous respiration increases. Depending on the use of the effluent, there may be additional energy demand for nitrification. This will be discussed in Section 8.5.

The potential of energy generation depends directly on the methane production and the efficiencies of the biogas collection and electric energy generation. The production of methane depends on the applied anaerobic sludge age and the operational reactor temperature. In Section 3.9.2 expressions were derived for the fractions of the influent COD discharged in the effluent f_s (Eq. 3.24), transformed into sludge f_x (Eq. 3.25) and converted to methane in an anaerobic digester:

$$f_d = 1 - f_s - f_x \tag{3.17}$$

Therefore for an applied organic load MS_{ti}, the generated mass of methane is:

$$MCH_4 = MS_{ti}(1 - f_s - f_x)/4 \tag{8.21}$$

However, not all generated methane will effectively be captured: a part will remain dissolved in the liquid phase and is discharged in the effluent.

As the methane fraction in the biogas produced from sewage will typically be high (>70–80%), to good approximation the solubility of methane in the digester is the same as that in pure water (Table 6.1), which is about 20 mg · L^{-1}, equivalent to 4 × 20 = 80 mg COD · L^{-1}. Hence, stated differently, the methane generated from a COD concentration of approximately 80 mg · L^{-1} will not be collected in the UASB reactor. Furthermore, a fraction of the produced methane may escape from the biogas collection system via several points of emission (diffusion of gas through concrete gas collection boxes, porosity of the collection tubes, small leakages in pressurized tubing etc.). The quantity of methane lost depends on the quality of the civil works and the maintenance applied, but it can add up to a considerable fraction of the biogas production. The electrical conversion efficiency of quality gas motors is about 40%, producing 5 kWh · kg^{-1} CH_4 (refer also to Chapter 6).

EXAMPLE 8.4

Calculate the potential of per capita electricity generation and compare this to the electrical consumption in an anaerobic–aerobic system. To do this calculate the production and collection of methane in the anaerobic sewage treatment system as a function of anaerobic sludge age for per capita contributions of 100 g COD · d^{-1} and 120 l · d^{-1}. Assume losses of 30% of methane in leaks and consider also the presence of 20 mg CH_4 · L^{-1} in the effluent.

For the energy consumption, calculate the aeration demand in the aerobic system operated at a sludge age of 8 days with and without nitrification (assume a per capita contribution of 56 g · L^{-1} ammonium nitrogen). Evaluate this for temperatures of 15°C and 25°C.

Solution

By using Eq. (3.27) to calculate f_d, the fraction of digested COD in the UASB reactor is calculated as a function of the anaerobic sludge age:

$$MCH_4 = f_d \cdot MS_{ti}/4 \quad (8.21)$$

Knowing that the electric energy production is 5 kWh · kg^{-1} CH_4, the electric power generation can be calculated. In Figure 8.10 the power production is plotted as a function of the anaerobic sludge age for the specified conditions. The solubility of methane of 20 mg · L^{-1} was taken into account, resulting in methane losses in the effluent equal to 120 × 0.02 = 2.4 g CH_4 · hab^{-1} · d^{-1}, while 30% of the methane collected in the gas phase is lost from the biogas collection system.

With regards to the energy demand for aerobic post-treatment, the daily oxygen demand required to convert the soluble biodegradable organic material in the anaerobic effluent can be calculated as follows. First the biodegradable soluble COD concentration in the UASB effluent is calculated:

$$S_{bs} = (f_s - f_{nsl})S_{ti} \quad (8.22a)$$

In addition, a fraction f_{di} of the sludge production (f_x) is not retained by the GLS separators and discharged in the UASB effluent (see Section 3.9.2):

$$S_{bp} = f_{di} f_x S_{ti} \quad (8.22b)$$

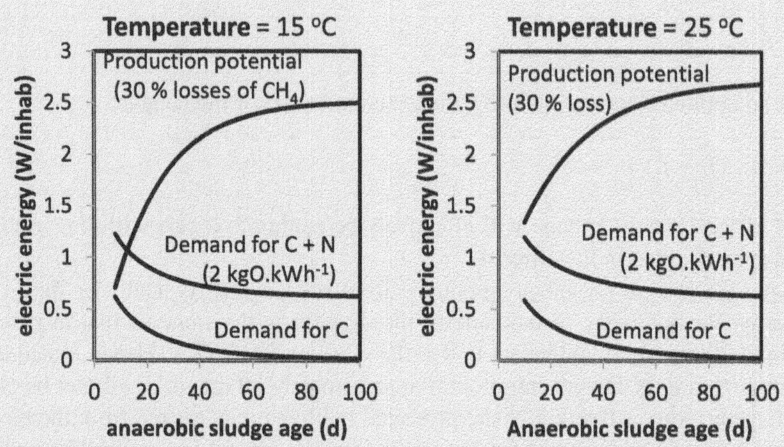

Figure 8.10 Per capita energy consumption for aeration and potential electrical power generation as function of the anaerobic sludge age for temperatures of 15 and 25°C. The applied aerobic sludge age is 5 days and cases with (C + N) and without (C) nitrification are considered.

If it is considered that this fraction is basically particulate material that can be degraded in an aerobic environment the total biodegradable concentration in the UASB effluent is

$$S_b = S_{bs} + S_{bp} = (f_s - f_{nst} + f_{di}f_x)S_{ti} \tag{8.22}$$

The per capita daily mass of organic material oxidized in the aerobic post-treatment can now be calculated with the aid of a slightly adapted version of Eq. (8.4), which defines oxygen demand in the steady-state model for activated sludge systems:

$$MS_{oc} = S_bQ_{hab}\,[1 - f_{cv}Y_{ae} + f_{cv}(1 - f)\,b_{ae}Y_{ae}R_{sae}/(1 + b_{ae}R_s)] \tag{8.23}$$

If nitrification is included the associated oxygen consumption is:

$$MO_n = 4.57MN_c = 4.57N_cQ_{hab} \tag{8.24}$$

where:

MO_n = daily oxygen demand for nitrification ($g\ O_2 \cdot d^{-1} \cdot inh^{-1}$)
MN_c = daily mass of nitrified ammonium ($g\ N \cdot d^{-1} \cdot inh^{-1}$)
N_c = concentration of nitrified ammonium ($g\ O_2 \cdot L^{-1}$)

The energy demand for aeration is directly dependent on the daily mass of oxygen consumed for COD oxidation and nitrification, and the oxygen transfer efficiency of the aeration system. In the example an aeration efficiency of 2 kg $O_2 \cdot kWh^{-1}$ was adopted for both temperatures, which requires a very efficient aerator. Figure 8.10 shows the electrical energy production potential and the consumption for aeration per inhabitant, with and without nitrification.

Figure 8.10 shows very clearly that the potential of energy generation is much higher than the energy demand for the aerobic post-treatment, except when the sludge age in the anaerobic reactor is very short and nitrification is required. In practice the energy demand may be greater than that in Figure 8.10 because (1) the efficiency of the aeration system may not reach the assumed transfer efficiency of 2 kg $O_2 \cdot kWh^{-1}$, and (2) in addition to the demand for aeration there is demand for pumping, illumination, laboratory etc. which may amount to an additional power demand equivalent to the requirement for aeration. However, even if the energy demand is twice the value calculated for aeration alone, there is still enough energy generation capacity for self-sufficiency, at least when the anaerobic sludge age is not very short. This does not necessarily mean that energy generation with the produced methane should be applied. In most cases energy production will be too expensive: only in large systems does it become economically feasible to use the potential energy production. Naturally the feasibility also depends on the price of electric energy and the stability of supply.

8.4.3 Potential for reduced sludge production

Anaerobic pre-treatment has a strong effect on the produced excess sludge mass, which is another important operational cost factor of wastewater treatment systems. With the aid of Eq. (3.25) the fraction of the influent COD that is converted into volatile sludge in the anaerobic pre-treatment can be calculated; Eq. (3.24) allows for calculation of the concentration of biodegradable COD in the effluent of the anaerobic pre-treatment, while the fraction of this biodegradable material that will be converted into excess sludge in the aerobic post-treatment can be calculated with Eq. (8.6) as a function of sludge age in this reactor with the aid of the steady-state model for activated sludge treatment (Van Haandel & Van der Lubbe, 2012).

EXAMPLE 8.5

For the same conditions as in Example 8.4, calculate the production of anaerobic and aerobic excess sludge as functions of the sludge age in the anaerobic pre-treatment reactor, maintaining a sludge age $R_s = 5$ days in the aerobic post-treatment, and compare this production with that of a purely aerobic system.

Solution

The production of volatile excess anaerobic sludge per inhabitant will be given by:

$$E_x = f_{di} f_x M S_{ti} / f_{cv}$$
$$= f_{di} M S_{ti} [f_{np35} \times 1.015^{(35-T)} + 0.16 \times 1.014^{(35-T)} \exp(-0.04(R_{san} - 3.8 \times 1.06^{(35-T)}))]$$

In Figure 8.11 the production of excess anaerobic sludge as function of the anaerobic sludge age is plotted for 15 and 25°C.

Knowing the fraction of biodegradable COD in the anaerobic effluent, the production of active sludge and endogenous residue in the aerobic system for the sludge age of 5 d is calculated. A fraction of $f_{np} = 0.03$ in the UASB effluent was used to estimate the inert sludge production. The sum of the three components (active sludge, endogenous residue and inert sludge) is also shown in Figure 8.11. It can be observed that the anaerobic excess sludge production is larger than the aerobic excess sludge production. The difference becomes larger as the anaerobic sludge age increases, as less and less COD is discharged to the aerobic system. In the same figure the production of sludge in an aerobic system without pre-anaerobic treatment was also plotted.

From Figure 8.11 it is clear that the production of excess sludge in the aerobic system is higher than in the anaerobic–aerobic treatment system, especially when the anaerobic sludge age is not very short, as will usually be the case in practice ($R_{san} \approx 50$–80 days; see Chapter 3).

However if aerobic sludge is produced at a short sludge age, the excess sludge must be stabilized, and this reduces the volatile sludge production. The stabilization of aerobic sludge will be considered in more detail in Section 8.4.4. In Figure 8.11 the production of stabilized sludge from the conventional aerobic system is also indicated. When the sludge production of the anaerobic–aerobic system is compared with the stabilized sludge production from a purely aerobic system, the difference is still considerable: in the order of 50–100%, depending on the anaerobic sludge age.

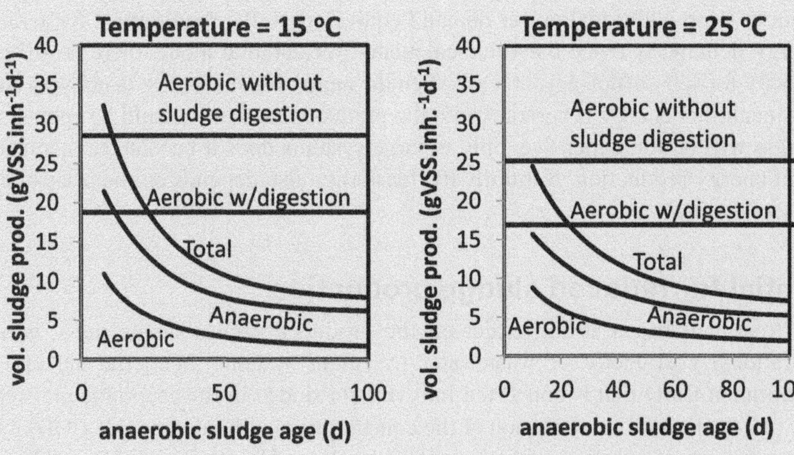

Figure 8.11 Production of organic anaerobic excess sludge as function of the applied sludge age in the anaerobic reactor for temperatures of 15 and 25°C (aerobic sludge age = 5 d).

8.4.4 Feasibility of aerobic sludge stabilization in the anaerobic reactor

In several sewage treatment systems in Brazil the aerobic post-treatment of anaerobic effluent with activated sludge systems is applied at full-scale. In these systems the UASB reactor is used not only for the pre-treatment of the influent COD, but also to stabilize the aerobic excess sludge. In this case the UASB reactor has two functions: degradation of the influent organic material and stabilization of the aerobic sludge of the post-treatment. These anaerobic–aerobic systems with simultaneous stabilization of excess aerobic sludge in the UASB reactor experience severe operational problems. This section describes an experimental study that explains why these problems exist and what can be done to solve them. It is very well known that in anaerobic digesters the digestion of the volatile solids from aerobic excess sludge is only partial. A considerable fraction of the solids persists as a non-biodegradable residue after digestion. Van Haandel and Van der Lubbe (2012) presented a model of the digestion efficiency of sludge conversion in digesters, in which it was considered that the active sludge (the live bacterial mass) is much more amenable to digestion than the non-active volatile solids (inert sludge and endogenous residue). Based on experimental results, empirical expressions were derived for the digestion efficiency of the active and inactive fractions:

$$\eta_{da} = 0.67T + 36 \tag{8.25a}$$

$$\eta_{dn} = 0.19T + 10 \tag{8.25b}$$

where:

η_{da}, η_{dn} = digestion efficiency of the active and inactive sludge fractions
T = temperature (°C)

Thus for example, at a digestion temperature of 25°C, the digestion efficiencies of active and inactive sludge are estimated at 53% and 15% respectively. The fraction of the organic solids in aerobic excess sludge that will be digested in a sludge digester can be expressed as the weighted average of the active and inactive sludge fractions:

$$\eta = f_{av}\eta_{da} + (1 - f_{av})\eta_{dn} \tag{8.25}$$

where:

η = overall digestion efficiency of aerobic excess sludge
f_{av} = active sludge fraction = ratio between active and volatile sludge concentrations

The active fraction in aerobic sludge can be calculated directly as a function of the sludge age using Eqs. (8.8) and (8.11):

$$\begin{aligned} f_{av} &= X_a/X_v \\ &= [(1 - f_{ns} - f_{np})YR_s/(1 + b_hR_s)]/[(1 - f_{ns} - f_{np})(1 + fb_hR_s)YR_s/(1 + b_hR_s) + f_{np}R_s/f_{cv}] \end{aligned} \tag{8.26}$$

During the research study excess sludge was generated in an aerobic system fed with raw sewage with non-biodegradable fractions $f_{ns} = 0.10$ and $f_{np} = 0.08$, at a constant sludge age of 3 days. In Figure 8.12 the active sludge fraction is plotted as a function of the sludge age for the average temperature during the study of 25°C. The fraction of the volatile aerobic excess sludge that can be anaerobically digested (Eq. (8.25)) is also indicated. For the applied aerobic sludge age of 3 days, the digestible sludge fraction is calculated at 41% of the volatile solids.

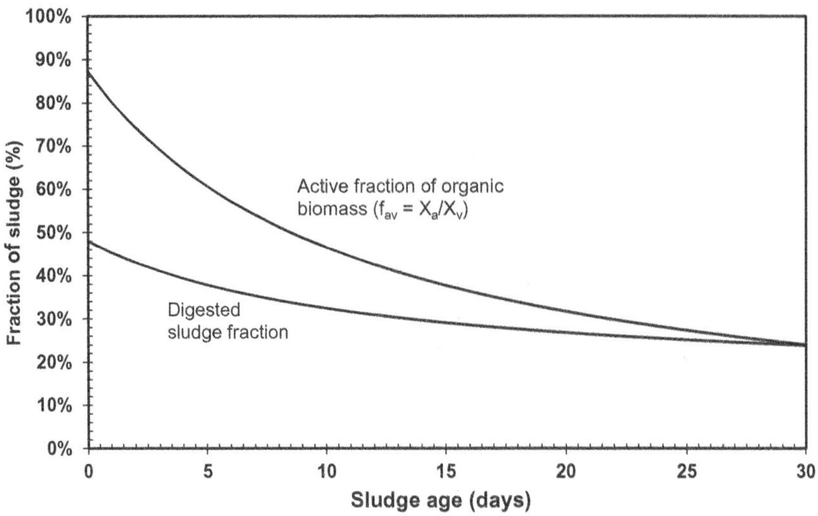

Figure 8.12 Effect of the aerobic sludge age on the composition and digestibility of aerobic sludge in conventional digesters.

The essence of the experimental investigation was to verify whether a UASB treating municipal sewage can also be used simultaneously to digest aerobic excess with the same efficiency as observed in conventional sludge digesters (at the same temperature). For this a series of 4 UASB reactors were operated. The first reactor was fed only with sewage, but incremental quantities of aerobic sludge were added to the other reactors. The additional organic loads were respectively 20%, 40% and 60% of the organic load of the sewage. Figure 8.13 illustrates the set-up of the experiment.

In practice the contribution to the COD load of the excess sludge from aerobic post-treatment reactors will never be as high as 40% or 60% of feed COD. Typically a UASB reactor removes two-thirds of the biodegradable organic load while the aerobic post-treatment removes the remaining one-third. Assuming that half of the latter COD load is oxidized and the other half is transformed into aerobic sludge, then the fraction of the feed COD ending up as aerobic excess sludge can be estimated at one-sixth. However, the high load contributions from aerobic excess sludge in the experiment were selected to evaluate whether the UASB reactor can be used to digest aerobic sludge even when the performance of the UASB reactor is poor and when therefore the production of aerobic sludge in the post-treatment system will be large.

Each UASB reactors had a height of 4 m, a diameter of 0.2 m and a useful volume of 120 litres. All reactors had sampling valves at various heights, so that it was possible to determine the profiles of the reactor sludge concentration as a function of height and to calculate the retained sludge mass. The reactors were operated without intentional sludge discharge. Under stationary conditions the sludge mass in the reactor remained essentially constant and therefore the production rate of sludge was equal to the discharge rate of solids in the effluent and could thus easily be measured. Knowing the daily discharge rate of sludge solids and the sludge mass present in the UASB, the anaerobic sludge age R_{san} was calculated as:

$$R_{san} = MX_t/ME_t \tag{8.27}$$

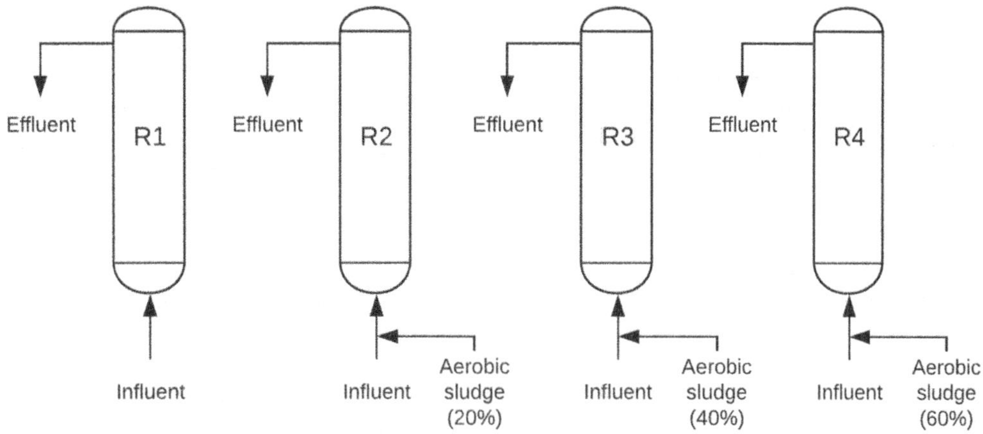

Figure 8.13 Schematic representation of the test set-up used in the research experiment.

where:

MX_t = sludge mass in the UASB reactor (g TSS)
ME_t = daily sludge mass discharged with the effluent (=excess sludge) (g TSS $\cdot$ d^{-1}).

The first UASB reactor (R_1) was operated without addition of aerobic sludge at a hydraulic retention time of 4 hours and at environmental temperature (±25°C). The sewage flow of 480 L $\cdot$ d^{-1} with an average COD of 700 mg $\cdot$ L^{-1} resulted in a daily COD load of 336 g COD $\cdot$ d^{-1}. Therefore the specific volumetric load was 336/120 = 2.8 g COD $\cdot$ L^{-1} $\cdot$ d^{-1}. Previous research (Derks, 2007) had shown that under these conditions a COD removal efficiency of 75–80% could be expected.

The other three UASB reactors were operated at the same combined (sewage + excess sludge) flow rate of 480 L $\cdot$ d^{-1}, but with COD additions in the form of aerobic excess sludge from the aerobic system in the range of 20% (R_2), 40% (R_3) and 60% (R_4) of the COD load in the sewage. The sludge was added proportionally to the sewage flow using a dosing pump. The 480 L $\cdot$ d^{-1} effluent of reactors R_1 to R_4 was discharged into 500 L tanks, so that it was possible to obtain daily average samples of the raw (non-settled) effluent by mixing the contents in these tanks. Samples of the settled effluent were obtained by settling the raw effluent in an Imhoff cone for 1 hour. After steady-state conditions had been obtained with essentially a constant sludge profile in the reactors, a 27 week investigation was carried out, during which the quality of the feed and the effluent of the UASB reactors and the sludge characteristics were monitored.

The average sewage characteristics are listed in Table 8.4. For all reactors the alkalinity was always sufficient to maintain a stable reactor pH around the neutral value, even when the maximum fraction of aerobic excess sludge was added to the total feed, and there was never any increase in VFA observed. Table 8.5 shows the applied flows and loads of both the aerobic sludge and the sewage, as well as that of the combined feed. The quality of the effluent and the sludge characteristics of the reactors were determined as well. The COD load and composition of the four reactor effluents are shown in Table 8.6.

The data summarized in Table 8.6 are calculated in the following way:

(1) Line 1: the daily total COD discharged into the effluent is the product of the raw effluent COD and the flow rate of 480 L $\cdot$ d^{-1}.

Table 8.4 Average values of the sewage fed to the reactors during the experimental study.

Variable	Suspended Solids mg · L^{-1}			VSS/TSS	COD	pH	Total Alkalinity	VFA
	TSS	VSS	ISS	%	mg O_2 · L^{-1}	–	mg $CaCO_3$ · L^{-1}	mg Hac · L^{-1}
Average	326	252	74	77	689	7.57	413	183
Maximum	462	372	116	85	841	7.77	462	227
Minimum	206	154	52	68	415	7.40	312	107
St. dev	62	50	21	4.6	105		33	36
Var. coef.	0.19	0.20	0.28	0.06	0.15		0.08	0.20

Table 8.5 Feed flow and load contributions (TSS/VSS/COD) from the sewage, the aerobic sludge and the combined feed into UASB reactors R_1 to R_4.

Variable	Aerobic Sludge				Sewage				Total			
	R_1	R_2	R_3	R_4	R_1	R_2	R_3	R_4	R_1	R_2	R_3	R_4
Q_i (l · d^{-1})	0	2.9	5.8	8.7	480	477	474	471	480	480	480	480
g TSS · d^{-1}	0	29	59	88	156	155	154	153	185	214	242	309
g VSS · d^{-1}	0	23	47	70	121	120	120	119	144	167	190	234
g COD · d^{-1}	0	35	70	106	181	180	179	178	216	251	285	359

Table 8.6 Composition of the average daily COD load in the effluent of the UASB reactors.

No	COD Loads (g · d^{-1})	R_1	R_2	R_3	R_4
1	Total COD	122.8	144.5	168.9	189.1
2	Soluble COD	71.4	73.8	75.1	77.9
3	Settleable COD	51.4	71.7	93.8	112.2
4	COD of anaerobic sludge	51.4	51.4	51.4	51.4
5	Part. COD not anaerobic sludge	–	20.3	42.4	60.8

(2) Line 2: the daily settled COD discharged into the effluent is the product of the settled COD concentration and the total flow. It can be noted that its value increases from R_1 to R_4 in line with the increase in the applied load of aerobic sludge, but this increase is very small and not significant considering the large variability of the experimental results, with a coefficient of variation in the range of 30%. This result indicates that the sewage COD load was digested efficiently in the reactors, independent of the applied sludge load and that no soluble COD accumulated in the UASB due to hydrolysis of the aerobic sludge.

(3) Line 3: the settleable COD is the difference between the total effluent COD (line 1) and the soluble COD (line 2). Its value depends greatly on the applied load of aerobic sludge, which is to be expected since only part of the aerobic sludge can be digested.
(4) Line 4: the settleable COD load of R_1 (51.4 g COD · d^{-1} or 28% of the COD load of the sewage) can be considered as the anaerobic sludge production. As the efficiency of removal of organic material from the sewage in the other reactors is only marginally smaller, the assertion is justified that in in reactors R_2 to R_4 the anaerobic digestion of organic material from the sewage develops with the same efficiency. Therefore it can be expected that in these the production of anaerobic sludge from the sewage is also approximately equal to that of R_1, i.e. 51.4 g COD · d^{-1}.
(5) Disregarding the small quantity of anaerobic biomass production from growth on digested aerobic sludge, the difference between the settleable COD (line 3) and the COD of the anaerobic sludge in the effluent (line 4) must be attributed to residual solids remaining after digestion of the aerobic sludge in the reactors R_2 to R_4. In fact in good approximation this COD mass is proportional to the COD mass added as aerobic sludge.

Figure 8.14 shows the daily COD load applied in the form of aerobic waste sludge to the four UASB reactors as well as the COD masses leaving the reactors in particulate form that can be attributed to non-biodegradable organic solids of digested aerobic sludge. On average the non-biodegradable particulate COD corresponded to 59% of the COD load applied in the aerobic sludge. Hence 41% of the volatile solids in the aerobic waste sludge were digested, which is equal to the theoretical value of the digestion efficiency in sludge digesters calculated previously and shown in Figure 8.12. Therefore it can be concluded that the UASB reactors digested the aerobic sludge with the same efficiency as observed in conventional sludge digesters that are operated at the same temperature, while the digestion of the organic material in the sewage was not adversely affected. However, it is important to stress that the experimental set-up during the research study was different from the usual combined anaerobic–aerobic configuration, because the

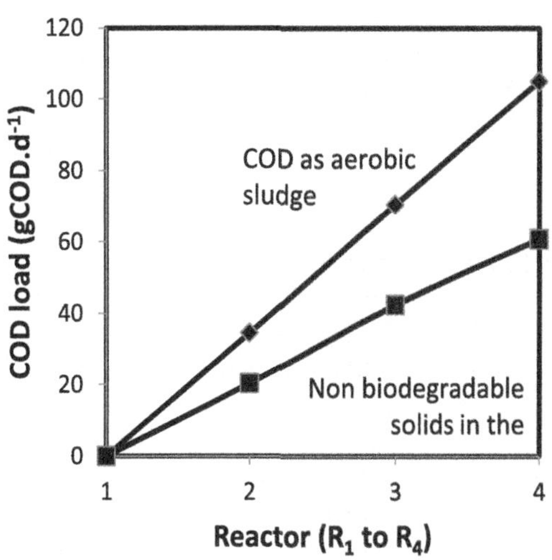

Figure 8.14 Relationship between daily COD loads fed as aerobic sludge and discharged as non-biodegradable solids in the four UASB reactors.

Table 8.7 Total and volatile sludge mass in the UASB reactors and their effluent. The value of the anaerobic sludge age is shown as well, calculated based on total and volatile suspended solids.

Parameter	Unit	R_1	R_2	R_3	R_4
Total sludge mass in reactor	g TSS	3047	3023	2801	2739
Volatile solids mass in reactor	g VSS	1706	1844	1708	1698
Total sludge discharge	g TSS · d^{-1}	45.7	63.7	83.4	99.7
Volatile sludge discharge	g VSS · d^{-1}	34.3	47.8	62.5	74.8
Sludge age (based on X_t)	days	66	47	34	27
Sludge age (based on X_v)	days	49	39	27	23

aerobic excess sludge did not originate from a system fed with the UASB effluent. This difference proved to be of vital importance, as will be discussed later in this section.

Table 8.7 shows the mass of total and volatile solids in the UASB reactors and in the effluent. From these data the sludge age can be calculated, either as the ratio of TSS mass and TSS discharge or VSS mass over VSS discharge. The values of the sludge age are shown in Table 8.7 as well. The data show two peculiarities: (1) the sludge age based on the total solids is systematically higher than the value based on the volatile solids, which means that there was selective discharge of volatile solids; and (2) the anaerobic sludge age becomes very short when the aerobic sludge mass is more than 40%, although the digestion efficiency of both the sewage COD and the aerobic sludge remains high. A possible explanation is that the non-biodegradable residue of the digested aerobic sludge is preferentially discharged from the UASB reactors.

Every week the physical and biological properties of the sludge were analysed. The tests were carried out both on samples of settled sludge in the effluents of R_1 to R_4 and the contents of the reactors themselves. To characterize the sludge, the specific methanogenic activity (SMA), the sludge stability (the fraction of volatile sludge mass that is digested) and the sludge settleability were determined. The results are summarized in Table 8.8. The activity tests were carried out according to the procedure presented in Appendix A1, using sodium acetate as a source of COD. The tests were carried out at 25°C, the same temperature at which the pilot reactors were operated.

Table 8.8 Results of tests for specific methanogenic activity (SMA), sludge stability and sludge settleability of the biomass in reactors R_1 to R_4 and in their effluents.

Test	Sample	R_1	R_2	R_3	R_4
SMA	Effluent solids	0.08	0.18	0.10	0.09
(mg COD · mg^{-1} X_v · d^{-1})	Sludge in reactor	0.14	0.17	0.18	0.19
Sludge stability	Effluent solids	11%	13%	15%	16%
(loss of volatile sludge mass)	Sludge in reactor	16%	17%	19%	19%
Sludge settleability:	Effluent solids	0.25	0.24	0.28	0.21
compressibility: k (l · g^{-1} TSS)	Sludge in reactor	0.04	0.04	0.05	0.04
Settleability:	Effluent solids	7.3	6.1	8.4	7.0
unhindered settling rate v_0 (m · h^{-1})	Sludge in reactor	6.5	6.5	8.4	5.2

The results in Table 8.8 indicate that the SMA of the sludge samples from the UASB reactors was certainly not adversely affected by the addition of aerobic sludge to UASB reactors: in fact the SMA tended to increase with increased addition of aerobic sludge to the reactors. Furthermore the SMA of the sludge expelled with the reactor effluent tended to be lower than that of sludge from the reactors. The results indicate that methanogenic sludge is selectively retained in the reactor. It is concluded that it is indeed the non-digestible fraction of the aerobic sludge that is preferentially expelled from the reactor. It is interesting to compare the experimental SMA results for the UASB reactors treating sewage + aerobic sludge with results from other authors. Silva and van Haandel (2014) and Pontes *et al.* (2003) investigated the SMA of sludges in UASB reactors treating sewage and aerobic sludge from systems with fixed biomass (submerged aerobic filters and biofilters respectively) and reported SMA values of 0.12 and 0.10 mg COD · mg^{-1} VSS · d^{-1} respectively. Leitão (2004) and Van Haandel and Lettinga (1994), using their own data and those published by other authors, published reported SMA values of sludges from UASB reactors treating sewage: 0.18 and 0.08–0.30 mg COD · mg^{-1} VSS · d^{-1} respectively. It is concluded that the results of this study are within the range obtained by other authors.

In the stability test as described in Appendix A1, sludge samples were incubated at ambient temperature (25°C) and it was determined experimentally how much of the initial organic solids in the sample were converted into methane, when left for a period of 28 days, by measuring the methane production over the period. The results from Table 8.8 indicate that the sludge tend to be less stable as more aerobic sludge is added, although the difference is small. Furthermore the effluent solids are more stable than the sludge from the reactors. By the criteria suggested by the EPA (2003) sludge is considered stable when the loss of volatile solids during the test is less than 20%. By this criterion the expelled sludge can be considered stable, i.e. it can be dewatered and disposed without additional treatment. Also the sludges in the reactors have a digestible fraction marginally below 20%.

Finally, sludge settleability tests were performed with sludge from reactors R_1 to R_4 as well as with the solids present in the effluents. It was established that the settling velocity in all cases conformed well to the Vesilind equation (see Appendix A2). The experimental values of the obtained constants v_0 and k are listed in Table 8.8. The settleability of the effluent solids was much lower than that of the sludge in the reactor. Figure 8.15 shows a very clear example of this different settling behaviour for reactor R_1. It can

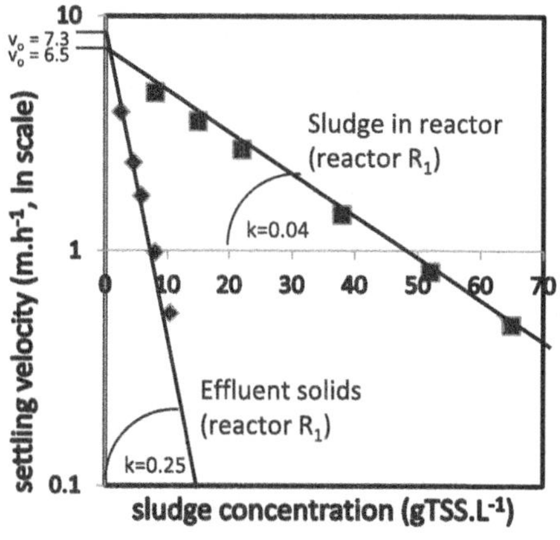

Figure 8.15 Semi-log plot of settling velocity against sludge concentration to determine the settleability constants for R_1 sludge.

be concluded that during the research project there was a selective loss of less settleable sludge from the reactor.

It is interesting to compare the calculated settling velocities to the upward velocity of the liquid in the reactors R_1 to R_4. Since these reactors have a height of 4 m and a residence time of 4 hrs, the upward liquid velocity was $1\ m \cdot h^{-1}$. Taking as example reactor R_1, it can be calculated that according to Vesilind's equation the highest concentration of sludge that can be retained by zone settling in the reactor is given by:

$$v = v_0 \exp(-kX_t) = 1\ m \cdot h^{-1} \tag{8.28}$$

For the settling characteristics determined for the effluent solids ($k = 0.25\ L \cdot g^{-1}$ and $v_0 = 7.3\ m \cdot h^{-1}$) and assuming that conditions for zone settling apply, the corresponding maximum solids concentration in the reactor can be calculated as $8\ g\ TSS \cdot L^{-1}$ for the upflow velocity of $1\ m \cdot h^{-1}$. This value can now be compared with the actual values obtained in the reactors. In Figure 8.16 the typical sludge concentration profile as a function of the height of reactors R_1 to R_4 is shown. It can be observed that for heights below 3 m (i.e. in almost the entire digestion zone), the actual sludge concentration was much higher than the maximum concentration calculated for the effluent solids, which means that it would have been impossible to retain sludge with such relatively poor settleability. So from the data on sludge settleability alone it can already be deducted that this sludge would be lost in the effluent.

In contrast, when considering the settleability constants of the sludge samples taken from reactor R_1 ($k = 0.04\ L \cdot g^{-1}$; $v_0 = 6.5\ m \cdot h^{-1}$), the maximum concentration that can be retained at $1\ m \cdot h^{-1}$ was 38 g $TSS \cdot L^{-1}$, which is higher than the concentration in most parts of the reactor. In fact, only well below 1 m of reactor height was the reactor sludge concentration higher than the value of $38\ g\ TSS \cdot L^{-1}$. Possibly in this bottom section the sludge settleability was higher owing to heavy particles introduced there. Another factor to consider is that the zone settling test is designed (and executed) for settling of sludge in absence of aeration (and biogas production!). The production of biogas has a significant disturbing effect on biomass settling, like that of nitrogen released in final settlers if denitrification is incomplete. Furthermore, as organic material is converted is converted while the feed travels upwards in the UASB reactor, the biogas velocity increases. This might be – in addition to the natural tendency for heavier solids to accumulate in

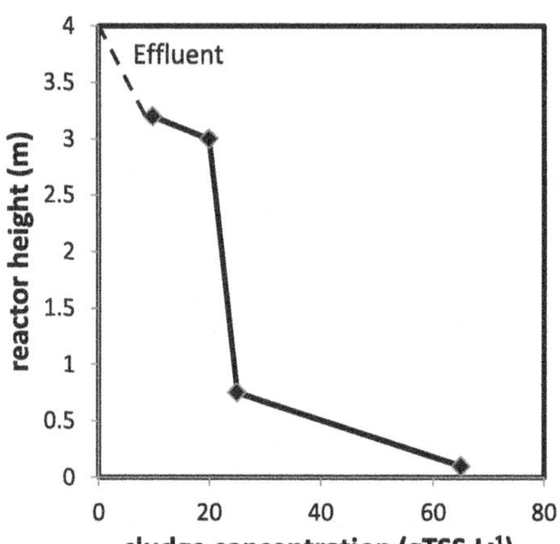

Figure 8.16 Typical concentration–height profile observed in reactors R_1 to R_4 during the experimental investigation.

the bottom part of the reactor – a second reason why at higher level in the reactor the biomass concentration is lower. One might argue that above the GLS separator biogas will be absent, but the solids concentration at this point is so low that zone settling no longer applies, and discrete particle settling proceeds at lower velocities.

From the results of this study discussed above, it can be concluded that it is indeed feasible to use the UASB sewage treatment reactor, not only for pre-treatment of raw sewage, but also for simultaneous stabilization of excess sludge from an aerobic system. This feasibility continues to exist even when if the COD fraction added in the form of aerobic sludge (evaluated at a maximum fraction of 60%) exceeds the maximum value that can expected in practice for anaerobic–aerobic systems (15–20%).

However, this does not mean that the configuration of the conventional activated sludge system as shown in Figure 8.1a can directly be replaced by an anaerobic–aerobic system as shown in Figure 8.2. The problem is that anaerobic stabilization of excess aerobic sludge by anaerobic digestion will convert only a fraction of the volatile solids into methane (about 40% at 25°C and under the conditions of the investigation). Hence most of the aerobic sludge remains after stabilization as non-biodegradable volatile solids in the UASB. As these solids have a relatively poor settleability, they will be expelled from the UASB reactor together with the reactor effluent. As these solids are not retained in the UASB reactor, they cannot be discarded together with the normal excess sludge discharge of the UASB reactor as tacitly assumed in Figure 8.2. Normally, there is no additional solids separation unit present between the anaerobic reactor and the aerobic post-treatment units, and consequently the solids expelled from the UASB reactor will be introduced in the aerobic system. However, being composed largely of non-biodegradable material, these solids will not be degraded in the aerobic system and thus end up being discarded again as excess sludge to the anaerobic reactor. Thus the serious problem is created that the non-digestible fraction of the aerobic sludge has no regular outlet from the system and therefore will accumulate in both the anaerobic and aerobic reactors. This will continue up to the point that the concentration of these solids exceeds the separation capacity of the final settler, after which the solids will be discharged through the only exit available in the system: i.e. together with the final effluent.

This vicious cycle due to endless recirculation of non-degradable solids is observed in all anaerobic–aerobic systems and is responsible for frequent operational problems. Naturally, the presence of these solids in the final effluent will affect its quality and therefore is undesirable.

The discharge of anaerobic sludge from UASB reactor does not offer a solution for this problem: the non-digested aerobic sludge in the UASB reactor has a settleability which allows only for a low concentration and therefore it will be expelled from the reactor, even if the reactor is not "full" of sludge.

In full-scale systems, with a large daily variation of the sewage flow, there is sometimes a practice of introducing the aerobic sludge into the UASB reactor at times of low flow and consequently at low upward liquid velocity. However, inevitably this sludge will be expelled as soon as the upward velocity of the liquid increases again during times of peak load. It can be concluded that, if aerobic sludge is discharged to the UASB, the inclusion of a gravity settler between the anaerobic and aerobic treatment units as shown in Figure 8.17 is unavoidable to maintain good operating conditions in both systems and a high quality of final effluent.

To make sure that the poorly settling solids are removed, low upflow velocities should be applied, at least well below the overflow velocities applied in the anaerobic unit, which results in a large settling area being required. In fact it might thus very well be more attractive to discharge the aerobic sludge not to the UASB, but rather thicken it in a dedicated gravity thickener before final disposal. Considering that the solids concentration in the UASB reactor effluent will be low (TSS concentrations in the range of 100–300 mg $\cdot$ L^{-1} were measured in R_4), the settler need not necessarily be mechanized. The settled sludge in the intermediate settler can be dehydrated without further treatment, since it consists mainly of non-biodegradable solids.

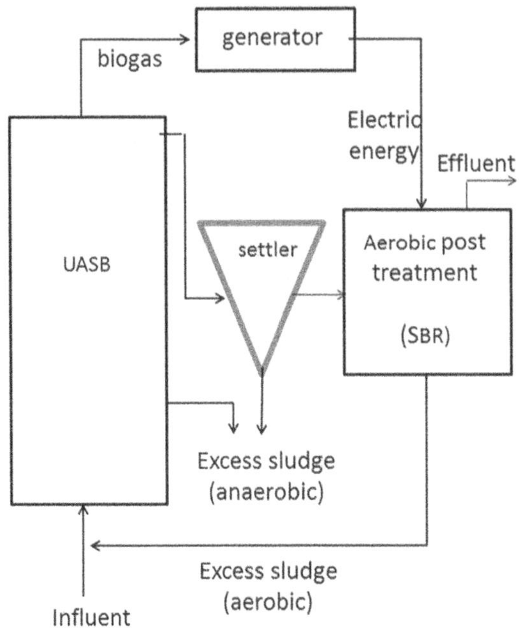

Figure 8.17 Intermediate settler to remove non-biodegradable solids from the UASB effluent before discharge in the post-treatment reactor.

8.5 NUTRIENT REMOVAL AFTER ANAEROBIC PRE-TREATMENT

In the previous sections it has been shown that an efficient pre-treatment process for COD removal as applied in a UASB reactor drastically reduces the concentration of biodegradable organic material and thus renders the traditional biological removal processes of nitrogen and phosphorus from its effluent virtually impossible. Relatively recently, however, alternatives have been developed that can remove these nutrients efficiently and thus in principle should allow an effluent of a quality compatible with legal standards. At the time of writing (2019), these alternatives are not mature enough for application on a low-strength, low-temperature effluent like that from a municipal sewage UASB, but development work on so-called mainstream application of these new processes is ongoing in many research institutes. The most promising of these alternative processes will be discussed in the following sections.

8.5.1 Nitrogen removal from effluent with low content of organic material

A process with potentially great possibilities for the removal of nitrogen from (pre-treated) municipal wastewater with a deficit of organic material was developed at the University of Delft (Mulder *et al.*, 1995): the Anammox process. It utilizes a species of autotrophic bacteria, which obtains energy from the oxidation of ammonia by nitrite, transforming these forms of nitrogen into molecular nitrogen, which desorbs from the liquid phase. Autotrophic denitrification, without the mediation of heterotrophic bacteria and the consumption of organic material, in its simplest form can be expressed as:

$$NH_4^+ + NO_2^- \rightarrow N_2 + 2H_2O \tag{8.29}$$

The Anammox process has two important advantages over the traditional processes of nitrification to nitrate and denitrification with organic material: (1) it greatly reduces the oxygen demand, and (2) it does not require organic material. The discovery of the Anammox process allowed the development of several systems to treat effluents rich in nitrogen and poor in organic material, such as the effluent from sludge digesters. Equation (8.28) represents only the catabolic part of the metabolic process. The Anammox bacteria, being autotrophic, need CO_2 for their growth. Vlaeminck *et al.* (2012) has proposed that Anammox bacterial mass can be represented as $CH_2O_{0.5}N_{0.15}$ and suggested the following overall equation:

$$NH_4^+ + 1.32\ NO_2^- + 0.066\ HCO_3^- + 0.13\ H^+ \rightarrow 1.02N_2 + 0.26NO_3^- + 0.066CH_2O_{0.5}N_{0.15} + 2.03H_2O \quad (8.30)$$

From Eq. (8.30) it follows that 0.066 mmol or 1.6 mg of bacterial mass is produced per $2.32 \times 14 = 32$ mg of substrates (NH_4^+ and NO_2^-), so that the yield coefficient is calculated as $1.6/32 = 0.05$ mg VSS $\cdot$ mg^{-1} N. In Eq. (8.30) it is assumed that CO_2 is reduced by NO_2^-, which in the process is oxidized to NO_3^-. Thus 0.26 mol of nitrate is produced per 2.32 mol of removed nitrogen, so that the maximum efficiency of the Anammox process is $0.26/2.32 = 89\%$. In a perfectly balanced process there will be 11% of the original ammonium load present as nitrate in the effluent, owing to the partial CO_2 reduction required for biomass synthesis.

As nitrogen in wastewaters is normally present as ammonium, especially after anaerobic pre-treatment, the catabolic reaction of Eq. (8.29) actually consist of two separate steps: (1) oxidation of part of the ammonia to nitrite, and (2) the redox process of the produced nitrite with the residual ammonia. These processes can be carried out in systems with two reactors: an aerobic zone for nitritation and an anoxic zone for Anammox, or the processes can be carried out alternatingly, separated in time. If the Anammox process is carried out in two separate stages, this implies that both stages of the process must be optimized and controlled individually. The investment costs and the difficulty of balancing the two stages are reduced when operating a single-stage system, where the processes develop simultaneously (Vlaeminck *et al.*, 2012). Due to these advantages, full-scale applications now tend to have a single-stage configuration, although this alternative will not necessarily prevail when treating waters with a low nitrogen concentration.

The challenge for the application of the Anammox system is that nitrite must be oxidized to nitrite (the first step in the nitrification process), but that the nitrite must not be further oxidized to nitrate (the second step). The kinetics of nitritation and nitratation are such that at temperatures lower than about 15°C the growth rate of nitrite oxidizers is greater than that of ammonium oxidizers, so that nitrite tends to be converted to nitrate and application of the Anammox system is not feasible. At temperatures above 15°C the situation is inverted: ammonium oxidizers grow faster than nitrite oxidizers and by careful management of the sludge age and the DO concentration, nitrite can be produced without further conversion to nitrate. For this reason the full-scale systems that are operational today all treat waste streams with ammonium at a high temperature, normally around 30–35°C, the optimal temperature for anaerobic sludge digestion. In regions with a cold climate, sludge digesters are normally heated and the liquid phase of the digesters will have a high temperature. Apart from the high temperature, another reason for the successful application of Anammox systems to the effluent of anaerobic sludge digesters is that the ammonium concentration is high, in the order of 1–2 g N $\cdot$ L^{-1}. If it is considered that the sludge digester must have a high alkalinity to maintain a stable pH, it is concluded that the partial CO_2 pressure in such a digester is high and if its effluent is released to the atmosphere, dissolved CO_2 will be removed, and the pH will rise.

So far, the application of the Anammox system to remove ammonium directly from pre-treated municipal sewage (for example UASB effluent, effluent from high-rate activated sludge systems with only BOD removal and nitrification) have been unsuccessful, mainly because it has not yet been possible to have stable (long-term) conversion of the ammonium into nitrite without producing any nitrate. In much of the world the sewage temperature will be well below 15°C for at least part of the year. A further complicating factor is that the ammonium concentration in municipal sewage is often quite low to start with, typically only 40–60 mg $N \cdot l^{-1}$. However, in regions with a hot climate and with year-round sewage temperatures well above 15°C, controlled nitrite production should be feasible, so that the system could be used for ammonium removal from anaerobically digested sewage.

Several pilot scale studies have produced promising results, but so far no full-scale Anammox systems for nitrogen removal from digested sewage have been built. However, there are some issues that should be considered:

- UASB effluent tends to have residual solids in its effluent (40–80 mg $TSS \cdot L^{-1}$ for a well designed and constructed reactor). If these solids are non-biodegradable, they might accumulate in the Anammox system and reduce the sludge age, which reduces the mass of Anammox bacteria, necessary for autotrophic nitrogen removal.
- A more significant problem is that biodegradable solids as well as dissolved biodegradable matter in the UASB effluent will be used for heterotrophic bacterial growth and that these bacteria easily outgrow the Anammox bacteria, which have a small yield coefficient and a low growth rate.
- To mediate the two above issues, it could be considered to install the Anammox after the aerobic post-system, which would then be designed only for the removal of organic material, and operated at low sludge age to suppress nitrification.
- H_2S will be present in the anaerobic effluent, but is toxic for both Anammox bacteria and nitrifiers. However, this can easily be removed by having a small aerated tank up front of the Anammox.

8.5.2 Phosphorus removal from effluent with low content of organic material

When the concentration of biodegradable organic material is low, as in the UASB reactor effluent, biological removal of phosphorus becomes impossible, but instead phosphorus removal by precipitation with metal ions can be considered. Metal ions of low-solubility salts such as $FePO_4$, $AlPO_4$ are used, which requires the addition of Fe^{3+} or Al^{3+} cations, generally in the form of $FeCl_3$ or $Al_2(SO_4)_3$.

Precipitation of struvite (also known as MAP, magnesium-ammonium-phosphate), the co-precipitate of magnesium, ammonia and phosphate at a pH value of 8–8.5 in flocculent or (preferably) granular form, which can be sold as fertilizer for agriculture, has received significant attention in the last 20 years and has by now become a well established technology. The reaction proceeds as per the equation shown below:

$$Mg^{2+} + NH_4^+ + PO_4^{3-} + 6H_2O \rightarrow MgNH_4PO_4 \cdot 6H_2O(s) \tag{8.31}$$

Either $MgCl_2$ or MgO can be used to supply the deficit in magnesium. Struvite precipitation can be readily applied after anaerobic treatment processes and there are several full-scale applications on reject water from anaerobic digesters and on specific anaerobically pre-treated wastewaters with high P-content. However, it is not suitable for treatment of effluent from municipal sewage UASB reactors, as the minimum residual PO_4-P that can be obtained is between 15 and 25 mg $P \cdot L^{-1}$, which is higher than the total phosphorus concentration in the feed for most municipal sewage. For this reason this technology will not be discussed here further.

The general experience with precipitation of phosphates with metal salts directly in the wastewater or from treated effluent is that the demand for these salts is much higher than can be expected based on reaction stoichiometry i.e., a large proportion of the cations is consumed by other processes, such as the formation of metal hydroxides. Figure 8.18 shows the relation between the molar ratio of the added metal ion and the phosphate concentration in the wastewater versus the range of residual phosphorus concentration that can be obtained.

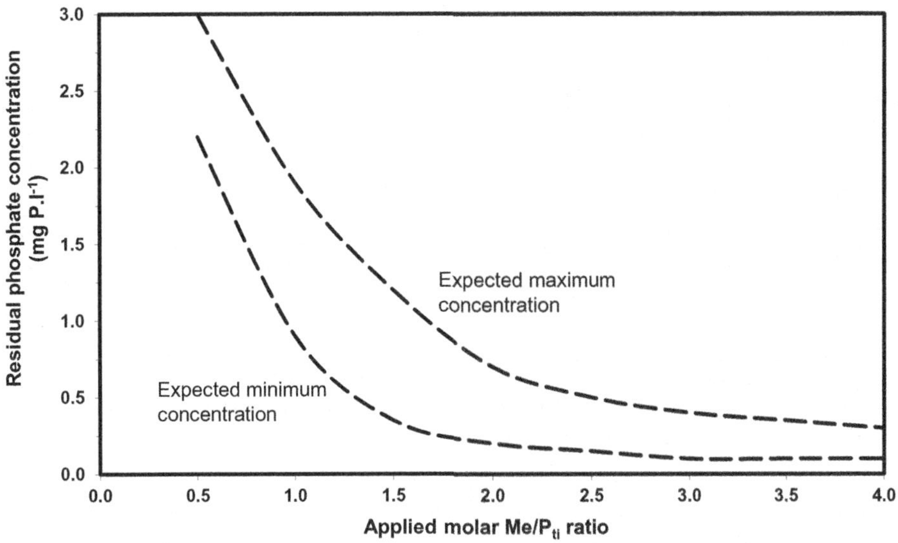

Figure 8.18 Residual PO_4-P concentration as function of molar Me/P dosing ratio for co-precipitation (de Jong *et al.*, 1993).

The figure is based on data collected in full-scale aerobic plants where co-precipitation was applied. The process configuration is shown in Figure 8.19. There are three important aspects in the diagram:

(1) For an equimolar Me/P ratio an effluent PO_4-P concentration of 2–3 mg $P \cdot L^{-1}$ is to be expected;
(2) For the lowest range of residual PO_4-P concentrations, a minimum Me/P ratio of 3 is required;
(3) However, even then the minimum residual *P* concentration is still in the range of 0.2–0.6 mg $P \cdot L^{-1}$, which may not be sufficiently low.

EXAMPLE 8.6

$Al_2(SO_4)_3$ is used at a molar Al:P dosing ratio of 3 to reduce the phosphorus concentration of UASB effluent from 10 to 0.5 mg $\cdot L^{-1}$. What will be the amount of inorganic sludge mass that is produced and what will be the sulphate concentration in the effluent?

Solution

To remove 9.5 mg $P \cdot L^{-1}$ or 9.5/31 = 0.31 mmol $P \cdot L^{-1}$, the addition of $3 \times 0.31 = 0.92$ mmol $\cdot L^{-1}$ of aluminium ions is required, of which one-third (0.31 mmol) will precipitate as $AlPO_4$ and two-thirds (0.61 mmol) as $Al(OH)_3$. Thus the production of inorganic sludge will be $0.31(27 + 31 + 64) + 0.61(26 + 3 \times 17) = 84.6$ mg $\cdot L^{-1}$. Compared to the biological sludge production, this is a very significant amount. In Figure 8.11 it was established that the biological sludge production in an aerobic system treating UASB effluent is in the order of 10 g $\cdot$ inh^{-1} or 10 g per 120 L = 83 mg $\cdot L^{-1}$. Hence, it may therefore be expected that the physical–chemical removal of phosphate will double the excess sludge production. According to the structural formula of the metal salt used, for each mole of Al, 1.5 mole of SO_4 is added as well. Hence the concentration of sulphate will increase with $0.92 \times 1.5 \times 96 = 132.4$ mg $SO_4 \cdot L^{-1}$ or 132.4/3 = 44.1 mg $S \cdot L^{-1}$. This is a very serious deterioration of the effluent quality that may lead to odour problems and reduces the applicability of agricultural reuse.

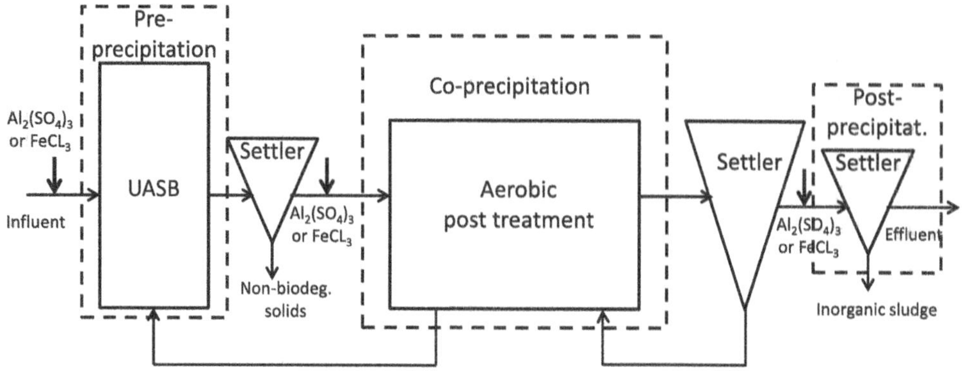

Figure 8.19 Flow sheet of anaerobic–aerobic treatment system with possible dosing locations to feed metal salts for chemical phosphate precipitation.

Lime ($Ca(OH)_2$) can also be used for precipitation of phosphorus. The main reactions for lime precipitation are the formation of calcium hydrogen phosphate $CaHPO_4$ and hydroxyapatite $Ca_{10}(OH)_2(PO_4)_6$. As calcium is usually present in sufficient quantity in the wastewater, lime addition is only necessary to increase the pH. The following reactions are relevant:

$$4Ca^{2+} + H^+ + 3PO_4^{3-} \rightarrow Ca_4H(PO_4)_3(s)\,(pK_s = -46.9) \qquad (8.32)$$

$$5Ca^{2+} + 3PO_4^{3-} + OH^- \rightarrow Ca_{10}(OH)_2(PO_4)_6(s) \quad (pK_s = -114) \qquad (8.33)$$

The values of the solubility constants indicate that for any pH value $Ca_{10}(OH)_2(PO_4)_6$ is much less soluble than $CaHPO_4$, but in practice oversaturated phosphate salt may form (see also Section 9.8). Another problem with lime treatment is that the bicarbonate concentration in anaerobic effluents is relatively high (to maintain a stable and neutral pH), and when lime is added, carbonate will tend to form and this will lead to precipitation of calcium carbonate (see also Chapter 5), so that the phosphate remains in solution:

$$Ca^{2+}(aq) + CO_3^{2-} \rightarrow CaCO_3(s)(\text{at pH} > 9.5) \qquad (8.34)$$

In large treatment plants installation of a lime recovery system can be considered from the thermal decomposition of calcium carbonate, but generally the produced sludge is used in agriculture to correct the pH of acid soils. As the dosage of lime is used only to raise the pH and is not directly linked to the concentration of phosphorus, its application can be considered in concentrated phosphorus effluents, such as reject water from the sludge digester or when the pH of the liquid phase is already high, such as in the case of polishing ponds treating anaerobic effluent (see Section 9.7). However, as discussed earlier, in this case struvite precipitation can also be applied, which is a product with much higher value. In general the use of lime for phosphorus removal is in decline due to the production of large amounts of inert sludge (lime impurities besides $CaCO_3$) and because lime is a difficult substance to handle, in terms of storage, preparation and dosing.

In practice there are three opportunities to apply metals salts to the treatment system: (1) in the anaerobic pre-treatment, (2) in the aerobic post-treatment, or (3) in a separate physical–chemical unit

Table 8.9 Advantages and disadvantages of pre-, co- and post-precipitation of phosphate.

Method	Advantages	Disadvantages
Pre-precipitation	• Flocculation of organic material may increase digestion efficiency • Less risk of heavy metal discharge in the final effluent • May improve sludge settleability	• May reduce dewaterability of the sludge • High consumption of chemicals • High inorganic fraction may reduce biological treatment capacity • Polymers may be required
Co-precipitation	• Less cost • Will improve sludge settleability	• Loss of treatment capacity due to high inorganic fraction • High inorganic sludge production • pH adjustment may be required
Post-precipitation	• Lowest residual P concentration • Can be combined with other post-treatment methods	• Additional treatment units are required • High investment costs • Higher risk of phosphate discharge

installed especially for phosphate precipitation from the aerobic effluent, shown schematically in Figure 8.19. Table 8.9 summarizes the positive and negative aspects of each alternative. It is concluded that in any configuration that is used, physical–chemical phosphorus removal has important disadvantages:

- The residual P concentration may be too high to be compatible with the existing legal standards for effluent quality, or alternatively at low residual P concentrations the required molar dosing ratio may be very high.
- The amount of metal salt required to precipitate phosphate is very considerable and will lead to a large increase of the sludge production.
- The salinity of the water increases considerably, which reduces the value of the final effluent for agricultural reuse, particularly if a metal sulphate salt is used.

In view of these important disadvantages, it is advisable that phosphate precipitation is avoided whenever possible. A particularly interesting alternative is agricultural reuse of the effluent, transforming the high concentration of phosphate into an asset.

Chapter 9

Post-treatment of digested sewage in polishing ponds

9.1 INTRODUCTION

Waste stabilization pond systems (WSP) have been widely used in municipal sewage treatment in most tropical countries, with generally satisfactory results in terms of effluent quality, at least when the project execution is technically adequate and there is a minimum of operation and maintenance. In this context "satisfactory" means that there is a considerable reduction of organic material and suspended solids as well as pathogens, but the nutrient concentrations remain almost unchanged in the treatment process. This effluent is suitable for irrigation, where the presence of nutrients can be considered an advantage, but not for discharge into surface waters as in practice usually occurs. In reuse the nutrients can replace fertilizer that would otherwise have to be applied to increase productivity. When the effluent is released onto surface waters, it can cause serious eutrophication problems.

As the name says, the main objective of a WSP is to stabilize organic material by transforming it into stable mineral products. To achieve this goal, a treatment system is used that is based on the metabolic activity of micro-organisms, particularly bacteria and algae. Algae produce oxygen through photosynthesis and this oxygen can be used by bacteria to oxidize biodegradable organic material. Alternatively, in the absence of oxygen, anaerobic bacteria can transform the organic material into biogas through the anaerobic digestion process (Mara & Pearson, 1987; Yañez, 1993). The result of the biological treatment is that the organic material concentration decreases drastically during the treatment process, leaving a final effluent with a low BOD_5 value. However, the required retention time (R_h) to produce enough oxygen to complete the treatment is long, even in places like Brazil, where the conditions of the tropical climate (high temperature, high incidence of solar irradiation for photosynthesis) are favourable for the process.

The long retention time required for the stabilization of the organic material has an important indirect advantage: the liquid remains in the lagoon system for a period sufficient for the complete removal of helminth eggs and a high efficiency of removal of thermotolerant (TT) or coliform bacteria, so that a final effluent with good hygienic quality is automatically guaranteed (Von Sperling, 1999; Yañez, 1993). It is concluded that conventional systems of WSP can produce a final effluent with low concentrations of BOD_5

Editor(s): Adrianus van Haandel and Jeroen van der Lubbe
doi: 10.2166/9781780409627_0335

and TSS and good hygienic quality, but for this good performance these ponds require a long retention time and, consequently, a large volume and area.

In order to assure good performance and at the same time minimize the retention, the WSP system must consist of several ponds operated in series (Mara, 1976) as shown in Figure 9.1. The first lagoon is an anaerobic lagoon (AL), which receives raw sewage, and the high BOD_5 load applied establishes a predominantly anaerobic environment.

In this pond the mechanisms of organic material removal are settling and anaerobic digestion (Marais & Shaw, 1961); algae cannot develop in this environment. In the second lagoon of the series, which is called a facultative lagoon or pond (FP), there is an anaerobic environment in the lower layer and an aerobic one in the upper layer, due to photosynthetic oxygen production by algae. In the FP two removal processes of the organic material occur simultaneously: oxidation in the upper part of the pond and anaerobic digestion in the lower part. Generally in a pond series there are one or more ponds after the FP, with a predominantly aerobic environment, that are called maturation ponds (MP). These MPs serve for complementary aerobic treatment and improvement of the hygienic quality of the effluent.

It is important to stress that the division of the WSP system into a series of ponds with AL, FL and ML is necessary when reducing retention time is desired. The total hydraulic retention time of WSP systems containing the three pond types in regions with a hot climate is in the order of 30 days. The anaerobic lagoon (AL) is small but relatively efficient: although it has a short retention time (2–5 days) it removes a large fraction of the organic material (50–60% of BOD_5 according to Mara (1976), particularly when the temperature is high enough (>20°C)). Therefore, the AL is instrumental in accomplishing much of the main function of the WSP system: the removal of the organic material. On the other hand the AL has several operational problems: strong odour emissions due to the production of sulphide, accumulation of solids from the influent, and methane release to the atmosphere, which is undesirable, since methane contributes significantly to the greenhouse effect. Usually more than one anaerobic lagoon is constructed, so that at least one lagoon is operational while the accumulated solids from the other lagoon are removed.

To overcome the serious operational problems associated with an AL, there is a tendency in Brazil to build systems without AL, releasing the raw sewage into a larger facultative pond. To avoid this FP

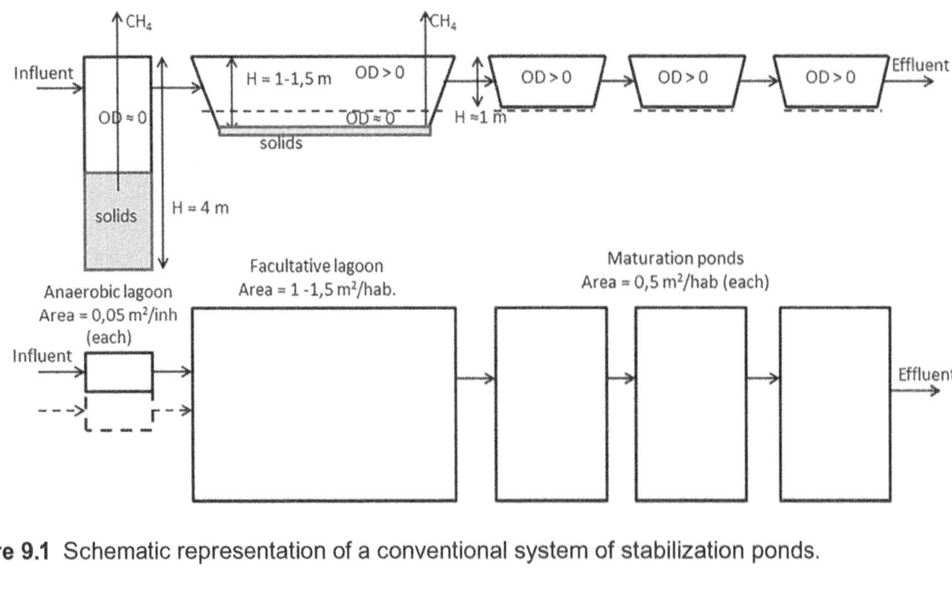

Figure 9.1 Schematic representation of a conventional system of stabilization ponds.

becoming predominantly anaerobic, its retention time must be increased by at least 50%, so that the total pond system will now have a retention time of much more than 30 days. However the accumulation of solids and the release of methane to the atmosphere will persist, although the layer of accumulated solids at the bottom will increase more slowly because the FL is much bigger. Only the emission of foul odours will be reduced.

The substitution of the AL by a UASB reactor maintains – and in fact increases – the removal of organic material through anaerobic digestion and eliminates the problems of the AL. The substitution of an AL by a UASB is an excellent measure as, due to its high efficiency, the UASB reactor replaces not only the anaerobic lagoon but also the facultative lagoon, while the effluent can be discharged directly to the MPs. Under these conditions the removal of residual organic matter is a process that takes relatively little time and the limiting factor that determines the minimum retention time of a pond system will normally be the removal of pathogenic organisms and not the stabilization of organic matter. In order to avoid confusion about terminology, ponds that are used for the post-treatment of effluents from efficient anaerobic systems will be called polishing ponds (PP), to distinguish them from WSP that treat raw sewage.

The ample experience with WSP shows that the most suitable configuration for the removal of the organic material from raw sewage is the anaerobic–facultative–maturation series. However, the design criteria must be modified if the removal of pathogenic micro-organisms is the primary goal of treatment in a polishing pond. The theory of reactors design applied to bacterial decay in ponds shows that for optimum efficiency the most suitable configuration is a single pond where there is no mixing of the liquid phase. This pond may be operated as a pond with continuous plug flow or in batch mode. As it has been experimentally established that it is impossible to avoid longitudinal mixing in a continuous-flow lagoon (Cavalcanti, 2003), the best alternative is to construct several lagoons with sequential batch regime, operating in parallel. The hygienic quality of treated wastewater is usually evaluated by the concentrations of thermotolerant bacteria (formerly called faecal coliforms) and helminth eggs. Normally the time to reduce the thermotolerant bacteria is longer than the time to remove helminth eggs (Saqqar & Pescod, 1995) and thus becomes the limiting retention time in polishing ponds. As the decay of thermotolerant bacteria is a first-order process (Marais & Shaw, 1961; van Haandel & Lettinga, 1994), the shortest possible retention time is obtained when the sequencing batch regime is applied (SBP), rather than flow-through ponds (FTP). SBP ponds may be designed as a series of ponds, each with a retention time such that the volume of each SBP is equal to the daily flow of sewage. The number of SBP depends on the time required to reduce the thermotolerant bacteria concentration in the effluent of the UASB reactor to a value below the maximum concentration imposed by the legal standard in force. In this chapter the performance of sequential batch ponds is described, and criteria are developed for the optimization of their design. The optimized design of a UASB reactor followed by polishing ponds will be compared to the optimal design of WSP.

Table 9.1 presents the main differences between WSPs and PPs. The most important aspect is the absence of an anaerobic pond prior to the polishing pond, which has the important practical advantages already mentioned. Table 9.1 can be summarized by stating that conventional WSP systems have major drawbacks when they are compared to the combination of a UASB reactor for pre-treatment and polishing ponds:

- The retention time is much larger, which leads to a huge area requirement.
- Conventional ponds at least occasionally spread an unbearable odour, making it impossible to operate them near urban areas.
- In anaerobic lagoons non-biodegradable solids accumulate, which have to be removed every few years, which requires the pond to be emptied.
- The methane from the anaerobic lagoons (and the other lagoons) is released into the atmosphere and contributes to the greenhouse gas effect (see Chapter 6).

Table 9.1 Differences between conventional waste stabilization (WSP) and UASB followed by polishing ponds (PP).

Parameter	Waste Stabilization Pond	UASB + Polishing Pond
Influent	Raw sewage	Digested sewage
Main objective(s)	COD and TSS removal Pathogen removal	Pathogen removal Nutrient removal is possible
Configuration	Series of lagoons (AL–FL–ML)	Series of sequential batch ponds operated in parallel
Bottom sludge	Rapid accumulation in AL ($250\ mg \cdot L^{-1}$)	Slow accumulation ($70\ mg \cdot L^{-1}$)
Desired flow regime	Continuous flow with completely mixed ponds	Ponds with a sequential batch regime (SBP)
Retention time (in hot climate)	In the range of 1 month	In the range of 1 week
Gas emissions	Biogas released to the atmosphere, causing foul odours	Biogas is generated in UASB reactor and can be flared off or used for energy generation
Methane generation	Methane released to the atmosphere contribution to GHG emission	Methane is generated in UASB reactor and can be flared off or used for energy generation
Areas of application	Far from urban regions requiring long outfalls	Proximity of population is no problem
Nutrient removal	Little (<20%)	N and P removal is feasible
Sewage treatment system	Centralized	"Segmented" network is feasible

- Due to the long retention time of conventional WSP, there will be more water losses by evaporation, which is important in the case of agricultural reuse: not only does this decrease quantity but also quality due to increased salinity of the final effluent.
- WSPs are necessarily constructed at a single location in a city, so there is no possibility of segmenting the collection network.

Table 9.1 clearly shows that WSPs have important shortcomings and disadvantages which can be reduced or eliminated when the combination of UASB + polishing ponds is used, which will always be far superior to a conventional WSP system.

An important factor in the decision on the configuration of lagoon treatment systems is that in the case of applying polishing ponds, it becomes more acceptable to operate treatment systems near urban areas, as the polishing pond works without emitting bad smells and thus avoids being noticed by the population. In this respect, it is noted that a very important advantage of being able to design systems near or even within urban areas is that the cost of the sewerage network – by far the most expensive part of a sewage collection and treatment system – is drastically reduced. Furthermore, in many cases it is possible to reduce costs when a single treatment plant for a city (which cannot be avoided for a WSP because of the foul odours) can be substituted by two or more local UASB reactors followed by polishing ponds. There are examples where the reduction in investment costs of the sewerage system from segmentation is so great that the difference with a single sewer network covers the cost of the treatment systems. Of course there is a limit to segmentation: having a large number of treatment systems becomes unfeasible as a result of increasing treatment costs, both investment (loss of economies of scale) and operational (more operating and maintenance personnel).

An aspect of great importance, indirectly linked to the anaerobic treatment system prior to the discharge of pre-treated sewage in the ponds, is the possibility of controlling the pH in the polishing pond. In the lagoon the stabilization of the organic material by aerobic or anaerobic bacteria generates CO_2, a weak acid that tends to lower the pH. Unlike stabilization, photosynthesis consumes CO_2 and, as such, results in an increase in pH. Due to the pre-treatment applied in the UASB reactor the concentration of organic material in the feed to the polishing pond will be low and therefore the production of CO_2 from the oxidation of this material will be reduced. In addition, in the polishing ponds there will be more photosynthesizing activity and consequently CO_2 consumption, because the UASB reactor effluent is clear and allows greater penetration of sunlight. Thus, there may be a significant reduction of the CO_2 concentration in the pond, resulting in a substantial increase in pH. This effect will be more pronounced if the pond is shallower which facilitates the penetration of sunlight over the whole depth. The increase of the pH in turn can have several beneficial effects. It tends to accelerate the decay rate of the thermotolerant bacteria, which allows for a reduction of the retention time. Also a high pH value in the liquid phase allows for possibility of removal in the polishing ponds.

When the pH increases, the NH_4^+ ion is progressively transformed into molecular ammonia, a gas that will desorb from the liquid phase, which allows for nitrogen removal through ammonia desorption. Furthermore, a high pH value changes phosphate equilibrium and results in a higher concentration of PO_4^{3-} ions, which in turn stimulates the precipitation of salts such as calcium phosphate ($Ca_4H(PO_4)_3$ or apatite ($Ca_{10}(OH)_2(PO_4)_6$). Removal of nutrients in WSP is more difficult, because the presence of organic material inevitably inhibits the pH increase and hence desorption of ammonium and precipitation of phosphate.

It is important to note that it is not always convenient to remove nutrients from the sewage system. Removal of nutrients is advantageous only when the effluent is released into surface water, especially if it is then stored in dams, making it important to reduce nutrient content to a minimum to avoid the exaggerated growth of algae, which impairs water quality (eutrophication). If the effluent is used for irrigation the opposite is true: there is an interest in conserving the nutrients, as these can then replace the nutrients from chemical fertilizers. The depth of the pond can be an instrument to control the concentration of nutrients in the effluent: when the pond is deeper, there will be less photosynthesis and consequently the pH increase will be reduced and with it the removal of the nutrients.

9.2 POST-TREATMENT IN POLISHING PONDS

The objective of post-treatment of digested sewage in polishing ponds is to improve the quality of the pre-treated sewage to the quality required by the legal standards for surface water discharge or agricultural reuse. Table 9.2 shows the typical composition of raw sewage, UASB effluent and the effluent from a polishing pond, as well as the desired final effluent quality. Although anaerobic digestion is efficient in the removal of organic matter and suspended solids, the concentrations of BOD_5 and TSS in digested sewage will generally still exceed the legal limits imposed. On the other hand, anaerobic digestion has little effect on other undesirable constituents of sewage, such as pathogenic organisms (quantified by the concentration of faecal coliforms and helminth eggs) and nutrients (notably nitrogen and phosphorus concentrations). There may a need to correct other variables, such as the settleable solids concentration, but in general the main objectives of digested sewage post-treatment in polishing ponds are, in order of importance: (1) reduction of pathogens, (2) nutrient removal, and (3) reduction of the BOD_5 and TSS concentrations.

In a polishing pond, the concentration of the constituents of the digested sewage changes over time, because of biological, chemical, and physical processes that develop. The most important biological

Table 9.2 Typical quality of raw and digested sewage and of the effluent of a polishing pond compared with the required quality by legal standards for unrestricted irrigation.

		Municipal Sewage			
Parameter	**Unit**	**Raw**	**Digested**	**Effluent PP**	**Desired Quality**
COD	$mg \cdot L^{-1}$	600	120	75	
BOD_5 (1)	$mg \cdot L^{-1}$	250	50	30	≤5
TSS	$mg \cdot L^{-1}$	400	100	50	
Nitrogen	$mg\ N \cdot L^{-1}$	50	45	10	
Phosphorus	$mg\ P \cdot L^{-1}$	10	9	7	
Thermotolerants (TT), *E.coli* (1,2)	CFU per 100 mL^{-1}	10^8	10^7	10^3	≤10^3
Helminth eggs(2)	$N^o \cdot L^{-1}$	100	50	0	<1
pH	(–)	7.5	7.0	9.0	
Settleable solids	$mL \cdot L^{-1}$	20	1	0.1	

Standards based on (1) CONAMA 18/06/86 and (2) WHO: Helmer *et al.* (1991).

processes are: (1) photosynthesis, (2) oxidation of organic material by bacteria that use oxygen, and (3) fermentation of organic material by anaerobic digestion. These three processes directly affect the removal of organic material. Figure 9.2 shows a schematic representation of a pond and the interaction between the different biological processes taking place in it. During photosynthesis, the algae use solar energy to reduce carbon dioxide, transforming it into cellular organic material.

In a simplified way this process can be represented by Eq. (9.1a):

$$CO_2 + H_2O \rightarrow COD + O_2 \tag{9.1a}$$

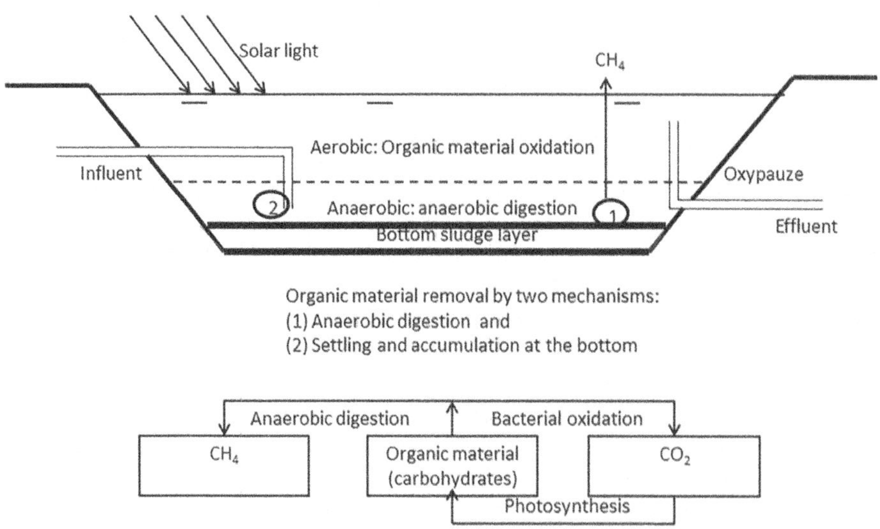

F]igure 9.2 Schematic representation of a polishing pond and of the main (biological) processes.

It is important to note that the generation of organic material (in terms of mass of COD) is accompanied by an equal amount of oxygen generated. In other words, stoichiometrically the mass of oxygen generated by photosynthesis is exactly enough to oxidize the organic material that is generated in the same process. However, the produced oxygen is generally not used by bacteria for oxidation of the cellular material of algae, but rather to oxidize the organic material present in the sewage. This bacterial oxidation can be represented by the following reaction:

$$COD + O_2 \rightarrow CO_2 + H_2O \qquad (9.1b)$$

In Figure 9.2 it can be seen that the processes of photosynthesis and bacterial oxidation are complementary, in the sense that the products of one process are the reactants of the other.

This shows that, in principle, there can be no net removal of organic material through oxidation if there is no source of oxygen other than from photosynthesis, such as atmospheric diffusion of oxygen into the liquid phase. However, the combined action of photosynthesis and oxidation may result in organic material removal by other mechanisms: the net result of the two processes is to transform organic matter in the sewage into a mass of algae. These algae can spontaneously form flocs, which subsequently settle out and become part of the bottom sludge layer of the lagoon, where they are digested or accumulated as organic non-biodegradable mass.

The relative rates at which the two biological processes occur directly affects important water quality parameters such as dissolved oxygen (DO), carbon dioxide (CO_2), biochemical oxygen demand (BOD_5), chemical oxygen demand (COD), suspended solids (SS) and pH, while others are indirectly affected, such as alkalinity and the nutrient concentrations (N & P). The rate of photosynthesis is determined mainly by the conditions of transparency, solar irradiation, temperature and depth of the pond, while the rate of bacterial oxidation is mainly determined by the concentration of the biodegradable organic material.

If the concentration of organic material is high, the dissolved oxygen concentration in the pond will be negligible because of the high demand for it (proportional to the concentration of the organic material). Under these conditions, the environment in the lagoon turns anaerobic, which will result in decay of algae at night when they need oxygen for respiration. In the anaerobic pond a black colour tends to develop (due to ferrous sulphide precipitation and production of coloured organic substances), which makes photosynthesis even more difficult.

If the organic matter concentration is lower, an upper aerobic layer forms in the lagoon, with a green colour, indicative of the presence of algae, while the lower layer remains anaerobic. This is the typical situation of a facultative pond. The oxypause indicates the level (liquid height) in the lagoon where there is a transition from the aerobic top layer to the anaerobic bottom layer. Due to the cyclic nature of photosynthesis (which requires sunlight), the position of the oxypause oscillates: rising daily as the sun sets and reaching a maximum level at dawn.

If the organic matter concentration decreases even further, most of the pond will have an aerobic environment, except in the sludge layer at the bottom of the pond. In this case the oxypause is located just above the sludge layer that always forms at the bottom of the pond due to flocculation of algae. Several authors (Mara & Pearson, 1987) have shown that in the case of raw sewage treatment the type of lagoon that develops can be related to the applied surface organic load. In tropical climates, the recommended BOD_5 load for anaerobic ponds is >400 kg $BOD_5 \cdot m^{-3} \cdot d^{-1}$, for facultative ponds <300 kg $BOD_5 \cdot m^{-2} \cdot d^{-1}$, and for maturation ponds: <150 kg $BOD_5 \cdot m^{-2} \cdot d^{-1}$.

In the case of polishing ponds, the load of organic material from the digested sewage is low and the ponds will have characteristics typical of maturation ponds, especially if the ponds are shallow (larger area and thus more photosynthesis per unit pond volume). Under these conditions it is possible to develop a high oxygen concentration, and there may even be oxygen desorption from the liquid phase into the atmosphere. In principle, the production of oxygen should be accompanied by an equal production (in COD) of algae mass (according to Eq. (9.1a), but in practice this increase is usually not observed because most algae flocculate and eventually settle at the bottom of the lagoon.

In addition to photosynthesis and bacterial oxidation, the third important biological process is anaerobic digestion, which develops at the bottom of the pond, where the oxygen from the photosynthesis does not reach or, if it arrives, is immediately consumed. The process can be represented schematically as:

$$\mathrm{COD} \rightarrow \mathrm{CH_4} + \mathrm{CO_2} \tag{9.1c}$$

It can be noted that anaerobic digestion also does not result in the destruction of organic material, it merely turns it into a gas (methane) that is released from the liquid phase and therefore from the effluent. In the case of WSP, the removal of organic material (expressed as COD) is typically of the order of 80%, while an important part (±50%) of the organic material present in the pond effluent consists of algae. At the bottom of the pond, sludge will accumulate with a high organic fraction: typically about 67% and with a ratio of 1.5 mg COD per mg VSS (Cavalcanti, 2003; Gonçalves *et al.*, 1999). Figure 9.3 (left) shows a typical division of the organic material in the raw sewage in WSP, where removal of organic material by absorption of atmospheric oxygen is not considered.

When efficient anaerobic treatment (e.g. in a UASB reactor) is applied prior to launching the sewage into a pond (which would then be a polishing pond), the concentration of organic material is substantially reduced, in the range of 65–75% (see Chapter 3), so that the COD concentration discharged into the polishing pond is less than one-third of the raw sewage COD. However, the COD in the final effluent (UASB + polishing pond) is not much lower than that of a WSP system: around 20% of the raw sewage COD (Cavalcanti, 2003). On the other hand, the accumulation of sludge at the bottom of a polishing pond (about 10% of the raw sewage COD) is much smaller than in a WSP (Cavalcanti, 2003).

Figure 9.3 (right) shows the typical division of the raw sewage COD in the combination UASB + polishing ponds, while Table 9.3 shows typical values of the division of raw sewage COD in both WSPs and polishing ponds, for a raw sewage concentration of 600 mg · L^{-1}.

In the polishing pond a large decrease of the organic load is observed after an efficient anaerobic pre-treatment in a UASB reactor, which will reduce the demand for oxygen. Furthermore, the transparency of the digested sewage is high, because the UASB reactor removes much of the colloidal material that is responsible for the turbidity of the raw sewage. Therefore, sunlight will be able to penetrate deeper into the polishing pond and subsequently there will be more photosynthesis, producing more oxygen per unit pond area.

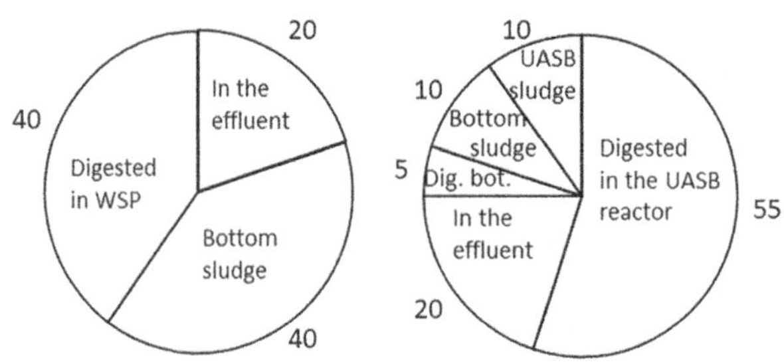

Figure 9.3 Typical division of raw sewage COD in waste stabilization ponds (left) and the combination UASB + PP (right).

Table 9.3 Typical division of the organic material present in the raw sewage over different COD fractions in WSP and UASB + PP combination.

COD Concentration	WSP	PP
Digested in pre-treatment	–	330
Sludge in pre-treatment	–	60
Digested in the lagoon	240	30
Sludge in the lagoon	240	60
In the effluent (including algae)	120	120
Total	600	600

The combination of a reduced oxygen demand with increased oxygen production will result in the establishment of a predominantly aerobic environment, like that in a maturation pond. Under these conditions, the anaerobic environment is restricted to the sludge layer at the bottom of the pond. In fact, the surface organic load applied to polishing ponds that receive effluent from UASB reactors is usually lower than the typical load of a maturation pond (150 kg $BOD_5 \cdot ha^{-1} \cdot d^{-1}$), even when the retention time in the pond is short. For this reason, the objective of polishing ponds is generally not the stabilization of the organic material, but rather the removal of the pathogens. It is interesting to note that there would be an increase of COD in the polishing pond if it were not for flocculation, settling and subsequent digestion of algae: the prevalence of photosynthesis leads to eutrophication!

9.3 PATHOGEN REMOVAL IN POLISHING PONDS

9.3.1 Kinetics of thermotolerant bacteria decay

Domestic sewage contains a huge variety of pathogenic micro-organisms, like viruses, bacteria, protozoa but also macro-organisms such as helminth eggs. As it is impossible to determine all these types on a routine basis, two types of organisms were adopted as indicators of the hygienic quality of sewage: the concentration of faecal or thermotolerant coliforms (expressed in colony forming units – $CFU \cdot 100\ mL^{-1}$) and helminth eggs (expressed in number of eggs per litre). The choice of these organisms as indicators of hygienic quality is mainly due to the fact that they demonstrate, compared with other pathogens, a higher survival rate in wastewater treatment systems. Thus, when their concentrations are below a certain limit, this indicates that the treated effluent has a good quality. The World Health Organization (WHO, 1989) established the following limits for unrestricted irrigation (which includes growing vegetables that are eaten raw, such as lettuce): a thermotolerant bacteria concentration below 1000 CFU per 100 mL and helminth eggs less than one unit per litre (Helmer *et al.*, 1991)

Removal of helminth eggs and thermotolerant bacteria (TT) from sewage occurs by entirely different mechanisms. Helminth eggs can survive for years in environments such as sewage treatment systems. Hence only physical removal is effective, for example adsorption to sludge flocs or simple settling, as the eggs have a slightly higher density than water. The retention time for efficient removal of helminth eggs from ponds is only a few days (Cavalcanti, 2003; Saqqar & Pescod, 1995). As the retention time in ponds is usually very long (in the order of weeks), the removal of helminth eggs is guaranteed whenever efficient bacterial removal is accomplished, so that there is no need to consider this aspect separately for the purpose of pond design. In contrast, the reduction in the faecal coliform concentration is a biological process, resulting from the bacterial decay (death) of the thermotolerant bacteria, which is a slow process and therefore should be carefully considered in the design of lagoons. Reporting on the pioneering work on kinetic modelling of decay processes (Marais & Shaw, 1961), Marais (1974) showed that this process can be described as a first-order reaction, with the reduction rate of TT bacteria proportional to their concentration:

$$r_d = (dN/dt) = k_b N \tag{9.2}$$

where:

r_d = decay (death) rate of the TT;
N = concentration of the TT (CFU · 100 mL^{-1});
t = decay time (d);
k_b = decay constant for TT (d^{-1}).

In chemical engineering, particularly in reactor design calculations, first-order processes have been thoroughly studied, because they are a type that occurs very frequently in practice.

It is possible to use theoretical models developed for reactor calculation to describe the process of TT decay in ponds. When a pond is considered to be a reactor operating at constant flow rate and volume, it can be shown that, for a given retention time in the pond, the efficiency of faecal coliform removal depends on the mixing regime of the pond (Levenspiel, 1999). Three different cases can be distinguished: (1) complete mixing, (2) no mixing (plug flow), and (3) partial mixing. In reality the first two cases are extreme cases, as in actual fact it is impossible to have either complete or no mixing. The extent of mixing in a pond will always be partial but depending on the conditions of the flow regime it may be approaching complete mixing or no mixing.

Table 9.2 shows that digested sewage can have a concentration of 10^6–10^7 CFU · 100 mL^{-1} (i.e. 10% of the concentration in raw sewage), whereas the WHO standard for unrestricted irrigation specifies a maximum of 10^3 CFU · 100 mL^{-1}. Hence to comply with the WHO recommendations, the reduction of the number of TT must be in the order of 10^3–10^4 CFU · 100 mL^{-1}, which requires a removal efficiency of at least 99.9–99.99%.

When a constant flow is applied to a completely mixed lagoon, the solution of the differential equation of Eq. (9.2) is given by (Marais, 1974) as:

$$N_e/N_i = 1/(1 + k_b R_h) \tag{9.3a}$$

When a series of M completely mixed lagoons is applied Eq. (9.3a) is modified to:

$$N_e/N_i = 1/[(1 + k_b R_{h1})(1 + k_b R_{h21})\ldots(1 + k_b R_{hM})] \tag{9.3b}$$

Equation (9.3b) can be simplified if the ponds in the series all have the same retention time so that $R_{hM} = R_h/M$):

$$N_e/N_i = 1/(1 + k_b R_h/M)^M \tag{9.3c}$$

where:

N_e = TT concentration in the pond effluent
N_i = TT concentration in the pond influent
R_h = total retention time in the pond system
R_{hM} = retention time in pond "M" of a series
M = number of ponds in series.

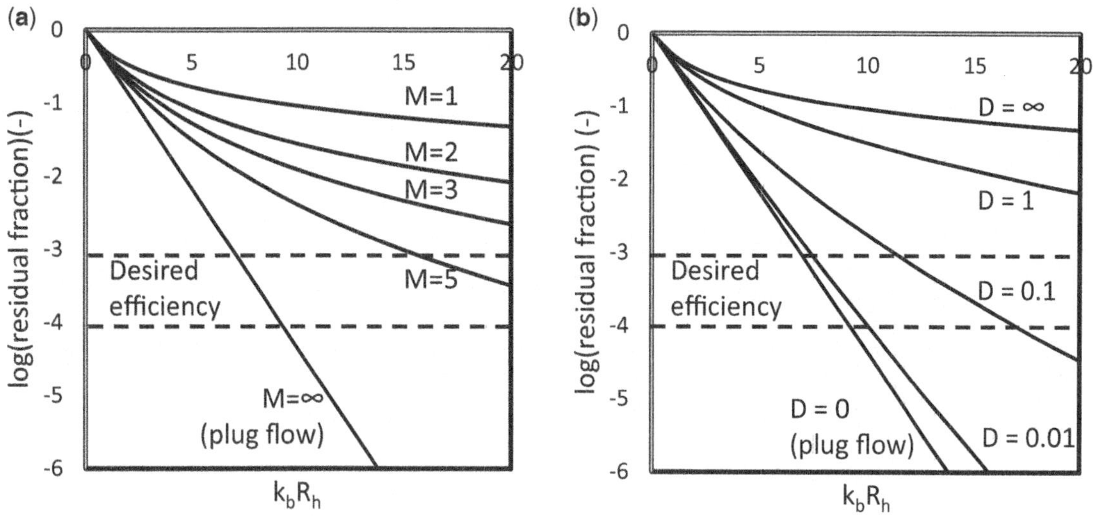

Figure 9.4 (a) Relation between the residual fraction of TT and the adimensional product k_bR_h, for different number of lagoons ($M = \infty$ represents a pure plug-flow system). (b) Relation between the residual fraction of TT and the adimensional product k_bR_h, for different values of the dispersion number "*D*" ($D = 0$ represents pure plug-flow).

Figure 9.4a is a graphical representation of Eqs. (9.3a) to (9.3c): the removal efficiency of TT (log scale) is plotted as function of the adimensional product k_bR_h. Figure 9.4a is very useful for the design of completely mixed ponds, as will be demonstrated in Example 9.1. If there is no longitudinal mixing in the pond the solution of Eq. (9.2) is given by an expression of exponential decay:

$$N_e/N_i = \exp(-k_bR_h) \tag{9.4}$$

Equation (9.4) is also indicated in Figure 9.4a. The flow regime in a pond without longitudinal mixing is called plug flow, since the mass of water moves along the pond as if it were a solid block (plug).

It can be observed that for any value of the total pond retention time, the TT removal efficiency in a plug-flow pond is higher than in a series of completely mixed ponds of equal total volume. For example, for a 99.9% removal efficiency (a log 3 decrease) of TT in a plug-flow pond, the value of $k_bR_h = 6.9$. For a typical value of the decay rate constant k_b of 1.5 d^{-1}, the required retention time $R_h = 6.9/1.5 = 4.6$ days, less than half the time that would be required in a series of five ponds (see Example 9.1). It is concluded that in principle it would be advantageous to have plug-flow ponds as, for the same retention time, these ponds are much more efficient than any other solution. Longitudinal mixing in ponds decreases when they are constructed as shallow units with a large length to width ratio (Agunwamba *et al.*, 1992; von Sperling, 1999; Yañez, 1993). This can be achieved by means of constructing a very long pond or, more practically, a rectangular pond partitioned by dividing walls, artificially inducing a high length to width ratio ("tubular" pond).

EXAMPLE 9.1

A 99.9% removal efficiency of TT bacteria is desired in a series of five completely mixed ponds (i.e., it is desired that of every 1000 TT in the influent only 1 survives after passage through the ponds). Find the retention time in each pond if the decay constant is $k_b = 1.5\ d^{-1}$.

Solution

Equation (9.3c) and Figure 9.4a show that the adimensional product $k_b R_h$ has a value of 14.6 for the specified conditions (i.e. five completely mixed ponds in series). For the value of the decay constant $k_b = 1.5\ d^{-1}$, the minimum retention time to achieve the required removal efficiency will be 14.6/1.5 = 9.7 d.

It can be observed that completely mixed ponds in series have a significant increased TT removal efficiency compared to that in a single pond with the same retention time. For comparison, a single pond with a retention time of 9.7 d has a removal efficiency of only 90%, while a pond in perfect plug-flow configuration has an efficiency of almost 99.9999% (remaining fraction $\approx 10^{-6}$).

Although WSP do not have an efficient configuration, the hygienic quality of the final WSP effluent treating raw sewage in a series of ponds is generally good, because the retention time for the removal of organic material is long enough to have an efficient removal of TT. In contrast, when digested sewage is subjected to post-treatment in a polishing pond, the removal of organic material is of secondary importance while that of TT becomes the prime objective. Hence it is very important that a plug-flow regime is approached as close as possible, so that the required retention time (and volume) is minimum. Figure 9.4a shows that there is a huge potential for reducing the retention time and therefore the volume and area of polishing pond by the adoption of a plug-flow regime and that the difference becomes more pronounced as the desired efficiency increases.

In addition to the applied flow regime, mechanical mixing due to environmental factors also must be considered. One of the most important ones is that resulting from the action of the wind, which not only induces the appearance of waves in the lagoons, but also a general mixing when the upper layer of the lagoon is dragged by the wind, causing a backflow in the lower layer. Another important source of mechanical mixing is the rise of biogas bubbles, formed mainly in the sludge layer at the bottom of the pond. Also the activity of insects and other macro organisms in the water, which when moving cause mixing of the liquid, must also be taken into account. Furthermore, the temperature variations that occur during the day can cause thermal stratification in the ponds, with a warm (and therefore lower-density) layer in the upper part of the pond during the day, whereas in the evening the upper layer tends to cool down until it is colder than the bottom layer (thermal inversion), resulting in density currents, when the heavy top layer sinks and the bottom layer rises. All these factors contribute to mixing in continuous-flow ponds, even when measures are taken to avoid this as much as possible.

Wehner and Wilhelm (1956) developed an expression that allows calculation of the efficiency of a first-order process in a continuous-flow reactor with partial mixing. They defined a dimensionless parameter called the dispersion number. To quantitatively express the influence of mixing on the efficiency of TT removal in a partially mixed polishing pond, this dispersion number must be experimentally determined. Once the value of the dispersion number is established, the fraction of TT remaining in the effluent will be given by the following equation:

$$N_e/N_i = 4a\exp(1/2D)/[(1+a)^2\exp(a/2D) - (1-a)^2\exp(-a/2D)] \qquad (9.5)$$

where:

$a = (1 + 4k_bR_hD)^{1/2}$
D = dispersion number.

Figure 9.4b shows the ratio N_e/N_i as a function of the dimensionless product k_bR_h for different values of the dispersion number: $D = 0$ (true plug-flow), $D = 0.01$ (little mixing), $D = 0.1$ (moderate mixing), $D = 1$ (intense mixing) and $D = \infty$ (complete mixing). It can be clearly observed that the increase in mixing intensity affects the removal efficiency of TT, especially when the desired efficiency is high. For example, for a 99.99% efficiency (4 log units), the required value of k_bR_h is 9.2 for a true plug-flow ($D = 0$), but for $D = 0.1$ the value of the product increases to 17. Therefore, the retention time R_h in the pond with little mixing ($D = 0.1$) will be 17/9.2 = 1.9 times higher than in the plug-flow pond without mixing ($D = 0$). It is concluded that even very little mixing can already lead to a large increase of the necessary R_h and hence of the pond area. The dispersion number in Eq. (9.5) is a quantitative measure for the mixing intensity in the pond and can be calculated from the experimental determination of the retention time distribution in the pond. After adding a tracer semi-instantaneously to the influent, this distribution is determined by measuring its concentration in the effluent as a function of time.

The mean retention time can be calculated from the retention time distribution using standardized calculations (Levenspiel, 1999). The same calculations also allow determination of the dead volume, i.e. the stagnant volume of the reactor that is not actively included in the treatment processes.

The data by Yañez (1993), obtained from South American experiments, and by Von Sperling (1999), with a specific focus on the situation in Brazil, show that in full-scale lagoons the dispersion number is large, even when pond has a high length to width ratio. Usually the value is greater than 0.2 and often even greater than 0.5. Furthermore, experimental investigations at pilot-scale ponds carried out in Brazil show that, even in ponds carefully designed to approximate the plug-flow regime as closely as possible and with limited influence of wind, the value of the dispersion number was at least 0.12 (Cavalcanti, 2003). It is concluded that due to imperfections in the flow regime, the dispersion number in polishing ponds with continuous flow is at least 0.1–0.2.

For this reason, the retention time required for efficient removal of TT will be more than double that of the theoretical minimum value of a plug-flow pond (Figure 9.4b). It is concluded that in practice it is not possible to obtain a plug-flow regime in continuous-flow ponds and therefore the maximum rate of TT removal in these ponds cannot be obtained.

9.3.2 Decay of thermotolerant bacteria in sequential batch ponds

An alternative to the continuous-flow regime is the Sequential Batch Pond (SBP). In sequential batch mode a pond is filled with UASB reactor effluent and the liquid is kept in the pond for the time required for a given removal efficiency, whereupon the lagoon contents are discharged. As in plug-flow ponds, the TT removal in sequential batch ponds as function of time can also described with a slightly modified version of Eq. (9.4):

$$N_n/N_i = \exp(-k_bt) \qquad (9.4a)$$

where t = time elapsed (days) after pond filling.

Therefore, the maximum rate of exponential decay is automatically guaranteed in batch-fed or batch-flow reactors (Levenspiel, 1999). Cavalcanti (2003), evaluating the performance of shallow lagoons (0.4–0.65 m) at an average temperature of 25°C, fed in sequential-batch mode (SBP) with digested sewage, and showed that to good approximation the decay of TT proceeded according to the kinetics of a first-order process until the TT concentration was below 10^3 per 100 ml, which occurred in 4–5 days at the pond temperature of 25°C. The author also reported that, during the period of exponential decay of the TT, there was also elimination (by settling) of the helminth eggs from the liquid phase.

Furthermore, because a predominantly aerobic environment was established, there was also a considerable reduction of BOD_5 and TSS concentrations. Thus, according to Cavalcanti (2003), in SBP with a retention time of only 4–5 days, it was possible to obtain a final effluent with quality compatible with that recommended by WHO (1989) for unrestricted irrigation.

This retention time is only half the minimum time required in a continuous-flow polishing pond, even after minimizing longitudinal mixing (Cavalcanti, 2003). Therefore the application of SBP reduces the area required for treatment by a factor of two compared with continuous-flow ponds. In comparison, in conventional WSP systems installed in tropical regions the retention time for a 99.99% TT removal efficiency is of the order of 25 days, which is two times greater than in a series of five continuous-flow polishing ponds and even five times more than in the SBP polishing ponds with the same depth as the maturation pond.

EXAMPLE 9.2

Compare the hydraulic retention time and the required area of three configurations for the treatment of 1000 $m^3 \cdot d^{-1}$ of sewage (10,000 inhabitants): a conventional WSP system; a UASB reactor with a series of five continuous-flow PP (minimal mixing assumed) and a UASB reactor system followed by SBP. Use the data from Table 9.4. The value of the decay constant $k_d = 1.5\ d^{-1}$ while the desired TT removal efficiency is 99.99%.

Solution

(1) *Waste stabilization pond*: Each of the two anaerobic ponds with a retention time of $R_h = 2$ d has a volume of 2000 m^3. Hence, for a depth of 4 m the required area is calculated as 2000/4 = 500 m^2. With respect to TT removal it should be considered that only one pond is operating at any time.

$$N_1/N_i = 1/(1 + k_d R_1) = 1/(1 + 1.5 \times 2) = 0.25 \text{ (i.e., 75\% removal)} \qquad (9.3a)$$

The volume of the facultative lagoon (with continuous flow) for $R_h = 15$ d is 15 × 1000 = 15,000 m^3. For a depth of 1.2 m the area is 15,000/1.2 = 12,500 m^3. The removal of TT is again given by $N_2/N_1 = 1/(1 + 1.5 \times 15) = 0.042$ (i.e., 95.2% removal).

As the overall efficiency of the TT removal is given as 99.9% it follows that $N_e/N_i = 10^{-4}$. Hence with Eq. (9.3c) the required retention time in the maturation ponds is calculated, considering the removal already affected in the stabilization and facultative ponds:

$$N_e/N_2 = N_1/N_i \cdot N_2/N_1 \cdot N_e/N_2 = N_e/N_a = 0.25 \times 0.042 \times 1/(1 + k_d R_m/3)^3 = 10^{-4} \text{ or } (1 + k_d R_m/3)^3 = 106$$

This equation can be solved for $R_m = 7.5$ d, i.e. each maturation pond has a retention time of $R_h = 2.5$ d. Each of the three maturation ponds has a volume of 2.5 × 1000 = 2500 m^3, which for a depth of 1 m yields a required area of 2500 m^2. Hence the total retention time of the WSP pond system is 2 + 15 + 3 × 2.5 = 24.5 d and the required area is equal to 500 + 12,500 + 7,500 = 20,500 m^2, which is slightly more than 2 m^2 per inhabitant. If the system had only two maturation ponds instead of three, the retention time would have to be:

$$(1 + k_d R_m/2)^2 = 106 \text{ and } R_m = 12.4 \text{ d, i.e. two ponds with a retention time of 6.2 d each.}$$

(2) *UASB + polishing ponds with continuous flow*: The UASB reactor with a retention time of 6 h requires a volume of 10,000/4 = 2500 m^3, possibly two units of 1250 m^3 with a height of 5 m. Therefore the required area would be 250 m^2 or units of 16 m long and wide.

As to the polishing pond: in this configuration a dispersion number of $D = 0.2$ is assumed and a series of five ponds of equal size. Also it is assumed that the TT removal in the UASB reactor is already 90%, so that in the polishing ponds the removal must be 99.9% to reach an overall efficiency of 99.99%. Hence:

$N_e/N_i = 0.001 = 1/(1 + k_k R_p/5)^5$ and $R_p = 10$ d.

Hence a series of five polishing ponds, with a retention time of 2 d each, will be able to achieve the desired effluent quality. The volume of the ponds is 10,000 m^3 and the total area 10,000 m^2 for an adopted depth of 1 m. The required area for this configuration would be 1 m^2 per inhabitant.

(3) *UASB + SBP*: In this configuration a series of sequential batch ponds is used, to be operated in parallel. Consequently, in each pond the TT removal efficiency will have to be 99.9% so that:

$N_e/N_i = \exp(-k_k R_b) = 0.001$ and $R_b = 6.9/1.5 = 4.6$ d.

If a series of six ponds is adopted, each with a volume of 1000 m^3, one such pond could treat the sewage flow of 1 day. After receipt of the batch is completed, a retention time of 4.6 d is required to achieve the desired TT removal efficiency. After that the final effluent is discharged and the pond is available again to receive feed. Thus, each pond needs a cycle time of 5.6 d to process a batch of UASB effluent, so that a minimum of six ponds is required for sequencing batch operation. The total volume of this series of ponds is 6000 m^3, with 6000 m^2 area for a depth of 1 m, while the per capita area is 0.6 m^2.

Table 9.4 Basic data for the design of WSP, UASB + continuous-flow PP, and UASB + SBP.

WSP	UASB + PP (Continuous Flow)	UASB + SBP
Anaerobic lagoon (2)	UASB	UASB
Number: 2	Volume < 2000 m^3 per unit	Volume < 2000 m^3 per unit
Depth: 4 m	Depth: 5 m	Depth: 5 m
Retention time: 2 d	Retention time: 6 h	Retention time: 6 h
Facultative lagoon	Polishing pond	Sequential batch ponds
Number: 1	Number: 5	
Depth: 1.2 m	Depth 1 m	
Retention time:15 d	Dispersion number: 0.2	
Maturation ponds		
Number: 3		
Depth: 1.0 m		

9.4 TRANSFER PONDS

Figure 9.5 (left) shows the layout as discussed in Example 9.2 (configuration C) with a series of six ponds sequentially receiving the UASB effluent. After the time required for TT decay of 4.6 d has passed, the final effluent is discharged, so that the pond is ready to receive another batch of UASB effluent, with a volume equal to the daily flow. However, this is not the only configuration in which batch ponds are used to produce the desired effluent quality, as is shown in Figure 9.5 (right). The effluent can also be discharged into a central pond, from where it can be distributed to the polishing ponds. Thus the effluent of the UASB reactor (operated under continuous-flow conditions) would accumulate in this pond and feed the polishing ponds operated as sequential batch reactors. This pond, denominated transfer pond, should be able to transfer effluent quasi-instantaneously to the different polishing ponds so that in principle one polishing pond less would be required. On the other hand, the transfer pond must have at least the same volume as a polishing pond.

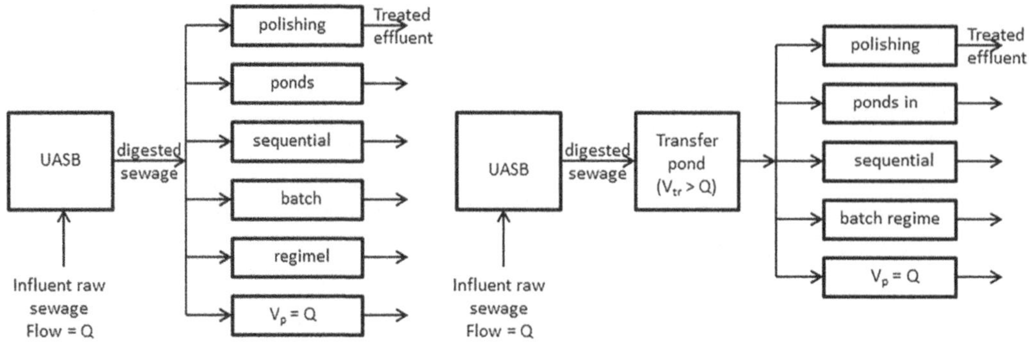

Figure 9.5 Flow scheme of sequential batch polishing ponds without (left) and with (right) intermediate transfer pond.

The advantage of having a transfer pond is that, apart from transferring digested effluent to the polishing ponds, it can serve several other functions, the most important being:

- The transfer pond also functions as a settling tank for solids discharged from the UASB reactor. Part of the settled solids could then be returned to the UASB reactor and the remainder be processed to produce a sludge cake or dry solids.
- If the transfer pond also acts as a settler, there is no need to have a GLS separator on top of the UASB reactor. If this structure is abandoned, the operation of UASB reactors becomes very easy. The principal operational problem of UASB reactors is the scum layer that forms at the surface of the interface under the separator elements. If the separator is abandoned, the scum problem no longer exists.
- If the retention time of the transfer pond is sufficiently high, photosynthesis starts, which establishes a population of algae in the discharge to the polishing ponds, effectively acting as seed so that photosynthesis can start immediately upon feeding the PP.
- If enough oxygen is produced by photosynthesis in the transfer pond, it can oxidize sulphides that may have been formed in the UASB reactor, thus suppressing the bad odours that would otherwise arise.
- Helminth eggs from the UASB reactor may be retained by settling and thus not discharged into the polishing ponds, guaranteeing that these pathogens will not be present in the final effluent. At any rate, a large fraction of the helminth eggs will be retained in the transfer pond and will end up in the excess sludge.
- The concentration of carbon dioxide dissolved in the UASB reactor effluent can be reduced by desorption to the atmosphere, reducing the acidity and facilitating the increase of pH by photosynthesis, thus accelerating the removal of ammoniacal nitrogen from the liquid phase in the subsequent polishing pond by desorption. Eventually this desorption can be stimulated by gentle agitation of the upper layer of the transfer pond.

The basic equation for TT decay, Eq. (9.2), can also be used to calculate the TT removal efficiency in a transfer pond. This efficiency will be low, because there is continuous introduction of digested sewage with a high concentration of TT (N_i). For this reason the removal efficiency in a continuous flow pond will be much lower than in a SBP. Cavalcanti (2003) has shown that, for practical purposes, the removal of bacteria in transfer ponds can be neglected, so that the transfer pond does not contribute to the removal of TT.

The important operating variables of transfer ponds are the retention time and the depth of the unit. As for the retention time, it should be at least 1 d, because the transfer pond should at least be able to store one batch of UASB effluent, equal to the daily flow. The advantage of applying a higher residence time is that the transfer pond will progressively perform other tasks as discussed above such as (a) sludge settler for the UASB reactor effluent, (b) development of photosynthesis, (c) sulphide oxidation, (d) reduction of pathogens, and (e) desorption of CO_2.

The pond depth is important because it defines the available area for receiving solar energy, needed for photosynthesis. Also, the depth determines the settling velocity that can be retained, that is, determines the settling efficiency. A practical consideration in relation to the transfer pond depth is that both the receipt of effluent from the UASB reactor and the distribution of this effluent over the polishing ponds should preferably be done gravity, i.e. without pumping. In general the UASB reactor can be built supported and not buried, so that the available discharge head is in the range of 4–6 m of water, depending on the height of the reactor. On the other hand, polishing ponds can be built on the ground with lateral slopes, with a depth of less than 1 m. Thus, there is possibility of operating the transfer pond with an available head of 3 m or more, without need for excavation.

An important aspect of transfer pond design is that the retained solids are effectively returned, that is, that there must be a mechanism that directs the settled solids to a certain sludge abstraction point on the bottom. One possibility is to construct a sloped bottom as in activated sludge settlers, where bottom scrapers can be placed to accelerate the convergence of the solids to the lowest point, from which they can be returned to the UASB or sent to excess sludge processing.

To optimize the post-treatment in polishing ponds, the dominant criterion for design must be identified. Cavalcanti *et al.* (2000b) have shown that for anaerobically digested sewage the removal of the hygienic quality indicator Faecal Coliform (TT) is slower than that of other undesirable constituents, such as residual BOD and TSS and helminth eggs. Thus, the post-treatment system should be designed to remove TT. The main variable to be determined is the retention time required to produce the desired effluent quality. This parameter can be calculated theoretically when adopting the kinetics of a first-order process as the basic expression of the decay of the bacteria, defined in Eq. (9.2). The retention time depends basically on two factors: (1) the flow regime that is used in the ponds, i.e. continuous flow-through or sequential batch regime; and (2) the value of the decay constant. In Section 9.4 the factors that influence the value of the decay constant are discussed.

9.5 FACTORS THAT INFLUENCE THE VALUE OF THE DECAY CONSTANT

A factor that hitherto has not been discussed is the value of the decay constant k_b. Research has shown that the value of this constant depends on several factors, with temperature, pond depth and pH being the most important. In the case of municipal sewage the temperature of the pond is an external parameter that cannot be influenced. On the other hand, the depth of the polishing pond is an important design variable, because it determines the area required for a given retention time. The depth has an indirect effect on the pH: if the depth is considerable the area for photosynthesis is relatively small. Consequently the biological consumption of CO_2 and the subsequent increase of the pH will be reduced. In contrast, in a shallow polishing pond there can be a considerable pH increase even at short retention time.

To evaluate the influence of the temperature, most researchers use an Arrhenius type of expression, usually having the value of k_{bt} at 20°C as a reference value:

$$k_{bt} = k_{b20} \cdot \theta^{(T-20)} \tag{9.7}$$

where:

k_{bt} = value of the decay constant for temperature T
k_{b20} = value of the decay constant at the reference temperature (20°C)
θ = temperature dependency coefficient (adimensional).

Table 9.5 Influence of the temperature on the value of the TT decay constant (k_b) in ponds.

Reference	Temperature (°C)	k_b Value (d^{-1})	θ Value
Marais (1974)	20	2.60	1.2
Sherry and Parker (1979)	20	1.5	1.06
Klock (1971)	20	1.1	1.07
Yañez (1993)	20	0.84	1.07

Table 9.5 summarizes k_b and θ values determined by various researchers. The available data seem to indicate a temperature dependency coefficient of $\theta = 1.07$. This means that the value of k_b doubles for a temperature increase of 10°C.

With respect to the influence of depth on the k_b value, few data are available in literature. Van Haandel and Lettinga (1994) suggested a hyperbolic relationship, with the constant k_b inversely proportional to depth:

$$k_{bh} = k_{b1}/H \tag{9.8}$$

where:

k_{bh} = decay constant in a pond with depth H
k_{b1} = decay constant in a pond with a depth of 1 m
H = pond depth (m)

From a large data set from ponds in Campina Grande, Brazil, obtained at a temperature of 25°C, a value of $k_{b1,25} = 1.9\ d^{-1}$ was determined for ponds with a depth of 1 m and a neutral pH. Corrected to the standard temperature of 20°C the value would be: $k_{b1,20} = 1.35\ d^{-1}$, which corresponds well to the data in Table 9.5, which were mostly obtained at a depth over 1 m.

To incorporate the influence of both temperature and depth, the expressions of Eqs. (9.7) and (9.8) can be combined:

$$k_{bht} = (k_{b1.20}/H)\theta^{(T-20)} \tag{9.9}$$

where:

k_{bht} = decay constant for a temperature T and a pond depth H
$k_{b1,20}$ = decay constant for a temperature of 20°C and a pond depth H of 1 m
θ = Arrhenius temperature dependency = 1.07.

With respect to the influence of the pH on the value of the decay constant, several researchers have shown that its value remains essentially constant in the neutral pH range. However, a significant increase in decay rate is observed at high pH values (above 9.0–9.5), as noted by Curtis *et al.* (1992).

9.6 ORGANIC MATERIAL AND SUSPENDED SOLIDS REMOVAL IN POLISHING PONDS

In the previous sections the kinetics of thermotolerant bacteria removal in polishing ponds was discussed. It was shown that even under favourable conditions such as high temperature and high solar irradiation, and applying the sequential batch regime, the minimum retention time to obtain a satisfactory hygienic quality of the final effluent is of the order of 5–7 days. In this section it is shown that this time is also enough to significantly reduce the residual concentrations of suspended solids and organic material in the effluent of the polishing pond, so that a low concentration of these variables is automatically obtained.

Chapter 3 showed that the UASB reactor has a high efficiency for treatment of domestic sewage in regions with a hot climate (sewage temperature above 18°C throughout the year), provided two conditions are fulfilled: (a) a suitable retention time must be applied (6–10 h for the conventional UASB reactor according to the Brazilian Standard 12.209), and (b) the produced excess sludge must be regularly removed. Under these conditions a removal efficiency of organic material and suspended solids in the range of 70–80% can be expected. Therefore, for COD, BOD_5 and TSS values in the raw sewage in the range of 500–700 $mg \cdot L^{-1}$, 200–300 $mg \cdot L^{-1}$ and 300–400 $mg \cdot L^{-1}$ respectively, a UASB effluent can be expected with COD of 100–200 $mg \cdot L^{-1}$, BOD_5 of 50–90 $mg \cdot L^{-1}$ and TSS of 60–120 $mg \cdot L^{-1}$. An additional feature of UASB effluent is that the effluent has low turbidity, because the colloids present in the raw sewage are efficiently flocculated and removed, which is why the solar rays can penetrate deeply into the PPs. For these reasons, PPs that receive effluent from UASB reactors have a lower load and turbidity than maturation ponds that receive effluent from a facultative pond in the conventional WSP system.

Consequently, photosynthetic oxygen production is higher than the oxygen demand for oxidation of organic material along almost the entire liquid column, except in the bottom section of the pond containing the settled sludge, regardless of the depth and the mixing intensity in the pond. Thus, algae are always present in PPs and the smell associated with the presence of sulphide is absent, even if it is present in the digested sewage.

To establish the behaviour of polishing ponds in relation to the removal of organic material and total suspended solids, Cavalcanti (2003) operated several pilot-scale continuous-flow polishing ponds. Each pond had different depth, ranging from 0.65 to 3 m. The shallow lagoon of 0.65 m was divided into five lanes that could be operated either in series or in parallel. Table 9.6 lists the main dimensions of the experimental ponds. The ponds received effluent from a UASB reactor treating municipal sewage. Throughout the experiments there was no intentional disposal of sludge from the UASB reactor, so that the ponds received the raw (non-settled) effluent from the digester. For this reason there was always a presence of some sludge particles in the digested sewage, but these settled in the polishing pond. Figure 9.6 shows the performance of the different polishing ponds in terms of the decrease of BOD_5, COD and TSS as functions of the retention time at different depths. The water temperature was 25–26°C.

The total (raw) BOD of the digested sewage in the polishing pond reduced rapidly to a value of 60–70 $mg \cdot L^{-1}$, due to the settling of anaerobic sludge flocs present in the UASB effluent. After this initial physical process, the decrease in BOD continues much more slowly until it reaches a final value of approximately 20–30 $mg \cdot L^{-1}$.

From Figure 9.6 several interesting aspects with regards to BOD removal in polishing ponds can be highlighted:

- The first-order model generally used to describe the decrease of (total) BOD in WSP is totally unsuitable for polishing ponds: the decrease of BOD_5 in time is certainly not exponential. The experimental data indicate that the process is closer to a zero-order reaction, in which case the BOD decrease would be linear with time.

Table 9.6 Dimensions of the experimental polishing ponds used during the investigation.

Code	Number	Volume (m^3)	Height (m)	Area (m^2)	Material
PP_1	5	32.5	0.65	50 (5 × 1 × 10)	Brick
PP_2	1	5	1.65	3 ($d = 2$ m)	Fiberglass
PP_3	1	6	2	3 (1.75 × 1.70)	Brick
PP_4	1	9	3	3 (1.75 × 1.70)	Brick

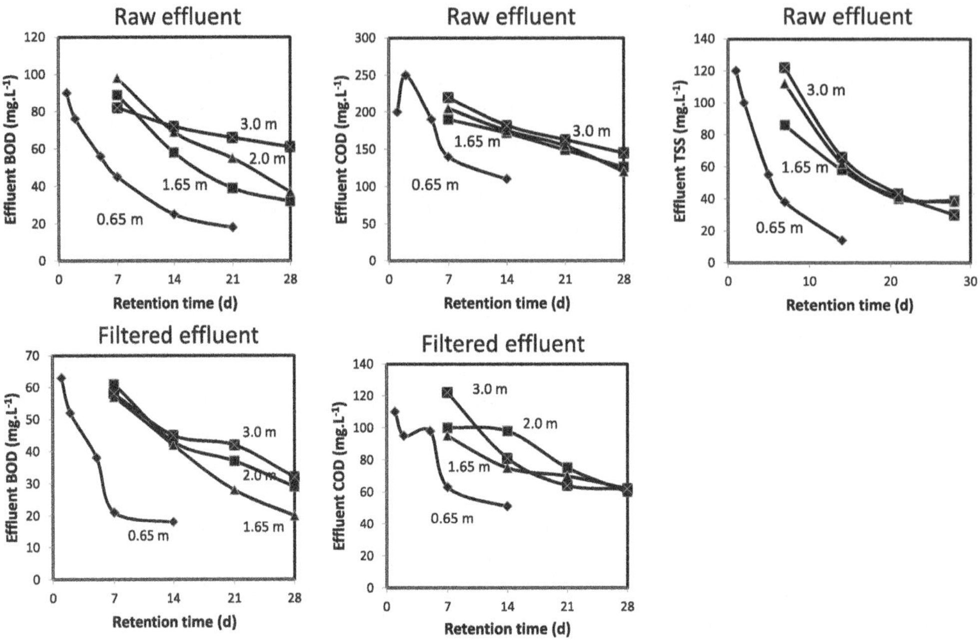

Figure 9.6 Performance in terms of removal of organic material and suspended solids in polishing ponds with a continuous-flow regime as functions of applied retention time and at various depths.

- The rate of decrease of BOD depends clearly on the depth: deeper ponds reduce the rate of decrease of BOD_5.
- The filtered BOD5 in UASB effluent constitutes 50–65% of the raw BOD5, indicating that a significant fraction (35–50%) can be attributed to the presence of particulate matter, initially predominantly faecal matter and UASB sludge, which is gradually replaced by algae during treatment in PPs.

COD has a qualitative behaviour very similar to that of BOD, with one exception: in the polishing pond COD tends to increase slightly or remain constant at the beginning of the polishing pond and then gradually decrease. The tendency of COD to increase is clearer in the shallow lagoon. This phenomenon must be attributed to algae production by photosynthesis, without a corresponding reduction of organic material by oxidation. The net oxygen production necessarily is accompanied by a corresponding increase of organic material (algae). Thus the production of organic material in terms of COD is continuously greater than its destruction by oxidation. The fact that, despite this, there is a decrease in COD can only be attributed to spontaneous flocculation and settling of the algae which thus withdraws the organic material from the liquid phase and adds it to the solid phase at the bottom of the pond, where it is partially digested and converted into biogas. In the ponds the COD decrease is approximately 100 mg $\cdot$ L^{-1} and should be attributed to flocculation and settling of algae. It is concluded that without algae flocculation it would be impossible to obtain a low COD concentration (less than 220 mg $\cdot$ L^{-1}). For the same reason the concentration of BOD in the effluent would also be much higher.

The settling of algae was confirmed by the analysis of bottom sludge in PP_1, which was shallow ($H = 0.65$ m) and consisted of five sequential lines. The lowest retention time in the first lane was 1 day, a time more than enough for settling of residual solids present in the effluent from the UASB reactors. Despite this, almost half of the bottom sludge was found in lanes 2–5 (Cavalcanti *et al.*, 2000c). The sludge that was present in lanes 2–5 therefore must have been due to the settling of solids that were formed in these ponds, presumably algae. Spontaneous flocculation and subsequent sedimentation can also be observed if a sample is withdrawn from the pond in a beaker and left in the sun: in less than 1 h a green precipitate is observed on the bottom of the beaker as a result of the algae settling.

Due to the abundant presence of algae in the effluent of polishing ponds, the COD value remains relatively high and a raw (non-settled) COD concentration of less than 100 mg $\cdot$ L^{-1} is rarely observed (Figure 9.6). In contrast, the filtered COD is much lower and values below 60 mg $\cdot$ L^{-1} (adopted as the effluent limit in some Brazilian states) can be expected for per capita areas of 1 to 1.5 m^2.

In terms of suspended solids removal, the polishing pond is not particularly efficient, again because of the presence of algae. In Figure 9.6, it is observed that the concentration of suspended solids decreases with time, but also depends on the depth.

9.7 PH VARIATION AND NITROGEN REMOVAL IN PONDS

Polishing ponds offer the great advantage of being able to remove nitrogen and phosphorus from digested sewage. This removal can be carried out by varying the pH in polishing ponds due to the physical and biological processes that occur. To analyse the variation of pH in ponds it is necessary to recall what was established in the Chapter 5 on the value and stability of pH in treatment systems. In this section the following important points are highlighted:

(1) The pH in treatment systems is mainly determined by the equilibria of the carbonic system, which in turn is influenced by the acidity and alkalinity
(2) The values of alkalinity and acidity are defined as:

$$\text{Alk} = 2[CO_3^{2-}] + [HCO_3^-] + [OH^-] - [H^+] \quad (5.10)$$

$$\text{Ac} = 2[CO_2] + [HCO_3^-] + [H^+] - [OH^-] \quad (5.12)$$

(3) Using these basic concepts the interrelationship between pH, alkalinity and acidity can be established:

$$\text{Alk} = [\text{Ac} + 10^{(\text{pH}-\text{p}K1)} - 10^{-\text{pH}}][1 + 2 \times 10^{(\text{pH}-\text{p}K2)}]/[1 + 2 \times 10^{(\text{p}K1-\text{pH})}] \quad (5.13)$$

(4) If the pH is in the neutral range ($5 < \text{pH} < 9$), the influence of the carbonic system is dominant and the influence of water hydrolysis can practically be ignored, so that Eq. (5.13) simplifies to:

$$\text{Alk} \approx \text{Ac}[1 + 2 \times 10^{(\text{pH}-\text{p}K2)}]/[1 + 2 \times 10^{(\text{p}K1-\text{pH})}] \text{ for } 5 < \text{pH} < 9 \quad (5.13a)$$

(5) Equation (5.13) can be used to construct the Deffeyes diagram (Figure 5.4) that can be used to evaluate variations of pH as a result of reactions that affect alkalinity or acidity.

In the specific case of polishing ponds, the processes affecting the pH value are (1) the removal of carbon dioxide, either physically by desorption processes or biologically by photosynthesis, (2) release of ammonia by desorption, and (3) precipitation of calcium carbonate. The effect of these processes can be evaluated by reaction stoichiometry.

- CO_2 desorption: This is a physical process and there is no reaction, but loss of CO_2 causes loss of acidity and thus increase of pH
- Photosynthesis: $CO_2 + H_2O \rightarrow \text{COD} + O_2$
- NH_3 desorption: $NH_4^+ \rightarrow NH_3\uparrow + H^+$
- $CaCO_3$ precipitation: $CO_3^{2-} + Ca^{2+} \rightarrow CaCO_3$

Table 9.7 Effect of different processes on alkalinity, acidity and pH in ponds.

Process	Unit	ΔAlk	ΔAc	ΔpH
CO_2 removal	$mmol \cdot L^{-1}$	0	−2	Increase
Desorption of NH_3	$meq \cdot L^{-1}$	−1	+1	Decrease
$CaCO_3$ precipitation	$mmol \cdot L^{-1}$	−2	0	Decrease

Now, using the definitions of acidity and alkalinity, the effect of the above processes on these two variables and on the pH can be calculated. Table 9.7 shows the stoichiometric variations in acidity and alkalinity. Section 5.3 showed how variations of the alkalinity and acidity of different processes can be used to calculate pH variation using the Deffeyes diagram.

Table 9.7 also indicates whether the pH tends to increase or decrease due to above processes. In practice, the precipitation of $CaCO_3$ is very low in the lagoons, even if the solutions are thermodynamically unstable, because it tends to form supersaturated solutions and indeed the precipitation of $CaCO_3$ is not actually observed in polishing ponds. For this reason precipitation of $CaCO_3$ will not be considered further in this text. When the concentration of calcium in polishing ponds decreases, this can generally be attributed to precipitation of phosphates rather than carbonates, which does not influence pH.

9.7.1 Carbon dioxide removal

CO_2 removal in polishing ponds occurs by two distinct mechanisms that take place simultaneously, one biological and the other one physical. Removal by physical desorption is due to supersaturation of this gas in the digested sewage. In the anaerobic pre-treatment system the concentration of CO_2 in the liquid phase is in equilibrium with the CO_2 partial pressure in the biogas, which is in the range of 5–20% for municipal sewage. However, in the lagoon the anaerobic effluent is in contact with air, with a much lower CO_2 partial pressure (0.03%), so that the liquid becomes supersaturated and CO_2 desorption will occur, until a new equilibrium between the concentration in the liquid phase and that in the air is established. The equilibrium concentration can be estimated by the Henry's Equation:

$$[CO_{2s}] = k_H p_{CO2} \tag{9.12}$$

where:

k_H = Henry constant = 0.034 at 25°C ($pk_H = 0.0138 \times T + 1.12$; Capri & Marais, 1974)
p_{CO2} = CO_2 partial pressure in the air = 31 mbar (0.031% of the atmospheric pressure)
CO_{2s} = dissolved CO_2 saturation concentration in water in equilibrium with air

The biological mechanism is the result of the three processes that occur in polishing ponds: photosynthesis, oxidation of organic material and anaerobic digestion. According to Eqs. (9.1a) to (9.1c), all three processes affect the CO_2 concentration in the liquid phase: photosynthesis consumes CO_2 and the processes of oxidation and anaerobic digestion produce CO_2. Therefore, depending on the relative rate of the three processes, there may either be biological production or consumption of CO_2. Invariably in polishing ponds the oxidation rate will be low because there is little organic material available while anaerobic digestion will be practically non-existent. Therefore in polishing ponds there the dominant process is biological CO_2 consumption for photosynthesis, which results in an increase of pH.

EXAMPLE 9.3

Estimate the equilibrium concentration of the carbonic system in water that is in equilibrium with air, as a function of the alkalinity.

Solution

At 25°C the CO_2 saturation concentration in pure water (Alk = 0) is calculated as:

$$[CO_2] = 0.034 \times 0.00031 \approx 10^{-5} \text{mol} \cdot L^{-1} \text{ or } 0.5 \text{ mg} \cdot L^{-1}$$

If the water has alkalinity, Eqs. (5.10 and 5.12) can be written to express the equilibrium CO_2 concentration in terms of acidity and alkalinity. By equating these expressions, it is possible to calculate for any pH the corresponding alkalinity and acidity values and use these to plot the Deffeyes diagram of Figure 9.7. The concentration of the carbonic system C_t is now calculated as the average of alkalinity and acidity: $C_t = (\text{Alk} + \text{Ac})/2$

Using the value of the total concentration of carbonic species C_t, the dissociation equations (5.6 to 5.8) can be used to calculate the concentrations of the individual species: $[CO_2]$, $[HCO_3^-]$ and $[CO_3^{2-}]$.

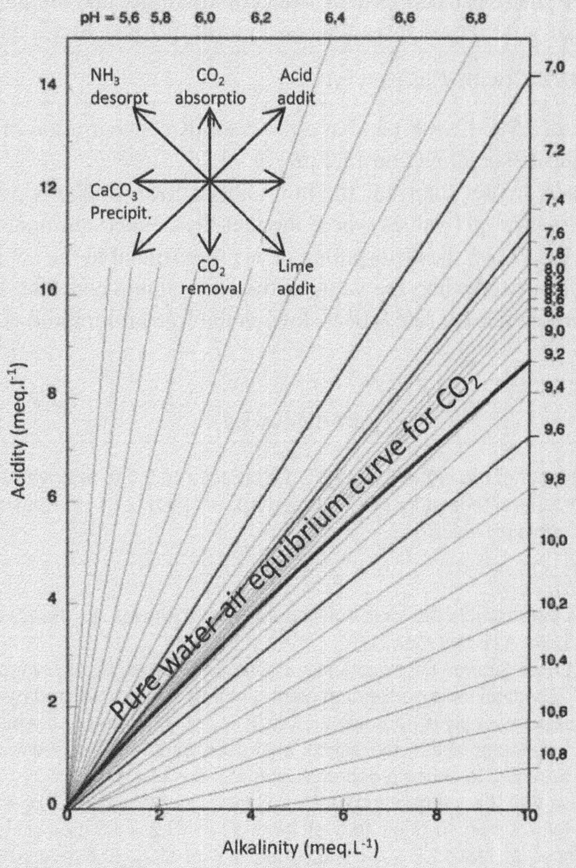

Figure 9.7 Deffeyes diagram for water–air equilibrium of CO_2.

In practice, the desorption rate of CO_2 is very slow when the pH becomes greater than about 7.5. Thus, for the pH to increase significantly above this value, it will be necessary that the photosynthesis develops at high rate, so that the rate of biological consumption of CO_2 by the algae exceeds that of the production of CO_2 by the bacteria. If the pH reaches a high value (pH > 9), the liquid phase may become sub-saturated with CO_2. Some CO_2 from the atmosphere may dissolve, however also at a low rate, since the inducing force (the difference between saturation and actual concentration) will always be small.

9.7.2 Ammonium removal

Ammonium removal is achieved by NH_3 desorption. Ammonium is formed from the dissociation of the NH_4^+ ion by the following reaction:

$$NH_4^+ \rightarrow NH_3 + H^+ \tag{9.13a}$$

$$[NH_3]/[NH_4^+] = K_a\,[H^+] \tag{9.13b}$$

Equation (9.13a) shows that the removal of ammonia is equivalent to the addition of a strong acid, i.e. per mmol of desorbed NH_3, there is a decrease of 1 meq of alkalinity and an increase of 1 meq of acidity. According to Emerson *et al.* (1975) the ionization constant pK_a is given by Eq. (9.13c):

$$pK_a = 0.09018 + 2729.92/T \text{ (with } T \text{ in Kelvin)} \tag{9.13c}$$

Thus, at a temperature of 25°C (298 K) and at a fixed solids concentration of 1 g · L^{-1} (i.e. an activity coefficient of about 0.9), the dissociation constant $pK_a = 9.1$.

Therefore, for pH values greater than 9.1, the non-ionized form (NH_3) is predominant, according to Eq. (9.13b), in contrast to neutral pH values where the non-dissociated ammonia fraction is small (1% at pH = 7.1). According to Fick's Law, the desorption rate is proportional to the concentration of non-ionized ammonia, which can be calculated using the value of the dissociation constant. It is concluded that in the neutral pH range the ammonia release rate will be low, as the free ammonium concentration NH_3 is low.

EXAMPLE 9.4

It is desired to increase the pH of the effluent from a UASB reactor to 9.6 to remove 42 mg · L^{-1} of ammonia by desorption. The effluent is characterized by pH = 7.0 and Alk = 8 meq · L^{-1}. How much CO_2 must be removed by photosynthesis to reach this pH value?

Solution

From the data provided in the example the ionic equilibrium in the Deffeyes diagram is characterized: point A in Figure 9.8. The acidity at point A is 11.8 meq · L^{-1}.

Removal of 42 mg N · L^{-1} or 3 meq · L^{-1} results in a decrease in alkalinity by 3 meq · L^{-1}, the final alkalinity being 8 − 3 = 5 meq · L^{-1}. The final ionic equilibrium point (point B) is characterized by the intersection of Alk = 5 meq · L^{-1} and pH = 9.6, where acidity is approximately 3.5 meq · L^{-1}. Therefore the acidity must decrease with 11.8 − 3.4 = 8.4 meq · L^{-1}. Remember that the acidity increases with 3 meq · L^{-1} due to ammonia release, so enough CO_2 needs to be removed to cause a decrease of 8.4 + 3 = 11.4 meq · L^{-1}

The diagram also shows that the intersection of Alk = 8 meq · L^{-1} with the dissolved CO_2 equilibrium curve with air is at an acidity of 7.2 meq · L^{-1}, so that at most 11.8 − 7.2 = 4.6 meq · L^{-1} can desorb. Therefore photosynthesis must remove at least 7.2 − 0.6 = 6.6 meq · L^{-1} of CO_2. (0.6 meq · L^{-1} is the residual acidity after photosynthesis = intersection point of the NH_3 desorption line with the photosynthesis line). In fact, photosynthesis will have to remove more CO_2, because desorption of CO_2 above pH = 7.5 becomes very slow.

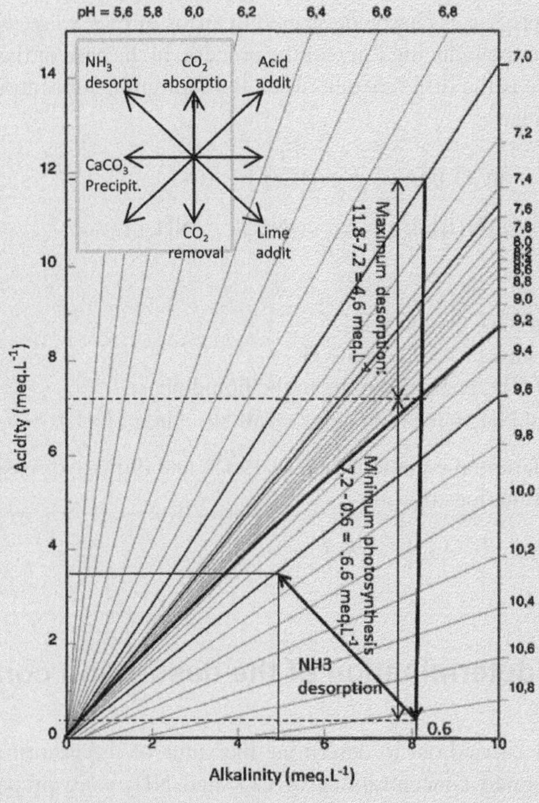

Figure 9.8 Deffeyes diagram, indicating the effects on pH of CO_2 and NH_3 desorption and biological CO_2 consumption.

To remove 6.6 meq · L^{-1} or 3.3 mmol · L^{-1} of CO_2, it will be necessary that there is a net production of 3.3 mmol of oxygen or about 100 mg · L^{-1} of oxygen. This is quite feasible, but the time it takes to achieve this depends on several factors: solar energy, depth and concentration of biodegradable organic material.

EXAMPLE 9.5

In the previous example the required removal of acidity was 11 meq · L^{-1}. Is this quantity normally available in the effluent from a UASB reactor?

Solution

The acidity removal requires that $11/2 = 5.5$ mmol $CO_2 \cdot L^{-1}$ is removed. The initially available carbonic species can be calculated from the expression that links the total concentration to the alkalinity and acidity: $C_t = (Alk + Ac)/2 = (11.8 + 8)/2 = 19.8/2 = 9.9$ mmol · L^{-1}, so that more than enough carbonic species are available for the pH increase. In general the availability of carbonic species is not a limiting factor for nutrient removal in polishing ponds.

9.7.3 Model of the kinetics of CO_2 and NH_3 removal by desorption

Desorption and absorption processes can be described in terms of Fick's law, which states that the rate of the process is proportional to the inducing force of the process. In the case of the volatile components (CO_2 and NH_3), the inducing force is the difference between the saturation concentration and the concentration that is present in the liquid phase:

For CO_2 desorption: $r_{dc} = (d[CO_2]/dt)_d = k_{dc} \cdot ([CO_2] - [CO_2]_s)$ (9.15)

For NH_3 desorption: $r_{dn} = (d[NH_3]/dt)_d = k_{dn} \cdot ([NH_3] - [NH_3]_s)$ (9.16)

where:

r_{dc}, r_{dn} = desorption rates for CO_2 and NH_3
k_{dc}, k_{dn} = desorption constants for CO_2 and NH_3
$[CO_2]$, $[NH_3]$ = CO_2 and NH_3 concentrations in the liquid phase
$[CO_2]_s$, $[NH_3]_s$ = CO_2 and NH_3 saturation concentrations, where $[NH_3]_s = 0$ (no ammonium in the air).

By integration of the differential equations in Eqs. (9.15) and (9.16), expressions are found for the CO_2 and NH_3 concentrations as functions of time:

$$\ln\{([CO_2] - [CO_2]_s)/([CO_2]_o - [CO_2]_s)\} = -k_{dc}t \quad (9.15a)$$

$$\ln([NH_3]/[NH_3]_o) = -k_{dn}t \quad (9.16a)$$

9.7.4 Experimental determination of the desorption constants for CO_2 and NH_3

An experimental study was carried out to determine the value of the desorption constants for CO_2 and NH_3. Solutions of oversaturated concentrations of CO_2 and NH_3 were prepared and the value of the concentrations was followed as a function of time in cylindrical pilot-pond models. Different liquid depths were evaluated, because desorption is a surface phenomenon, so it was expected that the value of the desorption constant depends on the depth.

Series P_1 was composed of three pilot ponds with depths of 0.4, 0.6 and 0.8 m respectively and contained a supersaturated CO_2 solution which was used to evaluate the CO_2 desorption kinetics. This solution was prepared by adding 4 meq $\cdot$ L^{-1} of sodium bicarbonate ($NaHCO_3$) to tap water (with an alkalinity of 2 meq $\cdot$ L^{-1}) and then to reduce pH to approximately 6 by addition of a strong acid (HCl). The initial composition of the contents of ponds P_1 is listed in Table 9.10 in Section 9.1.1.3.

Series P_2 also used three pilot ponds with depths of 0.4, 0.6 and 0.8 but these contained a supersaturated NH_3 solution, used to determine the ammonia desorption kinetics. An alkaline ammonium chloride solution was prepared by adding sodium hydroxide to tap water to raise the pH to 10, which resulted in the precipitation of calcium carbonate. To the supernatant, ammonium chloride (50 mg N $\cdot$ L^{-1}) was added. The initial composition of the contents of ponds P_2 is also shown in Table 9.10.

9.7.4.1 Determination of the CO_2 desorption constant

In the ponds of series P_1 with the supersaturated CO_2 solution (P_{1_40}, P_{1_60} and P_{1_80}), the only readily measurable parameter that changes with time is pH, because CO_2 desorption does not affect the alkalinity. The initial alkalinity and pH were 9.5 meq $\cdot$ L^{-1} and 6.13 respectively. The values of pH as a function of time are listed in Table 9.8.

Table 9.8 pH values as functions of time for the three pilot ponds of the series P_1.

Time (d)	P_{1_40}	P_{1_60}	P_{1_80}
0	6.13	6.13	6.13
1	6.45	6.26	6.26
2	6.96	6.53	6.37
3	7.48	6.88	6.59
4	7.63	7.13	6.82
5	7.97	7.44	7.05
6	8.29	7.62	7.17
7	8.6	7.89	7.54
8	8.63	8.45	7.62
9	8.69	8.47	7.75
10	8.85	8.5	8.02

The desorption constant was calculated with the aid of the proposed model, adopting a value of 0 for both the desorption rate of ammonium and photosynthesis, as these processes did not develop in the ponds of series P_1. One method to determine the value of the desorption constant is to calculate the variation of the CO_2 concentration with time. The desorption rate cannot be calculated directly from the change of the CO_2 concentration, because CO_2 is produced from bicarbonate present in the solution as desorption of CO_2 takes place. However, it is possible to determine the desorption rate from the variation of the total carbonic species concentration ($C_t = (Alk + Ac)/2$). Knowing that the alkalinity is constant during the experiment, the acidity can be calculated for the experimental values of the pH obtained as a function of time and from this the C_t values are calculated. As desorption was the only process taking place in the ponds of series P_1, the change of the total carbonic species concentration can be attributed to this process. The calculated CO_2 values are used in Figure 9.9 to calculate the desorption constant by plotting $\ln([CO_2] - [CO_2]_s)$ as function of the desorption time and determine the best straight line in accordance with Eq. (9.15a).

The results of the simulations showed that the expected inversely proportional relationship with depth was indeed observed. The best correlation between the experimental results of the desorption constant for the three investigated depths was obtained for $k_{dc} = 0.35/H$ (d^{-1}).

To establish the effect of temperature on the value of constant k_{dc}, similar experiments were carried out at 20°C and 30°C, using only ponds with a depth of 0.40 m. It was established that over the temperature interval the value of the constant increased by 58%. If an Arrhenius-type temperature dependency is assumed, the calculated increase per °C is calculated as 4.8%, so that the desorption constant of CO_2 can be expressed as a function of temperature T (in °C) and pond depth H (in m):

$$k_{dc} = 0.35/H \cdot (1.048)^{(T-25)}\ d^{-1}$$

9.7.4.2 Determination of the NH_3 desorption constant

In the case of ammonium desorption, Eq. (9.16a) can be used to determine the desorption rate constant. By determination of the values of the total ammonium concentration and the pH, the non-ionized concentration NH_3 can be calculated. The values of $\mathrm{Ln}[NH_3]/[NH_3]_o$ are plotted as function of time and the slope of the best straight lines through the experimental points are used to determine the desorption constant values. Figure 9.10 shows the experimental data.

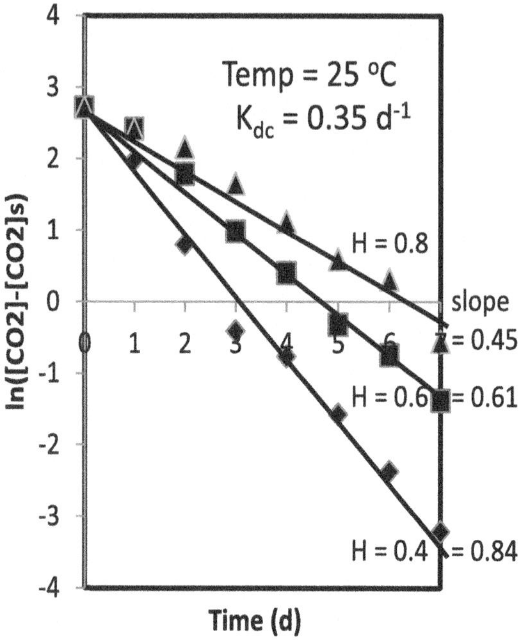

Figure 9.9 Experimental CO_2 concentration values during desorption of CO_2 in the pilot ponds of series P_1.

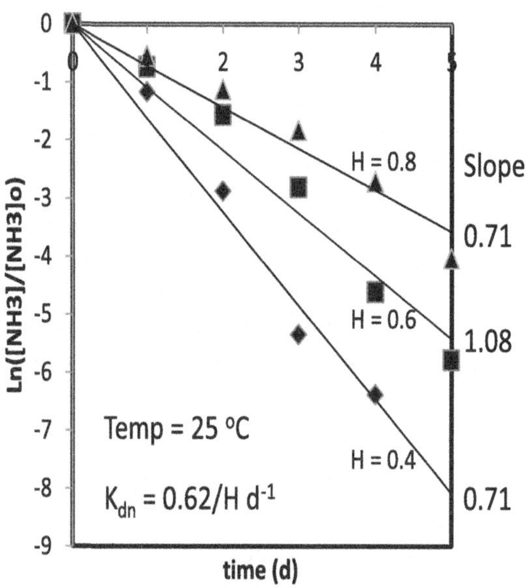

Figure 9.10 Experimental NH_3 concentration values during desorption in the pilot ponds.

As in the case of CO_2 desorption, the NH_3 desorption constant is clearly dependent on the pond depth. If an inversely proportional relationship is assumed, the dependency on the constant can be expressed as: $k_{dn} = 0.62/H$ (d^{-1}). To evaluate the temperature dependency, separate tests were carried out with ponds of 0.40 m at temperatures of 20°C and 30°C. An increase of the constant of 72% was determined for the temperature interval of 10°C.

If an Arrhenius temperature dependency is assumed, the increase per °C is 5.6% and the value of the ammonia desorption constant may be expressed as:

$$k_{dn} = 0.62/H\,(1.056)^{(T-25)}\ (d^{-1})$$

9.7.5 Model for pH in sequential batch polishing ponds

Having established expressions for the values of the desorption rate constants for CO_2 and NH_3, the rates of these desorption processes in sequential batch polishing ponds can now be calculated, when the pH is known. Equations (9.13a) to (9.13c) clearly show that a significant increase in pH is a prerequisite for high-rate ammonia desorption. This increase in pH is only possible if there is considerable biological removal of CO_2 in the pond. Since CO_2 desorption becomes very slow when the pH rises above a pH of 7.5, much of the CO_2 removal will have to come from photosynthesis.

The magnitude of the variations of alkalinity and acidity required to achieve a certain pH increase can be determined in the Deffeyes diagram. To establish the time required for this variation to take place, it is necessary to consider the kinetics of the relevant processes. These processes are the desorption of CO_2 and NH_3 that have been discussed above and biological CO_2 removal by photosynthesis which consumes CO_2 and produces dissolved oxygen. The rate of this process can be determined conveniently by the production of oxygen in the pond. The relationship between oxygen production and carbon dioxide consumption can be expressed as:

$$nCO_2 + nH_2O \rightarrow (CH_2O)_n + nO_2$$

The rate of oxygen consumption can therefore be expressed as:

$$r_{fo} = -(dCO_2/dt)_f = (dO_2/dt)_f = OPR \tag{9.17}$$

where:

$r_{fo} = (dO_2/dt)_f$ = photosynthesis rate in the pond (mol $O_2 \cdot L^{-1} \cdot d^{-1}$)
= rate of biological CO_2 consumption (mol $CO_2 \cdot L^{-1} \cdot d^{-1}$)
OPR = oxygen production rate (mol $O_2 \cdot L^{-1} \cdot d^{-1}$)

Actually Eq. (9.17) is an oversimplification of a much more complex expression that describes the production of organic material in the form of algae. A fraction of the algae mass is composed of nitrogen that is present as organic nitrogen and is produced by assimilation of ammonium, a process where 1 mol of H^+ is produced per mol NH_4^+ consumed. However, the fraction of nitrogen in algae is low so that the simplified equation of Eq. (9.17) is acceptable.

Equations (9.15)–(9.17) form the basis for a model that predicts both pH variation and ammonia removal in a sequential batch polishing pond. For this the differential equations must be written in terms of finite time elements that cause a finite change in CO_2 and NH_3 concentrations, as in Eqs. (9.18) and (9.19):

$$\Delta[CO_2]/\Delta t = -k_{dc} \cdot ([CO_2] - [CO_2]_s) - OPR \tag{9.18}$$

$$\Delta[NH_3]/\Delta t = [NH_3] - k_{dn} \cdot [NH_3]_o \tag{9.19}$$

These changes in turn result in variations of the alkalinity and acidity, as well as of the ammonia concentration and can be expressed as:

$$\Delta \text{Alk}/\Delta t = -k_{dn} \cdot [\text{NH}_3] \quad (9.20)$$

$$\Delta \text{Ac}/\Delta t = -2\ (k_{dc}([\text{CO}_2] - [\text{CO}_2]_s) - 2 \cdot \text{OPR} + k_{dn} \cdot [\text{NH}_3]) \quad (9.21)$$

$$\Delta \text{N}_{tot}/\Delta t = N_{tot} - k_{dn} \cdot [\text{NH}_3]. \quad (9.22)$$

On the other hand the pH variation can also be calculated from the variations of alkalinity and acidity. Considering that the H^+ and OH^- concentrations are so low that they do not influence the alkalinity and acidity one has:

$$\text{Alk} = \text{Ac}(1 + 2k_2/[\text{H}^+])/(1 + [\text{H}^+]/k_1). \quad (9.23)$$

This is an implicit quadratic expression that can be solved for $[H^+]$ resulting in:

$$[\text{H}^+] = (k_1 \cdot (\text{Ac/Alk} - 1)/2 + (k_1(\text{Ac/Alk} - 1)/2)^2 - 4 \cdot k_1k_2)^{1/2})/2$$

Furthermore $\text{pH} = v\log [\text{H}^+]$ so that:

$$\text{pH}_1 = -\log[(0.5(k_1(\text{Ac}_1/\text{Alk}_1 - 1)/2 + ((k_1(\text{Ac}_1/\text{Alk}_1/2)^2 - 4 \cdot k_1k_2)^{1/2})] \quad (9.24)$$

These equations form the basis for a model that will describe the variations of pH, alkalinity, acidity and ammonia concentration as a function of time. The model is a set of differential equations defined by Eqs. (9.15) to (9.24), which are presented in Table 9.9.

For a specific pond depth and temperature (that define the values of k_{dc} and k_{dn}), the only process that is not yet defined in Table 9.9 is the photosynthesis or CO_2 consumption rate. There seems to be no expression that links the rate of this process to the operational conditions and sewage and climate characteristics, so that it must be determined experimentally or estimated from previous experiments in similar ponds.

Table 9.9 Model summary of the equations to calculate the process prates and rate of change of concentrations in sequential batch polishing ponds.

Rate of photosynthesis Rate of CO_2 desorption Rate of NH_3 desorption:	$r_{fo} = -(dCO_2/dt)_f = (dO_2/dt)_f = \text{OPR}$	(9.17)
Alkalinity change:	$r_{dc} = (d[CO_2]/dt)_d = k_{dc}([CO_2] - [CO_2]_s)$	(9.15)
Acidity change	$r_{dn} = (d[NH_3]/dt)_d = k_{dn}([NH_3] - [NH_3]_s)$	(9.16)
pH change	$\Delta Alk/\Delta t = -k_{dn}[NH_3]$	(9.20)
CO_2 change	$\Delta Ac/\Delta t = -2(k_{dc}([CO_2] - [CO_2]_s) - 2OPR + k_{dn}[NH_3])$	(9.21)
Ammonium change	$pH_1 = -\log[(0.5(k_1(Ac_1/Alk_1 - 1)/2 + ((k_1(Ac_1/Alk_1/2)^2 - 4k_1k_2)^{1/2})]$	(9.24)
Total N change	$\Delta[CO_2]/\Delta t = -k_{dc}([CO_2] - [CO_2]_s) - OPR$	(9.18)
	$\Delta[NH_3]/\Delta t = [NH_3] - k_{dn}[NH_3]_o$	(9.19)
	$\Delta N_{tot}/\Delta t = N_{tot} - k_{dn}\ [NH_3]$	(9.22)

9.7.5.1 *Experimental verification of the validity of the model to predict pH changes*

An experimental study was carried out to establish if observed pH changes in polishing ponds could be predicted with the model of Table 9.9. The experiments were carried out at pilot scale, using ponds operated in batch mode. Ponds with three different water depths were operated: 0.40, 0.60 and 0.80 m. This range was chosen because ponds with a depth of less than 0.4 m are not practical in full-scale systems, and because it had been established before that for depths >0.8 m the pH increased only very slowly, even under the favourable conditions of Campina Grande, Brazil, where the experiments were carried out (high temperatures and abundant sunshine).

The study was divided into three parts, which were carried out simultaneously, so that the same weather conditions prevailed. In the first part, ponds were used with tap water solutions, which were oversaturated either with carbon dioxide (P_1) or with ammonium (P_2). The changes of the CO_2 and NH_3 concentrations were then used to determine the desorption rate constants as described in Section 9.6.4.

In the second part a series of ponds (P_3) were operated with effluent from a UASB reactor, but photosynthesis was limited artificially in such a way that the dissolved oxygen (DO) concentration remained approximately constant at a value near the saturation concentration of 7.8 $mg \cdot L^{-1}$. Therefore there was neither net biological oxygen production nor consumption. This was achieved by covering the pond when the DO concentration exceeded a maximum limit of 10 $mg \cdot L^{-1}$, and removing the cover when the DO concentration decreased below 6 $mg \cdot L^{-1}$. Hence, with the average rate of photosynthesis approximately equal to the rate of organic material oxidation, eventual changes in alkalinity and acidity could be attributed to physical CO_2 and NH_3 removal through desorption. Tests of VFA concentrations established that the oxidation of these acids was insignificant. It was evaluated whether it was possible to model the experimental desorption rates of CO_2 and NH_3 using the desorption constants determined previously in the first part of the study, when tap water was used instead of digested sewage with algae.

In the third part of the investigation (ponds P_4), photosynthesis was not limited, but the DO concentration was reduced chemically when it attained 10 $mg \cdot L^{-1}$ by automatically adding enough sulphite to reduce the DO value to 6 $mg \cdot L^{-1}$. Cobalt chloride ($CoCl_2$) was added to these ponds at a concentration of 0.2 $Co \cdot L^{-1}$ as a catalyst to accelerate the redox process. When sulphite was added, the contents of the pond was gently stirred to distribute the sulphite. Thus the accumulated DO concentration (and hence the accumulated biological CO_2 consumption) was calculated from the daily sulphite consumption.

The fifth series of ponds (P_5) was operated under identical conditions as the fourth one, but in this pond DO was not reduced, so that a high concentration built up and oxygen desorption took place. By comparing the pH changes in the fourth and the fifth series of ponds it could be established whether the DO reduction in the fourth pond influenced in any way the changes of pH in the pond (Figure 9.11, Table 9.10).

Thus three groups of five ponds each were operated simultaneously in the open air. The three groups had different depths: 40, 60 and 80 cm (see also Figure 9.13). Within each group of five ponds, the contents of the ponds and the objective of the operation differed. Ponds of the series P_3, P_4 and P_5 all had the same initial composition, but were operated in different ways. The ponds were filled with a mixture of settled UASB effluent (80%) and polishing pond effluent (20%) with a high algae concentration (about 800 $\mu g \cdot L^{-1}$). The initial composition of these ponds is also shown in Table 9.1.

9.7.5.2 *Evaluation of CO_2 and NH_3 desorption in polishing ponds*

Ponds of series P_3 (P_{3_40}, P_{3_60} and P_{3_80}) were used to evaluate the simultaneous processes of ammonium and carbon dioxide desorption in polishing ponds, using mixtures of UASB effluent and polishing pond effluent rich in algae. By limiting the rate of photosynthesis in the third pond, the pond was kept in an aerobic environment with the DO concentration near the saturation value, so that no net biological

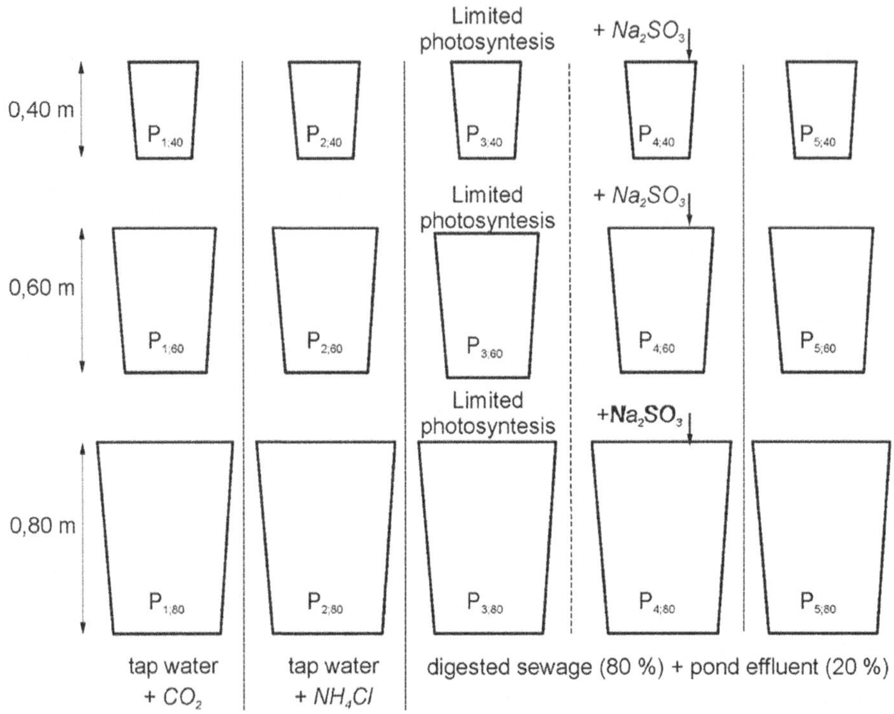

Figure 9.11 Representation of the experimental set-up of the batch ponds used in the experimental investigation: 5 series of 3 ponds (P_1–P_5) with different depths (40, 60 and 80 cm).

Table 9.10 Composition of the liquid phase in the contents of the different ponds.

Parameter	Units	Series P_1 ($CO_{2\ Desorption}$)	Series P_2 ($NH_{3\ Desorption}$)	Series P_3 to P_5 (Polishing Ponds)
pH	–	6.12	9.5	7.4
Alk	$meq \cdot L^{-1}$	6.5	12.3	7.5
$N\text{-}NH_3$	$mg\ N \cdot L^{-1}$	0	0	39
DO	$mg\ O_2 \cdot L^{-1}$	6.0	7.0	8.0
COD	$mg \cdot L^{-1}$	0	0	102
BOD_5	$mg \cdot L^{-1}$	0	0	34
TKN	$mg\ N \cdot L^{-1}$	0	0	43
P_{total}	$mg\ P \cdot L^{-1}$	0	0	6.5
P_{orto}	$mg\ P \cdot L^{-1}$	0	0	5.2
Chlorophyll a	$\mu g \cdot L^{-1}$	0	0	246
VFA	$mg\ HAc \cdot L^{-1}$	0	0	34
Ca^{2+}	$mg\ CaCO_3 \cdot L^{-1}$	104	52	95
Hardness	$mg\ CaCO_3 \cdot L^{-1}$	194	148	178

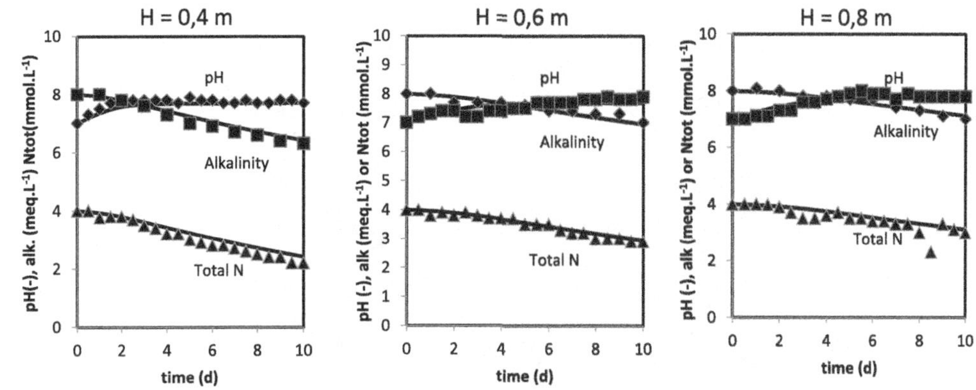

Figure 9.12 Experimental and simulated values of pH, total ammonium concentration ($NH_3 + NH_4^+$) and alkalinity as functions of time and for different depths in batch ponds with limited photosynthesis (series P_3).

consumption (or production) of CO_2 occurred in the ponds. For that reason the changes in alkalinity, acidity and pH were attributable only to desorption of CO_2 and NH_3. The data from the ponds could be compared to simulated profiles, using the model defined in Table 9.9 and the values of the desorption constants determined in the preceding Section 9.7.4.

Figure 9.12 shows the experimental data and the simulated profiles for the three readily measurable parameters in the ponds of series P_3: alkalinity, pH and ammonia, as function of time. It can be observed that in all cases there is a close correspondence between the experimental and the simulated data, so that it is concluded that the model can describe adequately both the stoichiometric and the kinetic aspects of the processes affecting the pH in polishing ponds, when the effect of photosynthesis is excluded.

One important conclusion that can already be drawn is that the removal of ammonium is indeed occurring through the mechanism of desorption: if it was due to nitrification and denitrification would take place, the rate of ammonium removal would be higher than the simulated one, which was not the case. Hence it is concluded that nitrification and denitrification do not take place to any appreciable extent.

9.7.5.3 Evaluation of pond behaviour without limitation of photosynthesis

The fourth (P_4) and fifth (P_5) series of ponds were operated without limitation of the photosynthetic activity of the algae. A difficulty in modelling the behaviour is the fact that the rate of photosynthesis is irregular because it depends directly on the actual weather conditions.

In the fourth series of ponds the rate of biological CO_2 removal was equal to the rate of oxygen production and this rate was determined by on line measurement of the DO concentration.

The DO concentration was controlled between 6 and 10 mg · L^{-1} by means of sulphite addition. Thus it was possible to determine the accumulated DO concentration over a day from the consumption of sulphite and this value was also attributed to the fifth series of ponds.

The rate of photosynthesis was taken as the average of the accumulated DO concentration over a day. Figure 9.13 shows experimental values of the accumulated DO concentration, which is equal to the rate of biological CO_2 removal for ponds P_{4_40}, P_{4_60} and P_{4_80}. Once the rate of biological CO_2 removal has been assessed, all information is available to simulate the behaviour of the ponds with respect to displacement of the ionic equilibrium point caused by the change of pH and alkalinity and ammonium desorption. Figure 9.14 gives the experimental values of the alkalinity, pH and the total ammonium concentration in

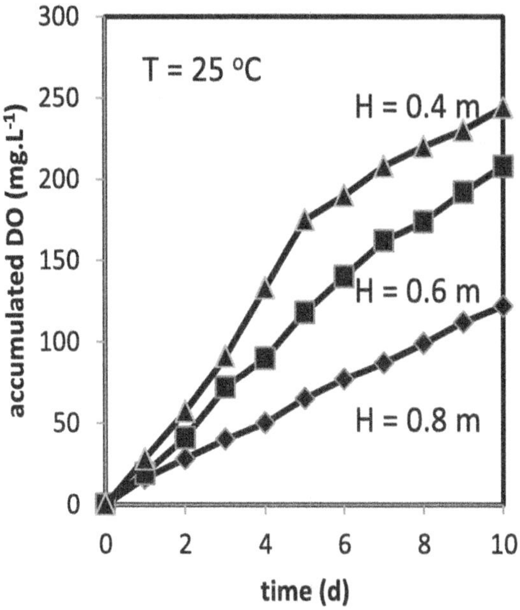

Figure 9.13 Experimental profiles of the accumulated oxygen production in batch ponds without photosynthesis limitation (series P_4).

ponds of series P_4. It can be noted that there is a close relationship between the experimental and simulated parameters in all ponds during the entire period of measurement.

When the curves of Figure 9.12 are compared to those of Figure 9.14, there is a very large difference in the behaviour with respect to the three variables: alkalinity, pH and total ammonium. This difference is clearly because photosynthesis was not limited in series P_4, so that extensive biological CO_2 removal

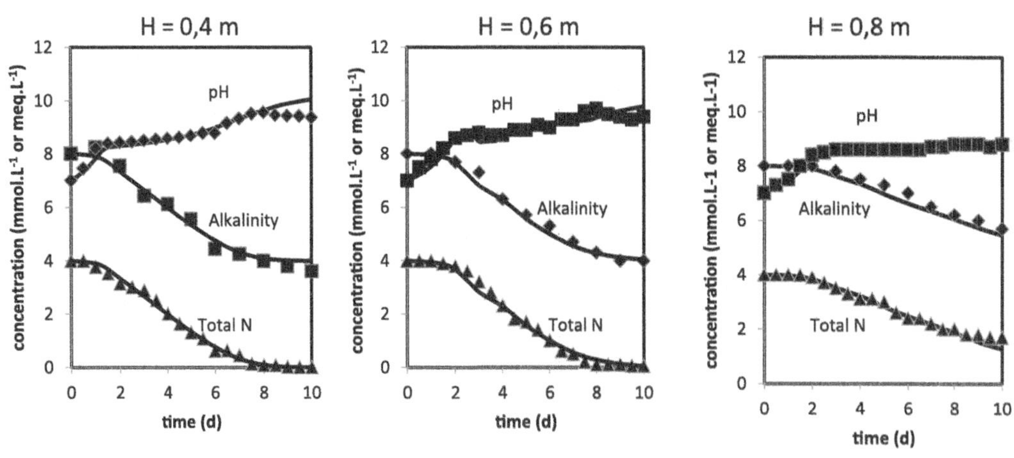

Figure 9.14 Experimental and simulated values of pH, total ammonium concentration ($NH_3 + NH_4^+$) and alkalinity as functions of time and for different depths in batch ponds with limited photosynthesis (series P_4).

took place in these ponds resulting in a much higher pH, more nitrogen removal and a greater decrease of alkalinity.

Parallel to the three ponds of series P_4, in which the accumulation of DO was determined by successive reductions of its concentration with Na_2SO_3, three ponds of series P_5 were also operated, but in these ponds the DO was not removed. It was observed that in the ponds of series P_5, without reduction of the produced DO, the pH was systematically higher than in the ponds of series P_4, with the sodium sulphite addition. This means that in the ponds of series P_4 the rate of photosynthesis was lower than in the ponds of series ponds P_5. The reason for this difference is not clear *a priori*, since the ponds were operated with the same digested wastewater plus algae seed and the operational conditions during the investigation were also identical. It is possible that either the cobalt catalyst (0.5 mg $\cdot$ L^{-1}) or the sulphite was slightly toxic for the algae, even though the latter was removed almost instantaneously by the DO in the ponds. It is concluded that the rate of biological CO_2 removal determined in the ponds of series P_4 was smaller than the value that could have been obtained if it had not been measured. Hence the rate of biological CO_2 removal that was determined is actually a conservative estimate of the value that could be obtained in practice.

Having established that the model of Table 9.9 is adequate to predict pH changes as well as changes of related variables (alkalinity, acidity and ammonium concentration) the model can now be used to evaluate the behaviour in polishing ponds under a variety of conditions, since the model permits calculation of the values of these variables as function of time. The next step is to carry out these calculations for a particular set of conditions and initial values, to show that it is indeed possible to predict what the behaviour of polishing ponds will be with respect to the displacement of the ionic equilibrium point and the extent of ammonia removal as a function of time, as function of the temperature and the pond depth, as long as a reasonable estimate of the rate of photosynthesis can be made. In Examples 9.6 and 9.7, these calculations will be demonstrated.

EXAMPLE 9.6

Evaluate the influence of temperature and pond depth on the rate and extent of CO_2 and NH_3 removal processes. Assume that the rate of photosynthesis is inversely proportional to the pond depth and is 24 mg $\cdot$ L^{-1} $\cdot$ d^{-1} in a pond with a depth of 0.4 m.

Solution

The depth of a pond has a strong influence on all NH_3 and CO_2 removal processes. The desorption rate constants are inversely proportional to the pond depth while the rate of photosynthesis also is roughly inversely proportional. In Figure 9.15 the rates of CO_2 and NH_3 desorption and the CO_2 consumption rate are shown for a temperature of 25°C. The rate of oxygen production and CO_2 consumption is also indicated for assumed values of 24, 16 and 12 mg O_2 $\cdot$ L^{-1} $\cdot$ d^{-1} for depths of 0.4–0.8 m. From the relative rates of CO_2 removal, it can be observed that desorption only has importance when the CO_2 concentration is high at the start of the treatment in the sequencing batch polishing pond. After the first day of operation the desorption rate is slightly negative (i.e. CO_2 absorption occurs).

Desorption of ammonium starts at a very low rate, because initially at the neutral pH the free ammonia concentration is low as well. As pH increases so does the rate of desorption, but as the total ammonia concentration decreases in time so does the NH_3 desorption rate. Figure 9.15 shows that the pond depth has a strong influence on the efficiency of nitrogen removal by desorption: whereas the time for substantially complete nitrogen removal is only one week in a pond of 0.4 m depth, the time required in a pond with 0.8 m depth is more than twice as high.

Temperature also has very strong influence on the performance of polishing ponds, but its influence cannot easily be expressed in a quantitative way as it affects many variables. The effluent composition – and notably

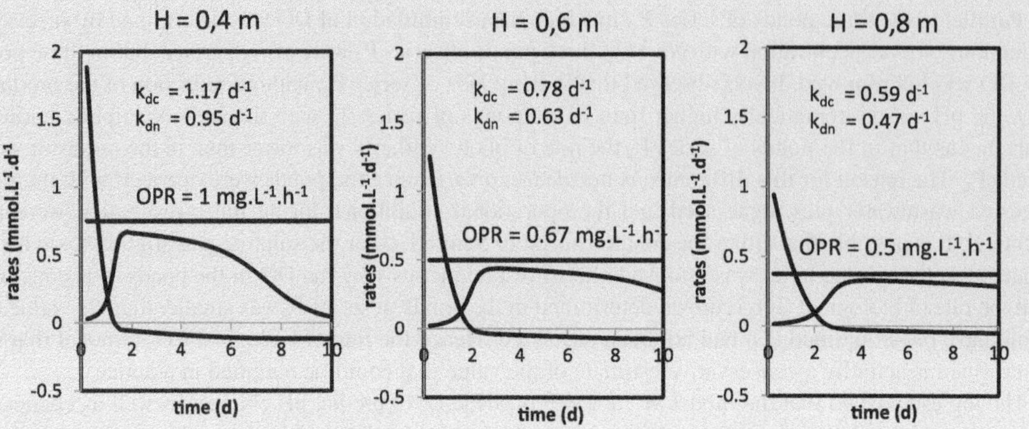

Figure 9.15 Rates of CO_2 and NH_3 desorption rates and CO_2 consumption rate at 25°C as function of time in ponds with depths from 0.4 to 0.8 m.

the biodegradable organic material of the pre-treatment system – is influenced by temperature as was shown in Chapter 5. The biodegradable organic material concentration influences the oxygen production rate, since it is oxidized in the pond, consuming oxygen and producing CO_2. The OPR is also influenced by temperature because it is related to photosynthesis. The desorption rate constants k_{dc} and k_{dn} are also a function of temperature.

Figure 9.16 shows the accumulated concentrations of the carbonic species, dissolved oxygen and total ammonium as function of time for three pond depths at a temperature of 25°C. The pond depth has a clear influence on the variation of each of these three variables in time.

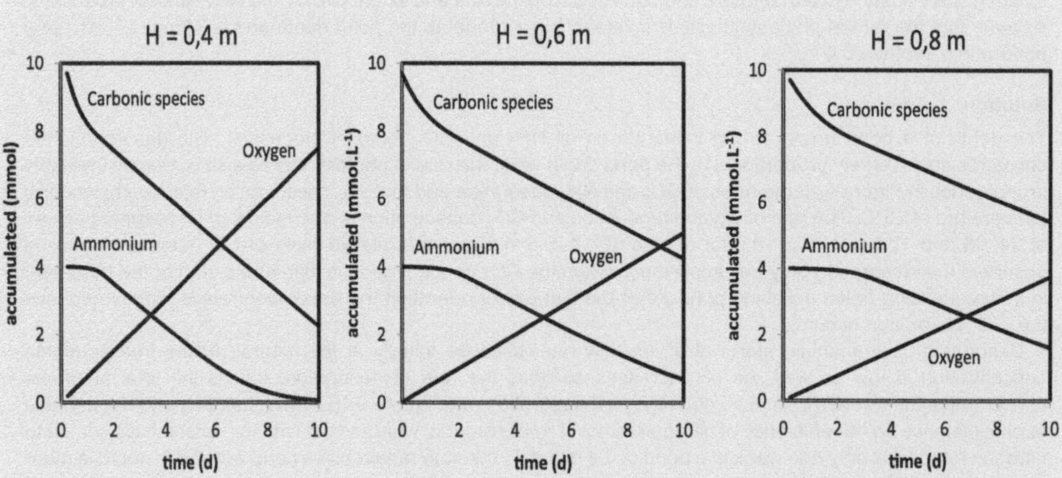

Figure 9.16 Accumulated values of carbonic species, ammonium and oxygen production (all in mmol · L^{-1}) as function of the treatment duration in a polishing pond at depths from 0.4 to 0.8 m.

EXAMPLE 9.7

Evaluate the influence of the photosynthesis rate on the removal of ammonium in a sequential batch polishing pond.

Solution

Using the model of Table 9.9, the rate of nitrogen removal was calculated for oxygen production rates of 0.5, 1 and 2 mg $O_2 \cdot L^{-1} \cdot h^{-1}$, which represent typical values for low, medium and high rates of photosynthesis in polishing ponds under tropical conditions. The results of the calculations are shown in Figure 9.17, showing the nitrogen profiles for the different rates of photosynthesis in ponds of 0.4–0.8 m depth. The profiles clearly show that the rate of photosynthesis is of crucial importance for the removal of nitrogen in polishing ponds.

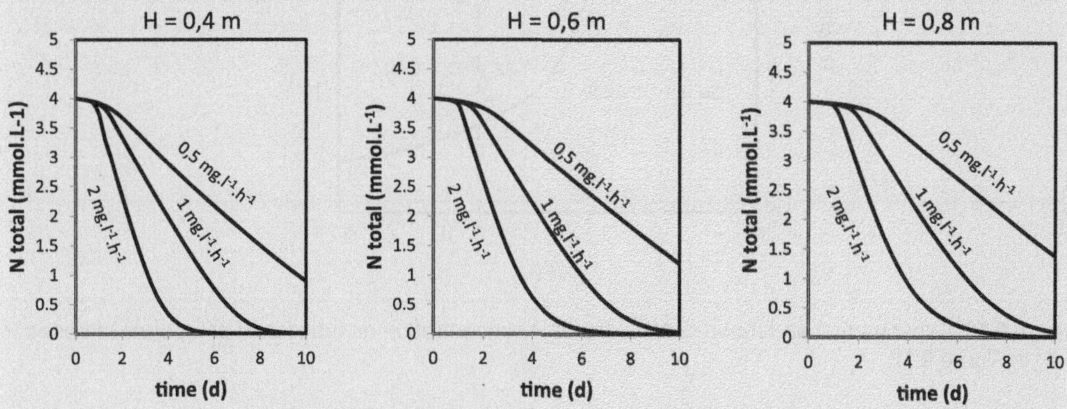

Figure 9.17 Profiles of total ammonium concentration as a function of time in sequential batch polishing ponds with different depths (0.4–0.8 m) for different rates of oxygen production and CO_2 consumption due to photosynthesis.

9.8 PHOSPHORUS REMOVAL IN POLISHING PONDS

An experimental study was carried out (Cavalcanti, 2003) to establish the relationship between pH and phosphate concentration in the effluent of polishing ponds, using sequential batch ponds with different depths. In all cases, the effluent was filtered through a fiberglass filter to eliminate the particulate phosphorus present in algae and organic material, as well as precipitated phosphate, thus determining only the residual soluble phosphate. The calcium concentration in the sewage used for these tests was about 1 mmol $\cdot L^{-1}$ (100 mg $CaCO_3 \cdot L^{-1}$), In Figure 9.18 the soluble phosphate concentration is plotted as a function of the pH in the ponds (dotted curve).

The thermodynamic solubility curves for different calcium salts are also plotted in Figure 9.18. The solubility of phosphates depends on the concentration of cations and the pH, which in turn determines the relative concentration of the phosphate species: $H_2PO_4^-$, HPO_4^{2-} and PO_4^{3-}. The solubility of phosphate was calculated from the equilibrium equations of the possible precipitates that could form: $Ca_4H(PO_4)_3$ and $Ca_{10}(OH)_2(PO_4)_6$. The equations of equilibrium reactions are according to Stumm and Morgan (1981):

$$Ca_4H(PO_4)_3 \rightarrow 4\ Ca^{2+} + 3\ PO_4^{3-} + H^+\ (pK = 46.9)$$

$$Ca_{10}(OH)_2(PO_4)_6 \rightarrow 10\ Ca^{2+} + 6\ PO_4^{3-} + 2\ OH^-\ (pK = 114)$$

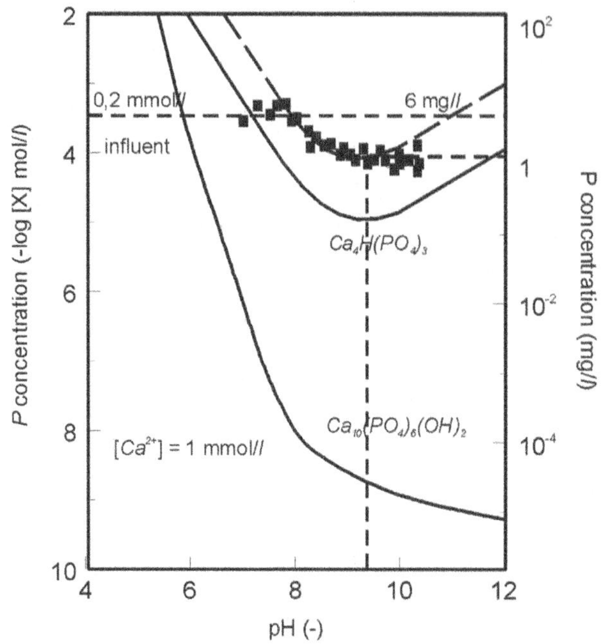

Figure 9.18 Experimental and theoretical values of P concentration as function of pH (ponds of series P_3 to P_5 in Figure 9.11).

The experimental results of phosphate solubility clearly show a trend similar to the solubility curve of $Ca_4H(PO_4)_3$ and not to that of the $Ca_{10}(OH)_2(PO_4)_6$ salt curve, which has a much lower solubility. However, comparing the experimental curve with the $Ca_4H(PO_4)_3$ solubility curve, there is a displacement of about two orders of magnitude in the logarithmic scale diagram, i.e., the solubility product is about 100 times greater than the thermodynamic equilibrium value for any pH value. The results indicate that for a significant removal of phosphate (e.g. a final concentration of 1 mg $P \cdot L^{-1}$), the pH in the polishing pond should be higher than 9, which coincides with the requirement for ammonia removal.

9.9 APPLICABILITY OF POLISHING PONDS

Waste stabilization ponds (WSP) have been amply applied in many regions, especially those with a hot climate. In this chapter the very serious shortcomings of WSP and particularly of anaerobic lagoons (AL) were clearly shown. Efficient pre-treatment of sewage in anaerobic units such as the UASB reactor, opens the way for the use of polishing ponds for post-treatment and can eliminate the problems inherent to WSP. Although the system consisting of UASB and polishing ponds has the same objectives as the conventional WSP system (i.e. stabilization of organic material and removal of pathogens) and uses the same biological processes (anaerobic digestion + photosynthesis + bacterial oxidation), adopting the first alternative has decisive advantages, as will be shown below.

9.9.1 Pre-treatment

The UASB reactor is very efficient in removing organic matter and solids in suspension, removing a large part of the colloidal material as well. Pre-treatment in this reactor decreases the organic load applied to the polishing pond, thus reducing the oxygen demand for oxidation. On the other hand, reducing the concentration of colloids reduces turbidity, which facilitates the penetration of sunlight into the polishing pond and therefore accelerates the photosynthetic production of oxygen.

In the UASB reactor the produced biogas is collected, allowing its use for the production of electric energy or heat or, alternatively, it can be burned in a simple flare, which can generate funds through carbon credits. Biogas combustion has two important advantages: (1) it eliminates the possibility of spreading odours, which is one of the major problems of conventional WSP, and (2) it reduces the release of CH_4, which contributes much more to the greenhouse effect than CO_2, the product of methane combustion. The combination of UASB + polishing pond also offers the possibility of a phased construction approach, first the anaerobic unit, then the pond. Many sanitation companies in developing countries have very limited financial resources to treat sewage up to the level compatible with legal regulations. In this case, the anaerobic treatment in the UASB reactor offers a possibility of partial treatment without excessive application of resources.

At a later stage, a post-treatment unit can be constructed when the necessary funds have been acquired. The UASB effluent does not comply with most of the quality standards, but it is far less damaging to the environment than the raw sewage. In the case of a WSP system, the construction of anaerobic ponds only is not realistic, as the effluent quality is much worse.

The high digestion rate in the UASB reactor allows for a much shorter retention time than in an anaerobic pond. This does not always mean that the cost of construction will also be lower, as this depends also on other factors: location and availability of the land, nature of the soil, topography etc. Perhaps the most important contribution to the total cost of treatment systems is from the construction of the sewer lines and final outfall. In the case of conventional WSP the odour problem requires that the treatment plant is built away from the urban region, increasing the costs of the sewer lines significantly. Anyway it will be difficult to obtain near a large urban region a large enough area with suitable characteristics to be used for a WSP system. Hence, this solution is in practice limited to small towns, but even in this case the combination UASB + polishing ponds is superior. In fact, it can be stated that there are no cases in regions with a hot climate that favour the application of conventional WSP.

The UASB produces an excess sludge with high concentration (50–70 g $TSS \cdot L^{-1}$), which is stable (low biodegradable organic matter, i.e. it does not putrify) and can be used as fertilizer in agriculture after simple drying to air by applying some method to improve hygienic quality (applying lime or heat as explained in Chapter 7). As the sludge can be discharged at regular (scheduled) intervals, the production tends to be constant over time, so that it is always available for sales. If there is no economic viability in the use of the dried sludge, it can be discharged to a landfill.

The operation and maintenance of the UASB reactor is relatively simple, so that in principle more than one unit can be installed in a single city. Thus it is possible that at least the pre-treatment is localized, which opens the possibility of making large reductions in the cost of the sewage network: instead of building a single sewer network with pumping stations, interceptors and high-cost outfalls, several local networks are constructed according to the drainage basins of the city, with a UASB reactor in each basin. After pre-treatment, the digested sewage (essentially raw water) can be transported much more easily than the raw sewage. This type of segmentation of the city wastewater treatment works is amply applied by SANEPAR in Paraná, and the experience obtained shows that there is a potential for a 40% reduction in costs of the construction of the network. In general, it is cheaper to build a distributed network with localized anaerobic pre-treatment than to build a single network with a centralized treatment plant. The applicability and financial feasibility depend greatly on the local topography and soil conditions.

For example, in Campina Grande, Brazil there is only one small pumping station, whereas in the rest of the sewer network gravity flow can be used to conduct the sewage to the central treatment plant. In this case, the advantages of localized treatment would be reduced or even non-existent.

9.9.2 Complementary treatment

The most important advantage of using the combination of UASB plus polishing pond is the reduction of the required area compared to the conventional WSP system, as was shown in Example 9.2. To obtain a filtered effluent BOD_5 of 30 mg · L^{-1}, the area is still considerable (0.8 m^2 per capita), but much smaller than the area required in a conventional WSP system, typically between 2 and 3 m^2 per capita (Mara, 1975; Van Haandel & Lettinga, 1994). The reduction of the installed pond area greatly increases the applicability of ponds as a post-treatment alternative.

Another advantage of the system of UASB + polishing pond is that is provides the definitive solution to the most serious operational problem of WSP, i.e. the odour that is generated and spreads in the surroundings (and sometimes well beyond them!). In the UASB + polishing pond system a (small) part of the sulphide is removed together with the biogas as hydrogen sulphide gas (see Chapter 6), but this biogas is collected and burned, transforming the sulphide into sulphur dioxide (odourless). In the case of diluted wastewaters such as municipal sewage, which produce little biogas, most of the sulphide actually remains in the liquid phase and in the polishing pond it will be oxidized by the oxygen generated in the photosynthesis process. As there are no problems with smells in polishing ponds, it is possible to construct the units much closer to urban areas, thus reducing the length of the connecting sewage network, which is generally a very large part of the investment costs.

An important aspect is that the performance of polishing ponds in terms removal of residual organic material and suspended solids as well as of hygienic quality is essentially independent of depth. This gives an extra degree of freedom in the operation of the polishing pond: by varying the depth, one can manipulate the residual concentration of nutrients in the effluent. In shallow ponds ($H < 0.4$ m) there is a tendency for nutrient removal, whereas in deep ponds ($H > 0.8$ m) they are retained. Thus, when the effluent of the system is to be used as irrigation water, the pond is operated with a depth of more than 1 m, thereby retaining the nutrients. However, in the rainy season (when no irrigation is applied), the ponds should be operated at a small depth in order to remove the nutrients and avoid eutrophication of the water bodies receiving the effluent.

The second most important operational problem of conventional WSP is the high rate of sludge production in the anaerobic lagoon. Gonçalves and Birth (1999) have shown that the deposition of solids in anaerobic lagoons is around 250 mg per litre. Hence, pond operation needs to be interrupted every 3 to 5 years in order to remove the accumulated sludge, which is an expensive, time-consuming and complicated operation. For this reason, it is necessary to have a spare anaerobic lagoon available that is taken into operation in the event of sludge removal. However, in the system of UASB + polishing pond, even when no waste sludge is discharged from the UASB reactor, the rate of solids deposition in the polishing ponds will be much lower (only 70 mg per litre of digested sewage). A large part of this settled sludge is the result of sedimentation of flocculated algae (Cavalcanti, 2003), which spreads over the entire pond area and accumulates so slowly that it allows decades of trouble-free operation.

It is concluded that anaerobic sewage pre-treatment in a UASB reactor or another efficient anaerobic reactor is an excellent measure when ponds are to be used for polishing treatment. The advantages of using pre-treatment in a UASB are economic (required area and especially possibility of segmentation of the sewage collection network), environmental (better quality of effluent, prevention of methane release), operational (removal of bottom sludge) and aesthetic (avoiding release of odours). These advantages are so clear and important that whenever construction of a conventional system of stabilization ponds is proposed, it should be mandatory to explain why the option to use efficient anaerobic pre-treatment was not considered.

Chapter 10

Spreadsheet design of anaerobic and complementary treatment plants

This chapter discusses the design of UASB reactors for sewage treatment. The design criteria are based on the experience obtained with reactors at both demonstration and full scale, at the Federal University of Campina Grande and at other research centres. The potential of the UASB reactor for sewage treatment was recognized early by the research group of Professor Lettinga in the 1980s, but the conditions under which they developed the first experimental work in Holland (low temperature and dilute sewage) led to poor results (De Man, 1990; Van der Last, 1991). Only when the UASB reactor was applied in tropical regions did the observed performance improve, for example in Cali, Colombia where the first demonstration-scale reactor was built operated in collaboration with the Lettinga Research Group (Schellinkhout *et al.*, 1985, 1986). Soon hereafter, UASB reactors were built by CETESB in Sao Paulo (Vieira & Souza, 1986) and in 1987 by the Municipality of Campina Grande. At SANEPAR in the subtropical state of Paraná, several systems similar to the UASB reactor were developed following research on the applicability of anaerobic digestion to sewage treatment (Gomes, 1985).

In the basic flow scheme presented in Figure 10.1, several elements of the anaerobic treatment system can be identified:

1. *Preliminary treatment.* Strictly speaking, the preliminary treatment is not part of the anaerobic treatment unit, but its application is necessary because the wastewater components to be removed would impair the operational stability and the performance of the UASB reactor if they were left in the sewage. These components include (a) coarse solids, (b) sand and silt, and (c) oils and greases. Preliminary sewage treatment will not be discussed in this text.
2. *The UASB reactor.* The most important elements of the basic structure of the reactor are size and shape. The size of the reactor depends directly on the flow to be treated and the retention time that is required. Key reactor components are the shape of the phase separator, influent distribution and

Editor(s): Adrianus van Haandel and Jeroen van der Lubbe
doi: 10.2166/9781780409627_0375

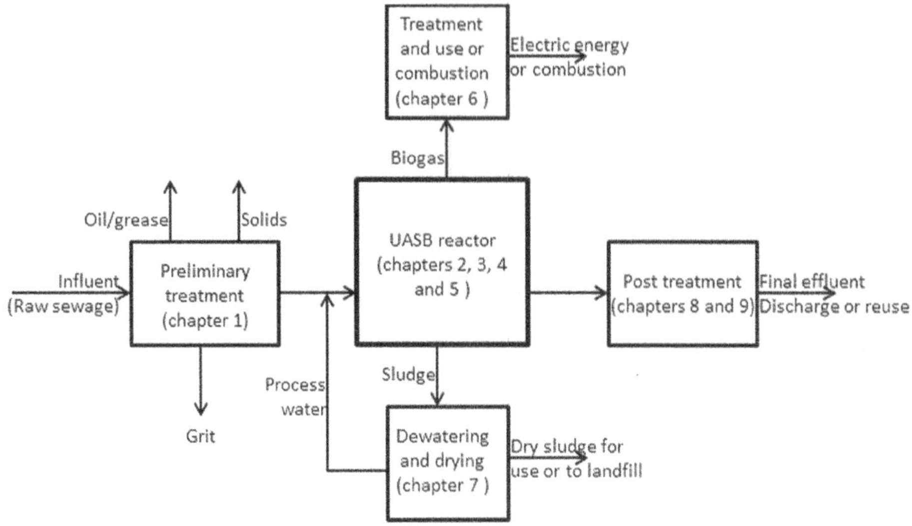

Figure 10.1 Basic scheme of the anaerobic treatment unit and its complementary processes, where the chapters in this book in which they are discussed are indicated as well.

effluent collection structures, devices to discharge excess sludge and sampling points to determine the sludge concentration profile.

3. *Use of biogas*. In many cases for sewage treatment systems this will be a simple flare, as the treatment and productive use of the generated biogas only has economic feasibility for very large units.
4. *Excess sludge handling*: sludge dewatering and drying units.
5. The quality of the anaerobic effluent does not allow its discharge into surface waters according to existing legislation, even when the anaerobic treatment is efficient. Therefore a post-treatment unit to improve the final effluent quality will always be required. The main alternatives for post-treatment are the activated sludge system and polishing ponds, which will be discussed in the last sections of this chapter.

10.1 DESIGN OF THE UASB REACTOR FOR SEWAGE TREATMENT

With regards to the design of the UASB reactor, the two key criteria that determine the success of any biological treatment system should be considered: (1) there must be adequate contact between the influent organic material and the biomass, and (2) the biomass must be sufficiently large through efficient solids retention. These two basic criteria lead directly to the definition of two fundamental parameters in the UASB reactor design: the upward velocity of the liquid in the digestion zone and in the settling zone, which in turn define the required area in the digestion and settling zones. The velocity in the digestion zone must be sufficient to avoid compaction of the sludge at the bottom of the reactor, which can lead to poor contact of the organic material with the sludge. Also, at low upflow velocities inorganic solids tend to be deposited in the reactor. The minimum allowed value is in the order of $0.3 \text{ m} \cdot \text{h}^{-1}$. On the other hand the upflow velocity may not be excessive, in order to prevent the sludge being carried along with the liquid. In the experiments presented in Chapter 5, there are clear indications that the characteristics of the sludge adapt to the applied upward velocity of the liquid: the application of a high upflow velocity increased the sludge settleability and the loss of sludge during moments of high hydraulic load was avoided.

As for the upflow velocity of the liquid in the settling zone, its value must be low in order to prevent the escape of sludge from the reactor. In general the ideal value of the upward velocity of the liquid in the sedimentation zone will be lower than in the digestion zone. Therefore in optimized systems the area of the digestion zone may be smaller than the settler area, although this is not feasible for large units.

In Chapters 3 and 4, three types of UASB reactor were discussed that differed by their geometry and configuration:

- The two-phase UASB reactor, $UASB_b$ (Figure 10.2a), is characterized by the fact that it has two biphasic separators instead of one three-phase separator. The physical layout of the $UASB_b$ reactor was introduced in Section 4.6.1.
- The UASB reactor with a single conical three-phase separator, $UASB_c$ (Figure 10.2a), was discussed in detail in Section 4.6.2.
- The conventional UASB reactor with rectangular geometry, $UASB_r$, was presented in Chapters 3 and 4 and is the most widespread of the three variants, with dozens of large systems operating in tropical or subtropical regions.

The three types of reactors are distinguished not only by their differences in physical layout, but also by the material of construction and the flow range they can handle. For example, the $UASB_b$ can be applied only for small flow, while the $UASB_c$ reactor is used for intermediate flow rates, and the $UASB_r$ reactor treats large flows. Table 10.1 summarizes the differences between the three reactor types.

Appendix A3 is a downloadable file that contains three spreadsheets, for $UASB_b$, $UASB_c$ and $UASB_r$ design. These spreadsheets will allow the user to make rational designs of the three types of UASB reactors. In Sections (10.2–10.4) the structure of each of the three spreadsheets is explained. In Section 10.5 several application examples of the spreadsheet are presented.

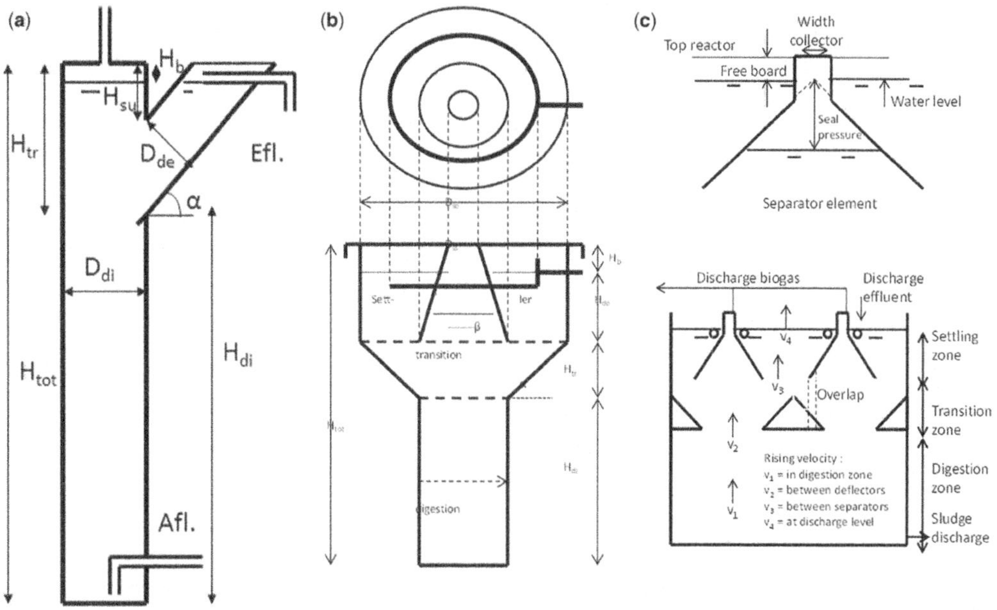

Figure 10.2 (a) Bi-phasic reactor with two-phase separators, $UASB_b$. (b) Reactor with a conical three-phase separator, $UASB_c$. (c) Rectangular reactor with trapezoidal three phase separators, $UASB_r$.

Table 10.1 Main characteristics and differences of the three basic configurations of UASB reactors for sewage treatment.

Characteristic	$UASB_b$	$UASB_c$	$UASB_r$
Material of construction	PVC tubes	Fibre glass	Concrete
Separator type	Two biphasic	Conical triphasic	Trapezoidal
Digestion zone	Cylindrical	Cylindrical	Rectangular
Transition zone	Cylindrical	Conical	Rectangular
Settling zone	Cylindrical (inclined)	Cylindrical	Rectangular
Volume range (m^3)	0.025–2.5	0.25–250	25–2500
Applied flow ($m^3 \cdot d^{-1}$)	0.1–10	1–1000	100–10,000
Height (m)	1–5	1.5–5	4–6

10.2 DESIGN OF THE BIPHASIC UASB REACTOR ($UASB_B$)

Figure 10.2a schematically shows the biphasic UASB reactor, which is characterized by extreme ease of assembly: for small reactors prefabricated pipes can be used with parts that can easily be inserted and glued within a few minutes. To rationally design the $UASB_b$ reactor it is necessary to define some parameters (basic data), which allows the calculation of other (derived) data. The basic data must be known or adopted by the designer. The derived data will completely define the physical dimensions of the reactor. The basic data for the $UASB_b$ reactor can be summarized as:

1. The sewage flow rate to be treated (Q).
2. Hydraulic retention time, which together with the flow rate sets the reactor volume.
3. Upward velocity in the digestion zone. This value must be enough to prevent the deposition of inorganic solids in the reactor ($>0.3\ m \cdot h^{-1}$) and avoid excessive sludge loss from the digestion zone ($<1\ m \cdot h^{-1}$).
4. Velocity in the settling zone. Preferably as small as possible, but it cannot be less than the velocity in the digestion zone, otherwise it is not possible to connect the settling zone to the reactor body.
5. Inclination of the settler angle. This angle should be high enough to allow the sludge accumulated in the decanter to slide into the digestion zone. The smaller the angle, the more efficient is the sedimentation of sludge (see Section 4.5).
6. Height of the transition zone above the settler, which determines the division of the reactor body between the digestion zone and the settling zone.
7. Free board. Determines the height of the gas chamber in the reactor. If desired an external hydraulic seal can be fitted, allowing to increase the gas chamber into the transition zone.

Having defined the basic data, the calculations can be carried out to determine the dimensions of the different parts of the reactor using a program like Excel. Table 10.2 shows the basic data and the derived variables to be calculated, as well as the expressions used to make the calculations. Table 10.3 shows the results of the calculations to determine the components of the $UASB_b$ reactor for different values of the flow to be treated. In the calculations the values of the liquid velocity in the digestion and settling zones were manipulated such that the diameters of these zones correspond to commercially available pipe diameters, which facilitates the construction of the unit and reduces costs. In the evaluation of the design three important parameters are considered, which are listed in **Bold** in Table 10.3. The fraction of the total volume allocated

Table 10.2 Design of the UASB$_b$ reactor. The symbols used refer to Figure 10.2a.

No.	Basic Data	Symbol	Expression	Units
1	Feed flow	Q		$m^3 \cdot d^{-1}$
2	Retention time	R_h		h
3	Velocity in the digestion zone	v_{di}		$m \cdot h^{-1}$
4	Velocity in the settling zone	v_{de}		$m \cdot h^{-1}$
5	Inclination of the settler	α		°
6	Height upper part	H_{su}		m
7	Free board height	H_b		m
No.	**Derived data**	**Symbol**	**Expression**	**Units**
8	Volume	V_{tot}	$Q/(24R_h)$	m^3
9	Digestion area	A_{di}	$Q/(24v_{di})$	m^2
10	Settler area	A_{de}	$Q/(24v_{de})$	m^2
11	Diameter digestion area	D_{di}	$(A_{di}/(3.14/4))^{0.5}$	m
12	Diameter settler	D_{de}	$(A_{de}/(3.14/4))^{0.5}$	m
13	Height of settler base	H_1	$A_{de}/\cos(\alpha)$	m
14	Height of settler top	H_{tr}	$H_1 + H_{su}$	m
15	Short part of settler	L_2	$H_{su}/\sin(\alpha)$	m
16	Long part of settler	L_3	$H_{tr}/\sin(\alpha)$	m
17	Average length of settler	L_4	$(L_2 + L_3)/2$	m
18	Volume of settler	V_{de}	$A_{de}L_4$	m^3
19	Volume of transition zone	V_{tr}	$H_{tr}D_{di}$	m^3
20	Volume of free board	V_b	H_bD_{di}	m^3
21	Volume of digestion zone	V_{di}	$V_{tot} - V_{de} - V_{tr} - V_b$	m^3
22	Height of digestion zone	H_{di}	V_{di}/A_{di}	m
23	Total height	H_{tot}	$H_{di} + H_{re} + H_b$	m
24	Fraction of digestion volume	f_{di}	V_{di}/V_{tot}	(-)
25	Fraction of transition volume	f_{tr}	V_{tr}/V_{tot}	(-)
26	Fraction of settler volume	f_{de}	V_{de}/V_{tot}	(-)
27	Fraction of free board volume	f_b	V_b/V_{tot}	(-)
28	Rising velocity in settler	v_{as}	$v_{de}\sin(\alpha)$	$m \cdot h^{-1}$

to the digestion zone is about 50–60%, which is a suitable value, although a higher value would be preferable. As for the upflow velocity of the liquid in the settler, it remains low for the entire range of flow rates, indicating that the retention of the solids will be efficient. This is accordance with the observation that the average sludge concentration in the UASB$_b$ reactor normally is higher than in the UASB$_c$ or UASB$_r$ variants.

It can be concluded that the UASBb reactor is extremely easy to build, will have a very low cost and a very short assembly time. However, the range of flow rates that can be treated with this reactor is limited. For flow rates above 10 $m^3 \cdot d^{-1}$, the application of the reactor is technically feasible, but not advisable for economic reasons: the height of the reactor becomes excessive and negates the simplicity of construction, which is precisely its strong point.

Table 10.3 Results of the design procedure for a $UASB_b$ reactor with a retention time of 6 hrs, a settler zone angle of 45°, a transition zone height H_{su} of 0.25 m and a freeboard of 0.1 m.

For Feed Flow Q =	0.1	0.3	0.5	1.0	3.0	10	20	$m^3.d^{-1}$
v_{dl}	0.23	0.4	0.42	0.59	0.64	0.94	1.05	$m \cdot h^{-1}$
v_{de}	0.53	0.4	0.42	0.59	0.64	0.94	1.05	$m \cdot h^{-1}$
H_{su}	0.25	0.25	0.25	0.25	0.25	0.25	0.2	m
H_b	0.10	0.10	0.10	0.10	0.10	0.10	0.10	m
$\mathbf{H_{tot}}$	**1.08**	**1.75**	**1.81**	**2.79**	**2.89**	**4.44**	**4.91**	**m**
H_{dl}	0.69	1.21	1.21	2.11	1.93	3.12	3.29	m
H_{tr}	0.39	0.53	0.61	0.67	0.96	1.31	1.62	m
D_{dl}	0.15	0.20	0.25	0.30	0.50	0.75	1.01	m
D_{de}	0.10	0.20	0.25	0.30	0.50	0.75	1.01	m
H_{su}	0.25	0.25	0.25	0.25	0.25	0.25	0.20	m
Short side settler	0.35	0.35	0.35	0.35	0.35	0.35	0.28	m
Long side settler	0.55	0.75	0.86	0.95	1.35	1.86	2.29	m
$\mathbf{f_{dl}}$	**0.50**	**0.51**	**0.48**	**0.60**	**0.50**	**0.55**	**0.52**	-
f_{de}	0.14	0.23	0.24	0.18	0.22	0.20	0.20	-
f_{tr}	0.28	0.22	0.24	0.19	0.25	0.23	0.26	-
$\mathbf{v_{as}}$	**0.37**	**0.28**	**0.30**	**0.42**	**0.45**	**0.66**	**0.74**	**$m \cdot h^{-1}$**

10.3 DESIGN OF THE UASB REACTOR WITH CONICAL SEPARATOR ($UASB_C$)

A convenient and easy way of constructing the $UASB_c$ is to secure the phase separator to the structure covering the unit as shown in Figure 10.2b. The phase separator has a base with a diameter larger than that of the digestion zone to prevent the entry of biogas bubbles into the settler zone. The difference between the two diameters is twice the overlap (line 6 in Table 10.4). To allow for maintenance, the cover and the separator unit must be removable. Therefore the effluent is collected in a peripheral channel, fixed to the settler wall. The width of this channel should allow passage of the separator, when the cover is placed on the reactor. The design of the $UASB_c$ reactor requires the definition of the basic parameters in Table 10.4. With the aid of the selected values of the basic data, all derived data including the dimensions of the $UASB_c$ reactor can now be calculated with the expressions that are also listed in Table 10.4.

Table 10.5 shows the results of calculations with the spreadsheet for various flows in the range of 1–300 $m^3 \cdot d^{-1}$. The output can be divided into several parts: (1) basic data (lines 1–9), diameters of various segments (10–15), tangent of phase separator (17), heights of various segments (18–22), volumetric fractions of the digestion, transition and settling zones (23–25) and upflow velocity at the base of the sedimentation zone (26). The values attributed to the basic parameters are dictated by the treatment experience or by the need to keep the derived data values adequate. The 0.5 m height of the phase separator (line 9) was adopted to ensure that the height of the settler ($H_{se} - H_b$) is large enough for efficient settling of the solids.

Line 17 is of particular interest: the angle of the slope must be greater than 45° to ensure that the solids deposited on the separator slide back into the digestion zone. However, when the flow is large, the calculated angle may be less than 45°. In this case there are two measures that can be taken to correct this: (1) increase the height of the phase separator and therefore of the settler, or (2) reduce the area of the settler, which is equivalent to increasing the upward velocity in it. In Table 10.5 it can be observed that for a flow rate of 300 $m^3 \cdot h^{-1}$ it was chosen to increase the height of the settler (line 18) as well as to increase the velocity in the settler, i.e. to

Table 10.4 Design of a UASB$_c$ reactor. The symbols used refer to Figure 10.2b.

No.	Basic Data	Symbol	Expression	Units
1	Feed flow	Q		m^3/d
2	Retention time	R_h		H
3	Rising velocity digestion zone	v_{di}		m/h
4	Rising velocity in settler zone	v_{de}		m/h
5	Inclination of conical settler	α		°
6	Overlap	O		M
7	Height of free board	H_b		M
8	Minimum settler height	H_{se}		M
No.	**Derived data**	**Symbol**	**Expression**	**Units**
9	Volume	V_{tot}	QR_h	m^3
10	Area digestion zone	A_{di}	$Q/(v_{di}24)$	m^2
11	Area settler zone	A_{de}	$Q/(v_{de}24)$	m^2
12	Diameter digestion zone	D_{di}	$(A_{di}/(3.14/4))^{0.5}$	m
13	Diameter separator	D_s	$1.1\ D_{di}$	m
14	Diameter upper part	D_{ss}	$>D_s/2$	m
15	Height of separator	H_{se}	>0.5	m
16	Tangent of separator angle	$\tan(\beta)$	$H_{se}/(D_s4)$	(–)
17	Separator angle	B	>45	°
18	Interface height	H_{in}	$0.5H_{se}$	m
19	Settler height	H_{de}	$H_{se} - H_b$	m
20	Area of upper part separator	A_{ss}	$3.14/4D_{ss}^2$	m^2
21	Separator diameter at surface	D_{sf}	$H_{se}/H_b(D_{se} - D_{ss}) + D_{ss}$	m
22	Separator area at surface	A_{sf}	$3.14/4D_{sf}D_{sf}$	m^2
23	External area	A_{ex}	$A_{sf} + A_{de}$	m^2
24	External diameter	D_{de}	$(A_{ex}/(3.14/4))^{0.5}$	m
25	Transition height	H_{tr}	$(D_{de} - D_{di})/2\tan(\alpha)$	m
26	Free board volume	V_b	H_bA_{ex}	m^3
27	Settler volume	V_{de}	$H_{de}A_{ex}$	m^3
28	Transition zone volume	V_{tr}	$3.14/24(D_{de}^3 - D_{di}^3)\tan(\alpha)$	m^3
29	Digestion zone volume	V_{di}	$V_{tot} - V_b - V_{de} - V_{tr}$	m^3
30	Effective volume	V_e	$V_{tot} - V_b$	m^3
31	Fraction of digestion zone	f_{di}	V_{di}/V_e	(–)
32	Fraction of transition zone	f_{tr}	V_{tr}/V_e	(–)
33	Fraction of settler zone	f_{de}	V_{de}/V_e	(–)
34	Height of digestion zone	H_{di}	V_{di}/A_{di}	m
35	Total height	H_{tot}	$H_b + H_{de} + H_{tr} + H_{di}$	m
36	Minimum settler area	A_{min}	$A_{ex} - D_s^2 3.14/4$	m^2
37	Maximum settler area	A_{max}	$A_{ex} - A_{sf}$	m^2
38	Maximum velocity in settler	v_{max}	Q/A_{min}	m · h^{-1}
39	Minimum velocity in settler	v_{min}	Q/A_{max}	m · h^{-1}

Table 10.5 Example of the calculations of the dimensions of a $UASB_c$ reactor for different flows and a retention time of 6 hrs for different flows (25°C).

For Feed Flow Q =		**1**	**3**	**10**	**30**	**100**	**300**	$m^3 \cdot d^{-1}$
Retention time	R_h	6	6	6	6	6	6	h
Velocity in the digester	v_{di}	0.3	0.3	0.35	0.45	0.6	0.9	$m \cdot h^{-1}$
Velocity in the settler	v_{de}	0.3	0.3	0.3	0.4	0.55	0.7	$m \cdot h^{-1}$
Desired inclination	α	45	45	45	45	45	45	°
Tangent	$\tan\alpha$	1.0	1.0	1.0	1.0	1.0	1.0	(–)
Overlap	O	10	10	10	10	10	10	%
Freeboard height	H_b	0.15	0.15	0.15	0.15	0.15	0.15	m
Height phase separator	H_{se}	0.5	0.5	0.5	0.5	0.5	0.5	m
Volume	V_{tot}	0.25	0.75	2.5	7.5	25.0	75.0	m^3
Diameter digestion zone	D_{di}	0.42	0.73	1.23	1.88	2.97	4.21	m
Diam. separator (base)	D_s	0.46	0.80	1.35	2.07	3.27	4.63	m
Diam. separator (top)	D_{ss}	0.23	0.40	0.68	1.03	1.64	2.31	m
Diam. sep. liquid surface	D_{sf}	0.30	0.52	0.88	1.34	2.13	2.49	m
External diameter	D_{de}	0.52	0.90	1.60	2.40	3.66	5.11	m
Separator tangent	$\tan\beta$	4.32	2.50	1.48	0.97	0.61	0.43	(–)
Angle separator	**B**	**78**	**69**	**56**	**45**	**45**	**45**	°
Height separator	H_{se}	0.50	0.50	0.50	0.52	0.80	1.10	m
Height settling zone	H_{de}	0.35	0.35	0.35	0.37	0.65	0.95	m
Height transition zone	H_{tr}	0.05	0.08	0.18	0.26	0.39	0.48	m
Height digestion zone	H_{di}	0.98	0.94	1.02	1.54	2.29	4.04	m
Total height	H_{tot}	**1.53**	**1.52**	**1.70**	**2.30**	**3.18**	**5.03**	**m**
Fraction digestion zone	f_{di}	**0.63**	**0.60**	**0.55**	**0.63**	**0.68**	**0.78**	**(-)**
Fraction transition zone	f_{tr}	0.04	0.07	0.13	0.14	0.15	0.12	(-)
Fraction settling zone	f_{de}	0.34	0.34	0.32	0.23	0.17	0.10	(-)
Max. velocity settler	v_{max}	**1.00**	**1.00**	**0.75**	**1.06**	**1.53**	**1.54**	$m \cdot h^{-1}$

decrease the diameter of the settler. For the flow rate of 300 $m^3 \cdot h^{-1}$ and a retention time of 6 hrs, the upward velocity is higher than 1 $m \cdot h^{-1}$, a very high value which may impair the performance of the reactor.

In addition to line 17, lines 22, 23 and 26 (all highlighted in **bold**) are also important for verifying that the design is adequate. Line 22 gives the total height of the reactor. As the unit will usually be constructed (partially) buried, it is advantageous that the height is small in order to minimize excavations. Another reason for having a reduced reactor height is that it increases the possibility of operating the reactor without the need to pump the raw sewage.

In line 23, the fraction of the effective volume of the reactor that is allocated to the digestion zone is calculated. It is suggested that this fraction should be at least maintained in the range of 60–70% to ensure that there is enough space to accommodate a large mass of sludge. Line 26 shows the upward velocity of the liquid at the inlet of the settler. If this value is too high, it makes the return of settled sludge to the digestion zone more difficult. A maximum value of 3 $m \cdot h^{-1}$ is suggested, but more research is needed to corroborate the suggested values for the required minimum volume fraction of the digestion zone and the maximum velocity at the settler inlet.

EXAMPLE 10.1

Design a $UASB_c$ reactor to treat the sewage of a population of 3000 inhabitants with a flow rate of 300 $m^3 \cdot d^{-1}$. A retention time of 6 hrs is required.

Solution

In Table 10.6 various solutions are presented: letters A, B, C, D, E and F in the first column indicate the different sets of basic data. Lines 15, 17, 22, 23 and 27 (in **bold**) show the data to be verified upon analysis of the results of the computer calculations.

In column A the first possible solution is presented, but when analysing the results one can see immediately that there are problems: in line 17 the angle of the separator is only 14°, where it should be at least 45°, in line 22 the total height is very small and in line 23 the fraction of the volume allocated to the digestion zone is practically non-existent. So the basic design data of solution A do not lead to a satisfactory design and must be changed.

Table 10.6 Different selections of basic data for the design flow of 300 $m^3 \cdot d^{-1}$ in order to find the optimal design for a $UASB_c$ reactor.

	Basic Data of Solution		A	B	C	D	E	F	Units
1	**Flow to be treated**	Q	300	300	300	300	300	300	$m^3 \cdot d^{-1}$
2	Retention time	R_h	6	6	6	6	6	6	h
3	Velocity in the digester	v_{di}	0.3	0.7	0.7	0.45	0.8	1	$m \cdot h^{-1}$
4	Velocity in the settler	v_{de}	0.3	0.3	0.7	0.4	0.8	0.8	$m \cdot h^{-1}$
5	Inclination (min)	A	45	45	45	45	45	45	°
6	Tangent desired angle	$\tan\alpha$	1.00	1.00	1.00	1.00	1.00	1.00	(–)
7	Overlap	O	1.1	1.1	1.1	1.1	1.1	1.1	m
8	Free board height	H_b	0.15	0.15	0.15	0.15	0.15	0.15	m
9	Separator height	H_{se}	0.5	0.5	0.5	0.5	0.5	0.5	
			Derived data						
10	Reactor volume	V_{tot}	75	75	75	75	75	75	m^3
11	Digestion zone diameter	D_{di}	7.29	4.77	4.77	5.95	4.46	3.99	m
12	Separator diameter	D_s	8.01	5.25	5.25	6.54	4.91	4.39	m
13	Diameter upper part	D_{ss}	4.01	2.62	2.62	3.27	2.45	2.19	m
14	Diameter at liquid surface	D_{sf}	5.21	3.41	3.41	3.58	2.91	2.49	m
15	**External diameter**	$\mathbf{D_{de}}$	**8.96**	**8.04**	**5.86**	**7.25**	**5.33**	**5.11**	m^2
16	Separator tangent	$\tan\beta$	0.25	0.38	0.38	0.98	0.65	1.00	(–)
17	**Arctangent**	β	**14**	**21**	**21**	**45**	**33**	**45**	°
18	Separator height	H_{se}	0.50	0.50	0.50	1.60	0.80	1.10	m
19	Settler height	H_{de}	0.35	0.35	0.35	1.45	0.65	0.95	m
20	Transition height	H_{tr}	0.83	1.64	0.55	0.65	0.43	0.56	m^2
21	Height digestion zone	H_{di}	0.00	−0.24	2.76	−0.48	3.14	3.46	m
22	**Total height**	$\mathbf{H_{tot}}$	**1.34**	**1.90**	**3.81**	**1.77**	**4.37**	**5.12**	**m**
23	**Fraction digestion zone**	$\mathbf{f_{di}}$	**0.00**	**−0.06**	**0.70**	**−0.20**	**0.68**	**0.60**	**-**
24	Fraction transition zone	f_{tr}	0.66	0.80	0.17	0.33	0.11	0.13	-
25	Fraction settler zone	f_{de}	0.34	0.26	0.13	0.87	0.20	0.27	-
26	**Max. velocity in settler**	$\mathbf{v_{max}}$	**1.00**	**0.43**	**2.32**	**1.63**	**3.69**	**2.32**	$m \cdot h^{-1}$

In column B the upflow velocity in the digestion zone was increased from 0.3 to 0.7 m · h^{-1}, but still the result is not satisfactory: the fraction of the volume that can be allocated to the digestion zone gives a negative value!!

In column C also the upflow velocity in the settling zone was also increased from 0.3 to 0.7 m · h^{-1}, but there is still a problem: the angle of the phase separator will only be 21°, which will still result in accumulation of sludge on the plate.

In column D it was decided to increase the height of the separator and therefore of the settling zone. A separator height of 1.6 m (line 18) gives the desired angle of 45°, but results in a negative volume of the digestion zone.

In column E the velocity was increased in both the digestion and settling zones but also this design is still not satisfactory: the phase separator angle is small (33°) and the velocity at the entrance of the sedimentation zone is high.

Finally in design *F* we find a solution that satisfies all the requirements when increasing the upflow velocity in the digestion zone to 1.0 m · h^{-1} and that in the sedimentation zone to 0.8 m · h^{-1}. The external diameter (5.11 m) has a reasonable value, the angle of the separator is 45°, the height is adequate (5 m), the fraction of the volume of the digestion zone is 60% and the velocity at the apertures of the sedimentation zone is 2.3 m · h^{-1}, which is acceptable. However the conditions that need to be imposed to achieve an adequate design are not favourable: the velocities of the liquid in the digestion and settling zones are high and can cause drag problems in the digestion zone and poor sludge retention in the settling zone and therefore performance drop, because the sludge age may be short. The magnitude of this problem depends fundamentally on settleability of the sludge and it is not known what it will be. In any case, it can be observed that 300 m^3 · h^{-1} is a large flow rate to be treated in the UASB$_c$ reactor and that yet larger flows may not result in the desired performance.

10.4 DESIGN OF THE CONVENTIONAL REACTOR (UASB$_R$) INCLUDING THE COMPLEMENTARY TREATMENT UNITS

In the case of the conventional UASB reactor the walls are vertical and the areas of both settling and overflow zones are the same. However, another variable is introduced, which is the number of phase separators of the UASB reactor, which has a strong impact on the physical design of the reactor. Spreadsheet UASB$_r$ allows to design the conventional UASB reactor with the assistance of a computer. In the spreadsheet the first column indicates the line number of the variable, parameter or constant in column 2. The third column indicates the nature of the data, which may be (1) data given by the nature of the design (in red), (2) data defined by the designer (green), (3) values calculated by the spreadsheet (black) and (4) values to be verified if they are in accordance with recommendations (blue). The spreadsheet will calculate the values of the black and blue cells, when the red and green cells are filled in.

10.4.1 Block 1: Wastewater characterization

Table 10.7 lists several comments for Block 1, located in lines 9–22 of the spreadsheet, which specifies the sewage flow and concentration, composition and load of the organic material to be digested as well as the temperature of the wastewater.

10.4.2 Block 2: Applied organic load and calculation of the anaerobic digestion efficiency

Table 10.8 lists several comments for Block 2, located in lines 23–45 of the spreadsheet, which specifies the load to be applied and calculates expected anaerobic digestion efficiency. As the applied COD loading

Table 10.7 Block 1 – Wastewater characterisation.

Line #	Considerations
10	Contributing population: the population considered will normally not be equal to the population of the city for several reasons: (1) Not everybody will have access to the sewerage network; (2) The design project is usually not done for the current population but for the population expected at a future date at the end of the useful life of the wastewater treatment plant, taking into account (a) the current population, (b) the population growth rate, (c) the fraction of the population currently served, (d) incorporation of parts of the population currently not served and (e) the expected lifetime of the design.
11	Per capita contribution: the ideal would be to have data on the per capita contribution and its temporal evolution for the different groups of the population as well as their socioeconomic habits. As this is not usually known, usually a value is taken according to the local standards, for example 120 $L \cdot d^{-1}$ though this may be too high for lower-income groups.
12	Average daily sewage flow rate: this is the average flow of sewage that reaches the treatment system. In principle this flow is equal to the product of the population and the contribution per capita. However, in practice several factors may interfere: (1) Owing to a lack of maintenance of the sewage network, part of the sewage may be discharged into the urban drainage network (rainwater collection) or conversely, rainwater may eventually enter the sewage network; (2) In an area of high groundwater level there may be considerable infiltration of groundwater into the sewage network, especially during rainy periods.
13	Influent COD concentration: this is the average of the products of COD concentration times flow rate at different times of the day, divided by the average daily flow (line 12). This value is typically much higher than the product of the average COD concentration and the average flow rate, because usually high COD concentration and high flow rate coincide.
14	Organic load: the COD load of the influent is not equal to the product of the average daily flow rate and the average COD concentration in the influent. Normally the flow and the COD have a quasi-sinusoidal behaviour with a frequency of once per day. Ideally, the values of the flow rate and COD concentration should be determined as a function of time with a dozen or more data points measured per day, to be able to draw a curve of the cumulative COD load as a function of time. If these data are not available (for example, because the sewerage network has not yet been built), the organic load must be estimated. Normally, in those cases the values of the average flow and COD concentration are adopted, and the organic load is calculated as the product of these two, multiplied by a factor that corrects for the variability of flow and COD concentration, for example $f_c = 1.25$, that is average organic load $= f_c \times$ average daily flow rate $\times$ average COD concentration. In the sheet the organic load MS_{ti} is calculated as the product of Q_iS_{ti} but this can be overwritten by the design engineer.
15	Sulphate concentration: it is assumed that sulphate is fully reduced to sulphide, oxidizing organic material in the process and reducing the methane production potential.
16	The concentration of organic material oxidized by sulphate (mg $COD \cdot L^{-1}$): stoichiometrically this is twice the concentration of reduced sulphate (mg $S \cdot L^{-1}$) (Chapter 6).
17	Available organic load for anaerobic digestion: the organic load in the influent (line 14) corrected for the oxidized organic load (line 16).

(Continued)

Table 10.7 (Continued) Block 1 – Wastewater characterisation.

Line #	Considerations
18	Temperature: this has a significant influence on the efficiency of COD removal. For the design it is necessary to consider the average sewage temperature in the coldest month of the year. It is important to note that the sewage temperature is different from the environmental temperature and varies much less. For example, in Campina Grande the average monthly air temperature ranges from a maximum of 29.9°C in January to a minimum of 17.8°C in August, but the sewage temperature is always in the range of 25 to 26°C, regardless of the month of the year. In Curitiba (one of the coldest places in Brazil), the average temperature varies between 26.7°C in February and 12.9°C in July. The sewage temperature in this capital varies between 16 and 22°C with an average of 19–20°C.
19	The ultimate soluble inert COD fraction: non-biodegradable under anaerobic conditions at a temperature of 35°C, for example $f_{ns35} = 0.08$.
20	The ultimate inert particulate COD fraction: non-biodegradable under anaerobic conditions at a temperature of 35°C, for example $f_{np35} = 0.07$.
21	The fraction of non-biodegradable and soluble COD (f_{ns}) under anaerobic conditions at the design temperature is calculated with the aid of Eq. (5.26).
22	The fraction of non-biodegradable and particulate COD (f_{np}) under anaerobic conditions at the design temperature is calculated with the aid of Eq. (5.27).

Table 10.8 Block 2 – Definition of the applied organic load and calculation of the anaerobic digestion efficiency.

Line #	Considerations
24	Applied volumetric organic loading rate (L) in kg COD · m^{-3} · d^{-1}: this is an important variable that has implications for all aspects of the project. A low VLR increases the conversion efficiency of organic material into methane, but it should not be too low, because this leads to a very large volume requirement of the system. The admissible value depends mainly on the temperature. Once the applied organic load has been defined, it is possible to calculate the sludge age, which in turn allows to establish the efficiency of the anaerobic digestion. In the spreadsheet several organic loading rates can be specified and the worksheet will then calculate the corresponding sludge age, the anaerobic digestion efficiency and the volume of the reactor. The designer will then select the applicable volumetric load based on the values of these variables.
25	Specific organic sludge production. In lines 25–35 it is attempted to establish the value of the specific organic sludge production by trial and error. In line 25 an initial estimate is specified by the designer for the specific sludge production, which in lines 26–35 is used to calculate the theoretical production of volatile sludge. This value is then filled again in line 25 and this is repeated until the calculated theoretical value is equal to the specified value.
26	VSS/TSS ratio: The ratio of volatile sludge to total sludge must be known in order to calculate the total sludge production in the system from the specific production of volatile sludge. The value of this fraction depends on the characteristics of the influent. The VSS/TSS ratio tends to increase when the temperature decreases, and the hydrolysis becomes less efficient. Its value is almost always in the range of 0.55–0.65.

(*Continued*)

Table 10.8 (Continued) Block 2 – Definition of the applied organic load and calculation of the anaerobic digestion efficiency.

Line #	Considerations
27	Average UASB reactor sludge concentration: this parameter is required to be able to estimate the sludge mass in the system. The concentration ranges from 60–100 g TSS · L^{-1} at the bottom of the reactor to 10–20 g TSS · L^{-1} at the top of the sludge blanket. The mean concentration is influenced by the quality of the phase separator and the sludge settleability. An average value of 20 g TSS · L^{-1} can be used as a conservative estimate.
28	Total volume: the total volume of the reactors is calculated as the ratio of applied available organic load (line 17) and the specified applied volumetric organic load (line 24).
29	Required sludge mass: calculated as the product of the average sludge concentration (line 27) and the volume of the reactors (line 28).
30	Hydraulic retention time: calculated as the ratio between the volume of the reactors (line 28) and the flow of wastewater to be treated (line 12).
31	Organic sludge production: the product of the available organic load (line17) and the specific production of volatile sludge (line 25).
32	Anaerobic sludge age: defined as the ratio between the organic sludge mass and daily production of organic sludge (line 31). The volatile sludge mass is the product of VSS/TSS ratio (line 26) and total sludge mass (line 29). Check if the sludge age is greater than the minimum sludge age required for methanogenesis, which is given by $R_{min} = 20 \times 1.1^{(20-T)} + 5$ (d) (Van Haandel and Van der Lubbe, 2012)
33	Fraction of feed COD ending up in the effluent: this is the sum of the soluble non-biodegradable fraction (line 21) and the soluble biodegradable fraction expressed in Eq. (5.20): $f_s = f_{ns35}\ 1.037^{(35-T)} + 0.20\ (1.011)^{(T-35)} \exp[-0.04(R_s - 3.8 \times 1.06^{(35-T)})]$
34	Fraction of feed COD fraction ending up in the excess sludge: this is the sum of the particulate non-biodegradable COD fraction (line 22) and the particulate biodegradable fraction plus produced anaerobic biomass expressed in Eq. (5.21) $f_x = f_{np35}\ 1.015^{(35-T)} + 0.16 \times 1.014^{(35-T)} \exp[-0.04(R_s - 3.8 \times 1.06^{(35-T)})]$
35	Verification of the theoretical organic sludge production: Now the theoretical volatile sludge production expressed as COD fraction (line 34) is divided by 1.5 mg VSS · mg^{-1} COD and this value is compared to the value originally adopted in line 25. If these two values are not equal, then the value of line 25 must be replaced by the last calculated theoretical value until they are.
36	Digested feed COD fraction: equal to unity less the fractions in the effluent (line 33) and in the sludge (line 34).
37	Minimum required *in situ* methanogenic activity: calculate the minimum value of the activity required to digest the applied available organic load. This is defined as the digested fraction times the available organic load divided by the mass of volatile sludge. If the calculated value is higher than 0.2 mg COD · mg^{-1} VSS · d^{-1} it is possible that the system will be overloaded due to lack of methanogenic capacity. However, in practice this hardly ever occurs: the limitation normally lies in the particle hydrolysis and not in methanogenesis.
38	Fraction of sludge in the effluent: defines the fraction of the produced excess sludge that cannot be retained in the reactor and is expelled with the effluent. The expression used in this spreadsheet to estimate this is $2/R_h$ (with R_h in hours).

rate to the reactor increases, the efficiency of the anaerobic digestion will decrease, whereas the size of the reactors is inversely proportional to the applied volumetric COD load. In this block the average sludge concentration in the UASB reactor must be specified, as at present it is not possible to predict the value of this variable. However, based on experience, in the case of raw sewage treatment the average concentration is between 20 and 30 g TSS · L^{-1} when the entire volume of the UASB reactor is considered. So for a conservative design, a value of 20 g TSS · L^{-1} would appear to be adequate.

Depending on the applied load, the efficiency of the anaerobic digestion is calculated, i.e. the division of the influent COD into its three fractions: effluent, sludge and biogas. This division basically depends on two variables: the sludge age and the temperature. Calculations are made to calculate the sludge age for any volumetric loading rate that is adopted.

It is considered that sludge ages of more than 100 days are counterproductive, as typically the maximum extent of digestion has been achieved (see Chapter 3), i.e. it will result in a reactor that is underloaded. The minimum sludge age that can be applied depends on the temperature and on the post-treatment that is given to the UASB effluent.

10.4.3 Block 3: Dimensioning of the UASB reactor and phase separator elements

Table 10.9 lists several comments for Block 3, located in lines 46–81 of the spreadsheet, which defines the basic geometry of the reactor, based on the organic load to be applied. Important variables are the reactor height, number of reactors, number of separators, and the inclinations of separator and deflector. In principle the reactor will have a square surface to minimize the length of the perimeter and therefore the area of the side walls. However, it can also be specified as a rectangular unit if this is advantageous, considering the terrain where the reactor is to be built and the size of the separators to be used. In Table 10.9 several variables are explained. It is assumed that the UASB reactor has a configuration like in Figure 10.3.

10.4.4 Block 4: Design verification (lines 83–97, Table 10.10)

Table 10.10 lists several comments for Block 4, located in lines 83–97 of the spreadsheet, in which the calculated values of important variables are compared with the values considered adequate for the operation of a well performing UASB reactor. In addition to the rising velocity of the liquid phase, important factors are the division of the reactor volume into zones of digestion, sedimentation and transition.

10.4.5 Block 5: Special devices in the UASB reactor (Lines 98–145, Table 10.11)

Table 10.11 lists several comments for Block 5, located in lines 98–145 of the spreadsheet, which defines the feed inlet and effluent outlet structures as well as the special devices required for discharge of excess sludge from the reactor and the collection of biogas. In this block the dimensions of inlet feed wells (final turndishes) with individual distribution boxes for each inlet feed point are calculated. This type of feed requires thorough pre-treatment to eliminate coarse solids. Figure 10.4 schematically shows the inlet structure.

For the collection of effluent, perforated pipes are used that are placed in parallel to the gas boxes, as shown in Figure 4.18. There may be one or two pipes per gas box. The liquid surface level in the UASB reactor is set in such a way that the effluent pipes are submerged, and the effluent enters into the perforations of the collection pipes below the water level (see also Section 4.3.5). The individual effluent collection pipes discharge into a common effluent channel where the water level is controlled by a central weir (Figure 4.31).

Table 10.9 Block 3 – Geometry of the UASB reactor. A GLS separator design as shown in Figure 10.3 is assumed.

Line #	Geometry of the UASB Reactor: Number and Size of the Reactors – Considerations
48	Volume of the individual reactors: the total required reactor volume has been calculated previously in line 28.
49	Number of reactors: to be defined by the designer, considering that generally more than one reactor is constructed and that a single reactor is not larger than about 2000 m^3.
50	Volume per reactor: ratio of lines 48 and 49.
51	Reactor height: the total reactor height is to be selected by the designer and has high repercussion on the resulting upflow velocity and the footprint and cost of the project.
52	Reactor area: ratio of reactor volume and selected depth. Typically not larger than 600 m^2.
53	Length. Usually a square geometry is chosen (resulting in the smallest perimeter), but within reason any length is possible (up to 30 m).
54	Width = reactor area/length (up to 30 m).
Line #	**Geometry of the phase separator – considerations**
56	Height of the biogas collector (gas box): typically 0.3–0.8 m depending on the configuration and whether it is desired to have the gas–liquid interface in the gas box or somewhere between the (rigid) gas collection plates
57	Width of the gas collector: choose >0.3 m for easy cleaning. In the spreadsheet this is the internal width.
58	Freeboard: difference between the liquid surface and the top of the reactor to avoid spilling of liquid and scum.
59	Number of separator elements: this has an important impact on the division of the volume in digestion, transition and settling zones and on the height of the total GLS separator.
60	Fraction of surface available for apertures between separators: ratio of area between separator elements and total area. This value determines the velocity at the base of the separator as well as at the base of the deflectors. Normally the ratio is chosen such that these velocities are equal.
61	Number of deflectors: necessarily equal to the number of separator elements.
62	Width of separators (at bottom of element): area covered by separators divided by the number of elements.
63	Width between separators (=aperture width): area not covered by separators divided by the number of elements.
64	Separator angle (gas collection plate): usually between 45 and 60°. A high value facilitates the sliding of settled sludge on the elements back to the digestion zone, but increases the height of the settling zone.
65	Deflector angle, same considerations as in line 72.
66	Tangent of the separator (calculated).
67	Tangent of the deflector (calculated).
68	Height of inclined part of separator element: projected height of gas collection plates.
69	Height separator element: sum of the projected height of the gas collection plates (line 68) and height of collector (gas box) (line 56). The height must be sufficient to accommodate the height occupied by the gas chamber (line 77).
70	Maximum allowed height between separator and top deflector: calculated as the extension of the gas collection plates to the deflectors.

(*Continued*)

Table 10.9 (Continued) Block 3 – Geometry of the UASB reactor. A GLS separator design as shown in Figure 10.3 is assumed.

Line #	Line # Geometry of the phase separator – considerations
71	Selected height between separator and top deflector: value between 0 and the maximum allowed value (line 70). Typically not higher than 0.2 m.
72	Overlap: adopt typical default value of for example 10% of the width of separator elements, or 10–15 cm.
73	Width deflector: width of aperture between two separator elements plus twice the overlap.
74	Area of the apertures between the deflectors: equal to the reactor area minus the area occupied by the deflectors. Normally this area is equal to the area between separator elements so the upflow velocity at the base of the two is equal.
76	Deflector height.
77	Height of gas chamber (i.e. the clearance between gas–liquid interface and the top of the gas box): choose >0.5 m to avoid clogging of the gas outlet.
78	Liquid height below gas–liquid interface: this is equal to the difference the height of the separator element (line 69) and the liquid height occupied by the gas chamber (line 77). If possible, this should be >0.3 m to avoid loss of solids at the bottom of the separator).
79	Width of gas–liquid interface: width of the liquid surface in each separator element.
80	Area of gas–liquid interface: width interface times length of the reactor.
81	Rate of biogas release at the interface: which is equal to the biogas production rate divided by the area of the interface.
82	Level in external hydraulic seal, which sets the pressure that is required in the separator to maintain the gas–liquid interface at the desired level: equal to the height of the separator element (line 69) – free board (line 58) – height below gas chamber (line 78) or even simpler height of gas chamber (line 77) minus freeboard (line 58).

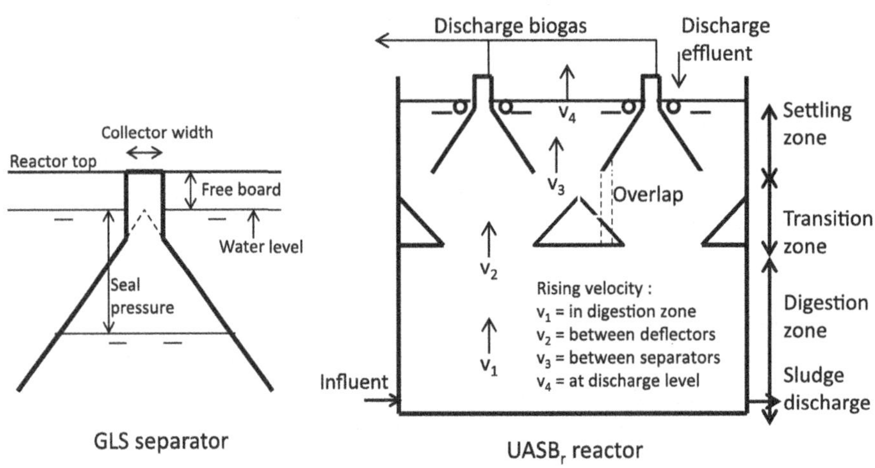

Figure 10.3 Details of $UASB_r$ reactor and the phase separator.

Table 10.10 Block 4 – Verification of the design of the UASB reactor.

Line #	Considerations
84	Velocity in the digestion zone: feed flow/reactor surface area.
85	Velocity at the deflector base: feed flow/available area aperture between deflectors.
86	Velocity at the separator base: feed flow/available area aperture between separators.
87	Velocity at the surface: flow/available overflow area (area of gas boxes is considered).
91	Height of settling zone: immersed part of separator element, which is equal to separator height – freeboard.
92	Height of transition zone: height between separator and deflector plus the deflector height.
93	Height of digestion zone: liquid reactor height – transition zone height – height of settling zone.
94	Total liquid height: sum of the heights of the three zones.
95	Liquid volume fraction for settling: the height of the settling zone divided by the total liquid height.
96	Liquid volume fraction of the transition zone: the height of the transition zone divided by the total liquid height.
97	Liquid volume fraction for digestion: the height of the digestion zone divided by the total liquid height.

Table 10.11 Block 5: Structures for influent feeding, effluent collection as well as special devices for excess sludge disposal and biogas collection.

Line #	Considerations for Influent Feed Distribution
100	Area per feed inlet point: traditionally 1 per 1–4 m^{-2}, experience suggests that the lower density is enough.
103	Number of feed wells (final turndishes) per reactor: this depends on the size of the reactor and number of feed inlet points. When the number of feed wells is low, the required diameter may be too large to fit in between two gas boxes, which complicates construction.
104	Diameter of feed pipeline to the well. The diameter should be sufficiently small to avoid settling of solids in the feed pipeline (velocity $>0.3\ m \cdot s^{-1}$)
106	Number of distribution points per feed well: equal to the number of feed inlet points (line 101) divided by the number of wells per reactor (line 103). In practice the number is always smaller than 100.
107	Width of distribution boxes: must be enough for maintenance.
108	Internal perimeter of feed well: sum of the width of distribution boxes.
109	Diameter of central well: calculate this from the internal perimeter (line 108).
110	Length of distribution boxes: must be enough for maintenance.
111	External diameter feed wells: equal to internal diameter + 2 × length of distribution boxes.
112	Height of overflow weir in distribution box: must be sufficient to handle feed flow.
113	Width of weir in distribution box (at base angle of 45°): equal to two times the height of the weir, while it should be smaller than the width of the distribution box (line 107).

(*Continued*)

Table 10.11 (Continued) Block 5: Structures for influent feeding, effluent collection as well as special devices for excess sludge disposal and biogas collection.

Line #	Considerations for Influent Feed Distribution
114	Depth of distribution box: equal to depth of feed well.
115	Height of feed well base above water level in reactor. The available head must be high enough to avoid blockages in the distribution tubes (>0.5 m).
116	Average sewage flow: calculated in $m^3 \cdot s^{-1}$.
117	Flow per feed inlet point: q_{av} = average sewage flow divided by number of feed inlet points times the number of reactors.
118	Average liquid level over the inlet weir: $h = (q_{av}/1.34)^{0.4}$.
119	Maximum flow per distribution point (with weir completely submerged): $q_{max} = 1.34\ h_{max}^{5/2}$. Note that h_{max} is defined as 0.1 m, which is the weir height (line 112).
120	Ratio between maximum and average flow (based on overflow weir configuration).
121	Diameter distribution tube (should be enough to avoid air entrainment).
122	Velocity in distribution tube: value should be $<0.3\ m \cdot s^{-1}$.
Line #	**Considerations for (submerged) effluent collection pipes**
124	Number of effluent collection points (orifices): equal to number of inlet feed points in order to ensure uniform flow distribution in the reactor.
125	Number of effluent collection pipes (per reactor): equal to twice the number of GLS separators.
127	Diameter effluent collection pipe (should be sufficiently small for self-cleaning).
128	Diameter of orifice: this should be small enough so that overall head loss at the orifice is much higher than in the rest of the effluent collection system, which ensures uniform effluent abstraction.
129	Velocity in effluent pipe (max): reactor flow per effluent pipe divided by the area of the effluent pipeline.
130	Velocity in effluent pipe (min): maximum velocity (129)/number of orifices in pipe (126).
131	Velocity in orifice: influent flow per reactor/number of points(124)/area orifice.
132	Distance between orifices: equal to length reactor/(number of orifices per tube + 1).
133	Distance between tubes (average): width of reactor/number of tubes (125).
Line #	**Considerations for excess sludge discharge from the reactor**
136	Diameter of sludge discharge pipes: should be large enough to avoid blockages but not too large as solids might settle, for example 0.15 m.
137	Low level of sludge discharge: near the bottom of the reactor, for example 0.2 m.
138	High level of sludge discharge: at higher level in the reactor, for example 1.5–2 m.
139	Number of discharge points per level: one per 50–75 m^2 for each level to allow for withdrawal over the whole surface of the reactor.
Line #	**Considerations for biogas collection (central flaring or utilization assumed)**
141	Collection points per gas box: two outlets are typically enough to avoid blockages).
142	Diameter of biogas outlets: 40 mm to prevent blockages.
143	Diameter common reactor biogas header: 50 mm.
144	Diameter to the external hydraulic seal (several reactors): 75 mm.

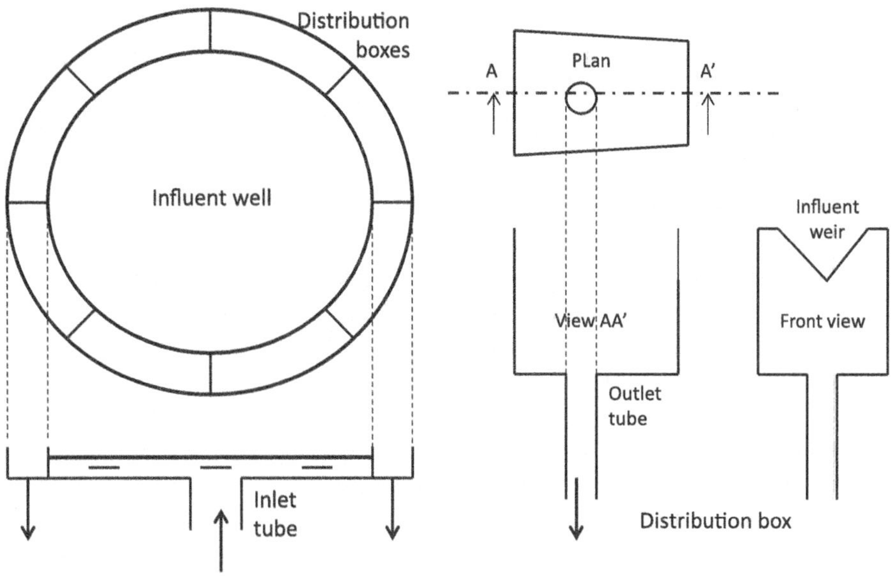

Figure 10.4 Details of the feed inlet structure with central feed to the influent well and discharge to the feed inlet points via distribution boxes located at the periphery of the centre well.

For the excess sludge collection it is assumed that the waste sludge can be discharged from two levels: one near to the bottom and the other higher in the reactor (refer to Section 4.3.6 and Figure 4.32).

10.4.6 Block 6: Flow and composition of the generated biogas (lines 146–166, Table 10.12)

Table 10.12 lists several comments for Block 6, located in lines 146–166 of the spreadsheet. For simplicity, it is considered that biogas is composed of only four components: methane, carbon dioxide, nitrogen and water vapour. The proportion of these components in the biogas is calculated, using the theory developed in Chapter 6. The potential for electrical power production is also estimated.

10.4.7 Block 7: Design of sludge drying beds

Table 10.13 lists several comments for Block 7, located in lines 167–203 of the spreadsheet, which deals with the design of sludge drying beds and is composed of two parts. The first part deals with the calculation of bed productivity, i.e. the mass of sludge that can be processed per unit area and per unit time. In the second part, the dimensions of drying beds are calculated for a specified final humidity.

10.4.8 Block 8: Post-treatment of anaerobic effluent in polishing lagoons

Depending on the desired final effluent quality, the limiting factor for polishing ponds may be the removal of thermotolerant coliforms (TT) or nutrients: nitrogen and/or phosphorus. If no nutrient removal is required, for example when the effluent is to be used as irrigation water, the retention time in a polishing pond is dictated by the requirement of efficient TT removal, which occurs at maximum rate when the pond is operated as a sequential batch unit. If nutrient removal is desired, the pH of the pond must increase to a

Table 10.12 Block 6 – Flow and composition of the generated biogas. It is assumed that biogas is composed of only four gases: methane, carbon dioxide, water, nitrogen and hydrogen sulphide.

Line #	Considerations
147	Distribution coefficient of methane (Table 6.1) – calculated based on temperature.
148	Distribution coefficient of nitrogen (Table 6.1) – calculated based on temperature.
149	Distribution coefficient of hydrogen sulphide (Table 6.1) – calculated based on temperature.
150	Daily mass of methane produced: digested COD load divided by four.
151	Production of methane gas: initial estimate not considering methane losses.
152	Ratio of gas flow/liquid flow (Q_g/Q_l): initial estimate.
153	Reactor pH: typically around 7.0 for municipal sewage.
154	Alkalinity in UASB effluent: typically around 8–10 meq · L^{-1}.
155	Partial CO_2 pressure (pCO_2) in the biogas: calculated based on pH and alkalinity.
156	Partial H_2O pressure (pH_2O) in the biogas: calculated based on temperature.
157	Partial N_2 pressure (pN_2) in the biogas: calculated based on Henry's law.
158	Partial CH_4 pressure in the biogas (p_{CH4}): equal to $1 - pCO_2 - pH_2O - pN_2$.
159	Daily losses of methane in the effluent: calculated based on methane solubility.
160	Methane losses (leakages etc): estimated, depends on quality of workmanship and materials used (e.g. concrete without proper coating or lining is not gas tight): 10–20% is possible.
161	Collected methane in the biogas (lines 150–159–160).
162	Production of methane gas (2nd estimate).
164	Energy dissipation: calculated from potential energy of biogas rising from the bottom.
165	Electric energy production potential: 40% electrical conversion efficiency is assumed.
166	Potential energy production per capita: energy production potential/population.

high level, for example about 8.5 for nitrogen removal and 9.5 for phosphate removal. Thus in this case the model must be used to calculate the minimum retention time for an effluent batch to acquire the desired pH value. This time depends mainly on three factors: temperature, pond depth and the rate of photosynthesis.

Another variable that must be considered is the presence of a transfer pond, that can function as an equalization basin between the UASB reactor operating under continuous flow conditions and the polishing ponds operating in sequential batch mode. The main variables for the design of the transfer pond are the retention time and the depth, along with the temperature and the photosynthesis rate. In the first part of this block the minimum residence time of polishing ponds operated in a sequential batch regime is calculated to reduce the concentration of thermotolerant and helminth eggs and thus produce a final effluent for unrestricted irrigation. The required retention time is calculated with the aid of Eq. (9.4), which can be rewritten as:

$$R_h = \ln(N_e/N_i)/k_{bh,t} \tag{10.1}$$

where:

$k_{bh,t} = k_{b1,20}/H\theta^{(T-20)}$ = decay constant at temperature T and depth H in the pond
$k_{b1,20} = 1.35\ d^{-1}$ = decay constant for a temperature of 20°C and a pond depth of 1 m

Table 10.14 lists several comments for Block 8, located in lines 206–229 of the spreadsheet, used to design polishing ponds with or without a transfer pond, if no nutrient removal is required.

Table 10.13 Block 7 – Design of sludge drying beds.

Line #	Design Data and Calculation of the Bed Productivity – Considerations
169	Time to discharge a sludge batch on a drying bed (T_1): <1 day.
170	Percolation constant: typical value is 320.
171	Applied solids load: typically in the range of 20–30 kg TSS · m^{-2}.
172	Sludge concentration: 60–80 g TSS · L^{-1} for bottom sludge, 15–25 g TSS · L^{-1} for sludge discharged with the top extraction system.
173	Percolation time (T_2): calculated with Eq. (7.1).
174	Water evaporation rate: obtained from meteorological data.
175	Initial humidity of evaporation: normally when percolation stops: $U_i \approx 0.8$.
176	Final (desired) sludge humidity: U_e typically between 0.5 and 0.8.
177	Relative evaporation rate: (Sections 7.4.2 and 7.4.3) = $1.25U_e$ as per Eq. (7.6).
178	Evaporation time (T_3): calculated with Eq. (7.5).
179	Time required to collect dry sludge (T_4): typically 2 days, depending on degree of mechanization used.
180	Total cycle time = time for discharge + percolation + evaporation + collection.
181	Drying bed productivity: ratio of applied solids load (171) and cycle time (180).
Line #	**Design of the drying beds – considerations**
183	Total anaerobic sludge mass (29).
184	Daily excess sludge production: COD mass ending up as sludge (kg COD · d^{-1}) divided by f_{cv} (1.5 g COD · g^{-1} VSS) and f_v (line 26).
185	Sludge age: calculated in line (32).
186	Maximum fraction of reactor sludge mass that may be discharged: to be defined (<50%). Note that it might very well be from all reactors simultaneously, as it is currently set up in the sheet.
188	Maximum period between discharges: sludge age times the maximum discharge fraction.
189	Selected interval between discharges: rounded value for convenience.
190	Mass of sludge discharged per batch: time between discharges (189) divided by the sludge age (185) times the sludge mass in reactor (183).
191	Minimum required bed area: daily sludge production (line 184) divided by productivity (181).
192	Minimum number of beds: should be >1, otherwise decrease the value of the interval between discharges (line 189).
193	Selected number of beds (select this so that the area per bed (line 194) is manageable.
194	Area of single drying bed: total area of beds divided by number of beds.
197	Reserve capacity: fraction of bed area.
198	Number of reserve beds: number of operational beds (193) times reserve capacity (197).
201	Height of discharged sludge batch on bed: applied solids load (171) divided by discharge sludge concentration (172).
202	Dried sludge mass (ton · d^{-1}): sludge production rate (184) times (1 – final humidity) (line 176).
203	Dried sludge mass per capita: dried sludge mass (202) divided by contributing population (10).

Table 10.14 Block 8 – Design of polishing ponds for production of final effluent for unrestricted irrigation, without nutrient removal and with or without a transfer pond.

Line #	General Data – Considerations
206	Decay constant at 20°C.
207	Arrhenius coefficient for temperature dependency of decay constant θ (see Chapter 9).
208	Temperature: average temperature of coldest month (given the long retention time).
209	Pond liquid depth: typically 1 m for a polishing pond.
210	Decay constant at temperature T and pond depth (see above).
211	Desired TT removal efficiency: normally between 99.9 and 99.99% (log 3 to log 4 removal), see Chapter 9.
212	Minimum required retention time R_h: calculated with Eq. (10.1).
Line #	**Design without a transfer pond – considerations**
214	Required time to fill a pond: normally 1 day.
215	Required time to discharge the final effluent from the pond: normally 1 day.
216	Number of polishing ponds required: as per line (214), typically each polishing pond receives the treated daily sewage flow. Hence the number of ponds is equal to the total retention time R_h for required TT removal efficiency (rounded up to next integer) + time for feeding (1 day) + time for discharge.
217	Volume of a polishing pond: numerically equal to daily sewage flow.
Line #	**Design with a transfer pond – considerations**
221	Required time to fill a pond: normally 0.5 day (gravity fed from transfer pond).
222	Required time to discharge the final effluent from the pond: normally 0.5 day.
223	Number of polishing ponds required: as in line (216).
224	Depth of the transfer pond: 2 m.
225	Retention time in the transfer pond: typically 2 days.
226	Volume of the transfer pond: sewage flow times retention time.
229	Total area ponds: area of transfer pond + area of polishing ponds.

EXAMPLE 10.2

Design a polishing pond system composed of a series of sequential batch ponds for post-treatment of 1000 $m^3 \cdot d^{-1}$ of UASB effluent, allowing one day for filling and one day for discharging the treated final effluent of the ponds. The objective is to remove 99.99% of the thermotolerant coliforms, but nutrient removal is not required. Use a pond depth of 1 m. Temperature is 25°C.

Solution

The required retention time in a single batch polishing pond is calculated with Eq. (10.1):

$$R_h = \ln(N_e/N_i)/k_{bh,t} = \ln(10^4)/(1.35 \times 1.07^{(25-20)}) = 9.2/1.9 = 4.85 \text{ d}$$

As the pond is filled in 1 day, its volume is equal to the daily design flow rate of 1000 m^3. If a rounded value of 5 days is used as required retention time for TT decay after filling the pond, each pond has an operational cycle of 7 days: 1 for feeding, 5 for TTC decay and 1 for discharge of the final effluent. Thus a minimum of 7 ponds would be required (as shown in Figure 10.5, top), while a spare pond is recommended as reserve capacity for maintenance. The area of each pond 1000 m^2 ($H = 1$ m). If it is assumed that the sewage is produced by 10,000 inhabitants, the required area per i.e. would be 7 × 1000/10,000 = 0.7 m^2. per habitant. As the decay constant is inversely proportional to depth, the required area remains the same for other depths.

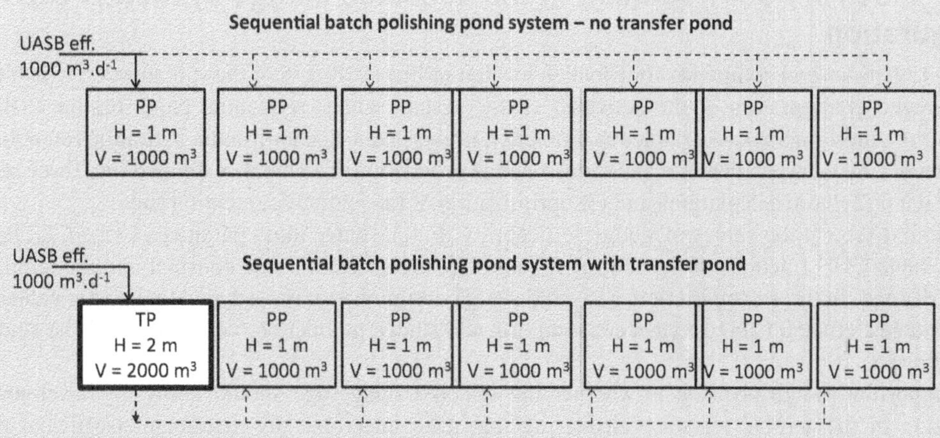

Figure 10.5 Schematic representation of polishing pond systems (with and without transfer ponds) for post-treatment of UASB effluent to produce final effluent suitable for unrestricted irrigation.

EXAMPLE 10.3

Design a polishing pond system for the same conditions as in Example 10.2 but with a transfer pond with a retention time of 2 d and a depth of 2 m.

Solution

The volume of the transfer pond is $2 \times 1000 = 2000$ m^3 and the area is $2000/2 = 1000$ m^2. The transfer to and from the polishing ponds will now be faster: it is assumed that the feed and discharge operations are completed in 1 d. Therefore the number of polishing ponds reduces to 6. Figure 10.5 (bottom) shows the polishing pond configuration to produce final effluent for unrestricted irrigation with a transfer pond. It can be seen that the structure of both pond systems is not very different, but the presence of a transfer pond can facilitate the operation of the system.

If nutrient removal is desired, then the model of Chapter 9 is used to estimate the required retention time in the ponds to reach the pH required for nutrient removal. The retention time for phosphorus removal will invariably be longer than the time for ammonium desorption and this time in turn will be higher than that for TT decay. Phosphorus removal is only feasible if the pH in the pond increases to a value well above 9. The residual phosphorus concentration reaches a minimum value of 1 mg $\cdot$ L^{-1} at a pH of 9.3. A further increase of the pH may actually be counterproductive because it becomes toxic and may lead to the decay of the algae, including those that after discharge are left in the pond as an inoculum for the next batch. As the increase of pH depends mainly on the rate of photosynthesis, which is an unpredictable variable, it is best to apply a variable retention time and consider the post-treatment concluded when a maximum pH is reached, for example pH = 9.3. A lower value for the maximum pH may be adopted, if it is considered that the retention time for pH = 9.3 is too long, but then probably the final phosphate concentration will be higher than the minimum value of 1 mg P $\cdot$ L^{-1}.

10.4.9 Block 9: Post-treatment in the activated sludge system in SBR configuration

Table 10.15 lists several comments for Block 9, located in lines 228–276 of the spreadsheet, which defines the optimized configuration of the activated sludge system with a sequential batch regime (SBR) and calculates the most important variables as a function of the sludge age: minimum hydraulic retention time, final effluent COD, oxygen demand and excess sludge production. The block is divided into three sections: (1) basic data, (2) basic calculations and (3) optimization of the hydraulic retention time.

In the first section the values of sludge settleability (k, v_0), sludge mass parameters (Y_{ae}, f, f_{cv}, b_h), non-biodegradable COD fractions in the sewage (f_{ns} and f_{np}), the utilization rate constant and the temperature must be defined. In the second section, activated sludge design theory is used to calculate the values of the main variables: effluent COD, oxygen consumption and sludge production, which requires the sludge age to be defined.

An important design decision is whether the selected sludge age should allow for development of nitrification in the aerobic reactor, considering that after anaerobic pre-treatment insufficient organic material remains in the reactor to allow for denitrification. Typically due to sulphide toxicity the required sludge age for nitrification is higher than for direct treatment of municipal sewage, unless a pre-aeration step is applied in which the sulphide is removed. If a sludge age is chosen that allows for nitrification to develop, then the oxygen required for this process must be considered: in fact the oxygen demand for nitrification is much higher than the demand for removal of biodegradable organic material.

It is important to remember that the aerobic sludge production can only be stabilized in the UASB reactor if the solids discharged from the anaerobic reactor are separated from the liquid phase so that they cannot enter the post-treatment unit (see Chapter 8). In the third section the optimal volume for the post-treatment system is established, by applying the optimization procedure presented in Chapter 8.

Table 10.15 Block 9 – Post-treatment in the activated sludge system in SBR configuration.

Line #	Basic Data for Aerobic Post-Treatment – Considerations
232	Vesilind compressibility constant (k): typically 0.2 L · g^{-1} TSS for anaerobic sewage.
233	Vesilind constant for unhindered settling velocity (v_0): typically 10 m · h^{-1} for anaerobic sewage.
234	Yield coefficient for aerobic processes: $Y_{ae} = 0.45$ mg VSS · mg^{-1} COD.
235	Temperature (20°C): same as for the UASB.
236	Decay constant $b = 0.24$ d^{-1}.
237	Endogenous residue fraction upon decay (f): 0.2.
239	f_{ns} of raw sewage: typically in range 0.10–0.15.
240	f_{np} of raw sewage: typically in range of 0.08–0.15.
241	COD utilization constant: 3 g COD · g^{-1} VSS · d^{-1}.
Line #	**Basic calculations – considerations**
243	COD load in UASB effluent (line 44).
244	Influent COD concentration S_{ti} (line 13).
245	COD concentration in UASB effluent S_{tu} (line 41).
247	Non-biodegradable soluble COD in UASB effluent S_{nsu}: not affected by anaerobic treatment. Its value is equal to $f_{ns}S_{ti}$.

(Continued)

Table 10.15 (Continued) Block 9 – Post-treatment in the activated sludge system in SBR configuration.

Line #	Basic calculations – considerations
247	Inert soluble COD fraction in UASB effluent: equal to $S_{nsu}/S_{tu} = f_{ns}S_{ti}/S_{tu}$.
248	Inert particulate fraction in UASB effluent (f_{np}): adopted value, typically low at 0.02.
249	Biodegradable COD concentration in UASB effluent: $S_{bu} = S_{tu}(1 - f_{ns} - f_{np})$.
250	Sludge age: main design parameter to be selected by designer. If only COD removal is required, a value of 6–8 days is appropriate at 20°C.
251	Fraction of feed COD consumed by exogenous respiration: calculated with Eq. (8.4).
252	Fraction of feed COD consumed by endogenous respiration: calculated with Eq. (8.4).
253	Oxidized COD fraction of UASB effluent (mS_o): sum of lines (251) and (252).
254	Oxygen consumption in post-treatment: $MS_o = Q_iS_{tu}mS_o$ (= Line 245 × line 12 × line 253).
255	Biodegradable load in UASB effluent: $MS_{bu} = Q_iS_{bu}$.
256	Final effluent COD load: $MS_{te} = Q_iS_{nsu}$ (presence of solids in the effluent is ignored).
257	COD discharged in excess sludge: $MS_{xv} = Q_iS_{tu} - MS_o - MS_{te}$.
258	Volatile excess sludge production: $ME_v = MS_{xv}/f_{cv}$.
259	Volatile fraction of sludge: typically higher than the value for raw sewage, f_v is expected to be in the range of 0.75–0.8.
260	Total aerobic sludge production: $ME_t = ME_v/f_v$.
Line #	**Optimization of the retention time**
262	Retention time: needs to be changed by the user until the value of line 264 (flow that is to be treated per unit reactor volume) is equal to line 277 (treatment capacity per unit volume) and therefore line 278, the difference between these two parameters, is zero.
263	Unit volume of the SBR reactor: always fixed value of one.
264	Flow rate per unit reactor volume: inverse of retention time.
265	Liquid height of the reactor: maximum liquid height typically not more than 6 m due to limitations in aeration equipment.
266	Batch size per unit reactor volume: to be decided by designer.
267	Active sludge concentration (X_a): calculated with Eq. (8.8).
268	Endogenous residue concentration (X_e): calculated with Eq. (8.9).
269	Volatile sludge concentration (X_v): calculated with Eq. (8.11).
270	Total sludge concentration (X_t): calculated with Eq. (8.12).
271	Minimum aeration time (t_{aerm}: calculated with Eq. (8.18).
272	Minimum settling time (t_{se}): calculated with Eq. (8.20).
273	Feed time (t_{feed}): value depending on local conditions and selected pump or gravity flow transfer capacity.
274	Discharge time (t_{disc}): value depending on local conditions and selected pump or gravity flow transfer capacity.
275	Batch cycle time: total time required for processing of a batch of UASB effluent, equal to sum of $t_{feed} + t_{aerm} + t_{se} + t_{disc}$.
276	Number of treated batches per day = 24 hrs divided by batch cycle time.
277	Treatment capacity: batch volume times number of batches per day.
278	Difference: treatment capacity – flow per unit volume (line 277 264), should be zero.

10.5 APPLICATION EXAMPLES

In this section some applications of the spreadsheet for $UASB_r$ reactor design, including the complementary units, presented in the previous section will be discussed.

EXAMPLE 10.4

Evaluate the optimal organic load to treat sewage for a population of 2,50,000 inhabitants (30,000 $m^3 \cdot d^{-1}$) with a COD of 600 mg $\cdot$ L^{-1} and a sulphate concentration of 20 mg SO_4-S $\cdot$ L^{-1}. Adopt a design temperature of 20°C and non-biodegradable fractions $f_{ns} = 0.15$ and $f_{np} = 0.11$.

Solution

In Table 10.16 the data of Block 2 are calculated for the specified conditions of Example 10.4, using the applied volumetric COD load as a variable. After establishing the value of the volatile excess sludge production using the iterative trial and error loop of lines 25–35, the feasibility of a applied volumetric loading rate on the performance of the UASB reactor can now be evaluated from two calculated variables: the sludge age (line 32) and the required minimum specific methanogenic activity (line 37).

For the evaluated applied volumetric loads, the sludge age is only 45 and 61 d respectively for the highest considered applied loads of 3.5 and 3 kg COD $\cdot$ $m^{-3} \cdot d^{-1}$, which may be considered too short for efficient anaerobic digestion at 20°C, resulting in the production of excess sludge that may not be well stabilized. Also the required minimum SMA for these loads is on the high sides: 0.14 and 0.13 g COD $\cdot$ g^{-1} VSS $\cdot$ d^{-1} respectively, although methanogenic activity normally is not the rate limiting process in anaerobic treatment of municipal sewage, but hydrolysis.

From the point of view of performance probably the best option would be to select a volumetric loading rate of 2.5 kg COD $\cdot$ $m^{-3} \cdot d^{-1}$, resulting in a sludge age of 81 d and a required minimum SMA of 0.113 g COD $\cdot$ g^{-1} VSS $\cdot$ d^{-1}.

Table 10.16 Performance of the UASB reactor as a function of the applied COD load for the conditions of Example 10.4.

Line #	Variable	Applied Volumetric COD Load in kg/($m^3 \cdot d$)				Unit
		3.5	**3.0**	**2.5**	**2.0**	
25	Specific organic sludge production (mE_v)	0.1062	0.0915	0.0823	0.0775	g VSS $\cdot$ g^{-1} COD
27	Average UASB reactor sludge concentration (X_{tu})	25	25	25	25	kg TSS $\cdot$ m^{-3}
28	Total volume (V_t)	4800	5600	6720	8400	m^3
29	Required sludge mass (MX_t)	12.0	14.0	16.8	21.0	ton TSS
30	Hydraulic retention time (R_h)	0.16	0.19	0.22	0.28	d
31	Organic sludge production (ME_v)	1784	1537	1383	1302	kg VSS $\cdot$ d^{-1}
32	Anaerobic sludge age (R_s)	45.06	61.02	81.41	108.06	d
33	Feed COD fraction ending up in effluent (f_s)	0.21	0.18	0.17	0.16	%
34	Feed COD fraction ending up in sludge (f_x)	0.1593	0.1372	0.1235	0.1163	%

(Continued)

Table 10.16 (Continued) Performance of the UASB reactor as a function of the applied COD load for the conditions of Example 10.4.

Line #	Variable	Applied Volumetric COD Load in kg/(m³ · d)				Unit
		3.5	3.0	2.5	2.0	
35	Theoretical specific organic sludge production (mE_v)	0.1062	0.0915	0.0823	0.0775	kg VSS · kg^{-1} COD
36	Digested feed COD fraction (f_d)	0.63	0.68	0.71	0.72	%
37	Minimum required methanogenic activity	0.141	0.130	0.113	0.093	g COD · g^{-1} VSS · d^{-1}
38	Fraction of produced excess sludge lost with effluent	0.346	0.297	0.247	0.198	%
39	Soluble and not retained sludge fraction	0.266	0.225	0.199	0.183	%
40	COD fraction discharged with the excess sludge from reactor	0.104	0.096	0.093	0.093	%
41	COD concentration in effluent ("raw COD")	0.160	0.135	0.119	0.110	g COD · L^{-1}
42	COD concentration in discharged sludge	0.062	0.058	0.056	0.056	g COD · L^{-1}
43	COD load for sludge production (MS_{xv})	1749	1621	1561	1567	kg COD · d^{-1}
44	COD load in effluent (MS_{te})	4474	3788	3340	3070	kg COD · d^{-1}
45	Digested COD load (MS_d)	10,577	11,391	11,899	12,163	kg COD · d^{-1}

EXAMPLE 10.5

For the same conditions as in Example 10.4 evaluate the influence of the applied COD load on the physical design layout of the UASB reactors and therefore on the investment costs.

Solution

In Table 10.17 the influence of the applied load on the size of the reactor and on the design of the phase separators is indicated. The table shows that the applied load has a very strong influence on the volume of the UASB system and therefore on the construction costs. The number of reactors and the number of separator elements in each reactor is also influenced by the applied load. The volume of each reactor in practice is not much larger than 2000 m³, whereas the number of separators is limited by the condition that the total height of a separator element should be more than 0.8 m (0.5 m for the gas chamber inside the separator and 0.3 m for the liquid phase beneath it). The numbers in Table 10.17 clearly show that the applied load has a decisive influence on the construction costs of the treatment system.

Table 10.17 Example 10.5 – Size of the GLS separators as a function of the applied COD load for the conditions of Example 10.4.

Line #	Variable	Applied Volumetric COD Load in kg/($m^3 \cdot d$)				Unit
		3.5	3.0	2.5	2.0	
48	Total volume of reactors	4800	5600	6720	8400	m^3
49	Number of reactors	3	3	3	4	(-)
50	Volume per reactor	1600	1867	2240	2100	m^3
51	Total reactor height	5	5	5	5	m
59	Number of GLS separators	8	8	10	10	(-)
69	Height of GLS separators	0.87	0.93	0.83	0.81	m

EXAMPLE 10.6

Evaluate the influence of the applied COD load on the rising velocity of the liquid and the division of the reactor volume in the fractions for digestion transition and settling.

Solution

In Table 10.18 the values of the upflow velocity at four critical levels of the reactor are calculated: (1) in the digestion zone, (2) at the base of the deflectors (3), at the base of the separators, and (4) at the overflow level. The values of the upflow velocities at different levels in the UASB reactors for the loads of 3.5 and 3 kg $COD \cdot m^{-3} \cdot d^{-1}$ are very high, so that these reactors may have a hydraulic overload, along with an organic overload as was observed in Example 10.4.

Table 10.18 Example 10.6 – Verification of the upflow velocities of the liquid at different levels and the division of the volume over digestion, transition and settling zones as a function of the applied COD load for the conditions of Example 10.4.

Line #	Variable	Applied Volumetric COD Load in kg/($m^3 \cdot d$)				UoM
		3.5	3.0	2.5	2.0	
84	Velocity in digestion zone	1.30	1.12	0.93	0.74	$m \cdot h^{-1}$
85	Velocity at deflector base	2.99	2.57	2.14	1.71	$m \cdot h^{-1}$
86	Velocity at separator base	2.99	2.57	2.14	1.71	$m \cdot h^{-1}$
87	Velocity at discharge (overflow) level	1.50	1.27	1.08	0.87	$m \cdot h^{-1}$
95	Liquid volume fraction for settling	14	15	13	13	%
96	Liquid volume fraction for transition	16	17	15	15	%
97	Liquid volume fraction for digestion	70	68	72	73	%

EXAMPLE 10.7

Design the special devices in the UASB reactors (feed distribution, effluent collection, biogas collection and sludge discharge).

Solution

Table 10.19 shows the results of the calculations as a function of the applied load under the conditions of Example 10.4 on the number of inlet and outlet points of the reactors for the different applied loads. It can be observed that the inlet and outlet structures are not heavily influenced by the applied load.

Table 10.19 Example 10.7 – Inlet and outlet structures and outlet devices for sludge discharge and biogas as functions of the applied COD load for the conditions of Example 10.4.

Line #	Variable	Applied Volumetric COD Load in kg/(m³ · d)				Unit
		3.5	3.0	2.5	2.0	
100	Area per feeding point	2.00	2.00	2.00	2.00	m^2
101	Number of inlet points per reactor	160	187	224	210	(–)
106	Number of inlet feed points per feed well	80	93	75	70	(–)
125	Number of collection pipes per reactor	16	16	20	20	(–)
127	Diameter effluent collection pipe	0.15	0.15	0.15	0.15	m
128	Diameter orifices	0.0125	0.0125	0.0125	0.0125	m

EXAMPLE 10.8

Evaluate the influence of the applied COD load on the flow and composition of biogas produced by the treatment system.

Solution

Table 10.20 shows the results of the calculations as a function of the applied load under the conditions of Example 10.4 on the flow and composition of the biogas produced in the treatment system. It can be observed that the flow of biogas (and particularly methane) is strongly influenced by the applied COD load, whereas the biogas composition is much less affected.

Table 10.20 Example 10.8 – Biogas production and composition as functions of the applied COD load for the conditions of Example 10.4.

Line #	Variable	Applied Volumetric COD Load in kg/(m³ · d)				Unit
		3.5	3.0	2.5	2.0	
150	CH_4 production	2644	2848	2975	3041	kg $CH_4 \cdot d^{-1}$
155	Partial CO_2 pressure	0.050	0.050	0.050	0.050	bar
156	Partial H_2O pressure	0.022	0.022	0.022	0.022	bar
157	Partial N_2 pressure	0.086	0.081	0.078	0.076	bar
158	Partial CH_4 pressure	0.841	0.847	0.850	0.851	bar
166	Per capita energy production	1.25	1.37	1.45	1.50	$W \cdot hab^{-1}$

Appendix A1

Anaerobic biodegradability, specific methanogenic activity and sludge stability

In this appendix three practical tests will be discussed that are very convenient to evaluate the feasibility of anaerobic digestion of a particular wastewater and to monitor anaerobic digester performance:

(a) *Anaerobic biodegradability test*: The BOD test is suitable for evaluating the feasibility and degree of degradation of organic material in an aerobic environment, but does not provide information on the possibility of applying anaerobic digestion to treat a specific wastewater. Anaerobic biodegradability can be evaluated experimentally by measuring the methane produced from organic material present in a sample of wastewater when it is brought into contact with an anaerobic sludge batch under previously selected and stable conditions.

(b) *Specific methanogenic activity test*: When placing a substrate that can be degraded by methanogens in contact with a known mass of an anaerobic sludge, then the specific methanogenic activity (SMA) of this sludge can be determined from the methane production rate. The SMA test provides the mass of COD that can be digested by the sludge per unit mass of sludge and per day under optimized conditions, previously chosen and it is of fundamental importance to evaluate the quality of a sludge;

(c) *Sludge stability*: This test measures the fraction of organic material that is converted into methane when a sludge sample is kept under suitable conditions and when enough time for digestion is allowed. The result of the test is indicative for the stability of the sludge: a high value of the residual degradable fraction indicates a poorly stabilized sludge with high content of biodegradable material that is likely to cause odour issues if disposed of without further treatment.

Editor(s): Adrianus van Haandel and Jeroen van der Lubbe
doi: 10.2166/9781780409627_0405

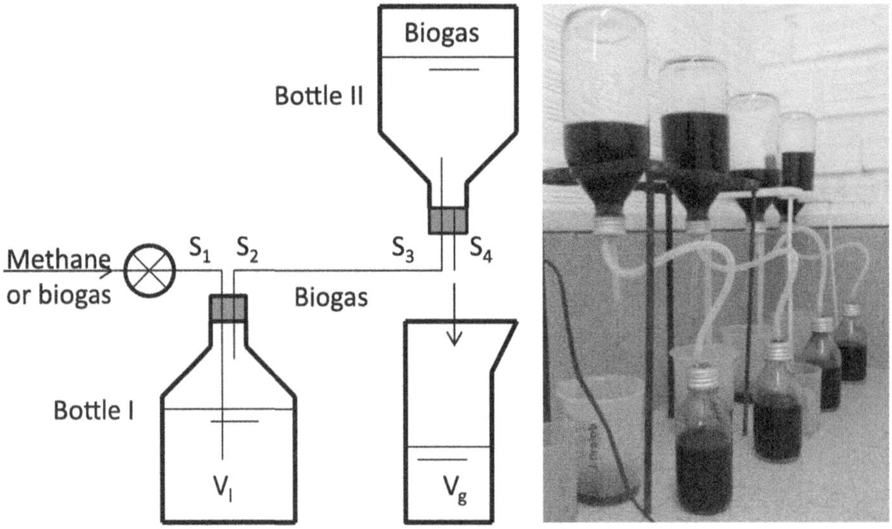

Figure A1.1 Schematic representation of the standard test set-up of the equipment to determine the anaerobic biodegradability of the organic material in wastewater, as well as sludge activity and sludge stability. On the right side a picture of an actual lab test set-up is shown.

A1.1 ANAEROBIC BIODEGRADABILITY OF ORGANIC MATERIAL

A schematic test set-up with the equipment required to perform the biodegradability test is shown in Figure A1.1. This set-up is generic and thus can also be used for the determination of specific methanogenic activity and for sludge stability tests.

In the case of a biodegradability test, a wastewater sample is placed in a serum bottle (bottle I), together with well stabilized anaerobic sludge. The rubber cap of the bottle is pierced by two syringes, the end of one syringe (S_1) is dipped in the liquid phase, while that of the other syringe (S_2) remains in the gas phase above the liquid. After replacing the cap, methane or biogas is blown through the syringe into the sample, in order to remove the air (oxygen) present in the sample and in the gas phase above it. This could be important when the concentration of oxygen (DO) is high, as its presence could be toxic for the methanogens. However, in the case of sewage treatment the anaerobic sludge is typically at least partially facultative aerobic, which means it has a great oxidative metabolic capacity, despite being anaerobic, and in a short time will eliminate the oxygen in the sample.

On the other hand, part of the methane produced during the test (its saturation concentration is around $1\ \text{mmol} \cdot L^{-1} = 16\ \text{mg}\ CH_4 \cdot L^{-1} = 64\ \text{mg COD} \cdot l^{-1}$) will remain dissolved in the liquid phase if it was not previously saturated with the methane or biogas mixture.

After flushing the liquid phase with methane, syringe S_1 is closed and syringe S_2 is connected through a thin flexible tube to a second bottle via a third syringe S_3, perforating the rubber cap. This second bottle II is placed with the lid down, creating a vacuum above the liquid level (the bottom has atmospheric pressure, the top liquid level less). During the test, biogas will be generated in bottle I, resulting in an increase of gas pressure, and when it is sufficiently high, the biogas will escape through the tube and accumulate above the liquid phase in bottle II. Consequently in bottle II the pressure will increase and as a result an amount of liquid equal to the volume of gas produced in bottle I will be expelled via a

Figure A1.2 Modern automated equipment to determine biodegradability and sludge activity or stability. The gas production of several simultaneous tests can be measured and recorded.

fourth syringe (S_4). Thus the volume of liquid thus expelled from bottle II can be measured by a simple volumetric flask or by weighing.

If the liquid in bottle II has a high pH, then CO_2 will be dissolved and only methane production will accumulate. If the pH is low, CO_2 will not dissolve, and all biogas produced will be determined. If desired, this feature can be used to estimate the methane fraction of the produced biogas, an important factor in the design of biogas treatment and utilization units

The same principle of measuring the methane or biogas production is used in modern test equipment, but the gas production is measured continuously by means of a pressure monitor and the results are stored in a computer. As the operation is automated, several tests can be carried out in parallel. Figure A1.2 shows an example of a modern test set-up, including an example of the graphs that can be generated by specific software. The sample bottles are slowly agitated to ensure that there is enough contact between sludge mass and substrate during the whole testing period.

Returning to the test set-up discussed in Figure A1.1, the concentration of digested COD in the sample can be calculated from the measured (produced) gas volume knowing that the anaerobic digestion of 4 g of COD results in the production of 1 g of methane. Furthermore, it is known that under standard conditions of pressure and temperature (1 bar, 273 K), 64 g of COD generates 1 mole of methane (16 g CH_4) which has a gas volume of 22.4 litres. Hence 1 litre of methane generated under standard conditions is obtained by digestion of 64/22.4 = 2.86 g of COD, so that:

$$S_{ban} = 2.86V_s/V_i = 2.86V_{CH_4}[P_{CH_4}\,273/(273 + T)]/V_i \qquad \text{(A1.1)}$$

where:

S_{ban} = COD digested in the sample,
V_s = volume of the produced methane under standard conditions;
V_{CH_4} = volume of produced methane under actual conditions of the test;
P_{CH_4} = methane pressure under test conditions in Bottle II;
V_1 = sample volume.
T = temperature (°C).

EXAMPLE A1.1

An 800 ml sample with a COD of 1000 mg · L^{-1} is subjected to a degradability test. The sample is incubated with a batch of seed sludge and during the production test a total volume of 270 ml of caustic soda solution is expelled from Bottle II. In the blank (control) test 800 ml of water without organic material is added to an equal mass of seed sludge: this test produces 30 ml. Estimate the biodegradable COD fraction under anaerobic conditions. The test was executed at a temperature of 30°C, while the local altitude was 700 m (i.e. $P_{atm} = 0.96$ Ba = 730 mm Hg).

Solution

At 30°C the water vapour pressure P_a is equal to 23 mm Hg, so the methane pressure P_{CH_4} in the test is 0.96(730 − 23)/730 = 0.93 Ba. Consider that, as caustic soda is used in Bottle II instead of water, methane production instead of biogas production is being measured. Using Eq. (A1.1) and correcting for the methane produced in the control bottle, one has:

$$
\begin{aligned}
S_{ban} &= 2.86 V_{CH_4} P_{atm}(760 - P_a)/760[273/(273 + T)]/V_1 \\
&= [2.86(270 - 30)0.96(730 - 23)/730 \times 273/303]/0.8 \\
&= 720 \text{ mg COD} \cdot l^{-1}
\end{aligned}
$$

Therefore the biodegradable COD fraction of the sample in an anaerobic environment is:

$f_b = 720/1000 = 72\%$.

EXAMPLE A1.2

If a 1 litre bottle is used for a 500 ml wastewater + sludge sample and the remaining headspace of the bottle is filled with air, calculate the COD that will be oxidized instead of being transformed into methane.

Solution

The molar volume of gases at 25°C is about 24 mL · $mmol^{-1}$. Thus, a volume of 500 ml air has 500/24 = 20.8 mmol with 21% oxygen. Therefore, it contains 0.21 × 20.8 mmol = 4.4 mmol of oxygen or 4.4 × 32 = 140 mg O_2. It can therefore oxidize 140 mg of COD. If the sample of wastewater + sludge has a volume of 0.5 L, the oxidized COD concentration will be 140/0.5 = 280 mg COD · L^{-1}. This is a considerable fraction of the COD concentration of sewage.

EXAMPLE A1.3

A biodegradability test on a solution of 1200 mg · L^{-1} of sodium acetate, with a sample size of 500 ml (Example A1.2), results in the production of 112 ml of methane. The atmospheric pressure during the test was 0.96 atmosphere and the temperature was 30°C. Is this result compatible with the theory?

Solution

The sample had a COD of 565 mg · L^{-1} and therefore 0.5 × 565 = 283 mg COD was incubated. As acetate is a biodegradable material in an anaerobic environment, the methane production potential is 283/4 = 71 mg CH_4.

On the other hand, at a temperature of 30°C the water vapour pressure is 23 mm Hg. Thus, at a pressure of 0.96 atm. in the methane accumulation bottle, the methane partial pressure is:

$$P_{CH_4} = (0.96 \times 760 - 23)/760 = 0.93 \text{ atm.}$$

On the other hand, the volume of 1 mole (16 g) of methane at 30°C and atmospheric pressure is 22.4(273 + 30)/273 = 24.9 L. The volume of one mole of methane at a pressure of 0.93 atm. is 24.9/0.93 = 26.7 L. Under these conditions the volume of 71 mg of CH_4 has a volume of 71 × 26.7/16 ml = 118 ml. When comparing experimental (112 ml) and theoretical (118 ml) methane production there is a difference of only 5%. Part of the difference can be attributed to the anabolic formation of sludge in the process of anaerobic digestion consuming organic material. Other possible sources of error would be (1) dissolution of methane in the liquid phase of bottle I and II, (2) gas losses due to diffusion through the wall of the tube from bottle I to bottle II, (3) sulphide reduction with simultaneous consumption of organic material, and (4) incomplete anaerobic digestion of the substrate. Taking into consideration these factors, the obtained result can be considered as satisfactory.

A1.2 SPECIFIC METHANOGENIC ACTIVITY

The specific methanogenic activity test (SMA) can be carried out with the same equipment as used for the anaerobic degradability test shown in Figures A1.1 and A1.2. In the SMA test, a sludge sample is added to a buffer solution, which ensures that the pH remains at 7.0 during the test, while micro- and macro-nutrients are added. Thus, the methane production rate from a substrate mixture is determined under ideal conditions for methanogens. If the maximum acetogenic methane production is to be determined, sodium acetate is used, whereas if the maximum total methanogenic production is to be tested, both acetotrophic and hydrogenotrophic archaea must have adequate substrate. This is achieved by using a solution with equimolar concentrations of acetic, propionic and butyric acid. The difference between methanogenic production with the VFA mixture and with acetate as sole substrate indicates the hydrogenotrophic activity. The SMA test is performed at a constant temperature, usually the optimum range for mesophilic methanogenesis: 30–37°C or alternatively the temperature in the reactor from which the sample was taken.

To minimize the risk of test failure, adequate operating conditions must be selected. In order to have a fast and reliable response, both the substrate and the sludge sample must be inoculated in sufficient quantity to produce in a short period of time an amount of methane that can be measured accurately. As stoichiometrically there is a production of 1 g of methane per 4 g of COD digested, under the usual test conditions (30–37°C and close to atmospheric pressure), this mass of methane has a volume of approximately 1.5 litres, as 16 g of methane have a volume of about 24 litres, i.e. the methane density is 16/24 = 0.67 g $CH_4 \cdot L^{-1}$. Therefore 1.5 L of methane is produced from the digestion of 4 g of COD, or 0.375 L $CH_4 \cdot g^{-1}$ COD.

Hence, to prevent bottle II from being drained completely during the test, it is good practice to incubate not more than 2 g COD $\cdot L^{-1}$ liquid volume of bottle II. On the other hand the incubated COD mass must also not be too small, as it will result in a small methane production and consequently potentially a large measurement error.

It is also desirable that the test is finished in a reasonable time, for example a maximum of five days. If there is a previous estimate of the sludge SMA, this gives an idea of how much of the sludge mass should be incubated. For example, consider the case when a sludge activity of 0.1 mg COD $\cdot mg^{-1}$ VSS $\cdot d^{-1}$ is expected (which is in the typical range expected for sludge generated in anaerobic sewage treatment) and it is desired to digest around 1 g of incubated COD in 5 days. The rate of digestion is equal to 1/5 = 0.2 g COD $\cdot d^{-1}$, and hence the inoculated sludge mass should be at least 0.2/0.1 = 2 g of organic sludge mass (VSS). If the expected SMA of the sludge is higher (for example 0.5 g COD $\cdot g^{-1}$ VSS $\cdot d^{-1}$), then the inoculated sludge

Table A1.1 Required sludge mass for inoculation as a function of the estimated specific methanogenic activity (SMA_e) and the desired duration of the test to digest 1 g of substrate COD.

SMA_e (g COD · g^{-1} VSS · d^{-1})	Required Sludge Mass for the Duration of the Test				
	1 d	2 d	3 d	4 d	5 d
0.04	25.0	12.5	8.3	6.3	5.0
0.06	16.7	8.3	5.6	4.2	3.3
0.08	12.5	6.3	4.2	3.1	2.5
0.1	10.0	5.0	3.3	2.5	2.0
0.15	6.7	3.3	2.2	1.7	1.3
0.2	5.0	2.5	1.7	1.3	1.0
0.25	4.0	2.0	1.3	1.0	0.8
0.3	3.3	1.7	1.1	0.8	0.7
0.4	2.5	1.3	0.8	0.6	0.5
0.5	2.0	1.0	0.7	0.5	0.4
1.0	1.0	0.5	0.35	0.25	0.2

mass may be smaller: 0.2/0.5 = 0.4 g VSS. For a desired maximum test duration (T) and an estimated activity SMA_e, the mass of sludge to be used per g of COD to be digested will be equal to:

$$MX_v = 1/(T\,SMA_e) \tag{A1.2}$$

Using Eq. (A2.2), in Table A1.1 the required sludge mass is calculated as a function of the assumed methanogenic activity and the desired maximum test duration, for 1 g · L^{-1} of incubated substrate COD.

If the test time is excessive, then alternatively the incubated substrate COD concentration can be decreased, but care must be taken not to use a COD concentration that is too low. The Monod equation shows that if the COD concentration is not much higher than the value of the half-saturation constant, the anaerobic digestion does not develop at the maximum rate. For the presumed substrate addition of 1 g of COD, the maximum methane production is 375 ml, which can be measured conveniently with bottle II, if its volume is greater than 0.5 L. The volume of bottle I is in principle irrelevant, but it is advantageous that it is not too large, so that the gas production results in additional mixing of the bottle contents and facilitates contact between the sludge and the substrate. If this mixture is insufficient, it is possible that the gas production will finish before all the biodegradable organic material has been digested, only using the substrate present in the lower part of the bottle, i.e. where the sludge is located. In any case it is highly recommended to induce gentle mechanical agitation, for example by means of a shaker.

In conclusion the preparation procedure of the SMA test can be summarized as follows:

(a) The volume of bottle II and bottle I is between 0.5 and 1 L;
(b) Organic material to be incubated: 1 g of COD (equivalent to 1.3 g of $CH_3COONa \cdot 3H_2O$);
(c) For acetotrophic methanogenic activity use acetate, for total methanogenic activity use a VFA mixture of acetic, propionic and butyric acid;
(d) Minimum sludge mass: $1/(T\,SMA_e)$ for a desired maximum test duration of T days;
(e) Slow agitation prevents insufficient contact between sludge and substrate.

EXAMPLE A1.4

A 50 ml anaerobic sludge sample with a concentration of 40 g VSS · L^{-1} is incubated with a solution of micro- and macronutrients and sodium acetate at a pH of 7.0. The temperature and pressure during the methane production test are 30°C and 0.96 atm respectively. Estimate the specific methanogenic activity (SMA) from the observed methane production curve, which is shown in Figure A1.3.

Solution

The gas production figure shows that after a brief period the methane production to good approximation was linear with time up to 3.5 days, after which methane production continued, but at a much lower rate, and one attributable to the endogenous production of methane by the sludge. The maximum methane rate is around 200 mL · d^{-1} while the endogenous rate is 30 mL · d^{-1}. It is concluded that methane production due to substrate utilization is equal to the difference between these two production rates: 200 − 30 = 170 mL · d^{-1}, which corresponds to a COD of:

$$S_d = 2.86 \times 170 \times 0.96\ (76023)/760 \times 273/303 = 408 \text{ mg COD} \quad \text{(A1.1)}$$

Therefore the specific methanogenic activity of the incubated sludge (0.05 L at 40 g VSS · L^{-1} = 2 g VSS) is calculated as:

$$\text{SMA} = 408/2000 = 0.20 \text{ mg COD} \cdot \text{mg}^{-1} \text{ VSS} \cdot \text{d}^{-1}.$$

The SMA test can also be used to compare the potential of methane production of acetotrophic and hydrogenotrophic archaea, when using a substance that produces the substrates for both: acetate and hydrogen. Such a substrate may for example be propionate, which in the acetogenic process decomposes as.

$$C_3H_6O_2 + 2H_2O \rightarrow C_2H_4O_2 + CO_2 + 3H_2$$

The two substrates thus generated can be converted into methane:

Acetotrophic reaction: $C_2H_4O_2 \rightarrow CH_4 + CO_2$
Hydrogenotrophic reaction: $3H_2 + {}^3/_4CO_2 \rightarrow {}^3/_4CH_4 + {}^3/_2H_2O$.

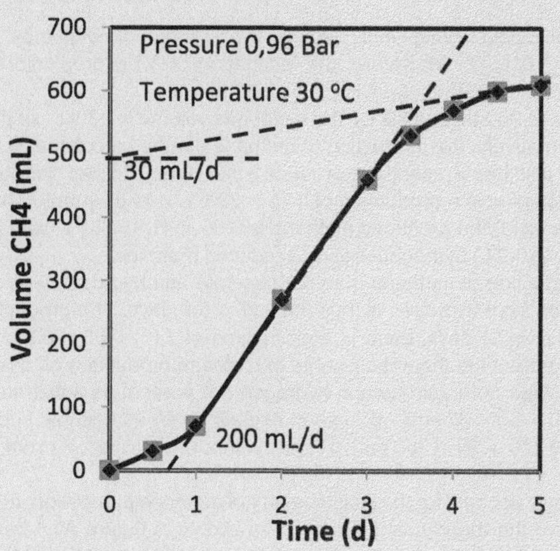

Figure A1.3 Production of methane as function of time of a sample with 5 g VSS inoculated and CH_3COONa as substrate.

EXAMPLE A1.5

When using propionate instead of acetate, it is observed that the maximum methane production increases from 170 to 340 mL · d^{-1}, whereas the endogenous production remains at 30 mL · d^{-1}. Compare the hydrogenotrophic and acetotrophic production.

Solution

Methane is produced by both acetotrophic and hydrogenotrophic archaea at equal rates. From the above decomposition equation it can be observed that upon acetogenesis of propionate, the substrate generation for hydrogenotrophic archaea is only three-quarters of the substrate generation for acetotrophic archaea.

As the methane production rates of both were equal in the experiment, then one-quarter of the generated acetate is not converted into methane and accumulates during the test as long as hydrogen is being produced. Hence after hydrogen is already depleted, one-quarter of the acetate that was generated by acetogenesis will have accumulated and is digested at the maximum rate of acetotrophic methanogenesis, which is 170 mL · d^{-1} as defined in the previous example. As the accumulated acetate is one-quarter of the generated acetate from propionate, the time to consume it will be one-third of the time during the phase of high-rate methane production.

EXAMPLE A1.6

Construct the theoretical methane production curve from the sample of Example A1.4 when 3.7 g · L^{-1} propionic acid is applied in 1 L of water with sufficient macro- and micronutrients and pH adjusted buffer.

Solution

The molar weight of propionic acid is 74 g · mol^{-1}, so that 3.7 g represents 0.05 mol. The COD of the propionic acid is 1.51 mg COD · mg^{-1} $C_3H_6O_2$. Therefore, the incubation of 3.7 g propionic acid or 5.6 g COD has a methane production potential of 5.6/4 = 1.4 g CH_4.

Under the test conditions (0.96 atm and 30°C), the molar gas volume is 23.9 L, so that 1.4 g CH_4 or 0.087 mol has a volume of 2.1 L. Alternatively, the production of methane can be calculated as follows: from the reaction equation 1 mol of propionic acid has a potential to produce 3 mol H_2 or 0.75 mol hydrogenotrophic CH_4 and 1 mol acetotrophic CH_4, so the potential is a production of 1.75 × 23.9 L = 41.8 L · mol^{-1} CH_4. Hence, for 0.05 mol of substrate (3.7 g of propionic acid) the produced methane is 0.05 × 41.8 = 2.1 L, with a fraction of 1/1.75 = 57% (1.2 L) acetotrophic and 43% (0.9 L) hydrogenotrophic produced methane.

In Example A1.5 the production of methane from both acetate and hydrogen was equal to 170 mL · d^{-1}. The methane production duration from hydrogen is thus 0.9/0.17 = 5.2 days. The production duration from acetate is 1.2/0.17 = 7.1 days. In these 7.1 days, there is a production of 7.1 × 0.03 = 0.2 L of endogenous methane. Now all the elements to construct the theoretical curve of methane production as a function of time are known, as shown in Figure A1.4. When both acetate and hydrogen are present as substrate (period from 0 to 5.2 d), the production is 170 + 170 + 30 = 370 mL · d^{-1}; when acetate is left as a single substrate (period from 5.2 to 7.1 d), the rate decreases to 170 + 30 = 200 mL · d^{-1}; and without substrate left (from 7.1 d onwards) production is 30 mL · d^{-1}.

The calculations above only account for the stoichiometry of methane production. In reality the actual methane production may deviate from the theoretical production as shown in Figure A1.4 because of two reasons: (1) when the substrate concentrations become low the actual methane production rate decreases (following the Monod equation); and (2) there may be inhibition of methane production due to an inadequate partial hydrogen pressure (see Figure 2.10).

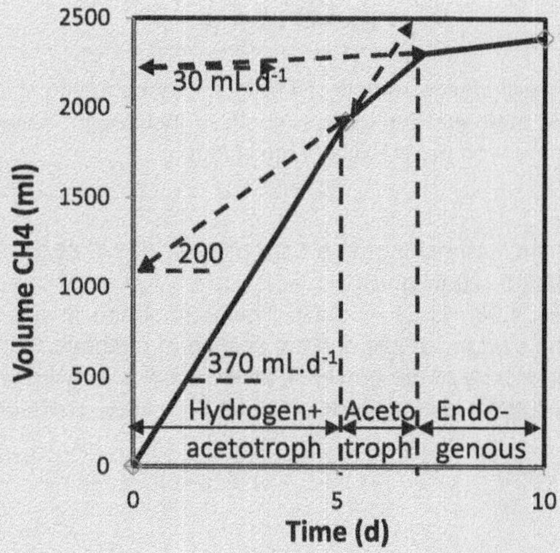

Figure A1.4 Theoretical methane production curve (Example A1.6).

A1.3 SLUDGE STABILITY TEST

A sludge stability test is carried out to assess the mass of organic material in a sludge sample that can still be digested. It reflects the fraction of the volatile solids remaining in the sludge that is formed by particulate biodegradable organic material which is settling onto the sludge. This material can be digested to methane through the normal sequence of processes.

This parameter is important when it is desired to discard sludge on a drying bed. According to EPA (2003), the fraction of biodegradable organic material remaining in the sludge should not exceed 20% of the sludge mass to avoid decomposition in contact with air, resulting in bad odours and a slimy liquid phase with difficult treatment. Normally, if the applied anaerobic sludge age is high enough (as has been discussed in Section 3.9), it may be expected that the produced UASB sludge is stable, but even then, it is possible that the scum discharged from the reactor surface or from within the phase separator is unstable. For this reason it is important that the scum is mixed with the excess sludge, so that the mixture has a biodegradable fraction that is low enough.

The stability test follows the same procedure as the SMA test, but no substrate is added. Instead, the production of biogas or methane from the biodegradable organic material already present in the sludge sample is observed. As this material is particulate and must be hydrolysed, the duration will be longer than in the case of the SMA test. A period of 30 days is prescribed for a temperature of 30–35°C (EPA, 2003).

EXAMPLE A1.7

What is the maximum allowed methane production per kg of sludge for the sludge to be considered stable?

Solution

If 20% of the sludge mass is digested during the test, this means a loss of 200 g VSS per kg of TSS. As 200 g VSS has a methane generation potential of $1.5 \times 200/4 = 64$ g $CH_4/16 = 4$ mol CH_4 or approximately 110 L methane, this means that a methane production of more than 0.11 $m^3 \cdot kg^{-1}$ or $L \cdot g^{-1}$ indicates an unstable sludge.

EXAMPLE A1.8

To evaluate the stability of a waste sludge, a sample of 4.08 g VSS is incubated at 30°C at atmospheric pressure. The sludge produces 320 ml of methane during one month of incubation, while the mass of volatile sludge decreases by 3.54 g VSS. Comment on the stability of the sludge.

Solution

The methane production of 320 mL corresponds to a COD of 320/375 = 0.85 g COD which in turn corresponds to a volatile sludge mass of 0.85/1.5 = 0.57 g VSS.

The observed decrease was 4.08 − 3.54 = 0.54 g. Therefore, there is a good agreement between the decrease in organic mass of the sludge sample and the volume of methane produced. The difference can be attributed to experimental error or else to the presence of some soluble hydrolysed organic material that was not yet digested. The result indicates a loss of solids of 0.54/4.08 = 14%. Therefore the sludge can be qualified as stable.

Appendix A2

Sludge settleability

Particles of sludges at concentrations above about 1 $g \cdot L^{-1}$ no longer settle individually: the particles are all interlinked and settle all together at the same rate, forming a distinct interface between the settling sludge and a clarified supernatant substantially free from solid particles. The velocity of the settling particles is called the zone settling velocity (ZSV). In the UASB reactor, equilibrium is established between the ZSV of the solids and the ascending velocity of the liquid phase, thus forming a stationary layer of sludge, whose concentration depends on the settleability of sludge and the upward velocity of the liquid phase.

Zone settling may be observed in the batch settler set-up described by White (1975) and shown schematically in Figure A2.1. The apparatus consists of a transparent vertical cylinder, in which a batch of sludge is placed. A stirrer is connected to the central axis, which in turn is driven by a low-rotation motor, and this gently stirs the sludge. After placing the sludge batch in the cylinder, the following behaviour can be observed:

- A short time (a few minutes) after placing the sludge in the cylinder, a sharp interface is formed, separating the clear supernatant not containing suspended solids in the upper part and settling sludge in the lower part.
- In the region below the interface all sludge particles settle at the same rate, so that the interface is also displaced at the same rate.
- Simultaneously at the bottom of the cylinder, sludge with a higher concentration accumulates. With time, an ever larger fraction of the sludge particles become part of this concentrated sludge.
- After some time, the sludge–supernatant interface approaches the region of concentrated sludge and its rate of displacement starts to decrease gradually.

Figure A2.1 also shows a typical curve of the interface displacement with time. The zone settling velocity is defined as the gradient of the linear (or linearized) part of this curve. Several cylinders may be used in parallel, in order to determine the zone settling velocity simultaneously for different sludge concentrations. The use of different concentrations allows the experimental determination of the relationship between the

 Anaerobic Sewage Treatment: Optimization of Process and Physical Design of Anaerobic and Complementary Processes
Editor(s): Adrianus van Haandel and Jeroen van der Lubbe
doi: 10.2166/9781780409627_0415

sludge concentration and the zone settling velocity. It is of great practical importance that an experimental method is available to determine the value of ZSV experimentally.

Among several empirical relationships (for example, Dick, 1972; Vesilind, 1968), the most widely applied is the Vesilind equation, which relates the zone settling velocity to the sludge concentration in the following manner:

$$v = v_0 \exp(-kX_t) \quad \text{(A2.1)}$$

where:

v = zone settling velocity ($m \cdot h^{-1}$)
X_t = sludge concentration ($g\ TSS \cdot L^{-1}$)
k = compressibility factor ($L \cdot g^{-1}\ TSS$)
v_0 = unhindered settling velocity ($m \cdot h^{-1}$).

In the Vesilind equation the constant k reflects the compressibility of the sludge, which is the ease with which the sludge can be concentrated. The constant v_0 is the unhindered settling velocity, which is the extrapolated settling velocity of the sludge for a concentration of zero. The Vesilind equation has the decisive advantage over other expressions that it can quantify the sludge settleability and that the settling velocity is not affected by the initial sludge concentration.

To calculate the zone settling velocity v for any concentration, it is necessary to experimentally determine the constants k and v_0. Figures A2.1 and A2.4 schematically show two test alternatives: the static and the dynamic test.

In the static test, a series of transparent test tubes of the same size are filled with sludge samples with different concentrations. The displacement velocity of the interface between supernatant and

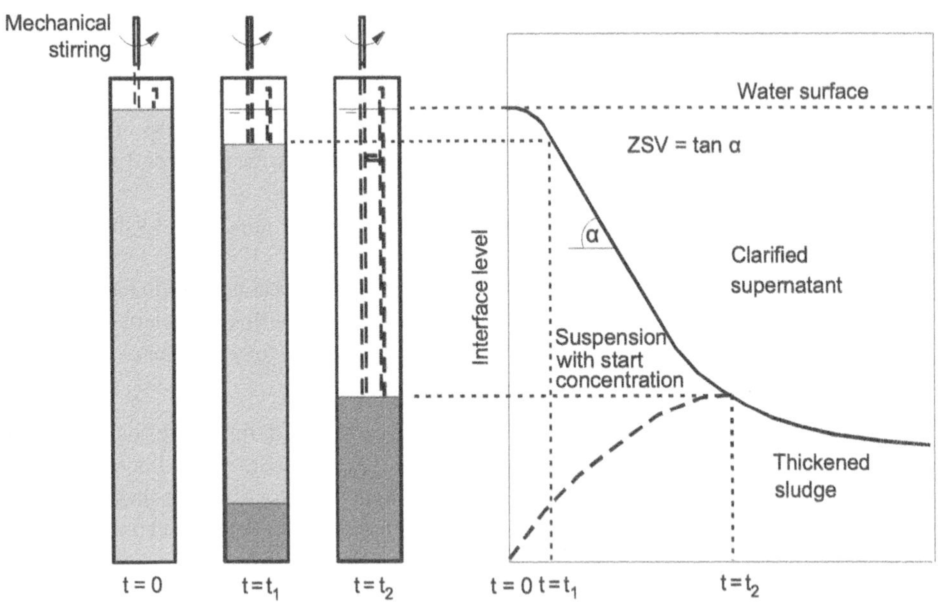

Figure A2.1 Equipment for the static test to determine the ZSV (left) and a typical curve of the interface displacement as a function of time (right).

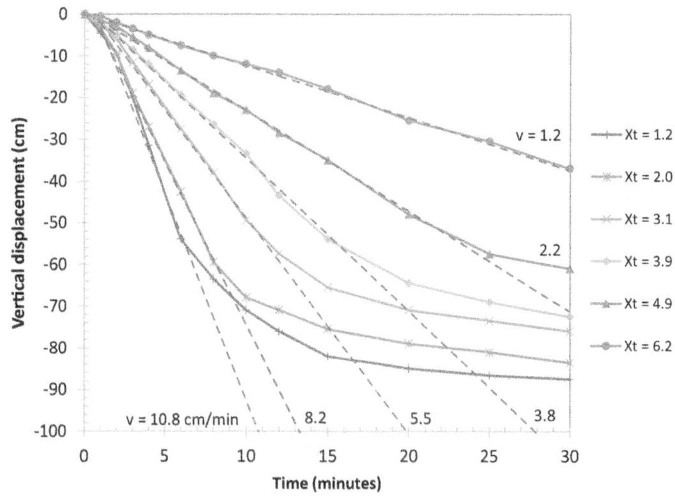

Figure A2.2 Graphical interpretation of the static settleability test (from Table A2.1).

settling sludge is observed as a function of time and plotted as shown in Figure A2.2 (Example A2.1). The inclination of the linear part of the curves of Figure A2.2 is equal to the zone settling velocities at different concentrations. The logarithm of these zone settling velocities is plotted as a function of the sludge concentration as shown in Figure A2.3. According to the Vesilind equation (Eq. (A2.2) one has:

$$\ln(v) = \ln(v_0) - kX_t \tag{A2.2}$$

Therefore, the slope of the straight line passing through the experimental points of a semi-log plot of v versus the sludge concentration X_t is equal to the value of the constant k, while the intersection with the y-axis provides the value of v_0 (see Figure A2.3).

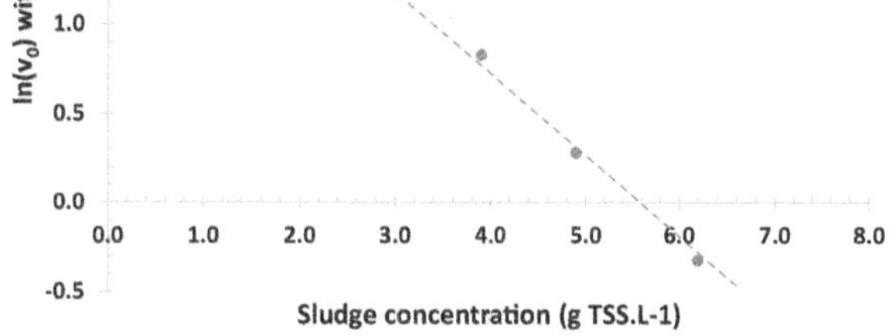

Figure A2.3 Semi-log plot of ZSV against sludge concentration to determine k and v_0.

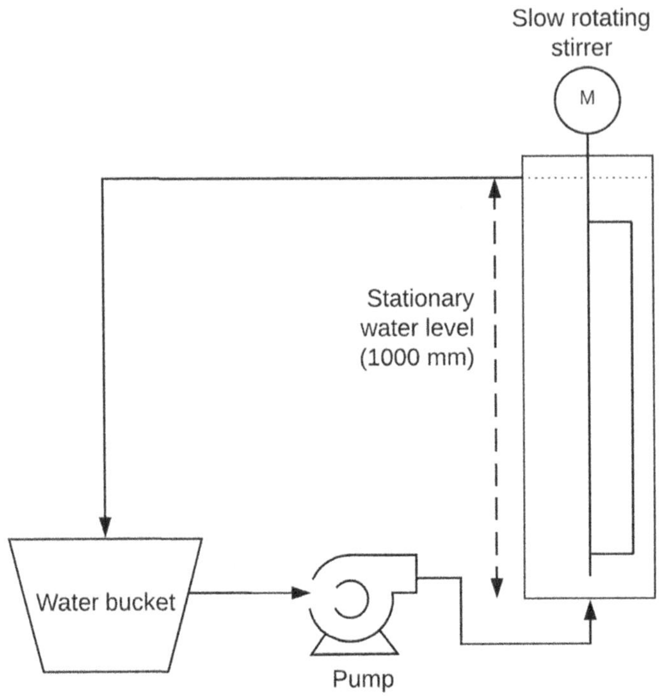

Figure A2.4 Schematic representation of the equipment to determine the zone settling velocity (dynamic test).

EXAMPLE A2.1

A static settling test is carried out with six sludge batches of different sludge concentrations. Table A2.1 shows the displacement of the interface as a function of time for 30 minutes. Determine the values of the Vesilind constants k and v_0.

Solution

To determine the values of the settleability constants, follow the procedure as described above:

(1) Plot the position of the interface as a function of time at different concentrations. Figure A2.2 shows the data points and associated curves.
(2) Find the gradient ($\Delta Y/\Delta X$) of the linear trendlines, using the straight sections of the displacement time curves (displayed as bold numbers in Table A2.1). The gradient is equal to the zone settling velocity v for these concentrations.
(3) Plot the values of $\ln(v)$ as function of the initial sludge concentration as indicated in Figure A2.3 and draw the best-fit straight line through the data points.
(4) The value of k is equal to the gradient of the straight line ($\Delta Y/\Delta X$) and v_0 is equal to the intersection of the straight line with the vertical axis. In this example $k = 0.44\ \text{L} \cdot \text{g}^{-1}$ and $v_0 = \exp(2.5) = 20.3\ \text{cm} \cdot \text{min}^{-1} = 12.2\ \text{m} \cdot \text{h}^{-1}$.
(5) These are typical values for aerobic sludge with relatively poor settleability. In general aerobic sludge has a much better compressibility (lower k value).

Table A2.1 Displacement of the sludge interface in the cyclinders with different starting sludge concentrations as a function of settling time.

Time (min)	Sample Sludge Concentration X_t (g TSS L^{-1})					
	1.2	2.0	3.1	3.9	4.9	6.2
0	0.0	0.0	0.0	0.0	0.0	0.0
1	−4.5	−3.5	−2.5	−2.0	−1.5	−0.5
2	−9.5	−9.5	−6.0	−5.0	−3.0	−2.0
3	−20.0	−19.0	−11.5	−8.5	−5.5	−3.5
4	−31.5	−27.0	−17.0	−12.0	−8.0	−5.0
6	−54.0	−42.5	−27.5	−19.5	−13.5	−7.5
8	−63.5	−59.5	−38.0	−26.5	−19.0	−10.0
10	−71.0	−68.0	−49.5	−33.5	−23.0	−12.0
12	−76.0	−71.0	−57.5	−43.5	−28.5	−14.0
15	−82.0	−75.5	−65.5	−54.0	−35.0	−18.0
20	−85.0	−79.0	−71.0	−64.5	−48.0	−25.5
25	−86.5	−81.0	−73.5	−69.0	−57.5	−30.5
30	−87.5	−83.5	−76.0	−72.5	−61.0	−37.0

In the dynamic test, a known mass of sludge is placed in a transparent cylinder and water is introduced in upward direction, which results in expansion of the sludge blanket in the cylinder until equilibrium is reached between the upward velocity v_l of the water and the settling velocity v of the sludge, so that:

$$v = v_0 \exp(-kX_t) = v_l = Q/A \quad \text{(A2.3)}$$

where:

Q = water flow rate passing through the transparent cylinder ($m^3 \cdot min^{-1}$)
A = area of the transparent cylinder (m^2).

By using different water flow rates, different upward velocities are imposed, and the sludge mass expands accordingly until the zone settling velocity becomes equal to the upward velocity. When equilibrium is reached, the level of the sludge blanket is read-off from the cylinder and the actual sludge concentration is calculated from the known initial mass:

$$M_b = V_b X_t = HAX_t$$

where:

M_b = sludge mass in the cylinder (g TSS)
H = height of the sludge blanket (m)
A = area of the cylinder (m^2)
V_b = sludge volume
X_t = sludge concentration (g TSS $\cdot L^{-1}$)

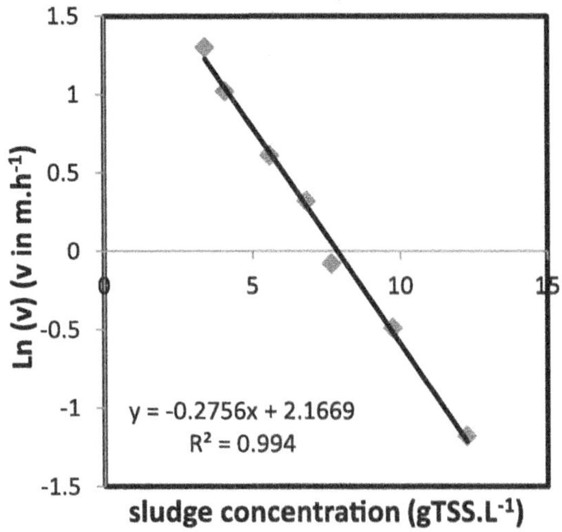

Figure A2.5 Graphical determination of the zone settling in the dynamic settleability test.

When several values of the water velocity (=ZSV) have been determined and the corresponding sludge concentration has been determined, the experimental data are plotted in a ln(ZSV) $-X_t$ diagram as in Figure A2.5.

EXAMPLE A2.2

In a transparent cylinder of 5 cm diameter and with 100 cm height of liquid a dynamic sludge settleability test is carried out. A sludge sample with a concentration of $12\ g \cdot L^{-1}$ and a volume of 0.5 L was placed in the cylinder, and therefore the sludge mass was $12 \times 0.5 = 6$ g TSS. An upward flow rate was applied to expand the sludge blanket and the height of the expanded sludge was measured at various flow rates. The results are shown in columns 1 and 2 of Table A2.2 below. Determine the settling constants of this sludge.

Solution

The solution proceeds according to the following steps (see Table A2.2 below):

(1) Calculate the area and volume of the cylinder: area = 19.6 cm^2 and volume = 1.96 L.
(2) For the applied flow rates (column 1), calculate the flows in $m^3 \cdot h^{-1}$ (column 3) and, dividing by the area of the cylinder, calculate the upward liquid velocity of the water (column 4); a conversion from $mL \cdot min^{-1}$ to $m^3 \cdot h^{-1}$ is made to obtain the result of the constants k and v_0 in the usual units of $L \cdot g^{-1}$ and $m \cdot h^{-1}$ respectively.
(3) From the measured heights of expanded sludge blanket, calculate the sludge volume and knowing that the mass of the sludge sample is 6 g TSS, calculate the sludge concentration (column 5).
(4) Calculate the natural logarithm of the upward velocities (column 6).
(5) Plot the data of ln(v) as function of the sludge concentration (X_t) and establish the gradient of the straight line through the experimental points (Figure A2.5).
(6) For the data presented, the calculated value of k follows from the inclination of the straight line as $0.28\ L \cdot g^{-1}$ and the v_0 value from the intersection of the line with the ordinate axis as: $\ln(v_0) = 2.17$, i.e. $v_0 = 9\ m \cdot h^{-1}$.

Table A2.2 Applied flow rates and corresponding height of the expanded sludge blanket during dynamic tests for the conditions in Example A2.2.

1	2	3	4	5	6
Flow **($ml \cdot min^{-1}$)**	**Height** **(cm)**	**Flow** **($m^3 \cdot h^{-1}$)**	**ZSV** **($m \cdot h^{-1}$)**	**X_t** **($gTSS \cdot l^{-1}$)**	**ln(v)** **(–)**
10	25	0.0006	0.307	12.27	−1.18
20	35	0.0012	0.613	9.76	−0.49
30	40	0.0018	0.920	7.67	−0.08
45	45	0.0027	1.380	6.82	0.32
60	55	0.0036	1.840	5.58	0.61
90	75	0.0054	2.760	4.09	1.02
120	90	0.0072	3.681	3.41	1.30

The values of the obtained Vesilind constants are typical for a well settling aerobic sludge or an anaerobic sludge with very poor settling characteristics.

References

Agunwamba J. C., Egbuniwe N. and Ademiluyi J. O. (1992). Prediction of the dispersion number in waste stabilisation ponds. *Water Research*, **21**, 85–89.

Aisse M. M. and Bollmann H. A. (1989). Avaliaçâo da aplicaçâo de reatores anaeróbios defluxo ascendente no tratamento de egots domêsticos de pequenas comunidades (Evaluation of the application of upflow anaerobic reactors for domestic sewage treatment in small communities). Proceedings of 15th Brazilian Congress on Sanitary and Environmental Engineering, 17–22 September 1989, Belém, Brazil. ABES, Rio de Janeiro, p. 140. (In Portuguese).

Andrade Neto, C. O. de (2004). Anaerobic Filter Applied for the Treatment of Domestic Sewage. PhD thesis, Federal University of Campina Grande, Brazil.

Angelidaki I. and Sanders W. (2004). Assessment of the anaerobic biodegradability of macropollutants. *Reviews in Environmental Science and Bio-Technology*, **3**, 117–129.

APHA, AWWA and WEF (2017). Standard methods for the examination of Water and wastewater. Washinton D.C., USA.

Araújo M. L. de A. (2000). Configuration of the UASB Reactor: In Series or Parallel? MSc Thesis, Federal University of Campina Grande, Brazil. (In Portuguese)

Ardern E. and Lockett W. T. (1914). Experiments on the oxidation of sewage without the aid of filters. *Journal of the Society of Chemical Industry*, **33**, 523.

Associação Brasileira de Normas Técnicas (ABNT) (2011). Elaboration of hydraulic-sanitary design of sewage treatment stations, 2nd edn. Rio de Janeiro Brazil (in Portuguese).

Barbosa R. A. and Sant'Anna G. L. Jr. (1989). Treatment of raw domestic sewage in a UASB reactor. *Water Research*, **23**, 1483–1490.

Bezerra S. M. da C. (1997). Optimization Strategy for Start-up of a UASB Reactor. MSc thesis, Federal University of Campina Grande Brazil (in Portuguese).

Bitton G. (2005). Wastewater Microbiology. Wiley, Hoboken, NJ, USA.

Buisman C., IJspeert P., Janssen A. and Lettinga G. (1990). Kinetics of chemical and biological sulphide oxidation in aqueous solutions. *Water Research*, **24**, 667–671.

Butler J. N. (1964). Ionic Equilibrium: A Mathematical Approach. Addison-Wesley, Reading, UK.

Cakir F. Y. and Stenstrom M. K. (2005). Greenhouse gas production: a comparison between aerobic and anaerobic wastewater treatment technology. *Water Research*, **39**, 4197–4203.

Capri M. G. and Marais G. v. R. (1974). Anaerobic Filter Treatment of Wine Distillery Waste, Research Report W1, Department of Civil Engineering, University of Cape Town, South Africa.

Cavalcanti P. F. F. (2003). Integrated Application of the UASB Reactor and Ponds for Domestic Sewage Treatment in Tropical Regions. PhD thesis, University of Wageningen, The Netherlands.

Cavalcanti P. F. F., Medeiros E. J. S., Silva J. K. M. and van Haandel A. C. (1999). Excess sludge discharge frequency for UASB reactors. *Water Science and Technology*, **40**(8), 211–219.

Cavalcanti P. F. F., van Haandel A. C. and Lettinga G. (2000a). Polishing ponds for post treatment of digested sewage. Part 1: Flow-through ponds. *Water Science and Technology*, **44**(4), 237–245.

Cavalcanti P. F. F. van Haandel A. C. and Lettinga G. (2000b). Polishing ponds for post treatment of digested sewage. Part 2: Sequential batch ponds. In: Proceedings of the VI Latin-American Workshop and Seminar on Anaerobic Digestion, 5–9 November 2000, Recife, Brazil, pp. 352–359.

Cavalcanti P. F. F., van Haandel A. C. and Lettinga G. (2000c). Accumulation of bottom solids in polishing ponds treating digested sewage. In: Proceedings 1st Latin American Conference on Stabilization Ponds and Reuse, 24–27 October 2000, Santiago de Cali, Colombia, pp. 236–244.

Chernicharo C. A. L. (2006). Post treatment options for the anaerobic treatment of domestic wastewater. *Reviews in Environmental Science and Bio-Technology*, **5**, 73–92.

Chernicharo C. A. L. (2007). Biological Wastewater Treatment Series: Volume 4 – Anaerobic Reactors. IWA Publishing, London, UK.

Chernicharo C. A. L., Almeida P. G. S., Lobato L. C. S., Couto T. C., Borges J. M. and Lacerda Y. S. (2009). Experience with the design and start up of two full scale UASB plants: enhancements and feedback. *Water Science and Technology*, **60**(2), 507–515.

Coelho A. L. S. S., do Nascimento M. B. H., Cavalcanti P. F. F. and van Haandel A. C. (2003). The UASB reactor as an alternative for the septic tank for on-site sewage treatment. *Water Science and Technology*, **48**(11–12), 221–226.

Collazos C. J. (1990). Operation of the La Rosita pilot plant 2nd Stage. Report for the defence corporation of the Bucaramanga catchment area, Bucaramanga, Colombia. (In Spanish).

Consentino P. R. S. (2002). Physical chemical post treatment of anaerobically pre treated effluents. MSc thesis, Federal University of Campina Grande, Brazil. (in Portuguese).

Curtis T. P., Mara D. D. and Silva S. A. (1992). Influence of pH, Oxygen, and Humic substances on ability of sunlight to damage faecal coliforms in waste stabilization pond water. *Applied and Environmental Microbiology*, **584**, 1335–1343.

Deffeyes K. S. (1956). Carbonate equilibria: a graphic and algebraic approach. *Limnology and Oceanography*, **10**, 412.

De Jong P., Voors E. H. and van der Pluijm J. L. P. M. Manual of Chemical Phosphorus Removal (in Dutch), STOWA Report 93-06, 1993.

De Man A. W. A. (1990). Anaerobic Purification of Raw Sewage with the Aid of Granular Sludge in UASB Reactors, Final Report, Department of Water Pollution Control, Agricultural University of Wageningen, Netherlands (In Dutch).

Derks Y. M. (2007). Respirometry for the Determination of the Metabolic Capacity of Sludges. MSc thesis, Federal University of Campina Grande, Brazil (in Portuguese).

De Vrieze J., Hennebel T., Boon N. and Verstraete W. (2012). Methanosarcina: The rediscovered methanogen for heavy duty biomethanation. *Bioresource Technology*, **112**, 1–9.

Dick R. I. (1972). Gravity thickening of sewage sludges. *J. Water Pollution Control Federation*, **44**, 368–380.

Dold P. L., Ekama G. A. and Marais G. v. R. (1980). A general model for the activated sludge process. *Progress in Water Technology*, **12**(6), 47–77.

Draaijer H., Maas J. A. W., Schaapman J. E. and Khan A. (1992). Performance of the 5 MLD UASB reactor for sewage treatment at Kanpur, India. *Water Science and Technology*, **25**(7), 123–133.

Emerson K., Russo R. C., Lund R. E. and Thurston T. T. (1975). Aqueous ammonium calculations: effect of pH and temperature. *Journal of the Fisheries Research Board of Canada*, **23**, 2379–2383.

EPA 2003. Control of Pathogens and Vector Attraction in Sewage Sludge. EPA/625/R-92/013, Revised 2003. U.S. Environmental Protection Agency, Washington, DC, USA.

Ferguson J. F. and McCarty P. L. (1969). The precipitation of phosphates from fresh waters and waste waters, Technical Report 120, Stanford University, Stanford, CA, USA.

Genung R. K., Donaldson T. L. and Reed G. D. (1985). Pilot-scale development of anaerobic filter technology for municipal wastewater treatment. In: Proceedings of the Seminar on Anaerobic Treatment of Sewage, June 1985, M.S. Switzenbaum (ed.), University of Massachusetts, Amherst, MA, USA.

Ghasimi D., de Kreuk M., Maeng S. K., Zandvoort M. and van Lier J. B. (2016). High-rate thermophilic bio-methanation of the fine sieved fraction from Dutch municipal raw sewage: cost-effective potentials for on-site energy recovery. *Applied Energy*, **165**, 569–582.

Gomes C. S. (1985). Research at SANEPAR and State of Parana, Brazil, with anaerobic treatment of domestic sewage in full scale and pilot plants. In: Proceedings of the Seminar on Anaerobic Treatment of Sewage, June 1985, M.S. Switzenbaum (ed.), University of Massachusetts, Amherst, MA, USA.

Gonçalves R. F., de Araujo V. L. and Bof V. S. (1999). Combining upflow anaerobic sludge blanket (UASB) reactors and submerged aerated biofilters for secondary domestic wastewater treatment. *Water Science and Technology*, **40**(8), 71–79.

Guimarães P., Melo H. N. S., Cavalcanti P. F. F. and van Haandel A. C. (2003). Anaerobic–aerobic sewage treatment using the combination UASB-SBR activated sludge. *Journal of Environmental Science and Health, A*, **38**, 2633–2641.

Gujer W. and Zehnder A. J. B. (1983). Conversion processes in anaerobic digestion. *Water Science and Technology*, **15**(8–9), 127–167.

Harned H. S. and Davis R. Jr. (1943). The ionization constant of carbonic acid in water and the solubility of carbon dioxide in water from 0 to 50°C. *Journal of the American Chemical Society*, **65**, 2030–2037.

Harned H. S. and Hamer W. J. (1933). The ionization constant of water and the dissociation of water in potassium chloride solutions from electromotive forces of cells without liquid junction. *Journal of the American Chemical Society*, **55**, 2194–2206.

Harned H. S. and Scholes S. R. Jr. (1943). The ionization constant of HCO_3^- from 0 to 50°C, *Journal of the American Chemical Society*, **63**, 1706–1709.

Haskoning (1989). Anaerobic treatment of domestic wastewater under tropical conditions. Final Report, Royal Dutch Consulting Engineers and Architects, Amersfoort, Netherlands.

Haskoning and Euroconsult (1990). Monitoring Report of the 5 MLD UASB Treatment Plant at Kanpur. Technical Report, May 1990\ Royal Dutch Consulting Engineers and Architects, Amersfoort, Netherlands/Euroconsult, Paris, France.

Heffernan B., Otten M. and van der Lubbe J. G. M. (2010a). Design and operational aspects of UASB sewage treatment plants. Presented at Singapore International Water Week, 28 June – 2 July 2010, Singapore.

Heffernan B., Van der Lubbe J. G. M. and Van Lier J. B. (2010b) Performance Review of Large Scale Up-flow Anaerobic Sludge Blanket Sewage Treatment Plants. Proceedings of Bi-annual IWA World Water Congress - Montreal, Canada, 19th – 24th September 2010.

Heffernan B., van Lier J. B. and van der Lubbe J. G. M. (2011). Performance review of large scale up-flow anaerobic sludge blanket sewage treatment plants. *Water Science and Technology*, **63**(1), 100–107.

Helmer R., Hespanhol I. and Saliba L. J. (1991). Public health criteria for the aquatic environment: recent WHO guidelines and their application. *Water Science and Technology*, **24**(2), 35–42.

Henze M. and Harremoës P. (1983). Anaerobic treatment of wastewater in fixed film reactors – a literature review. *Water Science and Technology*, **15**(8–9), 1–101.

Henze M., Grady L. Jr, Gujer W., Marais G. v. R. and Matsuo T. (1987). Activated Sludge Model No 1. IAWPRC Scientific and Technical Report No 1. IAWPRC, London, UK.

Holmes D. E. and Smith J. A. (2016). Biologically produced methane as a renewable energy source. *Advances in Applied Microbiology*, **97**, 1–61.

IPCC (2007). III.I. Avoidance of methane production in wastewater treatment through replacement of anaerobic systems by aerobic systems (version 6).

Janssen A. J. H., Lettinga G. and Keizer A. d. (1999). Removal of hydrogen sulphide from wastewater and waste gases by biological conversion to elemental sulphur: colloidal and interfacial aspects of biologically produced sulphur particles. *Colloids and Surfaces A: Physicochemical and Engineering Aspects*, **151**, 389–397.

Jeris J. S. (1983). Industrial wastewater treatment using anaerobic fluidised bed reactors. *Water Science and Technology*, **15**(8–9), 169–176.

Jeris J. S., Ehlers R. and Witkowski P. (1985). Anaerobic Treatment of municipal wastewater using the fluidized bed process. In: Proceedings of the Seminar on Anaerobic Treatment of Sewage, June 1985, M. S. Switzenbaum (ed.), University of Massachusetts, Amherst, MA, USA.

Jewell W. J. (1982). Development of the attached microbial film expanded bed process for anaerobic waste treatment. In: Biological Fluidized Bed Treatment of Water and Wastewater, P. E. Cooper and B. Atkinson (eds.), Ellis Horwood, Chichester, UK.

Jewell W. J., Switzenbaum M. S. and Morris J. W. (1981). Municipal wastewater treatment with the anaerobic attached microbial film expanded bed process. *J. Water Pollution Control Federation*, **53**, 482–490.

Klock W. J. (1971). Survival of coliform bacteria in wastewater treatment lagoons. *J. Water Pollution Control Federation*, **43**, 2071.

Kotsyurbenko O. R., Chin K., Glagolev M., Stubner S., Simankova M. V., Nozhevnikova A. N. and Conrad R. (2004). Acetoclastic and hydrogenotrophic methane production and methanogenic populations in an acidic West-Siberian peat bog. *Environmental Microbiology*, **6**, 1159–1173.

Lakshminarayana J. S. (1972). Prevention of sewage pollution by stabilization ponds. *Environmental Letters*, **8**, 121–134.

Leitão R. C. (2004). Robustness of UASB Reactors Treating Sewage under Tropical Conditions. PhD thesis, Wageningen University, Wageningen, Netherlands.

Leite L. A. (2016). Aerobic Post-treatment of Digested Sewage in an Activated Sludge System with a Sequential Batch Regime. MSc thesis, Federal University of Campina Grande Brazil. (in Portuguese)

Lens P. N. and Hulshof Pol, L. (2000). Environmental Technology to Treat Sulphur Pollution: Principles and Engineering. IWA Publishing, London, UK, 550pp.

Lettinga G., van Velsen A. F. M., Hobma S., de Zeeuw W. and Klapwijk A. (1980). Use of the upflow sludge blanket (USB) reactor concept for biological wastewater treatment, especially for anaerobic treatment. *Biotechnology and Bioengineering*, **22**, 4699–4734.

Levenspiel, O. (1999). Chemical Reaction Engineering, 3rd edn. Wiley, New York, USA.

Loewenthal, R. E. and Marais, G. v. R. (1976). Carbonate Chemistry of Aquatic Systems: Theory and Application. Ann Arbor Science, Michigan, USA.

Mahmoud N., Zeeman G., Gijzen H. and Lettinga G. (2004). Anaerobic sewage treatment in a one-stage UASB reactor and a combined UASB–Digester system. *Water Research*, **38**, 2348–2358.

Mara D. D. (1975). Proposed design for oxidation ponds in hot climates *Journal of the Environmental Engineering Division. ASCE*, **101**, 296–300.

Mara D. D. (1976). Sewage Treatment in Hot Climates. Wiley, Chichester, UK.

Mara D. D. and Pearson H. W. (1987). Waste Stabilisation Ponds: Design for Mediterranean Europe. World Health Organisation, Geneva, Switzerland.

Marais G. v. R. (1974). Faecal bacteria kinetics in waste stabilisation ponds. *Journal of the Environmental Engineering Division, ASCE*, **100**, 119–139.

Marais G. v. R. and Ekama G. A. (1976). The activated sludge process. Part I – Steady state behaviour. *Water SA*, **2**, 163–200.

Marais G. v. R. and Shaw V. A. (1961). A rational theory for the design of sewage stabilization ponds in Central and South Africa. *Transactions of the South African Institute of Civil Engineering*, **3**(11), 205–227.

McCarty P. L. (1981). One hundred years of anaerobic treatment. Presented at the 2nd Int. Conf. on Anaerobic Digestion, 6–11 September 1981, Travemünde, Germany.

Menar, A. B. and Jenkins, D. (1972). Calcium Phosphate Precipitation in Wastewater Treatment, Report EPA-R2-72-064, US EPA, Washington DC, USA.

Merill D. T. and Jordan R. N. (1975). The high rate treatment of raw domestic sewage by lime precipitation and dissolved air flotation. *Progress in Water Technology*, **7**(3/4), 379–389.

Metcalf and Eddy Inc. (2013). Wastewater Engineering – Treatment and Reuse. McGraw Hill, New York, NY, USA.

Monod J. (1950). La technique de culture continue: theorie et applications. (Continuous culture technique: theory and applications). *Annales de l'Institut Pasteur*, **79**, 390–410, (in French).

Moreira E. A., van Haandel A. C., Cavalcanti P. F. F. and Sales de Melo A. (2006). Determination of the percolation rate of rainwater in sludges drying beds. Italian-Brazilian Symposium, Fortaleza-Ce, Brazil. ABES, Rio de Janeiro, Brazil. CD-ROM.

Mulder A., van de Graaf A. A., Robertson L. A. and Kuenen J. G. (1995). Anaerobic ammonium oxidation discovered in a denitrifying fluidized bed reactor. *FEMS Microbiology Ecology*, **16**, 177–184.

Nobre C. A. and Guimaraes M. O. (1987). Experiments of anaerobic digestion of urban wastes. *Revista DAE*, **47**, 75. (In Portuguese)

Oliveira M. G. (1990). Evaluation of the Performance of UASB Reactors Treating Sewage in North-east Brazil. MSc thesis, Federal University of Campina Grande, Brazil. (In Portuguese).

O'Rourke J. T. (1968). Kinetics of Anaerobic Treatment at Reduced Temperature, PhD thesis. Stanford Univ., Stanford, USA.

Pacheco E. J. and Pessoa C. A. (2017). Domestic Sewage Treatment, 8th edn. ABES, Rio de Janeiro, Brazil. (In Portuguese).

Pessotto B. (2015). Design and Performance Evaluation of an Aerobic Sludge Blanket Reactor with Mechanical Mixing and a High-Rate Settler Treating Domestic Wastewater. DSc thesis, University of Sao Paulo, Sao Carlos, Brazil, 172pp.

Phelps E. B. (1944). Stream Sanitation. Wiley, New York, USA.

Pontes P., Chernicharo C. A. L., Frade E. C. and Porto M. T. R. (2003). Performance evaluation of an UASB reactor used for combined treatment of domestic sewage and excess aerobic sludge from a trickling filter. *Water Science and Technology*, **48**(6), 227–234.

Rabinowitz B. and Marais G. v. R. (1981). Chemical and Biological Phosphorus Removal in the Activated Sludge Process. Research Report W32. Department of Civil Engineering, University of Cape Town, Rondebosch, South Africa.

Rinzema A. (1988). Anaerobic Treatment of Waste Water with High Concentrations of Lipids or Sulphate. PhD thesis, University of Wageningen, Netherlands.

Rinzema A. and Lettinga G. (1988). Anaerobic treatment of sulfate containing waste water. In: Biotreatment Systems, D. L. Wise III (ed.), CRC Press, Boca Raton, FL, USA, pp. 65–109.

Sales V. dos S. (2013). Evaluation of the methane production capacity in UASB reactors treating vinasse under different operational conditions. MSc thesis at the Federal University of Campina Grande Brazil (in Portuguese).

Santos L. S., Morais S. R. and van Haandel A. C. (2016). Influence of phase separator design on the performance of UASB reactors treating municipal wastewater. *Water SA*, **42**(2), 176–182.

Santos S. L. Simoes J. P. O. and van Haandel A. C. (2016). Design optimization of a simple single family anaerobic sewage treatment system: Part 1: Influence of the configuration. *Water Practice & Technology*, **12**, 1.

Saqqar M. M. and Pescod M. B. (1995). Microbiological performance of multi-stage stabilization ponds for effluent use in agriculture. *Water Science and Technology*, **23**(7–9), 1517–1524.

Sastry C. A. and Mohanrao G. J. (1975). Waste stabilization pond design and experience in India. In: Ponds as a Wastewater Treatment Alternative, Water Resources Symposium No 9, 23–25 July 1975, University of Texas, Austin, Texas, USA.

Schellinkhout A. and Collazos C. J. (1992). Full scale application of the UASB technology for sewage treatment. *Water Science and Technology*, **25**(7), 159–166.

Schellinkhout A., Lettinga G., van Velsen L. and Rodriguez G. (1985). The application of the UASB reactor for direct treatment of domestic wastewater under tropical conditions. In: Proceedings of the Seminar on Anaerobic Treatment of Sewage, June 1985, M. S. Switzenbaum (ed.), University of Massachusetts, Amherst, MA, USA.

Schellinkhout A., Lettinga G., Van Velsen L., Louwe Kooijmans L. and Rodrigues G. (1986). The application of the UASB reactor for direct treatment of domestic waste water under tropical conditions. In: Schwitzenbaum M. S. (ed.) Proceedings of the seminar of anaerobic treatment of sewage, University of Massachusetts, Amherst, USA.

Schroepfer G. J., Fullen W. J., Johnson A. S., Ziemke N. R. and Anderson J. J. (1955). The anaerobic contact process as applied to packinghouse wastes. *Sewage and Industrial Wastes*, **27**, 460–486.

Sheen R. T. and Woodruff E. B. (1944). Selective calcium softening for industrial water supplies. *Industrial Engineering Chemistry*, **36**, 971–979.

Sherry G. P. and Parker C. D. (1979). Development of an improved quantitative relationship between bacterial die off, design and operational factors for anaerobic, aerobic and maturation type lagoon systems. *Progress in Water Technology*, **11**(4/5), 427.

Silva Filho J. de A. and van Haandel A.C. (2014). Stabilization of aerobic sludges from post treatment in the anaerobic pre treatment unit. *Revista Dae*, **194**, 86–102.

Silva S.A. (1989). Stabilization pond series with UASB reactors for the treatment of domestic waste in NE Brazil. In: Proceedings of 15th Brazilian Congress on Sanitary and Environmental Engineering, 17–22 September 1989, Belém, Brazil. ABES, Rio de Janeiro, p. 86. (In Portuguese)

Silva S. (2016). Feasibility of sludge return from settled solids from UASB effluent. MSc thesis, Federal University of Campina Grande, Brazil. (In Portuguese)

Song Z., Williams C. J. and Edyvean R. G. J. (2004). Treatment of tannery wastewater by chemical coagulation. *Desalination*, **164**, 249–259.

Souza M. E. (1986). Criteria for the utilization, design and operation of UASB reactors. *Water Science and Technology*, **18**(12), 55–69.

Souza F. P. (2000). Influence of the Area/Depth Ratio of a UASB Reactor on its Performance. MSc thesis, Federal University of Campina Grande, Brazil. (In Portuguese).

Souza C. L., Chernicharo C. A. L. and Aquino S. F. (2010a). Quantification of dissolved methane in UASB reactors treating domestic wastewater under different operating conditions. Proceedings of the 12th IWA Anaerobic Digestion Conference (AD12). Guadalajara, Mexico.

Souza C. L., Chernicharo C. A. L. and Melo G. C. B. (2010b). Methane and hydrogen sulphide measurements in UASB reactors treating domestic wastewater. Proceedings of the 12th IWA Anaerobic Digestion Conference (AD12). Guadalajara, Mexico.

Souza C. L., Chernicharo C. A. L. and Aquino S. F. (2011). Quantification of dissolved methane in UASB reactors treating domestic wastewater under different operating conditions. *Water Science and Technology*, **64**(11), 2259–2264.

Speece R. E. (1983). Anaerobic biotechnology for industrial wastewater treatment. *Environmental Science and Technology*, **17**, 416A–427A.

Speece R.E. (2008). Anaerobic Biotechnology and Odor/Corrosion Control for Municipalities and Industries. Archaea Press, Nashville, Tennessee, USA.

Stumm W. and Morgan J. J. (1981). Aquatic Chemistry – An Introduction Emphasizing Chemical Equilibria in Natural Waters. Wiley, New York, USA.

Switzenbaum M. S. and Jewell W. J. (1980). Anaerobic attached-film expanded-bed reactor treatment. *Journal of the Water Pollution Control Federation*, **52**, 1953–1965.

Truesdale A. H. and Jones B. J. (1973). WATEQ, a Computer Program for Calculating Chemical Equilibria of Natural Waters. NTIS Report PB 20464, NTIS, Springfield, VA, USA.

Uemura S. and Harada H. (2010). Application of UASB technology for sewage treatment with a novel post-treatment process. In: Environmental Anaerobic Technology, H. H. P. Fang (ed.), Imperial College Press, London, UK, pp. 91–112.

van der Last A. R. M. (1991). Anaerobic Treatment of Settled Sewage with the EGSB and the FB Processes. Final Report, Department of Water Pollution Control, Agricultural University of Wageningen, Netherlands. (In Dutch)

van der Lubbe J. and Heffernan B. (2012) Method and installation for recovery and utilisation of methane from an anaerobic liquid effluent. WO2012078044A1.

van der Lubbe J. G. M., Heffernan, B. and van Lier, J. B. (2010a). Engineering and operational issues in UASB reactors treating municipal sewage. Presented at Anaerobic Digestion Conference (AD12), 31 October–4 November 2010, Guadalajara, Mexico.

van der Lubbe J. G. M., Otten M. and Heffernan B. (2010b). UASB – How To Get It Right. *Asian Water*, **26**, 6.

van Haandel A. C. and Lettinga G. (1994). Anaerobic Sewage Treatment: A Practical Guide for Regions with a Hot Climate. Wiley, Chichester, UK, 226pp.

van Haandel A. C. and van der Lubbe J. (2012). Handbook Biological of Wastewater Treatment: Design and Optimisation of Activated Sludge Systems, 2nd edn. IWA Publishing, London, UK.

van Lier J. B. (2008). High-rate anaerobic wastewater treatment: diversifying from end-of-the-pipe treatment to resource-oriented conversion techniques. *Water Science and Technology*, **57**(8), 1137–1148.

van Lier J. B., Vashi A., van der Lubbe, J. G. M. and Heffernan B. (2010). Environmental anaerobic technology – applications and new developments. In: Anaerobic Sewage Treatment Using UASB Reactors: Engineering and Operational Aspects, H. W. Fang (ed.), Imperial College Press, London, UK.

Vavilin V. A., Fernandez B., Palatsi J. and Flotats X. (2008). Hydrolysis kinetics in anaerobic degradation of particulate organic material: an overview. *Waste Management*, **28**, 939–951.

Vaxelaire J., Bongiovanni J. M. and Puiggali J. R. (2010). Mechanical dewatering and thermal drying of residual sludge. *Environmental Technology*, **20**, 29–36.

Vesilind P. A. (1968). Theoretical considerations: design of prototype thickeners from batch settling tests. *Water and Sewage Works*, **115**, 302–307.

Vieira S. M. M. and Garcia A. D. Jr. (1992). Sewage treatment by UASB reactor. Operation results and recommendations for design and utilization. *Water Science and Technology*, **25**(7), 143–157.

Vieira S. M. M. and Souza M. E. (1986). Development of technology for the use of the UASB reactor in domestic sewage treatment. *Water Science and Technology*, **18**(12), 109–121.

Vlaeminck S. E., De Clippeleir H. and Verstraete W. (2012). Microbial resource management of one-stage partial nitritation/anammox. *Microbial Biotechnology*, **5**, 433–448.

von Sperling M. (1999). Performance evaluation and mathematical modelling of coliform die-off in tropical and subtropical waste stabilization ponds. *Water Research*, **33**, 1435–1448.

Wehner J. F. and Wilhelm R. H. (1956). Boundary conditions of flow reactor. *Chemical Engineering Science*, **6**, 89–93.

White M. J. D. (1975). Settling of Activated Sludge. Technical Report TR11, Water Research Centre, Swindon, UK.

Wiechers H. N. S. (1977). Calcium carbonate and hydroxide solubility product values for heterogeneous systems. *Progress in Water Technology*, **9**, 531–545.

World Health Organization (1989). Health Guidelines for the Use of Wastewater in Agriculture and Aquaculture. Technical report series 778, World Health Organisation, Geneva, Switzerland.

Yañez F. (1993). Stabilization Ponds. CEPIS, Lima, Peru. (In Spanish)

Yoda M., Hattori M. and Mijaji Y. (1985). Treatment of municipal wastewater by anaerobic fluidised bed: behaviour of organic suspended solids in anaerobic reactor. In: Proceedings of the Seminar on Anaerobic Treatment of Sewage, June 1985, M. S. Switzenbaum (ed.), University of Massachusetts, Amherst, MA, USA.

Young J. C. (1991). Factors affecting the design and performance of upflow anaerobic filters. *Water Science and Technology*, **24**(8), 133–155.

Young J. C. and McCarty P. L. (1969). The anaerobic filter for waste treatment. *Journal of the Water Pollution Control Federation*, **41**(5, Research Supplement), R160–R173.

Yvon-Durocher G., Allen A. P., Bastviken D., Conrad R., Gudasz C., St-Pierre A., Thanh-Duc N. and del Giorgio P. A. (2014). Methane fluxes show consistent temperature dependence across microbial to ecosystem scales. *Nature*, **507**, 488–491.

IWA Publishing's authorised EU representative for General Product Safety Regulations is Diane D'Arras, 15 rue Duret, 75116 Paris, France, e-mail: safety@iwap.co.uk.

Printed and bound by CPI Group (UK) Ltd, Croydon, CR0 4YY
16/06/2026
02140373-0002